TUBER CROPS

THE AUTHORS

Dr. Sanyat Misra is a Forestry graduate from UAS, Bangalore; M.Sc. (Ag.) Hort. from GBPUAT, Pantnagar (Udham Singh Nagar), Uttarakhand; and Ph.D. from Dr. YSP UHF, Nauni, Solan (H.P.). He served ICAR-NBPGR (New Delhi and Shimla) for 3 years and in 2004 joined BAU, Kanke (Ranchi) as Assistant Professor *cum* Jr. Scientist. At present, he is Head, Deparment of Horticulture looking after administrative, extension, teaching and research activities of the Department. He had been Chairman in case of 7 M.Sc. students. To his credit, he has some 45 research publications, 11 books and 3 bulletins.

Sanjay Mishra is a Lucknow-based botany graduate *cum* software professional who to his credit has 4 books (2 books on softwares, 1 as 'Soilless Crop Production' and one book 'Herbs, Spices and Medicinal Plants' in 2 volumes). It is his untiring efforts that this book has reached to this stage.

Dr. R.L. Misra got retired from Division of Floriculture and Landscaping, ICAR-IARI, New Delhi-110012 in May 2008 as Principal Scientist *cum* Project Coordinator (Floriculture), All India Coordinated Research Project on Floriculture. He also worked as Officer-in-Charge, Indo-Israel Preject on Research and Development on Farms, IARI, New Delhi. As Adjunct Faculty he continued further teaching works to M.Sc. and Ph.D. students at IARI from 2010 to 2014. As Chairman, he guided 6 M.Sc. and 16 Ph.D. students in IARI. To his credit, he developed 30 gladiolus varieties, has more than 500 publications: 11 books, 9 technical bulletins, 15 Review articles, 169 research publications, 120 technical articles, 57 symposium papers and more than 140 book chapters. He is Fellow of (i) **Indian Society of Genetics and Plant Breeding** (1980), (ii) **Indian Society of Ornamental Horticulture** (2009), (iii) **Horticultural Society of India** (2018), and (iv) **Promotion of Horticultural Science, Bengaluru** (2022). Also he is HSI Gold Medal Awardee (2013). **Indian Society of Ornamental Horticulture**, New Delhi in 2018 on his outstanding floricultural works, and **The Orissa Horticultural Society**, Bhubaneswar in 2023 on his outstanding horticultural works conferred Lifetime Achievement Awards. In 2007, he was also conferred with Urivi Vikram Charitable Trust Award in New Delhi.

TUBER CROPS

Authors

Sanyat Misra

Sanjay Mishra

R.L. Misra

2025

Daya Publishing House®

A Division of

Astral International Pvt. Ltd.

New Delhi – 110 002

ISBN 978-93-5919-546-9

Published by : **Daya Publishing House®**
A Division of
Astral International Pvt. Ltd.
– ISO 9001:2015 Certified Company –
4736/23, Ansari Road, Darya Ganj
New Delhi-110 002
Ph. 011-43549197, 23278134
e-mail: info@astralint.com
Website: www.astralint.com

Printed at : **Replika Press**

Preface

Indian continent is rich in flora and fauna of various kinds and there are enormous number of plant species endemic to this region. In agricultural field, now we are self-sufficient though up to 1960s this country used to import grains, oils and various other edibles to feed our peoples. At present we have 63 State/Central/Deemed-to-be Agricultural Universities including 6 exclusive horticultural universities, and every university with dozens of regional colleges to solve regional problems. Indian Council of Agricultural Research is the apex agricultural body in the country coordinating various regions as well as universities across the country. At present we are in a position to export food grains, and horticultural produce of all kinds including vegetables. India is the leading producer of vegetables at global level after China, and at present we are producing more than 100 million tonnes of vegetables annually.

In this book, we are dealing with various vegetables having their underground parts modified in the form of corms, tubers, tuberous roots and rhizomes to store food materials for perennation. We could found only 35 such vegetables to be tuberous though in certain other literature sources certain root crops such as yam bean (*Pachyrrhizus erosus*), Japanese arrowroot (*Pueraria lobata*), *etc.* are also described as tuber crops due to wrong placing though these are root crops. All those described here are categorized as per their underground structures as given here under:

Cormous vegetables described here are: i) **arrowhead** (*Sagittaria* spp.), ii) **Chinese water chestnut** (*Eleocharis dulcis*), iii) **elephant-foot yam** (*Amorphophallus* spp.), and iv) **wild arum** (*Arum maculatum*).

Stem tubers (true tubers) described here are: i) **bitter potato** (*Solanum* × *curtilobum* and *S.* × *juzepczukii*), ii) **Chinese potato** (*Coleus parviflorus*, syn. *C. tuberosus*), iii) **Indian arrowroot** (*Tacca leontopetaloides*), iv) **lady's sorrel** (*Oxalis tuberosa*), v) **nasturtium** (*Tropaeolum tuberosum*), vi) **potato** (*Solanum tuberosum*), vii) **ullucu** (*Ullucus tuberosus*), viii) **yams** {greater yam, *Dioscorea alata*; lesser yam, *D. esculenta*); white yam, *D. rotundata*; *etc.*}, and ix) **yellow nutsedge** (*Cyperus esculentus*).

Root tubers (tuberous roots described here are: i) **African yam bean** (*Sphenostylis stenocarpa*), ii) A**merican groundnut** (*Apios* spp.), iii) **cassava** (*Manihot esculenta*), iv) **coontie palm** (*Cycas integrifolia*),v) **Soh-phlong** (*Flemingia vestita*), vi) **sweet potato** (*Ipomoea batatas*), vii) **yacon** (*Smallanthus sonchifolius*), viii) **winged bean** (*Psophocarpus tetragonolobus*), and ix) **Zombi pea** (*Vigna vexillata*) .

Rhizomatous vegetables described here are: i) **arrowroot** (*Maranta arundinacea*), ii) **Bengal arum**(*Typhoniun trilobatum*), iii) **bugleweed** (*Lycopus* spp.), iv) **curcuma** (*Curcuma* spp.), v) **giant taro** (*Alocasia macorrhizos*), vi) **Guinea arrowroot** (*Goeppertia allouia*), vii) **Indian shot** (*Canna indica*), viii) **Jerusalem artichoke** (*Helianthus tuberosus*), ix) **lotus** (*Nelumbo* spp.), x) **swamp taro** (*Cyrtosperma merkusii*), xi) **tannia** (*Xanthosoma sagittifolium*), xii) **taro** (*Colocasia esculenta*), and xiii) **woundwort** (*Stachys* spp.).

Every important aspect of each and every edible tuber crop has briefly been covered in a quite comprehensive manner apart from their photographs in most of the cases. The book has been written for horticultural students (B.Sc., M.Sc. and Ph.D.) and those preparing for ARS Exams, for teachers and researchers in horticulture and vegetables, to development officers at district level, and more so for farmers of the country.

Authors have all the love to Master Vibhor Mishra (son of Mr. Sanjay Mishra) for helping in the proof-reading of this manuscript; and gratefulness to Mrs. Smita Mishra (wife of Mr. Sanjay Mishra) for going through the whole manuscript grammatically.

Sanyat Misra

Sanjay Mishra

R.L. Misra

Contents

Introduction

The word tuber originates from the Latin *tuber* meaning 'lump, bump or swelling' due to the underground organs being thick and fleshy. Tubers are enlarged, swollen and fleshy portions of underground stems (shoot tubers, stem tubers or true tubers), roots with a part of stem (fleshy roots, root tubers or tuberous roots) or sometimes even aerial tubers in the leaf axils (tubercles or aerial tubers) and corms but here all these organs will broadly be called as tubers for the sake of smooth understanding. Tubers are used for the plant's perennation (survival during rigours of the winter, dry or summer months) to store energy and nutrients for regrowth during next growing season, and as a means of asexual reproduction. Unlike corms or bulbs, tubers do not have a basal plate from which new shoots or roots grow; instead they are covered with nodes or buds (eyes) which grow as shoots or roots. Sweet potatoes bear 'root tubers', the potatoes and yams technically 'stem tubers', the elephant foot yam 'basal plate-less corms' and colocasias the 'rhizomes'. All these are modified underground stems.

Stem tubers (true tubers) are enlarged and thickened stolons to develop into storage organs. The tuber has all the parts of a normal stem and is covered by thin and tough skins for their protection, *vis-a-vis* latent buds (eyes) which sprout and grow up through the soil surface as shoots and stems when conditions are congenial as in case of potato. Tubers have indefinite bud (eye) arrangement over its surface and in all its internal part as in case of a typical stem. Internally, a tuber is filled with starch stored in enlarged parenchyma-like cells. The inside of a tuber has the typical cell structures of any stem including pith, vascular zones and a cortex. Tubers are generally formed annually and disintegrate once the food reserves are depleted, forming new tubers in the process . Examples of tubers are *Dioscorea*, potatoes (*Solanum tuberosum*), *Tropaeolum, Umbilicus, etc.* Many plants with tubers are short-lived and are usually grown only to harvest the fleshy edible tubers. Some perennial stem tubers being almost circular or flattened increasing in size each season ~ producing shoots from the upper side and roots from either side or both.

Stem tubers can be cut into many pieces each bearing a bud, and planted individually to create new plants that will be exact replicas of the parent plant. As tubers mature, new tubers may form from their roots and stems. Stem tuber generally starts as an enlargement of the hypocotyl section of a seedling but sometimes also include the first node or two of the epicotyl and the upper section of the root. The tuber with usually oblong-rounded shape has a vertical orientation with one or a few vegetative buds on the top and fibrous roots produced on the bottom from a basal section. Many tubers such as potatoes are grown as food because of their high nutrient content.

Some species such as *Ullucus tuberosus* and *Dioscorea batata* may have small **aerial tubers/tubercles** (aerial stem tubers/tubercles) on the leaf axils from the axillary buds nearly in the end of growing season when day lengths continue to shorten. They are easily removed from the plants, surface-dried and stored overwinter for spring planting.

Tuberous roots (fleshy roots, root tubers or storage roots) are genuine roots modified into fleshy underground storage organs but are devoid of their original absorbent function as in functional roots. As a rule, the eyes (buds) are found on the collar, *i.e.* proximal end where these typically form the crown at the base of the aerial stem where these roots are attached *e.g.* sweet potato. During sprouting, the tuberous root develops as a swelling on the emerging radicle of a seedling plant and increases in size to form a single structure with adventitious buds forming on its surface. Likewise, lateral roots arising from crown during the growing season also swell and make a cluster attached to a central stem. Besides ensuring the survival of the plant during adverse weather conditions or dormancy, together also provides energy to the developing plant and on depletion of the stored food in them these die and renew themselves annually producing multifold tubers in the process and so the cycle continues. Tuberous roots such as sweet potatoes usually form in clusters and may grow under soil surface vertically to laterally. The enlarged area of the tuber can be produced at the end or middle of a root or may involve the entire root. Though different in origin, it is similar in function and appearance to a stem tuber. The massive enlargement of secondary roots typically represented by sweet potato has the internal and external cell and tissue structures of a normal root; they produce adventitious roots and stems which again produce adventitious roots. In root tubers, there are no nodes and internodes or reduced leaves. The proximal end of the tuber which is attached to the parent plant has crown tissue that produces buds which grow into new stems and foliage on getting proper environment. The distal end of root tuber normally produces unmodified roots. In stem tubers the order is reverse, with the distal end producing stems. As the shoots of the new plant grow the stored reserves of the tuber are consumed in the process of producing the new roots, stems and reproductive organs; any remaining root tissue dies concurrently to the plant's regeneration of the next generation of tubers. Plants with tuberous roots may be long-lived. These plants are propagated in late summer to late winter by digging up the tubers and separating them making sure that each piece has some crown tissue for replanting.

Bulbs, Corms, Rhizomes, Fleshy Taproots and Stolons

Rhizomes, bulbs and corms are actually underground stems and not roots. Tubers can be either stem tuber, aerial tuber or a tuberous root (see above). These underground stems are storage containers for the plant.

Bulbs such as those of onions, usually have layers or scales of fleshy dormant leaves. The true bulbs as in case of onions are divided into layers including a papery outer layer. It is comprised of a plant's stems and leaves. The bottom of the bulb is a compacted stem with roots growing from it. Layers of nutrient-filled leaves sit at the bottom of the bulb and surround a bud that eventually becomes a flower. A bulb actually contains the embryo of a new plant. A lengthwise cross section would display a tiny stem and flower as well as fleshy modified leaves called scales encircling the embryo. These scales serve as food reserves for the tiny plant. There are two types of bulbs ~ one is like an onion with an onion-like skin called a tunic, and other has overlapping scales like garlic and no overall outer skin. Propagation is done by dividing bulbils (lateral buds on the base-plate) of bulb and replanting or storing until the next planting season.

Corms are solid units and generally have a rough husk-like layer of protection as leaf bases around them, and a basal plate where these were attached with mother corms. *Amorphophalus* produces corm-like structure but is devoid of basal plate, however, it is a structure in between the tubers and rhizomes but the shape being corm-like. Corms have the swollen underground stems for nutrient storage similar to those of stem tubers but they have a basal plate (the flat part of the plant where the root grows) while tubers don't. They are either rounded or slightly flattened at the top as well as at the bottom. A corm consists of one or more internodes, with at least one growing point, generally with protective leaves modified into skins or tunics though *Amorphophalus* does not have either internodes or shredded tunics on its body. Rightly, the *Amorphophalus* should be called only rhizome though its shape is not irregular albeit morphologically it matches with other visible characters of a rhizome. The tunic of a corm is formed from dead petiole sheaths ~ remnants of previous years leaves. The parent corm dies back but produces cormels or cormlets from buds on the top or side of the original corm. Large new corms may flower the following year but the smaller cormels may take several years to bloom. Cormous plants generally can be propagated by cutting the corms into sections each with at least one bud before replanting. Here also *Amorphophalus* lacks as it has visible bud only in the top. Each section with a bud can generate a new corm. Cormels can be separated from the shrivelled parent corm and planted separately.

Rhizomes are modified and swollen stem-like structures that grow horizontally forming roots from the bottom and shoots from the upper surface, sending shoots upwards. They grow underground and generally have a thickened stem that is used for storage. Rhizomes have eyes or buds that appear along the top and sides, which grow upward to produce new stems and foliage. Some rhizomes are fleshy as *Colocasia* while some are slender and elongated with internodes and often appear as nothing more than roots. These are multiplied through cut sections containing at least one eye. Rhizomes can grow in any direction. Both tubers and rhizomes can create new plants, but they do it in slightly different ways. Tubers have nodes (often called 'eyes') that appear anywhere on the flesh and sprout both new shoots and new roots while rhizomes sprout roots along the bottom and sides of the growth and stems along the top and sides.

Root crops have modified roots to function as storage organs while tuber crops have modified stems or roots to function as both storage and propagation organs. As such, the modified roots of root crops cannot propagate new crops whereas the modified stem or roots of tuber crops can propagate new crops. Tubers are frequently confused with plants having edible roots such as the carrot or radishes but they are not the same as carrots and radishes produce only tap roots. The fleshy portions of carrot or radish that we eat are actually a long and thick **taproot** and not a tuber.

Stolons are stems that creep horizontally aboveground. These stems or runners contain nodes or joints. Nodes are where the new roots and plants develop. Plants such as strawberry and spider plant have stolons. Division can be made by separating sections containing a node and then planting the section.

1

African Yam Bean [*Sphenostylis stenocarpa* (Hochst. ex A.Rich.) Harms]

Family: Fabaceae (the Pea family). **Subfamily**: Faboideae.

Common names: African yam bean, Wild yam bean and Yam-pea.

Synonyms: *Dolichos stenocarpus* Hochst. ex A. Rich., *Sphenostylis ornata* A. Chev., *Vigna ornata* Welw. ex Bak. (Anon., 2009).

Introduction and Origin

The word *Sphenostylis* originates from a Greek word *sphen*, meaning 'wedge shape' (Allen and Allen, 1981). Seven species are within the genus *Sphenostylis* but *S. stenocarpa* is the most popular and economically important among these species (Porter, 1992).

Three species of *Sphenostylis* are important food sources in Africa including the under-exploited *S. stenocarpa* (en.wikipedia.org/wiki/Sphenostylis) which originated in Ethiopia but is now widely cultivated in tropical Africa, especially West Africa, in Cameroon, Ghana and Nigeria (Porter, 1992). Both wild and cultivated types now occur in tropical Africa as far south as Zimbabwe, throughout West Africa from Guinea to southern Nigeria, being especially common in the latter country; and in Togo and the Ivory Coast, and in East

Africa from northern Ethiopia (Eritrea) to Mozambique including Tanzania and Zanzibar (Humanity development library 2.0).

African yam bean is an underutilized tropical African tuberous legume. Although it is grown for seeds and tuberous roots to be used as food (Okorie, 2018; en.wikipedia.org/wiki/Sphenostylis) in many parts of Africa (Klu *et al.*, 2001; Oshodi *et al.*, 1995) but the tubers which are consumed in most parts of East and Central Africa have not been given much attention in West Africa [Researchgate (a)]. Despite its nutritional benefits and potentials as a food security crop, there is low awareness of its productive and nutritional value [Researchgate (a)]. If the crop could be grown in large quantities, it may become the important source of protein needed by the people of sub-Saharan Africa (en.wikipedia.org/wiki/Sphenostylis).

Yam bean vines in field

Growing African yam bean has dual crop benefit of producing seeds as well as tubers at the same time as both are good source of protein, carbohydrate, dietary fibre (Ajibade *et al.*, 2005) and other nutrients. African yam bean with a good proportion of high protein and carbohydrate contents could be a promising food crop that can be useful in combating protein-energy malnutrition. Moreover, this useful hardy crop can thrive in extreme conditions including high rainfall, high acidity and infertile soils. The plant can tolerate an annual rainfall of less than 1,000mm (Okorie, 2018). It is also resistant to several major crop pests (en.wikipedia.org/wiki/Sphenostylis) and has high nitrogen-fixing ability (Klu *et al.*, 2001). At present, African yam bean has been relegated to an unimportant underutilized crop, predominantly grown by the older generation of farmers (Saka *et al.*, 2004). The genetic resources of this crop are in the hands of these farmers with little research effort concentrated on improving upon the quantity and quality of its yield thus threatening its survival (Ekpo, 2006).

The African yam bean legume is an important source of calcium and amino acids apart from being rich in protein and starch (Oshodi *et al.*, 1995). It contains amino acids that are important for development in early pre-school and school-going children as well as adults (Oshodi *et al.*, 1995). The yam bean is a useful source of nutrients for many African communities with a nutritional value comparable to that of the soybean though its cooking time is much longer (4–6 hours).

Botany

It is generally grown as an annual (Heuze and Tran, 2016) though is a perennial as it regenerates from its rootstocks at the commencement of the rainy season every year (Oagile *et al.*, 2007). African yam bean is a herbaceous but vigorous growing climbing vine (Humanity development library 2.0) reaching to a height of 1-3m or more depending on the length of the staking materials and cultivar. The crop produces primary and secondary branches akin to the varietal character which twine on the stakes or any nearby tree or other plant similar to cassava (Adewale and Odah, 2013). The leaves are trifoliate with oval leaflets that are 2.7-14cm long and 0.2-5.5cm broad (Heuze and Tran, 2016). The leaves of some cultivars are linear, ovate or elliptic (Oagile *et al.*, 2007) and could even be pigmented with pale green, green or dark green colouration.

African yam bean – with leaves, flowers, pods, seeds and tubers

Depending on the cultivar, climatic condition and cultural practices, the flowers emerge after 90–150 days of planting (Oagile *et al.*, 2007) which remain for 50-70 days. Four to ten flowers are arranged in racemes on long peduncles usually on the primary and secondary branches (Hutchinson and Dalziel 1958; Dukes, 1981). The flowers are obligate selfer and cleistogamous that open late in the evening till early morning and close before noon (Popoola *et al.*, 2011). Bright mauvish-pink, purple or greenish-white flowers are about 2.5cm in length, borne out on stout axilliary peduncles (Humanity development library 2.0). Standard petals twist slightly backwards on themselves at anthesis (Hutchinson and Dalziel 1958; Dukes, 1981). Flowers seem to exhibit both cross- and self-pollination (Adewale, 2011), where after fertilization up to 6 pods per peduncle may result (Hutchinson and Dalziel 1958; Dukes, 1981). The pollen grains produced by the crop possess tricolporate, fenestrate and scabrate exine with three colpus (Popoola *et al.*, 2011). The pollen size and fertility varies from one cultivar to the other.

Grain-bearing glabrous pods mature in 140-210 days after planting. According to Klu *et al.* (2001), the pods mature about 30 days after successful fertilization. Usually linear and long unicarpel pods when mature turn brown (Hutchinson and Dalziel 1958; Dukes 1981), hard and tough (Heuze and Tran, 2016). The pods are usually 20-30cm long (Heuze and Tran, 2016) and 1-1.5cm broad depending on the cultivar and growing conditions. Pods may or may not have the seed cavity ridges and may be of shattering or non-shattering types. Seed pods which may sometimes be flat or raised in a ridge-like form on both margins are usually prone to shattering, dehiscing along the dorsal and ventral sutures when dry (Oshodi *et al.*, 1995; Adewale *et al.*, 2010).

The seed pods generally contain 20-30 round, oval, oblong, rhomboid, ellipsoid or truncated seeds while the seed texture could be smooth, rough or wrinkled depending on the cultivar. Seeds show considerable variation in size and colour ~ the largest being usually 1cm long and 7mm wide (Oshodi *et al.*, 1995; Adewale *et al.*, 2010). The varieties with various seed sizes and colours (self or mosaic type) are available (Oshodi *et al.*, 1995; Adewale *et al.*, 2010). Mono coloured seeds may be white, grey, cream, light to dark brown and purple to black.

Some cultivars produce 1-4 potato- or sweet potato-like brownish-orange, cream or pink tubers while others do not produce any tuber (Adewale, 2011). The flesh is usually white and watery. Edible tubers which mature in about 210–240 days after sowing vary in shape and sizes akin to the cultivar group. They are about 5-7.5cm long with 50-150g weight though under favourable conditions they can weigh up to 300g and are usually spindle-shaped (Humanity development library 2.0). There is some evidence that yields of seeds and tubers are inversely related.

Genetics

Large genetic variability in traits among African yam bean has been reported by various workers. Based on the study of 21 accessions of African yam bean, Adesoye and Nnadi (2011) reported range of chromosome counts between 2n = 20 to 24 with 2n = 22 being the most frequent, and the chromosomes of the accessions studied were generally small in size ranging between 0.58 to 1.84μm, and on the basis of chromosome size ~ different accessions could be divided into three different groups. This information can be utilized for further progress in karyological and cytogenetic research towards improvement of this plant.

Varieties

This bean has a high level of genetic variability which may be useful in the hybridization programme in order to increase food production and sustainability. Though yam bean is in cultivation since long but there had been little selection work on specific traits (Klu *et al.*, 2001). The centre of diversity spreads from the west through to the east and southern parts of Africa and these areas are suspected to host the genetic resources of African yam bean. Different local varieties have been categorized on the basis of seed size and colour [mosaic types/multi-coloured or mono-coloured (white, grey, cream, light or dark brown, purple, black, *etc.*)].

Nutritional Value

Both the seeds and tuberous roots are good source of protein, carbohydrate, dietary fibre and other nutrients.

Approximate composition of African yam bean tubers in percentage as reported by various workers:

Range of Nutritional Values in Tubers of African Yam Bean as Reported by Various Workers

Component	*Values*	*References*
Caloric value (Kjg-1)	1,451	(Ojuederie and Balogun, 2019)
Carbohydrate (per cent)	54.47-79.24	(Ojuederie and Balogun, 2019; Ameh, 2007; Konyeme *et al.*, 2020)
Crude fat (per cent)	1.28-2.77	(Ojuederie and Balogun, 2019; Konyeme *et al.*, 2020)
Crude fibre (per cent)	2.4-9.8	(Ojuederie and Balogun, 2019; Ameh, 2007; Konyeme *et al.*, 2020)
Crude protein (per cent)	8.32-14.45	(Ojuederie and Balogun, 2019; Ameh, 2007; Konyeme *et al.*, 2020)
Nitrogen free extract (per cent)	71.18	(Ameh, 2007)
Ether extract (per cent)	0.6	(Ameh, 2007)
Minerals	*ppm*	*References*
Calcium	500-2680	(Ojuederie and Balogun, 2019; Ameh, 2007)
Iron	80-3,160	(Ojuederie and Balogun, 2019; Ameh, 2007; Konyeme *et al.*, 2020)
Magnesium	1,670-4,670	(Ojuederie and Balogun, 2019; Ameh, 2007)
Manganese	114.06	(Konyeme *et al.*, 2020)
Phosphorus	20-66,010	(Ojuederie and Balogun, 2019; Ameh, 2007)
Potassium	7,400-48,790	(Ojuederie and Balogun, 2019; Ameh, 2007)
Sodium	21,440	(Ameh, 2007)
Zinc	78.56-3,750	(Ameh, 2007; Konyeme *et al.*, 2020)
Copper	31.02-1,600	(Ameh, 2007; Konyeme *et al.*, 2020)

Other Constituents	*Quantity*	
Moisture (per cent)	39.4-55.5	(Ojuederie and Balogun, 2019; Ameh, 2007; Konyeme *et al.*, 2020)
Total ash (per cent)	2.44-7.69	(Ojuederie and Balogun, 2019; Ameh, 2007; Konyeme *et al.*, 2020)

Table source: Frontiersin.

Konyeme *et al.* (2021) through their studies (website: "dx.doi.org; ajol.info and Researchgate (b)") using standard analytical procedures to determine the proximate and mineral composition of fresh tubers of 17 African yam bean (AYB) accessions, reported approximate compositions in percentage as ash content 4.59-9.99 per cent, carbohydrate 46.59-66.52 per cent, crude fibre 6.93-12.13, fat 1.06-4.04 per cent, moisture content 11.36-21.91 per cent and protein 4.91-14.50 per cent. The range of mineral content evaluated were calcium 1.53-5.82, copper 10.59-44.93, iron 63.52-240.48, magnesium 0.59-2.26, manganese 42.25-160.01, nitrogen 0.75-2.23 per cent, potassium 1.34-5.08, sodium 0.05-0.22 and zinc 28.24-106.93. Approximate variables in the tubers at $P<0.05$ significantly distinguished the 17 AYB accessions.

Descriptive Statistics of the Proximate and Mineral Profile on 17 African Yam Bean Accessions

Variable	*Mean ± SE*	*Minimum*	*Maximum*
N	1.49±0.09	0.76	2.23
Protein	9.74±0.58	4.91	14.51
MC	15.50±0.77	11.37	21.92
Ash	7.69±0.37	4.59	9.99
Crude Fibre	9.80±0.42	6.94	12.14
Fat	2.77±0.21	1.07	4.05
CHO	54.47±1.25	46.60	66.52
Ca	4.245±0.30	1.54	5.83
Mg	1.66±0.11	0.60	2.26
K	3.58±0.26	1.34	5.09
Na	0.15±0.01	0.06	0.22
Mn	114.06±8.05	42.26	160.01
Fe	171.14±12.02	63.52	240.54
Zn	78.56±5.56	28.24	106.93
Cu	31.02±2.32	10.59	44.94

†N=Nitrogen, MC= Moisture, CHO= Carbohydrate, Ca= Calcium, Mg= Magnesium, K= Potassium, Na=Sodium, Mn= Manganese, Fe= Iron, Zn- Zinc, Cu= Coppe

The tubers are rich in starch and protein. Dry matter is approximately 35 per cent of which starch is about 80 per cent and protein about 14 per cent, ranging from 12.5 to 19 per cent for six varieties; however, some analysts have indicated that the non-protein nitrogen can be 50 per cent of the crude protein nitrogen (Humanity development library 2.0).

Nutritional Value of Seeds

African yam bean seeds have good nutritional profiles with high levels of protein, carbohydrate and other nutrients comparable with those of other common legumes. Its seed is only second to soybean in its protein content and the tuber is also rich in carbohydrate and protein. Moreover, seed processing by oven drying reduced the level of antinutrients with low effect on nutritional quality. Analytical figures for the seeds showed dry matter content about 90.5 per cent. The dry matter composition comprised of protein 24-28 per cent; fat 1.5-2 per cent; total carbohydrate 74.1 per cent; fibre 5.2-5.7 per cent; ash 2.8-3.2 per cent; calcium 61mg/100g; and phosphorus 437mg/100g. Amino acid content of the protein is very similar to that of soya bean though rather higher in histidine and iso-leucine. Energy content of the seeds per 100g dry matter is 1,640 kJ (Humanity development library 2.0).

Approximate composition of African yam bean (per cent) as obtained from various workers are ash 1.86–5.35, protein 19.53–29.53, carbohydrate 49.88–63.51, fat 1.39–7.53, fibre 2.47–9.57 and moisture 1.93–13.30 (Nwokolo, 1987; Adeyeye *et al.*, 1994; Nwosu, 2013; Abioye, 2015; Adamu *et al.*, 2015; Ade-Omowaye *et al.*, 2015; Baiyeri *et al.*, 2018; Anya and Ozung, 2019).

Phytochemicals in Tubers

As per websites "dx.doi.org; ajol.info and Researchgate (b)", Konyeme *et al.* (2021) followed standard lab methods to ascertain the presence *cum* quantity of some phytochemicals in the tubers of 17 African yam bean accessions. Result showed that variations existed among the studied accessions for 11 phytochemical constituents. The ranges found per 100g were 22.195-183g total alkaloid, 4.338-14.733g glycosides, 7.732-14.037g flavonoids, 16.448-32.287g phenols, 2.519-8.938ppm oxalate, 1.22-4.340ppm tannin, 1.475-5.232ppm saponin, 0.261-0.928ppm hydrogen cyanide, 1.532-5.435ppm phytate, 1.088-3.858ppm trypsin inhibitor and 11.537-23.904ppm organic acid.

Uses

The tubers are cooked and eaten in the same manner as potatoes (fried, boiled or roasted) which these resemble in flavour. Their protein recovery is higher than

Approximate Composition of African Yam Bean Seeds in Percentage as Reported by Various Workers

Region of Purchase of Sample	*Ash*	*Protein*	*CHO*	*Fibre*	*Fat*	*Moisture*	*References*
Nsukka	4.55	19.62	NR	8.45	3.00	NR	Nwokolo, 1987
Ekiti	2.60	20.51	50.24	6.09	12.20	8.36	Adeyeye *et al.*, 1994
Owerri	3.23	23.24	58.05	5.31	1.64	9.43	Nwosu, 2013
Southern Kaduna	2.64	21.78	63.51	9.29	2.84	9.23	Ndidi *et al.*, 2014
Southwest Nigeria	NR	19.53	NR	NR	1.39	12.04	Ade-Omowaye *et al.*, 2015
Southwest Nigeria	4.30	24.96	59.85	6.26	2.70	1.93	Adamu *et al.*, 2015
Enugu	3.66	27.66	49.88	9.57	4.29	4.94	
IAR and T and NACGRAB	3.47	29.53	52.59	2.75	2.80	9.03	Abioye, 2015
IITA Ibadan	1.86	24.31	58.64	2.47	1.88	13.30	Ajibola and Olapade, 2016
IITA Ibadan	4.28	22.46	53.68	6.47	3.59	9.52	Baiyeri *et al.*, 2018
Cross River	5.35	21.61	NR	7.00	5.12	NR	Anya and Ozung, 2019

NR: Not reported, CHO: Carbohydrate, IITA: International Institute of Tropical Agriculture, IAR and T: Institute of Agricultural Research and Training, NACGRAB: National Centre for Genetic Resources and Biotechnology.

Table source: Toyosi *et al.*, 2020.

seeds (Sanderson, 2005). The seeds are also eaten but must be soaked in water for about 12 hours before being cooked up to 4-6 hours.

Adverse Effects

Although no adverse effects have been found in literatures but some health problems have been discovered in relation to consumption of seeds including flatulence, stomach cramps, diarrhoea and dizziness. These problems result from the way they are usually cooked. These health problems can be resolved by exposing the yam bean seeds to pre-cooking treatments such as lactic acid fermentation using a low-level technological process. The fermentation method also reduces the amount of time and energy to produce a viable food product from the yam bean.

Propagation and Growth

Both seeds and tubers can be used for propagation of African yam bean. Tuber planting is done usually at the start of the rainy season whenever the rain becomes stable (Klu *et al.*, 2001). Seed germinates 5–7 days after sowing exhibiting hypogeal germination. The flowers emerge after 90–150 days of planting (Oagile *et al.*, 2007) and remain for about 50-70 days. After fertilization, 4 to 8 pods per peduncle are formed. Pods bearing the grains mature from 140 to 210 days after planting (Anon., 2009) or about 30 days after successful fertilization (Klu *et al.*, 2001).

Depending upon the cultivars the tubers usually mature in 210 to 240 days after sowing/planting (Anon., 2009) though can be harvested in 150-240 days.

Factors Effecting Growth

Climate and Soil

The plant is specially suited to lowland conditions but can be grown up to 1,800m agro-ecology ranging from savannah to rain forest provided there is a combination of adequate rainfall (1,000mm or more during the growing season) with reasonably good drainage (Ibirinde *et al.*, 2019; Humanity development library 2.0). Although African yam bean is a hardy crop and can thrive in extreme conditions including high rainfall, acidity and infertile soils, it grows well on deep, loose, sandy and loamy soils with good organic content and good drainage (Anon., 2009). It grows better in regions with annual rainfall between 800 and 1,400mm and where temperatures are between 19 and 27°C (Anon., 2009).

Field Preparation and Planting

Land should be prepared for planting by 3-4 deep ploughings, harrowings and through addition of organic fertilizers to make it fertile. Planting is usually done by hand by often putting 2-3 seeds (or one tuber) to one hole. In the east central state of Nigeria, yam beans are usually inter-planted with yams between April and June. Spacing may vary significantly, and often depends on the crop with which they have been inter-planted. A general spacing when inter-cropped with yam is 45cm apart alternated with yam in 120cm rows (Humanity development library 2.0).

African yam bean is usually grown as a minor crop along with yams, cassava and beans using same stakes as for yam for support though sometimes left to trail on the ground. It is sometimes stated that plants perform better when interplanted than when grown alone (Humanity development library 2.0). The seeds can be planted at the base of the heap or ridges of the main crop(s) or on a tilled land. The crop is also planted at the base of trees which serve as stake. Its sole cropping is rare because of the little importance attached to the crop by the farmers and consumers.

Staking and Weeding

About two weeks after planting, a 3-4m stake should be provided for the vine to climb and twine on. Delayed staking will result in vine spreading on the ground instead of climbing. The field should be free from weeds to avoid competition. Pests and diseases should be kept at a minimum. First weeding should be carried out 3 weeks after planting and every 2 months thereafter. The growing pattern of the crop can suppress weeds as it deprives the growing weeds of sunlight.

Fertilizers

Although African yam beans require little or no nitrogen fertilizer supplement because these plants can derive about 79–97.6 per cent of their nitrogen requirement from the atmosphere through nitrogen fixation. It has been observed that the plant performs maximally when NPK fertilizer is applied at the rate of 60 kg/ha (Togun and Olatunde, 1998; Olaposi and Adarabioyo, 2010). However, further increase in NPK fertilization beyond 60kg/ha proves detrimental to yield and yield-related parameters

Harvest

Pods: The pods are ready to harvest 140 to 210 days after sowing. Overstay of the dried pods on the field leads to seed wastage especially with shattering cultivars. For shattering cultivars it is recommended to start harvesting as soon as the matured pod is becoming dry.

Tubers: The tubers are ready to harvest 150 to 240 days after sowing/planting (Anon., 2009). Towards the end of the dry season, the crop is dug up by hand using hoe and spades. Care should be taken to avoid bruises on the tubers while harvesting so that these remain safe in storage.

Postharvest

Pods: If the harvested mature pods are not completely dry they should be dried before processing and storage. The dry pods should be threshed, seeds extracted and dried to store at 10–12 per cent moisture content. Threshing and willowing should be carried out in a clean environment to achieve a high level of purity in the seed.

Tubers: The wounded or bruised tubers should be separated and properly cured before storage.

Yield

The seed yield of African yam bean can be as high as 3,000kg/ha (Dukes, 1981). Per plant average seed yield is 100-200g and the tuber yield 500g (Anochili, 1984). At Nsukka, estimated yield based on at least a mean of 100 stands of two plants each, of four cultivars, alternated with yams on 120cm rows, amounted to about 3025kg/ha dry seeds and 1875kg/ha fresh tuber weight (Okigbo, 1973).

On a basis of 24,200 plants/ha, tuber yield has been calculated at 1,452-2,904kg/ha depending upon variety. Seed yield on the same basis ranged from 3461-3872kg/ha (Humanity development library 2.0).

Pests and Diseases

Although African yam bean is infested by various pests and diseases at various stages of development in the field and in storage though is less susceptible as compared to cowpea and soybean,

Reported fungal diseases are powdery mildew (*Oidium* sp.) which is parasitised by *Cincinnobolus cesati*. Leaf spot (*Phoma* sp.) and stem rust (*Aecidium* sp.) also infect this plant apart from mosaic virus (Humanity development library 2.0). Pests have not been defined in detail but include *Orthopterous* and *Lepidopterous* insects (Humanity development library 2.0). The larvae and adults of cutworms (*Agrotis* spp), nymphs and adults in case of aphids (*Aphis craccivora*), grasshopper (*Zonocerus variegatus*), *Maruca testulalis, Cydia ptychora*, leaf-rolling caterpillars (*Sylepta derogata*) and leaf miners have been described as causing serious damage to the foliage whereas larvae and adults of *Cydia ptychora, Heliothis armigera, Riptortus dentipes, Apion varium* and *Nezara viridula* have been described as causing damage to flowers (Ameh and Okezie, 2005). Nematodes may attack the root system leading to reduction in yield (Humanity development library 2.0).

Integrated pest management approaches should be adopted in keeping pests and diseases at a minimum to avoid crop failure or reduced yield. These pests and diseases should be timely controlled by adopting cultural, mechanical, biological, chemical methods, or a combination of two or more of these methods.

References

1. Abioye, V. 2015. Proximate composition and sensory properties of moringa fortified yellow maize-ogi. *Afr. J. Food Sci. Res.*, 3. www.sciencedirect.com/science/article/pii/S240584402032301X

2. Adamu, G., O. Ezeokoli, A. Dawodu, A. Adebayo-Oyetoro and L. Ofodile. 2015. Macronutrients and micronutrient profile of some underutilized beans in South Western Nigeria. *Int. J. Biochem. Res. Rev.*, **7**: 80-89.

3. Ade-Omowaye, B.I.O., G.A. Tucker and I. Smetanska, 2015. Nutritional potential of nine underexploited legumes in Southwest Nigeria. *Int. Food Res. J.*, **22**: 798-806.

4. Adesoye, A.I. and N.C. Nnadi, 2011. Mitotic chromosome studies of some accessions of African yam bean *Sphenostylis stenocarpa* (Hochst. Ex. A. Rich.) Harm. *Afr. J. Pl. Sci.* **5**(14): 835-41. DOI: 10.5897/AJPS10.100. ISSN 1996-0824. http: //www.academicjournals.org/ajps.

5. Adewale, B.D. 2011. Genetic Diversity, Stability and Reproductive Biology of African yam bean, *Sphenostylis stenocarpa* (Hochst. ex A. Rich.) Harms, p. 203. Ph.D Thesis, University of Agriculture, Abeokuta, Nigeria.

6. Adewale, B.D. and N.C. Odoh, 2013. A review on genetic resources, diversity and agronomy of African yam bean (*Sphenostylis stenocarpa* (Hochst. Ex A. Rich.) Harms): A potential future food crop. *Sustain. Agric. Res.*, **2**: 10-6.

7. Adewale, B.D., O.B. Kehinde, C.O. Aremu, J.O. Popoola, and D.J. Dumet, 2010. Seed metrics for genetic and shape determination in African yam bean. *Afr. J. Pl. Sci.*, **4**(4): 107–15.

8. Adeyeye, E.I., A.A. Oshodi and K.O. Ipinmoroti, 1994. Functional properties of some varieties of African yam bean (*Sphenostylis stenocarpa*) flour II. *Int. J. Food Sci. Nutr.*, **45**: 115-26.

9. Ajibade, S.R., M.O. Balogun, O.O. Afolabi, K.O. Ajomale and S.B. Fasoyiro, 2005. Genetic Variation in Nutritive and Anti-Nutritive Content of African Yam Bean (*Sphenostylis stenocarpa*). *Trop. Sci.*, **45**: 144-8. https: //doi.org/10.1002/ts.14.

10. Ajibola, G.O. and A.A. Olapade, 2016. Physical, proximate and anti-nutritional composition of African yam bean (*Sphenostylis stenocarpa*) seed varieties. *J. Food Res.*, 5: 67.
11. ajol.info/index.php/sa/cart/view/206274/194503.
12. Allen, O.N. and E.K. Allen, 1981. The Leguminous: A Source Book of Characterization, Uses and Nodulation. Macmillan Publishers Ltd., London, UK, pp. 420-22.
13. Ameh, G.I. 2007. Proximate and mineral composition of seed and tuber of African yam bean, *Sphenostylis stenocarpa* (Hochst. ex. a. rich.) harms. *ASSET*, Ser. B 6, pp. 1–10. doi: 10.4314/pprj.v10i1.35248.
14. Ameh, G.I. and C.E.A. Okezie, 2005. Pests and diseases of African yam bean, *Sphenostylis stenocarpa* (Hoechst. Ex. A. Rich) harms. *Bio-Research*, **3**(1): 14-20.
15. Anochili, B.C. 1984. Tropical Agricultural Handbook. Food Crop Production. Macmillan Publishers, London, pp.48-50.
16. Anon., 2009. Ecocrop database. FAO. http: //ecocrop.fao.org/ecocrop/srv/en/home. (https: //www.feedipedia.org/node/1660).
17. Anon. 2009. Genetic Resources Information Network (2009) Taxonomy for Plants. http: //www.ars-grin.gov/cgi-bin/npgs/html/taxon.pl?35250#dist.
18. Anon. 2009. Germplasm Resources Information Network. National Germplasm Resources Laboratory, USDA, Beltsville, Maryland. http: //www.ars-grin.gov (https: //www.feedipedia.org/node/1640).
19. Anya, M.I. and P.O. Ozung, 2019. Proximate, mineral and anti-nutritional compositions of raw and processed African Yambean (*Sphenostylis stenocarpa*) seeds in Cross River State, Nigeria. *Global J. agric. Sci.*, **18**, p.19.
20. Baiyeri, S., M. Uguru, P. Ogbonna, C.C. Samuel-Baiyeri, R. Okechukwu, F. Kumaga and C. Amoatey, 2018. Evaluation of the nutritional composition of the seeds of some selected African yam bean (*Sphenostylis stenocarpa*). *Agro-Sci. J. Trop. Agric. Food Environ. Ext.*, **17:** 37-44.
21. Dukes, J.A. 1981. Handbook of Legumes of World Economic Importance. Plenum Press, New York, USA, pp. 220-22.
22. dx.doi.org; ajol.info and Researchgate (a)
23. dx.doi.org; ajol.info and Researchgate (b)
24. dx.doi.org/10.4314/jasem.v24i10.12.

25. Ekpo, A.S. 2006. Changes in amino acid composition of African yam beans (*Sphenostylis stenocarpa*) and African locust beans (*Parkia filicoida*) on cooking. *Pakistan J. Nutr.*, **5:** 254-6. https: //doi.org/10.3923/pjn.2006.254.256

26. en.wikipedia.org/wiki/Sphenostylis.

27. Frontiersin: www.frontiersin.org/articles/10.3389/fsufs.2022.726458/full#: ~: text=George per cent 20et per cent 20al.,fiber per cent 20(2.47 per cent E2 per cent 80 per cent 939.57 per cent 25).

28. Heuze, V. and G. Tran, 2016. African yam bean (*Sphenostylis stenocarpa*). Feedipedia, a programme by INRAE, CIRAD, AFZ and FAO. https: //www.feedipedia.org/node/704. Last updated on March 16, 2016, 11: 56.

29. Humanity development library 2.0. Root crops: African yam bean (*Sphenostylis stenocarpa*). nzdl.org.

30. Hutchinson, J. and J.M. Dalziel, 1958. Flora of West Tropical Africa. Vol.1 (2), pp. 232-41. Crown Agents for Oversea Governments and Administrations, Millbank, London.

31. Ibirinde, D., C. Aremu, K. Balogun and L. Oladokun, 2019. Assessment of seed and tuber production potential in varieties of *Sphenostylis stenocarpa* (African yam bean). *Agric. Sci.*, **10**: 870-81.doi: 10.4236/as.2019.107066.

32. Klu, G.Y.P., H.M. Amoatey, D. Bansa and F.K. Kumaga, 2001. Cultivation and use of African yam bean (*Sphenostylis stenocarpa*) in the Volta Region of Ghana. *J. Food Technol.Africa*, **6**(3): 74–7. doi: 10.4314/jfta.v6i3.19292. hdl: 1807/2961.

33. Konyeme, T.E., B.L. Nyananyo and F.B.G. Tanee, 2020. Diversity in proximate analysis of tubers of some African yam bean (*Sphenostylis stenocarpa*) (Hochst Ex. A. Rich.) harms (Fabaceae) accessions. *J. Appl. Sci. Environ. Manage.*, **24**: 1787–93. doi: 10.4314/jasem.v24i10.12.

34. Konyeme, T.E., B.L. Nyananyo and F.B.G. Tanee, 2021. Phytochemical composition of tubers of selected accessions of African yam bean, *Sphenostylis stenocarpa* (Hochst.ex A. Richmond) harms (family fabaceae). *Sci. Africana*, **20**.1-16. 10.4314/sa.v20i1.1.

35. Ndidi, U.S., C.U. Ndidi, A. Olagunju, A. Muhammad, F.G. Billy and O. Okpe, 2014. Proximate, antinutrients and mineral composition of raw and processed (boiled and roasted) *Sphenostylis stenocarpa* seeds from Southern Kaduna, Northwest Nigeria. *ISRN Nutr.*, pp. 1-9.

36. Nwokolo, E. 1987. A nutritional assessment of African yam bean, *Sphenostylis stenocarpa* (Hochst ex A. Rich) Harms, and Bambara groundnut, *Vigna subterranea* L. *J. Sci. Food Agric.*, **41**: 123-9.

37. Nwosu, J.N. 2013. Evaluation of the proximate composition and antinutritional properties of African yam bean (*Sphenostylis sternocarpa*) using malting treatment. *Int. J. Basic Appl. Sci.*, **2:** 157-69.

38. Oagile, O., R. Michael, D. Davey and P.G. Alderson, 2007. African Yam Bean. *J Crop Improv.*,**7**(20): 53-71.

39. Ojuederie, O. B. and M.O. Balogun, 2019. African yam bean (*Sphenostylis Stenocarpa*) tubers for nutritional security. *J. Underutil. Legumes*, **1:** 56–68.

40. Okigbo, B.N. 1973. Introducing the Yam Bean. Proceedings of the 1st IITA Grain Legume Improvement Workshop, IITA, Ibadan, Nigeria, 29th October-2nd November 1973. https: //www.scirp.org/journal/paperinformation. aspx?paperid=93639#ref1.

41. Okorie, P.A. 2018. Assessing the anti-nutritional composition of four varieties of African yam bean in Afikpo town of Ebonyi state Nigeria. *Intern. J. Modern Res. in Engin. and Technol.* (IJMRET), **3**(6): 1–4.

42. Olaposi, A.O. and M.I. Adarabioyo. 2010. Flowering, pod formation and abscission and seed yield of African yam bean (*Sphenostylis stenocarpa*) under varied NPK application. *Agric. J.*, **5**: 273-6.

43. Oshodi, A.A., K.O. Ipinmoroti, E.I. Adeyeye and G.M. Hall. 1995. Amino and fatty acids composition of African yam bean (*Sphenostylis stenocarpa*) flour. *Food Chem.*,**53**(1): 1–6. doi: 10.1016/0308-8146(95)95778-5.

44. Popoola, J.O., A.E. Adegbite, O.O. Obembe and J.O. Agbolade, 2011. Reproductive mechanisms and pollen characterization in some accessions of an underutilized legume (*Sphenostylis stenocarpa* Hochst Ex. A. Rich) harms. *Int. J. Biodivers. Conserv.*, **3**: 185-92.

45. Porter, D. 1992. Economic botany of *sphenostylis* (leguminosae). *Econ. Bot.*, **46**(3): 262–75.

46. researchgate.net/publication/336086622_African_Yam_Bean_Sphenostylis_Stenocarpa_tubers_for_nutritional_security_1_2_Ojuederie_OB_and

47. researchgate.net/publication/351327695_Phytochemical_composition_of_tubers_of_selected_accessions_of_African_yam_bean_Sphenostylis_stenocarpa_Hochstex_A_Richmond_harms_family_fabaceae.

48. Saka, J.O., S.R. Ajibade, O.N. Adeniyan, R.B. Olowoyo and B.A. Ogunbodede, 2004. Survey of Underutilized Grain Legume Production Systems in the Southwest Agricultural Zone of Nigeria. *J. agr. Food Inform.*, **6**: 93-108. https: //doi.org/10.1300/J108v06n02_08.

49. Sanderson, Helen, 2005. The Cultural History of Plants (eds Prance, Ghillean and Mark Nesbitt). *Routledge*. p. 65. ISBN 0415927463.

50. Togun, A.O. and A.O. Olatunde, 1998. Effect of soil applied fertilizer on the growth and yield of African yam bean (*Sphenostylis stenocarpa*). *Niger. J. Sci.*, **32**: 43-8.

51. Toyosi, T. George, O. Obilana Anthony and Samson A. Oyeyinka, 2020. The prospects of African yam bean: past and future importance. *Heliyon.*, **6**(11): e05458.ISSN 2405-8440. https: //doi.org/10.1016/j.heliyon.2020.e05458. (www.sciencedirect.com/science/article/pii/S240584402032301X).

2

American Groundnut (*Apios* Fabr.)

Family: Fabaceae. **Subfamily**: Faboideae.

Synonym: *Cyrtotropis* Wall.

Apios is the Greek word for 'pear', referring to the pear-shaped tubers of some of its members. The species in the genus *Apios* are found in North America and Asia between latitudes of 50° and 20° (Seabrook, 1973). *Apios* consists of 9 accepted species, *viz. A. apios* (L.) MacMill, *A. americana* Medik., *A. carnea* (Wall.) Benth. *ex* Baker, *A. delavayi* Franchet, *A. fortunei* Maxim., *A. gracillima* Dunn, *A. marcantha* Oliver, *A. priceana* B. L. Rob., and *A. taiwanianus* Hosokawa. (Ildis). At least three members of this genus (*A. americana*, *A. fortunei* and *A. priceana*) are known to produce edible tubers (Seabrook, 1973; Hu and Tin, 2005).

I. *Apios americana* Medikus

Synonyms: *Apios tuberosa* Moench and *Glycine apios.*

Other common names: America-hodoimo, Cinnamon vine, Groundnut, Hodoimo, Hopniss, Indian potato and Potato bean.

Introduction and Origin

American groundnut is native to every eastern state of USA as well as Quebec, Ontario and Brunswick in southern Canada (Anon., 2008), natural range being from southern Canada down through Florida and west as far as the border of Colorado (Renolds *et al.*, 1990). *A. americana* can grow in the open sun to partial shade. It prefers low and moist woods, thickets, stream and riverbanks, ponds, marshes, meadows and wet ravines (Voss, 1985; McGregor, 1986). It can be found in cultivation up to 1,000m elevations (Elzebroek and Wind, 2008) from temperate to subtropical climates. Its tubers have traditionally been a staple food among most indigenous peoples of the Americas within the natural range of the plant (Beardsley, 1939).

The American groundnuts are commercially farmed in significant quantities in Japan (en.wikipedia.org). Before the American groundnut was introduced to Japan, the people on the main island of Honshu and the northern island of Hokkaido were already familiar with a native wild plant called hodoimo (*Apios fortunei*), which was occasionally eaten as an emergency food (Kinugasa and Yasuto, 1992). It was probably the effort of Dr. Kiyochika Hoshokawa which promoted the cultivation of this crop in Japan (Iwai and Matsue, 2007). Also there are reports of its cultivation in South Korea where it is grown for its nutritional values (Belamkar *et al.*, 2014).

Subfamily Faboideae is distinguished from other subfamilies of Fabaceae by its bilaterally symmetrical flower, with the uppermost petal largest as well as outermost, and the two basal petals fused together (Judd *et al.*, 1999) so making it looking like the butterfly wings.

Recognized Subspecies/Varieties

The International Plant Names Index identifies the following subspecies/varieties:

Apios americana f. *cleistogama* Fernald, *A.a.* f. *keihneri* Oswald, *A.a.* f. *mccculloughii* Oswald, *A.a.* f. *pilosa* Steyerm. and *A.a.* var. *turrigera* Fernald.

Louisiana State University (https://www.cultivariable.com) hosted a groundnut breeding project in the 1980s and 1990s that turned out a number of superior varieties. The budget was eventually pulled and the project ended without ever introducing the new varieties though some of the people involved took the new varieties with them and at least some of them have been preserved. The LSU varieties are both much larger and higher yielding than wild groundnut. There may be more than a dozen LSU varieties that still exist but tracking them down is a challenge.

Botany

American groundnut is a herbaceous perennial with 1–6m long vines bearing 8-15cm long, alternate and pinnately compound leaves having 4-7mm long deciduous stipules and 5-7 ovate to lanceolate leaflets per leaf, each leaflet with a round (not C-shaped) petiolule. The leaflets can be glabrous or pubescent and are ovate to lanceolate with a rounded base and an acute tip. They range from 2 to 10cm in length but typically 4-6cm long and 1.5-4cm wide (Voss, 1985; Gleason, 1963; McGregor, 1986; Fernald, 1950). Glabrous to lightly pubescent stem bears glabrous to pubescent petioles up to 1.8cm long. Rhizomes are slim and have 2 or more tubers up to 6cm in diameter in a series (Gleason, 1963). This plant shows considerable variation in the amount of pubescence, raceme density and the leaflet size among individuals.

Close-up of flowers

American groundnut usually flowers from July to September in the Great Plains. The flowers are violet-scented. In warm summer climates, groundnut may flower as early as July which is convenient for saving seed. In cooler climates, it may not flower until September. Some varieties may not flower at all. The pink, purple or red-brown flowers are produced in dense and 7.5–13cm long axillary racemes on 2-5cm long peduncles bearing 2-5mm long pedicels (Renolds Berthal *et al.*, 1990). The corolla is papilionaceous (butterfly-like) with retuse petals (bearing a notch at the top). The banner petal has a whitish back and a reddish-brown front. Wings are purplish-brown, keel is strongly curved (Voss, 1985) and sometimes coiled. There are 10 diadelphous (1+9) stamens. The pistil is stipitate with a disc that surrounds the stipe. The ovary is 5.5-7mm long and glabrous to tomentose with 7-16 ovules.

Seeds

The fruit (pod) is straight or little curved, 5–13cm long and 4-6mm wide (Renolds Berthal *et al.*, 1990; McGregor, 1986; Britton and Brown, 1970), glabrous to tomentose when young and can be glabrous or bearing striga (sharp-pointed appressed bristles) when mature; the tip is shortly-aristate to acuminate; the endocarp is silver to off-white (Woods, 2005); the dehiscent 5-13cm long pods may have up to 12 small seeds each (Bruneau and Anderson, 1988). This species experiences low pod and seed production in certain populations. This might be attributed to the many triploid individuals seen in the northern part of the geographic range of this plant.

The 4-5 mm long and glaucous seeds with wrinkled surface and an irregular shape are green when fresh and become dark brown when dry (McGregor, 1986; Woods, 2005).

Comparison with *Apios priceana* Species

The flower of *Apios americana* is purplish-brown and retuse, while *Apios priceana* has a whitish-green flower with a spongy appendage (Gleason, 1963) protruding from the top of each flower. *A. americana* produces multiple tubers compared to only one in *A. priceana*.

Genetics

Though the species is normally diploid with 2n=2x=22 chromosomes though there are even triploid forms (Renolds *et al.*, 1990). Only diploids are capable of producing seeds so can be propagated through seeds and tubers both whereas triploids will produce flowers but not seeds so will propagate only through tuber division (Bruneau *et al.*, 1988). Triploids are generally found in the northern part of American groundnut's range whereas diploids predominate in the southern part of the range (Bruneau *et al.*, 1988). One form of the species, *A.a.* f. *cleistogama* has flowers that do not open at all and form seeds through self-pollination (Fernald, 1950).

Nutritional Values

As per the fresh weight of a tuber of *A. americana*, the carbohydrate (primarily starch) is 36 per cent apart from being an excellent source of calcium and iron (Kinugasa and Watanabe, 1992). Tuber and flower also contain monosaccharides and oligosaccharides. The tuber has more of these types of carbohydrates than the soybean, potato and sweet potato. Calcium content is 10 times and iron 2 times greater than a potato though vitamin C is considerably less (Kinugasa and Watanabe, 1992). Its tubers contain 17 per cent protein by mass which is roughly three times greater than potato. The amino acid balance is good with the exception of cysteine and methionine (Wilson *et al.*, 1987). Fatty acid content of tubers is 4.2-4.6 per cent with linoleic fatty acids predominating (Wilson *et al.*, 1986). Some studies have indicated that tubers contain isoflavines that are linked to a lesser occurrence of prostate and breast cancer. However, in some instances people became sick after eating the tubers. Some studies state that *A. americana* had larger levels of genistein as compared to *A. carnea* and *A. fortunei* (Brown *et al.*, 2014). Genistein is a naturally occurring compound that structurally belongs to isoflavones and is described as an angiogenesis inhibitor and a phytoestrogen.

Nitrogen Fixation

Apios americana shares a symbiotic relationship with certain soil bacteria, *e.g. Rhizobium* (Zomlefer, 1994), forming nodules on the roots and fixing atmospheric nitrogen. A part of this nitrogen is utilized by the growing plant though remaining by other plants growing nearby. This could be a great advantage in comparison to other tuber crops (Putnam *et al.*, 1991).

Uses

All parts of *A. americana* (tubers, shoots, flowers and seeds) can be cooked and eaten. The highly palatable tubers have culinary characteristics of a potato with finer texture though with a little nuttier flavour. As the skin is little woody so its old tubers are usually peeled. The tubers can also be dried and ground into a powder for use as a thickening in soups, *etc.* or can be added to cereal flours when making bread. Small seeds are also a good source of protein and can be cooked and used as peas and beans. These can be ground into a powder and added to cereals when making bread, *etc.*

According to Hartwell (1967–1971), the tubers can be used in folk remedies for cancerous condition known as proud flesh in New England.

Propagation

The diploid species can be propagated using both seeds and tubers though triploids only by division of tubers. Both diploid and triploid can also be propagated from cuttings and tissue culture. The most adapted and convenient method of propagation is however from tubers.

Seeds are pre-soaked for 3-4 hours in tepid water before being sown in February/March in a cold frame. Seeds usually germinate in 1-3 months at 15°C (Rice, 1988). When seedlings become large enough to handle, these are taken out carefully and planted into individual pots. These are kept in light shade in the greenhouse for their first winter and then in late spring or early summer they are planted out in the main field.

Growing this crop from tubers ensures that variety remains true to type and saves a year of growing time over growing from seed. Vegetative propagation through tuber division can be carried out at almost any time though spring is the best time. After digging up the roots the tubers are harvested and generally replanted in spring to the area where plants are to be grown. If the tubers are harvested in winter they should be stored in a frost-free cool and dry place over the winter to plant them in the spring. The tubers lose moisture rapidly once they have been harvested hence are stored in a damp medium such as leafmold.

Growth and Development

The plant forms long thin roots which enlarge at intervals along their length to form the tubers ~ the effect is somewhat like a necklace. The species shares a symbiotic relationship with certain nitrogen-fixing soil bacteria. American groundnut takes several years to reach maximum yield. In a climate with warm summers, growers may probably get reasonable yield with good sized tubers after two growing seasons though it may take three years in milder climates. It rarely produces good yields in a single growing season though it may produce a reasonable crop in warm summer climates.

Factors Influencing Growth and Quality

Apios americana is a plant of the temperate zone where it is found at elevations up to 1,000m. It tolerates a range of climatic variations producing well in cool temperate zones as well as the subtropical conditions of South Florida (Duke, 1983). It can grow well in a sunny position, light dappled or partial shade. It does not like windy situations. The plant prefers a light and rich soil having pH range of 5-7.5 though can tolerate the pH range of 4.5-8.5. In a warm dry situation and in well-drained sandy soil the plants are long-lived with the tuberous roots increasing in size and number each year though aerial plant drying out during fall. It does not require pruning.

It grows best in areas where annual daytime temperatures are within the range of 10-20°C but can tolerate 8-30°C. When dormant, the plant can survive temperatures down to about -40°C, though young growth can be damaged by frosts. It prefers a mean annual rainfall in the range of 1,000-1,200mm but tolerates 700-1,500mm.

Season

The seedlings that are produced from seeds in the greenhouse should not be planted in winter. They are always planted out in main field in late spring or early summer. The tubers can be planted out at any time though spring is the best time.

Field Preparation and Planting

The tubers are quite cold-tolerant and can be planted at any time of year. After usual ploughing and harrowing, tubers can be planted 5-8 cm deep at a distance of about 30-45cm with trellising or 90-120cm without trellising. Mulching can be done to stop competition from weeds.

Intercropping

Though it can be grown without any support but yield recovery is more with something sturdy for its vines to climb. Though by nature it is a perennial plant, however, its vines die back at the end of the growing season, so tall annual crops can provide a good trellis for it. Some plants that have been used successfully are corn, sunflowers and Jerusalem artichokes.

Harvest

The crop may be harvested at any time of the year but two most preferred times are the fall after the vines are killed by frost or in the spring once frozen soil thaws out. The tubers are usually formed in the top 15cm of soil. The tuber is too fragile to come up when uprooting the plant. The tubers can be harvested in their first year but lifting after 2-3 years make this crop quite sizeable. Average yield is about 2-3kg of tubers per plant.

Storage

American groundnut tubers are stored much like potatoes in a cool, humid and dark environment and these will remain fresh for months together. Seed tubers can be stored for a surprisingly long time in the refrigerator. Freshly harvested tubers can be stored in plastic bags in the refrigerator for up to three years.

The most common way of preserving groundnut tubers is to dry and grind them into flour. Since this flour doesn't contain gluten therefore cannot be used to make bread without mixing other flours similar to that of potato flour. The flour also makes an excellent thickener for soups and stews much like arrowroot.

Diseases and Pests

It tends to be mostly pest-resistant. Only some forms of nematodes are likely to attack this crop. Beneficial nematodes can be added to the soil to kill these pests.

There haven't been significant disease issues in cultivation. However, the storage of tubers at inappropriate temperatures and humidity levels can cause severe fungal infection (*e.g.* tuber rot caused by *Fusarium acuminatum*) which adversely affects the tuber quality.

In Baton Rouge (Louisiana), cucumber mosaic virus and *Desmodium* yellow mottle virus are reported to be causal agents of diseases in *A. americana* (Valverde *et al.*, 1990). In case of virus diseases prevention is always better than cure.

II. *Apios fortunei* Maximowicz

Common names: Groundnut or Potatobean; in Japan: Hodo, Hodoimo.

Introduction and Origin

Apios fortunei, a native to Eastern China and Japan is distributed along the NE Coast of Asia, primarily in China, Japan and Korea. In the wild, *Apios fortunei* is often found near brooks (Seabrook, 1973). It is one of three species in the genus *Apios* that are known to produce edible tubers, although it has generally been considered an emergency *cum* famine food or as medicinal plant (Seabrook, 1973; Kinugasa and Yasuto, 1992).

Botany

Apios fortunei is a perennial climbing vine with pinnatifid ovate or lanceolate leaves some 3-7cm long having 3-7 leaflets. The plant produces a single large tuber. Flowers have been variously described as whitish-green, light yellowish-green with a red purple edge on the wing petal, or sulphur-green with rosy wing petals. The showy flowers appear form 6–26cm long pseudo-racemes or terminal panicles. Its legume-pods are linear, 7–8cm long and 5–6mm wide (en.wikipedia.org).

Comparison with *Apios americana*

Apios fortunei is very similar in form to *Apios americana* but has single tuber that grows larger than other *Apios* species but seems less promising for development since the yield is much lower. It may be a better performer in cool summer conditions than American groundnut which makes it an interesting possibility for hybridization.

Nutrition Values

The starch is found to be the predominant carbohydrate though smaller amounts of sucrose and glucose are found but there is no fructose (Seabrook, 1973; en.wikipedia.org).

III. *Apios priceana* B.L.Rob.

Common names: Price's potato-bean, Price's groundnut and Traveler's delight.

Introduction and Origin

Apios priceana, native to the SE United States (Alabama, Mississippi, Kentucky and Tennessee), is usually associated with openings in the forest canopy in mixed hardwood stands where ravine slopes grade into creek or stream bottoms. The plant is used as a food source by indigenous peoples of the Americas and early white settlers. The plant has only 25 known occurrences (Anon., 2010) and is federally listed as a threatened species. The threatened status of the species is primarily due to habitat destruction but other impacts such as disease, predation and historical tuber collection have also contributed substantially (en.wikipedia.org).

Botany

Price's potato-bean is a 3-5m long yellow-green vine that grows from a large and stout potato-like tuber that distinguishes it from other *Apios* species. Like *A. forunei*, it produces a single tuber which can grow larger than American groundnut tubers in the warm summer. Alternately arranged leaves comprising of 7 leaflets are 20–30cm long. Pale-pink whitish-green or greenish-yellow pea-like fragrant flowers appear during summer. The fruit is a long slender pod 10-15cm long. Its seeds are also somewhat different from *A. americana*.

Comparison of the Three Edible Species of *Apios*

	Apios americana	*Apios fortunei*	*Apios priceana*
Flower	Purplish-brown and retuse	Whitish-green, light yellowish-green with a red purple edge on the wing petal, or sulphurous green with rosy wing petals	Whitish-green flower with a spongy appendage protruding from the top of each flower
Tuber	Multiple tubers in series	Only one large tuber	Only one large potato-like tuber

References

1. Anon., 2010. Center for Plant Conservation. *Apios priceana. 2010-03-04. Archived from the original on 2015-09-23.*
2. Anon., 2018. Plants Profile for *Apios americana* (groundnut). plants.usda.gov. Retrieved 24 January 2018.
3. Anon., 2008. USDA, NRCS. The Plants Database. http: //plants.usda.gov/java/profile?symbol=APAM (Jan. 26th 2009).
4. Beardsley, Gretchen, 1939. The groundnut as used by the native peoples of Eastern North America. Papers of the Michigan Academy of Sciences Arts and Letters, **25**: 507–25.
5. Belamkar, Vikas; V.G. Bhattacharya; S.R. Kalberer; W.J. Blackmon; N.T. Weeks; A.D. Farmer; A. Wenger and S.B. Cannon, 2014. Identification of superior germplasm and development of genetic resources for *Apios americana*: a potential new legume crop. *Pl. and Anim. Genome XXII Meeting.*
6. Britton, N.L. and H.A. Brown, 1970. *An Illustrated Flora of the Northern United States and Canada: Vol. II.* Dover Publications, N.Y. Inc.
7. Brown, Kenneth; Jianhua Li; Katherine Wirth; Emily Wilkins; Jim Tufts and Fernando Bahena, 2014. Determination of Genistein and Protein Content in *A. carnea* and *A. fortunei* from China, and *Apios americana. Modern Appl. Sci.*, **9** (1).
8. Bruneau, A., J. Gregory and J. Anderson, 1988. Reproductive biology of diploid and triploid *Apios americana* (Leguminosae). *Amer. J. Bot.*, **75** (12): 1876–83.
9. Duke. J. 1983. Handbook of Energy Crops.

10. Elzebroeck, A.T.G. and K. Wind (eds), 2008. Guide to Cultivated Plants. Wallingford, UK: CAB International Publishing.
11. en.wikipedia.org. en.
12. Fernald, M. L. 1950. Gray's Manual of Botany (8th ed). American Book Company, N.Y., USA.
13. Gleason, H.A. 1963. The New Britton and Brown Illustrated Flora of the Northeastern United States and Adjacent Canada. Hafner Publishing Co., Inc, N.Y., USA.
14. Hartwell, J.L. 1967–1971. Plants used against cancer. A survey. *Lloydia*, pp. 30–34.
15. Hidemasa, S. 2013. Hodoimo America (USA Katamariimo). Archived from the original on 29 September 2013.
16. https: //en.wikipedia.org/wiki/Apios_priceana.
17. https: //www.cultivariable.com/instructions/root-crops/how-to-grow-groundnut/
18. Iwai, K and H. Matsue, 2007. Ingestion of *Apios americana* Medikus tuber suppresses blood pressure and improves plasma lipids in spontaneously hypertensive rats. *Nutr. Res.*, **27**(4): 218–24.
19. Judd, W.S., C.S. Campbell, E.A. Kellogg and P.F. Stevens, 1999. *Plant Systematics: A Phylogenetic Approach.* Sinauer Associates, Inc., Sunderland, Massachusetts, USA.
20. Kinugasa, H. and W. Yasuto, 1992. Nutritional Composition of the tubers of American groundnut (*Apios americana* Medikus). *Sonoda Women's College Studies*, **26**: 209–18.
21. McGregor, R.L. 1986. *Flora of the Great Plains.* The University Press of Kansas. Lawrence, Kansas, USA.
22. Putnam, D.H.; G.H. Heichel and L.A. Field, 1991. Response of *Apios americana* to nitrogen and inoculation. *HortSci.*, **26**(7): 853–5.
23. Reynolds Berthal, D.; J. William; E. Blackmon; E. Wickremesinhe; M.H. Wells; and R.J. Constantin, 1990. Domestication of *Apios americana. Adv. New Crops,* pp. 436–42.
24. Rice, G. (ed.), 1988. Growing from Seed (vol. 2). Thompson and Morgan, USA.
25. Seabrook, J.E.A. 1973. A Biosystematic Study of the Genus *Apios* Fabricius (Leguminosae) with Special Reference to *Apios americana* Medikus. M.Sc. thesis. University of New Brunswick, Fredericton, Canada.

26. Valverde, R.A., R. Provvidenti and C.A. Clark, 1990. Cucumber mosaic virus and *Desmodium* yellow mottle virus infections in wild groundnut (*Apios americana*). *Pl. Dis.*, **74**: 151-3.

27. Voss, E. G. 1985. *Michigan Flora Part II: Dicots.* Cranbrook Institute of Science. Ann Arbor, Michigan, USA.

28. Wilson, P.W.; F.J. Pichardo; J.A. Liuzzo; W.J. Blackmon and B.D. Reynolds, 1987. Amino acids in the American groundnut (*Apios americana*). *J. Food Sci.*, 52(1): 224–5.

29. Wilson, P.W.; J.R. Gorney; W.J. Blackmon and B.G. Reynolds, 1986. Fatty acids in the American groundnut (*Apios americana*). *J. Food Sci.*, **51**(5): 1387–8.

30. Woods, M. 2005. A Revision of the North American species of *Apios* (Fabaceae). *Castanea*, **70**: 85-100.

31. Zomlefer, W.B. 1994. *Guide to Flowering Plant Families.* The University of North Carolina, Chapel Hill, North Carolina, USA.

3

Arrowhead (*Sagittaria* L.)

Family: Alismataceae. **Type species**: *Sagittaria sagittifolia* L.

Synonyms: *Diphorea* Raf., *Drepachenia* Raf., *Hydrolirion* H.Lev., *Lophiocarpus* (Kunth) Miq., illegitimate, *Lophotocarpus* T.Durand, and *Sagitta* Guett.

Introduction and Origin

Sagittaria, a genus of aquatic herbs of temperate and tropical regions comprises species that are mostly native to South, Central and North America though some are also from Europe, Africa and Asia. In the wild, arrowhead plants grow in shallow aquatic habitats of swamps, marshes, pond edges and shallow rivers and streams throughout most of the southern half of Canada, the United States and Mexico (Angier, 1974). Its species go by a variety of common names including Arrowhead, Duck potato, Katniss, Omodaka (in Japanese), Swamp potato, Tule potato and Wapato/Wapatoo (en.wikipedia.org).

Arrowhead is a widely grown aquatic plant for ponds, swimming pools and aquaria. Almost all of its species or varieties that produce fleshy corms are edible and can be used in the same ways. They are a reliable source of starch and carbohydrates even during winter. Arrowhead was a favourite food of many native North American tribes.

Most of the species are aggressive grower and considered aquatic weeds in several areas. Several countries have officially declared some of arrowhead species as invasive due to their widespread root system. Their colonies can quickly take over areas with consistently wet soil. In the wild, individual plants can spread to

1-2m when the roots are submerged in shallow water or on the mud banks of swamps, streams, sloughs and ponds.

Botany

Sagittaria is a monoecious or dioecious herbaceous aquatic perennial that often has surface runners arising from the leaf axils and sometimes corms. Several species produce few to many small- to medium-sized subterranean edible corms at the base of the stem/rhizome. Its diploid chromosome number is 2n=22.

Sagittaria montevidensis Cham. & Schltdl.

It has three morphological growth forms that play an important role in the life cycle allowing the species to adapt to varying environmental conditions, and these are i) submerged with narrow, ribbon-, tape- or spatula-shaped leaves, ii) floating-long-stemmed with oval or egg-shaped leaves, and iii) aerial with arrow-shaped leaves (reflected in the name of the plant). The air and floating leaves are simple and entire as well as stalked. The petioles are 3-4-edged. The leaf blade is linear, lanceolate, ovoid, sagittate or hastate. The leaves are generally arrow-shaped and 1.5-6cm long (Clarke 1977; Medve and Lee Medve, 1990). The scapose flowers comprise of 3 sepals and 1-2cm long 3 deciduous white petals, standing generally above the water and are usually arranged in consecutive whorls (rarely in umbels). Whorled inflorescence may be branched further. Each flower-stalk is surrounded by a lanceolate bract that is shorter than peduncle. In monoecious plants, unisexual male and female flowers get developed in different parts of the plant; top of the inflorescence comprising only male flowers on long and slender stalks possessing 6, 7, 9 or up to 30 stamens; though female or bisexual flowers with up to 1500 spirally arranged superior and free carpels are present at the lower parts on short and thick stalks. Female flowers after being insect-pollinated produce aggregated fruits where each beaked nutlet (achene) that is strongly flattened, usually winged, thin and buoyant for several days.

Sagittaria guayanensis subsp. *lappula* (D.Don) Bogin

Nutritional Value

Every 100g of fresh arrowhead corm contains 103kcal food energy, 4.7g protein, 0.2g fat, 20g total carbohydrate, 0.8g crude fibre, 1.6mg thiamine, 0.25mg riboflavin, 1.4mg niacin, 5mg vitamin C, 1.5mg ash, 12mg calcium, 165mg phosphorus, 22mg sodium, 922mg potassium, 51mg magnesium, 0.7mg zinc and 6.6mg iron (Marles, 2000).

Uses

They are excellent plants for water purification. Small species are usually traded as low-maintenance aquarium plants. The starchy corms that are often formed at the end of axillary stolons of several species are edible raw, though are less bitter when cooked (Angier, 1974). These can be consumed raw, boiled, dried, baked, roasted, mashed, ground into flour, candied with maple sugar or can be cooked like potatoes. The plant stocks below the flower can be eaten raw or cooked after peeling (Moerman, 1998). Though its skin is edible but peeled corms are more palatable. They are high in starch and phosphorous (Medve and Lee Medve, 1990). Native Americans regularly harvest and consume the corms which they refer to as 'wapato'.

A number of tribes are known to have used the arrowhead plant for medicinal purposes. The Navajo use the plant to treat headaches, the Ojibwa eat the corms for indigestion and the Algonquin of Quebec use the root to treat tuberculosis. Some other uses of the arrowhead plant include a Cocopa gambling game and Iroquois corn fertilizer (Moerman, 1998).

Species

Opinion regarding number of species and subspecies differ in *Sagittaria* as per different authorities so there are some 28-41 species and some 12 subspecies out of which Wunderlin *et al.* (2022) have stated about 8 species and 5 subspecies.

Edible Species

Sagittaria cuneata E.Sheld.

Common names: Arumleaf arrowhead, Duck potato, Northern arrowhead or Wapato (en.wikipedia.org).

Synonyms: *Sagittaria arifolia* Nutt. ex J.G.Sm., *S.a.* var. *aquatilis* J.G.Sm. ex Norton, *S.a.* var. *cuneata* Lunell, *S.a.* var. *dimorpha* Lunell, *S.a.* var. *monomorpha* Lunell, *S.a.* var. *polymorpha* Lunell, *S.a.* var. *stricta* J.G.Sm., *S.a.* var. *tenuor* Blank., *S. cuneata* f. *cuneata*, *S.c.* f. *equiloba* Fernald, *S.c.* f. *hemicycla* Fernald, *S. hebetiloba* A.Nelson, *S. paniculata* Blank., *S. marioniae* R.Loxley, *S. sagittifolia* var. *minor* Pursh, and *S. suksdorfii* Gand (en.wikipedia.org; Anon., 2012; theplantlist.org).

Species Accepted by Anon. (2010a), Anon. (2010b), and/or by Wunderlin *et al.* (2022)

Sl.No.	Species	References	Common Names	Region
1.	*Sagittaria ambigua* J.G. Sm.	Anon. (2010a); Anon. (2010b)	Kansas arrowhead or Missouri Arrowhead	Oklahoma to Indiana
2.	*S. australis* (J.G. Sm.) Small	Anon. (2010a); Anon. (2010b); Wunderlin *et al.* (2022)	Appalachian arrowhead, Longbeak Arrowhead	SE US from Louisiana to Florida and as far north as Iowa and New Jersey
3.	*S. brevirostra* Mack. and Bush	Anon. (2010a); Anon. (2010b)	Short-beak arrowhead, Midwestern arrowhead	Central US (Great Plains, Mississippi and Ohio Valleys, Great Lakes); also Virginia and Saskatchewan; naturalized in California
4.	*S. cristata* Engelm.	Anon. (2010a); Anon. (2010b)	Crested arrowhead	Great Lakes region
5.	*S. cuneata* E. Sheld.	Anon. (2010a); Anon. (2010b)	*Sagittaire cuneaire*, Arum-leaf arrowhead, Nothern arrowhead, Wapato, Swamp Potato	Most of Canada including Yukon and Northwest Territories; Alaska; western and northeastern US
6.	*S. demersa* J.G. Sm.	Anon. (2010a); Anon. (2010b)	Chihuahuan arrowhead	New Mexico, northeastern Mexico
7.	*S. engelmanniana* J.G. Sm.	Anon. (2010a); Anon. (2010b)	Engelmann's arrowhead, Acid-water arrowhead	Eastern US from Mississippi to Vermont
8.	*S. fasciculata* E.O. Beal	Anon. (2010a, b)	Bunched arrowhead	North and South Carolina
9.	*S. filiformis* J.G. Sm.	Anon. (2010a, b); Wunderlin *et al.* (2022)	Threadleaf arrowhead	Eastern US from Alabama to Maine
10.	*S. graminea* Michx.	Anon. (2010a, b)	Sagittaire a feuilles de graminees, Grass-leaf arrowhead, Grassy arrowhead	Cuba; much of eastern and central US; eastern Canada; naturalized in Washington State and in Vietnam

Sl.No.	Species	References	Common Names	Region
11.	*S. guayanensis* Kunth	Anon. (2010a, b)	Guyanese arrowhead	Widespread across Latin America, the West Indies, China, India, Southeast Asia; introduced into Louisiana
12.	*S. intermedia* Micheli	Anon. (2010a, b)	Intermediate arrowhead	Greater Antilles, Colombia, southern Mexico
13.	*S. isoetiformis* J.G. Sm.	Anon. (2010a, b); Wunderlin *et al.* (2022)	Quillwort arrowhead	Cuba, from Florida, Georgia, Alabama, Mississippi, Carolinas
14.	*S. kurziana* Gluck	Anon. (2010a, b); Wunderlin *et al.* (2022)	Springtape or Strap-leaf *Sagittaria*	Florida; naturalized in Mariana Islands
15.	*S. lancifolia* L.	Anon. (2010a, b)	Bulltongue, Scythefruit arrowhead, Bulltongue arrowhead	Southeastern US from Texas to Delaware; West Indies; Latin America from southern Mexico to Brazil
16.	*S. latifolia* Willd.	Anon. (2010a, b); Wunderlin *et al.* (2022)	Wapato, Sagittaire latifoliee, Broadleaf arrowhead, Common arrowhead, Duck-potato	Widespread across most of North America, the West Indies and northern South America; naturalized in Hawaii, the western Himalayas and parts of Europe
17.	*S. longiloba* Engelm. ex J.G. Sm.	Anon. (2010a, b)	Long-barb arrowhead, Longlobe arrowhead, Longbarb arrowhead	Southern Great Plains, Arizona, New Mexico, California, Mexico, Nicaragua, Venezuela
18.	*S. macrocarpa* J.G. Sm.	Anon., 2010a, b)	Large-fruited Arrowhead	North and South Carolina
19.	*S. montevidensis* Cham. and Schltdl.	Anon. (2010a, b); Wunderlin *et al.* (2022)	California arrowhead, Giant arrowhead, Sagittaire spongieuse	Widespread across much of US, Mexico and South America
20.	*S. papillosa* Buchenau	Anon. (2010a, b)	Nipplebract arrowhead	Texas, Louisiana, Mississippi, Arkansas, Oklahoma
21.	*S. platyphylla* (Engelm.) J.G. Sm.	Anon. (2010a, b); Wunderlin *et al.* (2022)	Delta Arrowhead, Delta Duck-potato	South-central US with scattered populations in southeast, the Ohio Valley and Washington State; also Mexico and Panama; naturalized in South Australia, Italy, Java, Caucasus

Sl.No.	Species	References	Common Names	Region
22.	*S. rigida* Pursh	Anon. (2010a, b)	Canadian Arrowhead, Sagittaire dressee, Sessile-fruited arrowhead, Sessilefruit arrowhead, Sagittaire dressee	Canada from Quebec to Saskatchewan; common in northeastern and north-central US from Arkansas and Nebraska east to Virginia and New England; scattered populations in California, Idaho, Oregon, and Washington State; naturalised in Great Britain
23.	*S. sagittifolia* L.	Anon. (2010a, b)	Arrowhead, Hawaii arrowhead	Widespread across most of Europe; Siberia, Caucasus, Turkey
24.	*S. sanfordii* Greene	Anon. (2010a, b)	Valley arrowhead, Valley sagittaria	Endemic to California
25.	*S. secundifolia* Kral	Anon. (2010a, b)	Kral's water plantain, Kral's water-plantain, Little River arrowhead	Georgia and Alabama
26.	*S. subulata* (L.) Buchenau	Anon. (2010a, b); Wunderlin *et al.* (2022)	Awl-leaf arrowhead, Narrow-leaved Arrowhead	Eastern US from Louisiana to Massachusetts; naturalized in Great Britain, Azores and Java
27.	*S. teres* S. Watson	Anon. (2010a, b) (2010b)	Slender arrowhead, Quill-leaved sagittaria	Northeastern US
28.	*S. trifolia* L.	Anon. (2010a, b)	Threeleaf arrowhead	Widespread across much of Asia including Siberia, China, Japan, India, Iran, Indonesia, Philippines, *etc.*; also Ukraine and European Russia; naturalized in Fiji and Polynesia

Source: https://en.wikipedia.org/wiki/Sagittaria, Anon. (2010a, b), and/or by Wunderlin *et al.* (2022).

Other Accepted Species (Anon., 2010b)

1.	*S. chapmanii* (J.G.Sm.) C.Mohr (WCSP); from Texas to the Carolinas
2.	*S. aginashii* Makino; Japan, Korea, Primorye
3.	*S. altigena* Hand.-Mazz.
4.	*S. lichuanensis* J.K.Chen, X.Z.Sun and H.Q.Wang; from southern China
5.	*S. macrophylla* Zucc. – Papa de agua; from Mexico
6.	*S. natans* Pall.; From widespread across northern Europe and Asia from Sweden to Kamchatka; Russia, China, Kazakhstan, Japan, Korea
7.	*S. planitiana* G.Agostini; from Brazil, Venezuela
8.	*S. potamogetifolia* Merr.; from southern China
9.	*S. pygmaea* Miq. – Pygmy arrowhead; from China, Japan, Korea, Himalayas, Thailand, Vietnam
10.	*S. rhombifolia* Cham.; from Costa Rica; widespread across much of South America
11.	*S. sprucei* Micheli; from Colombia, Peru, Venezuela, northern Brazil
12.	*S. tengtsungensis* H.Li; from Tibet, Nepal, Bhutan, Yunnan
13.	*Sagittaria* × *lunata* C.D.Preston and Uotila – Sweden, Finland, northern Russia (*S. natans* × *S. sagittifolia*)

Source: Anon. (2010b) and https://en.wikipedia.org/wiki/Sagittaria

Subspecies under *Sagittaria* L. as accepted by Wunderlin *et al.* (2022) and Anon. (2010a)

Sl.No.	*Subspecies*	*References*
1.	*S. graminea* ssp. *chapmanii* (J.G. Sm.) R.R. Haynes and Hellq. ~ Chapman's Arrowhead	Anon. (2010a);Wunderlin *et al.* (2022) mentioned as: *S.g.* Michx. var. *chapmanii* J.G.Sm.; *S.g.* Michx. var. *graminea; and S.g. Michx. var. weatherbiana* (Fernald) Bogin, respectively
2.	*S. graminea* ssp. *graminea* Michx. ~ Grassy Arrowhead	
3.	*S. graminea* ssp. *weatherbiana* (Fernald) R.R. Haynes and Hellq. ~ Weatherby's Arrowhead	
4.	*S. guayanensis* ssp. *guayanensis* Kunth	Anon. (2010a)
5.	*S. guayanensis* ssp. *lappula* (D. Don) Bogin	Anon. (2010a)
6.	*S. lancifolia* ssp. *lancifolia* L. ~ Bulltongue Arrowhead	Anon. (2010a), Wunderlin *et al.* (2022)
7.	*S. lancifolia* ssp. *media* (Micheli) Bogin ~ Bulltongue Arrowhead	Anon. (2010a), Wunderlin *et al.* (2022)
8.	*S. montevidensis* ssp. *calycina* (Engelm.) Bogin	Anon. (2010a)
9.	*S. montevidensis* ssp. *montevidensis* Cham. and Schltdl.	Anon. (2010a)

Sl.No.	Subspecies	References
10.	*S. montevidensis* ssp. *spongiosa* (Engelm.) Bogin	Anon. (2010a)
11.	*S. trifolia* ssp. *leucopetala* (Miquel) Q. F. Wang	Anon. (2010a)
12.	*S. trifolia* ssp. *trifolia* L.	Anon. (2010a)

Source: https://en.wikipedia.org/wiki/Sagittaria, Anon. (2010a), and Wunderlin *et al.* (2022).

Arumleaf arrowhead is a perennial herb native to much of North America including most of Canada (every province and territory except Nunavut) as well as the western and northeastern United States (New England, Great Lakes, Great Plains, Rocky Mountain, Great Basin and Pacific Coast states; including Alaska but not Hawaii) (en.wikipedia.org; Flora of North America @ efloras.org, 2017; Anon., 2017). It is one of the smaller arrowhead species growing in slow-moving and stagnant water bodies such as ponds and small streams. The soil underneath the water can consist of mud or sand. Occasional dry spells without standing water are tolerated if the soil surrounding the root system remains moist. The plant prefers full or partial sun. Because of its emergent or floating leaves, this plant tolerates somewhat muddy water.

Botany

Arumleaf arrowhead, growing to 0.8 m, is mostly an emergent aquatic or floating aquatic (rarely submersed aquatic) plant. Like several other species, its submerged parts appear very different from those growing above the surface or on land. The plant grows from white or blue-tinged corms. Many of the variable-shaped leaves are sagittate (arrow-shaped) with two smaller, pointed lobes opposite the tip. The leaf blades are borne on very long petioles. Submersed leaves are ribbon-like, flat and linear with sharply pointed tips and may grow to 45cm long in deeper water. Flowering

Flowers and Immature seedheads

stems and leaf stalks are hairless. The inflorescence of this monoecious plant which rises above the surface of the water is a raceme made up of several whorls of flowers; the female flowers are in the lowest node whereas the male flowers appear in the upper nodes. The individual flower is up to 2.5cm wide with white petals. The male flowers have rings of yellow stamens at the centre and the female flowers have spherical cluster of pistils which ultimately develop into a group of tiny fruits (Gandoger, 1920). Fruit is a globular head of beaked seeds. The beak is erect at the top of the seed.

Uses

Corms can be eaten raw or cooked. Raw is little bitter but when roasted the taste becomes sweet. The plant has been traditionally used to treat headaches (Moerman, 1998). The corms have been eaten as a treatment for indigestion (Moerman, 1998). The Cheyenne give dried leaves to horses for urinary troubles and for a sore mouth (Hart, 1981). The Klamath use the rootstocks as food (Coville, 1897). The Ojibwe eat the corms against indigestion and as food. These are eaten fresh, boiled, dried or candied with maple sugar. Muskrat and beavers store these in large caches (Smith, 1932). The Northern Paiute use the roots for food (Fowler, 1989).

Sagittaria fasciculata E.O. Beal

Common names: Bunched arrowhead, Duck potato, Indian potato or Wapato.

Indian potato is only known to be found in wetlands in Henderson and Buncombe counties in North Carolina as well as in Greenville and Laurens counties in South Carolina. They need clean, slow-flowing or stagnant water on gently sloped landscapes or in deciduous forests. It prefers shaded areas on sandy loams. This is an endangered plant threatened by land-use conversion, habitat fragmentation and forest management practices. Most of its habitats have been drained (Weakley, 2012) and land conversion appears to be a predominant threat.

Botany

Indian potato grows to 35-38cm tall with long and narrow submerged leaves that are round in cross-section. Emerging leaves are flat, broadly ovate or lanceolate. It is probably the only member of its genus that has arrowhead in one or more of its common names but instead of sagitate leaves it has lanceolate. Submersed leaves are phyllodial, flattened and 9.5-16.5×0.6-1.5cm. Emersed leaves are with petiole nearly terete, 3-25cm, blade linear-lanceolate to ovate and 5.5-8.5×0.5-2.1cm.

Flowering is in spring—summer. Inflorescences are racemes of 2-5 whorls, 4.5-15×2-6cm, peduncles to 35cm, bracts connate, lanceolate, 2-4.5mm, delicate, non-papillose, fruiting pedicels spreading to recurved, cylindric and 1.5-4.5cm. Flowers are about 1cm in diameter, sepals appressed to spreading, not enclosing flower or fruiting head, filaments dilated ±equalling anthers, glabrous, and the pistils are pedicellate and without ring of sterile stamens. Fruiting heads are 5-6mm in diameter, achenes obovoid, abaxially keeled, 2.5-3×1.2-1.5mm, beaked, faces non-tuberculate, wings 1, ±entire, glands 0-1 and beak lateral and horizontal. This plant produces edible corms that are vigorously collected by the native Americans as a food source.

Sagittaria latifolia Willd.

Common names: Broadleaf arrowhead, duck-potato, Indian potato, katniss or wapato; in India: Jathipotia (Asomese) and Koukha (Bengali).

Broadleaf arrowhead is native to southern Canada and most of the contiguous United States as well as Mexico, Central America, Colombia, Venezuela, Ecuador and Cuba. It is also naturalized in Hawaii, Puerto Rico, Bhutan, Australia and much of Europe (France, Spain, Italy, Romania, Germany, Switzerland, the Czech Republic and European Russia) (Anon, 2017). It can be found in wet areas such as ponds and swamps (Spellenberg, 2001). It is marked by the dark green colour of the leaves, forming dense colonies on very wet soils forming long bands following the curves of rivers, ponds and lakes. It likes high levels of phosphates and hard waters.

This plant can be easily cultivated in 15-45cm deep water with no or little current but the soil should have decomposed manure. The corms are planted well-spaced in May with maximum of 12 plants per square metre at the water depth of 5-7cm. These can be multiplied through seeding or division. The corms are harvested from the ground in various ways: through feet adjustments, a pitchfork or a stick; and after digging up, the corms usually float to the surface. Ripe corms can be collected in the autumn.

Botany

Sagittaria latifolia is a variably-sized plant spreading from 2 to 20m having thin and white roots with white corms ranging from 0.3 to 1m long and 0.15 to 0.6m deep which are covered with a purplish skin. The plant has strong roots and can survive through wide variations of the water level, slow currents and waves. They grow in colonies and can cover large areas of ground. The plant produces rosettes of leaves and an inflorescence on a long rigid scape. Extremely variable light green leaves are spongy and solid, 10–50cm long and 1-2cm thin to wedge-shaped like those of *S. cuneata* with parallel venation.

Like several other species of the genus, its inflorescence is a raceme growing about 90cm above water. White flowers 2-4cm wide and whorled by threes appear from July to September (Spellenberg, 2001) with females being usually on the lower part and males on the upper though some specimens may also be dioecious. The flowers have three round white petals and three very short and curved dark green sepals. Male flowers have 25 to 50 yellow stamens and the female flowers have the sphere of green carpels (Radford *et al.*, 1968; Hickman, 1993).

Nutritional Values

Arrowhead, Raw

Nutritional Value per 100g

Energy	414 kJ (99 kcal)
Carbohydrates	20.23g
Fat	0.29g
Protein	5.33g
Vitamins	***Quantity per cent DV†***
Thiamine (B1)	15 per cent 0.17mg
Riboflavin (B2)	6 per cent 0.073mg
Niacin (B3)	11 per cent 1.65mg
Pantothenic acid (B5)	12 per cent 0.599mg
Vitamin B6	20 per cent 0.26mg
Folate (B9)	4 per cent 14μg
Vitamin C	1 per cent 1.1mg
Minerals	***Quantity per cent DV†***
Calcium	1 per cent 10mg
Iron	20 per cent 2.57mg
Magnesium	14 per cent 51mg

Manganese	17 per cent 0.36mg
Phosphorus	25 per cent 174mg
Potassium	20 per cent 922mg
Sodium	1 per cent 22mg
Zinc	3 per cent 0.28mg

Units: μg = micrograms; mg = milligrams; IU = International units.

†Percentages are roughly approximated using US recommendations for adults.Source: USDA Food Data Central

Source: en.wikipedia.org

Uses

This plant produces starchy edible corms that have traditionally been extensively used by native Americans. The corms can be eaten raw or cooked. The taste is similar to potatoes and chestnuts, and they can be prepared in the same fashions: roasting, frying, boiling, and so on. They can also be sliced and dried to prepare flour. Other edible parts include late summer buds and fruits.

Sagittaria macrophylla Zucc.

Common names: Papa de agua (water potato).

Sagittaria macrophylla is endemic to central Mexico primarily in the region close to the nation's capital. It grows in clean, shallow and slow-moving water. It is threatened by habitat destruction due to urbanization (Zepeda *et al.*, 2005). Underground starchy corms are a traditional food source for the people of the region. Its large and hastate (spearhead-shaped) leaves have up to 30cm long blades. Terminal lobe is large and broadly lanceolate while two basal lobes are much smaller and narrower (Tanaka and Nguyen, 2007; Lansdown, 2014).

Sagittaria sagittifolia L.

Common names: Arrowhead; In Japan: Kuwai; in China: Cígu.

Synonyms: *Alisma sagittaria* Stokes, *Sagitta aquatica* (Lam.) St.-Lag., *S. palustris* Bubani, *Sagittaria acuminata* Sm., *S. aquatica* Lam., *S. bulbosa* (Poir.) Donn, *S. heterophylla* Schreb., *S. minor* Mill., *S. tenuior* Gand., *S. vulgaris* Gueldenst., *Vallisneria bulbosa* Poir.

Arrowhead is native to wetlands of most of Europe from Ireland and Portugal to Finland and Bulgaria, and in Russia, Ukraine, Siberia, Japan, Turkey, China, India, Australia, Vietnam and the Caucasus. It is also cultivated as a food crop in some other countries. In Britain it is the only native *Sagittaria* (Tanaka and

Nguyen, 2007; Lansdown, 2014). It can be weedy or an invasive weed. It also appears on state noxious weed lists for 46 states of USA.

The plant will thrive in wet or marshy places in a range of climates from tropical to warm-temperate. Where cultivation is undertaken its requirement is often similar to rice. It is usually propagated using pieces of corm with an axilliary bud. In some countries the field preparation is similar to that used for rice or lowland tarot. This involves ploughing, discing and harrowing to level the surface and providing a well-puddled soil suitable for flooding, into which the pieces of corm are set by hand at a depth of about 20cm. The corms mature in about 6-7 months and are generally dug by hand.

Sagittaria sagittifolia can be weedy or invasive according to the USDA, Animal and Plant Health Inspection Service (APHIS) and Plant Protection and Quarantine (PPQ). It also appears on state noxious weed lists for 46 states (en. wikipedia.org).

Botany

Sagittaria sagittifolia grows in 10–50cm deep water with arrow-shaped leaves having 15-25cm long and 10-22cm broad leaf blades which are borne on a long petiole holding the leaf up to 45cm above water level. The plant also has up to 80cm long and 2cm broad narrow linear submerged leaves. The flowers are 2-2.5cm broad with three small sepals and three white petals and numerous purple stamens.

Uses

The taste of round corm is similar to a potato but somewhat crunchier even when cooked. In Japan, its corms are eaten particularly on the eve of New Year, and in China it is often used in winter hot pots. The young petiole leaves and corms are used for soups and vegetables (Tanaka and Nguyen, 2007). The corms are used as a starchy vegetable after boiling and are a constituent of several Japanese and Chinese meat dishes (Humanity development library 2.0). Formerly in USA, these were much used by Indians though now it is seldom employed as human food there except by some ethnic minorities (Humanity development library 2.0).

It is used in Chinese medicine, and in 2006 seven new ent-rosane diterpenoids and a new labdane diterpene were purified from the plant. Four of these compounds (sagittine A–D) exhibit antibacterial activity against *Streptococcus mutans* and *Actinomyces naeslundii* while another (sagittine E) was active only against *A. naeslundii* (MIC = 62.5 μg ml–1). Recently, the same group identified five new diterpenoids from *Sagittaria pygmaea*. None displayed activity against *A. actinomycetem comitans* while four of the others were active against *A. viscosus* and three against *S. mutans*, of which 18-β-D-3',6'-diacetoxyglucopyranosyl-ent-kaur-16-ene was found most active (Devi *et al.*, 2009).

Corms are sometimes employed as a source of starch in China or as pig-feed. The plant has also been used medicinally for skin diseases and in childbirth (Humanity development library 2.0).

Nutritional Values

The composition of the edible portion of the corms has been reported containing water 70.6 per cent, energy 448kJ/100g; protein 5 per cent; fat 0.3 per cent; carbohydrate 22.4 per cent; fibre 0.9 per cent; calcium 13mg/100g; iron 2.6mg/100g; phosphorus 165mg/100g; potassium 729 mg/100g; thiamine 0.16mg/100g; riboflavin 0.04mg/100g; niacin 1.4mg/100g; and ascorbic acid 5mg/100g (Humanity development library 2.0). Carbohydrate consists mainly of starch with about 2 per cent of sucrose. The starch grains are large, round, oval or rounded-angular with a diameter of up to 30-36 microns. Sagittariol, an anti-inflammatory principle occurs in its plants (Humanity development library 2.0).

Sagittaria trifolia L. syn. *Sagittaria sinensis* Sims

Common names: Chinese arrowhead or Threeleaf arrowhead.

Threeleaf arrowhead is native to Ukraine, European Russia, Siberia, the Russian Far East, Central Asia, China, India, Iran, Iraq, Japan, Korea, Pakistan, Indonesia, the Philippines and many smaller countries in between. It is also naturalized in Fiji, as well as in the Cook and Society Islands in the Pacific. Cultivation of this plant showed that growing in moist but less soggy soil yields more corm than those grown in soggy or aquatic location. It is cultivated by the Meitei community of Manipur.

Botany

It is a glabrous perennial herb growing to about 1m in height forming axillary stolons, often ending with a corm. The leaves are sagittate, apex acuminate to rounded, lateral lobes longer than the middle and the triangular lobe forming an arrowhead. Inflorescence is racemose with 3 to many whorls of 3 flowers, lower 1-3 whorls usually branched; bracts free or connate towards base; flowers unisexual, female flower on lower 1-8 whorls with short pedicels, male flower with pedicels; sepals reflexed, ovate; petals obovate; stamens numerous; and anthers yellow (Ci, 2010). Fruit forms an oblongate head which consists of numerous triangular obovate achenes with laterally bent beak (Lim, 2014).

Phytoconstituents and Nutritional Values

The species was reported to have a rich phytoconstituents like phenolic compounds and flavonoids which can scavenge free radicals in the body and can be a potential antioxidant. In a study, about 40 free radical scavenging compounds were reportedly detected in addition to the nutrients and minerals like carbohydrate, fat, proteins, fibre, calcium, magnesium, potassium, sodium, iron, zinc and copper (Ahmed *et al.*, 2019). The terpenoid, sandaracopimaric acid isolated from the tuber of *S. trifolia* showed a good immunosuppressive activity (Yuan *et al.*, 1993). Several diterpene trifoliones A, B, C and D were isolated which showed inhibitory effects on the histamine release during the *in-vitro* immunomodulatory test in rats (Yoshikawa *et al.*, 1993, 1996).

Uses

Due to its edible underground corms, it is cultivated as a food crop in parts of Asia. The corms are cooked and eaten as vegetables by the Meitei community of Manipur. Starchy corms have a distinct taste with a slight bitterness (Facciola, 1990). The dried corms are ground into flour to be used in making bread and other culinaries. The corm contains proteins, fats, carbohydrates, fibre, calcium, phosphorous, iron, potassium, thiamine, riboflavin, niacin and ascorbic acid (Duke and Ayensu, 1985; Suzuki *et al.*, 1993). The young petiole leaves and corms can

be used in soups and as vegetables. The presence of D-raffinose, D-verbascose, D-fructose and glucose was reported by Li (2008).

The plant is diuretic and is also used against scurvy. The leaf is used against skin diseases. Tuber has got galactofuge activity and discutient and can cause premature birth (Grieve, 1984; Duke and Ayensu, 1985). In Chinese traditional medicine, the essential oil of *S. trifolia* is used during childbirth and to cure skin diseases. In relation to it, the antimicrobial activity of this essential oil showed a significant result against several microorganisms (Zheng *et al.*, 2006). In Vietnam, it is used against pimples and in times of dizziness (Tanaka and Nguyen, 2007).

Propagation

Sagittaria can reproduce *via* several methods. It flowers mostly from summer to autumn and is a prolific seeder, with each plant having the ability to produce hundreds of thousand seeds. Seed production occurs from September to May. Germination can occur anytime when the conditions are favourable. It can reproduce vegetatively by stem or root fragment, and also from corms and lateral offshoots out from the mother plant. During spring the runners are produced that push through the substrate, sending out new plants as they go. These runners can be pinched off and planted. The corms with an axilliary bud are considered a good planting material.

Sexual Propagation

If seeds are being used, the growers should make sure to begin sowing them in late fall so that young plants can be moved outdoors in the following spring. Seeds may have to undergo cold treatment beforehand to ensure that they germinate within a shorter time-frame. Farmers can use a tray setup with a few low pots and approximately 2.5–5cm of water. Ideally, the soil should stay moist or wet throughout the germination and growing periods. The setup can be placed in a greenhouse or cold frame to ensure that the seedlings are protected from adverse elements. Once they are large enough to handle, individual seedlings can be transplanted separately into individual growing pots, or in one large pot maintaining proper spacing. Inter-plant spacing depends on the type of plants being grown.

Asexual Propagation

Similar setup that keeps the soil wet can also be used for vegetative propagation. Corms or offshoots can be removed from the mother plant in spring or fall but they can be planted at any time during the growing period (spring-summer). They are buried at a depth of approximately 5-10cm in fine organic soil. The pointed tip or hook-like part that extends from the corm should face upwards. Compost or manure can be mixed into soil to enhance its nutrient richness. Once the corms have produced new roots and leaves, they can be moved to their permanent positions. Arrowheads grown out of pots can be moved further into the pond as their leaf stalks lengthen.

Growth and Development

Arrowhead leaves form long shoots by autumn with corm formations at the ends. In the spring, new plants develop from these corms by producing runners that push through the substrate, sending out new plants as they go. Plants are fixed in the bottom with the help of these shoot-forming corms. Many species grow wild in ponds, streams, marshy areas and bogs with a few centimetres of water covering their roots. Other varieties of the plant are cultivated by gardeners and grown in bog gardens or areas with very poor drainage. Some species that grow beneath the water surface, especially the dwarf ones are popular for use by aquarium hobbyists. In aquarium, these plants appreciate the addition of CO_2 to the water, though they do well without CO_2 provided that the lighting is sufficient to encourage plant growth.

Most species are low-maintenance when grown in pots. Colonies that are planted directly into the ground may be challenging to control, especially as they may compete with the rest of the pond-edge plants. To prevent self-seeding, the flowers are removed before they develop into seed clusters. These potentially invasive plants are associated with many benefits, especially when colonies are well-maintained or when the spread is restricted to small areas.

Factors Influencing Growth and Quality

Arrowhead plants will thrive in wet or marshy places in a range of climates from tropical to warm temperate. This can grow well under diverse conditions tolerating even freezing temperature but the temperatures between the ranges of 18-28°C and moderate amount of light is suitable for most species. When growing in large containers or aquaria providing a moderate amount of light between the ranges of 2-3 watts per gallon will suffice the need. Regular addition of carbon dioxide and fertilizers rich in iron and trace elements will help in flourishing the growth and development of the plant in an aquarium. Perfect water pH is from 6.5 to 7.5. Wherever its cultivation is to be undertaken, it should be given conditions often similar to that used for rice.

Colonies of several species such as *S. latifolia* can survive outdoors through winter as long as the root systems are situated under a few centimetres of soil. They can withstand temperatures that drop to as less as -20°C. The plant's leaves will naturally die back as temperatures begin to decrease in fall. It is advisable to cut back any remaining leaves before the first frosts occur as the plant's rootstock will go dormant through winter.

Field Preparation and Planting

Arrowhead plants are cultivated in shallow pools or in very damp and wet soils without having any special growing requirements other than maintaining a cover of shallow water 10-12cm deep over the soil. The planting materials can be directly planted in shallow pools or along the side or margins of the large ponds similar to that of paddy planting. This involves ploughing, discing and harrowing to level the surface and providing a well-puddled soil suitable for flooding into which the pieces of corm are set by hand at a depth of about 10cm (Humanity development library 2.0).

If the growing area is a container like aquarium or bathtub, the entire container should be two-third filled with a mixture of 65-75 per cent good quality topsoil, preferably a loam containing some clay and 25-35 per cent well-rotted manure or compost. The soil mix should be added leaving enough head room to add the correct depth of water. The corms can be planted directly into this soil mix and water added until the surface is covered to the appropriate depth.

The growing location may be a large pond than a suitable large pots filled with the above mentioned soil mix, can also be used. After planting in these pots, a topping of gravel or pebbles as mulch is usually done to keep the soil in place. If necessary, bricks can be arranged under the pots to bring them to the correct depth from the surface of the water.

In ponds or shallow pools, it can be planted throughout the year though usually in spring or even in summer (if plant replacement is required).

Sagittaria grows best when provided with a nutrient-rich substrate. If that's not possible, the water column should be regularly fertilized. This plant is very sensitive to low iron levels in the water. Yellowing of the leaves usually indicates that there is an iron deficiency.

Interculture

Any dried or decaying plant parts should be removed. These will attract bacteria and may decrease the pond's oxygen levels as they decompose. If the pond freezes over during winter, growers may choose to bring marginally planted plants indoors by gently digging them out and storing them in a cool, dark area to keep them dormant. They can be replanted the following spring when the final frosts have thawed.

Harvesting

The corms are ready to harvest when the leaves turn yellow and die back. The corms are harvested from the ground in various ways: with the feet, a pitchfork or a stick and after digging up the corms, these usually float to the surface. Ripe corms can be collected in the autumn and are also often found there floating freely.

Pests and Diseases

Arrowhead isn't associated with serious diseases or pests. Aphids or spider mites that tend to infest other nearby plants may find their way to arrowhead colonies. The foliage and flower stalks should be monitored for these pests and can be manually removed as soon as they are sighted. If they begin to occur in large amounts, a detergent mixture or organic pesticide can be used.

A fungal leaf smut disease caused by *Doassansia horiana* has been observed in *S. trifolia* var. *edulis* and *S. latifolia* (Tanimoto and Kusakari, 2000). Black rot of *Sagittaria sagittifolia* caused by bacterium *Bacillus amyloliquefaciens* has been reported for the first time from China (Hwang, 2012). To keep such infections under check, such plants should be uprooted gently and destroyed.

References

1. Ahmed, M., M. Ji, A. Sikandar, A. Iram, P. Qin, H. Zhu and Z. Sun, 2019. Phytochemical analysis, biochemical and mineral composition and GC- MS profiling of methanolic extract of Chinese arrowhead *Sagittaria trifolia* L. from Northeast China. *Molecules*, **24**(17): 3025.
2. Angier, B. 1974. Field Guide to Edible Wild Plants, 26pp. Stackpole Books, Harrisburg, PA, USA.
3. Anon. 2010a. Integrated Taxonomic Information System (ITIS). Retrieved on 3rd October, 2022. https: ##www.itis.gov#servlet#Single RptSingleRpt?search topic=TSN and search value=38903#nulln.
4. Anon. 2010b. World Checklist of Selected Plant Families: Royal Botanic Gardens, Kew. apps.kew.org. Retrieved 2017-01-30.
5. Anon. 2012. World Checklist of Selected Plant Families: Royal Botanic Gardens, Kew. http: //www.theplantlist.org/tpl1.1/record/kew-287210
6. Ci, G.S. 2010. *Sagittaria* Linnaeus. *Flora of China*, **23**: 84-6 (Sp. Pl. 2: 993.1753).
7. Clarke, C.B. 1977. Edible and Useful Plants of California. California Natural History Guides. University of California Press, Berkeley, USA.

8. Coville, F.V. 1897. Notes on the Plants Used by the Klamath Indians of Oregon. *Contr. from the U.S. Natl. Herbar.*, **5**(2): 87-110.
9. Devi, B. Parimala and R. Ramasubramaniraja, 2009. Dental cares and medicinal plants ~ an overview. *J. Pharm. Res.*, **2**(11): 1669-75.
10. Duke. J.A. and E.S. Ayensu, 1985. Medicinal Plants of China. Reference Publication.
11. en.wikipedia.org
12. Facciola, S. 1990. Cornucopia – A Source Book of Edible Plants. Kampong Publications.
13. Fowler, 1989
14. Grieve, 1984. A Modern Herbal. Penguin.
15. Hart, J.A. 1981. The Ethnobotany of the Northern Cheyenne Indians of Montana. *J. Ethnopharmacology*, **4**: 1-55.
16. Hickman, J.C. 1993. The Jepson Manual: Higher Plants of California, 1400p. University of California Press, Berkeley, USA.
17. https: //en.wikipedia.org/wiki/*Sagittaria*.
18. https: //www.vocabulary.com/dictionary/Sagittaria.
19. Humanity development library 2.0. Root crops: Arrowhead (*Sagittaria sagittifolia*). nzdl.org.
20. Joseph, G.Z., 1832. Abhandlungen der Königlich Bayerischen Akademie der Wissenschaften. Mathematisch-naturwissenschaftliche Abteilung 1: 298, *Sagittaria macrophylla*
21. Lansdown, R.V. 2014. *Sagittaria sagittifolia*. IUCN Red List of Threatened Species. e.T167821A43126944.
22. Li, T.S.C. 2008. Vegetables and Fruits: Nutritional and Therapeutic Values, 304p. CRC Press, Boca Raton, USA.
23. Lim, T.K. 2014. *Sagittaria trifolia*. Edible Medicinal and Non-Medicinal Plants, pp. 96-102. doi: 10.1007/978-94-017-9511-1_2.
24. Marles, R.J. 2000. Natural Resources Canada, and Canadian Forest Service. Aboriginal Plant Used in Canada's Northwest Boreal Forest. UBC Press, Vancouver, Canada, 273p.
25. Medve, R.J. and Lee Medve Marry, 1990. Edible Wild Plants of Pennsylvania and Neighboring States. Pennsylvania State University Press, University Park, PA, USA.

26. Moerman, D.E. 1998. Native American Ethnobotany. Timber Press, Oregon, Portland, USA.
27. Radford, A.E., H.E. Ahles and C.R. Bell, 1968. Manual of the Vascular Flora of the Carolinas, 1183p. University of North Carolina Press, Chapel Hill, USA.
28. Smith, H.H., 1932, Ethnobotany of the Ojibwe Indians. *Bull. Public Museum of Milwaukee*, **4**: 327-525.
29. Spellenberg, R. 2001. National Audubon Society Field Guide to North American Wildflowers, 335p. Western Region. Knopf, Germany.
30. Suzuki, A., M. Kaneyama, K. Shibanuma, Y. Takeda, J. Abe and S. Hizukuri, 1993. Physicochemical properties of Japanese arrowhead (*Sagittaria trifolia* L. var. *sinensis* Makino) starch. *Denpun Kagaku*, **40**(1): 41-8.
31. Tanaka, Y. and V.K. Nguyen, 2007. Edible Wild Plants of Vietnam: The Bountiful Garden. Orchid Press, Bangkok, 175p, Thailand.
32. Tanimoto, T. and Si Kusakari, 2000. Leaf smut of *Sagittaria latifolia* caused by *Doassansia horiana. J. Gen. Pl. Pathol.*, **66**: 289–93.
33. Weakley, A.S. 2012. Flora of the Southern and Mid-Atlantic States. Working Draft of 28 September 2012. University of North Carolina Herbarium (NCU), North Carolina Botanical Garden, University of North Carolina at Chapel Hill. Online. Available: http: //herbarium.unc.edu/flora.htm (Accessed 2012).
34. Wunderlin, R.P., B.F. Hansen, A.R. Franck and F.B. Essig, 2022. Atlas of Florida Plants (http: //florida.plantatlas.usf.edu/). [Landry, S.M. and K.N. Campbell (application development), USF Water Institute.] Institute for Systematic Botany, University of South Florida, Tampa, USA.
35. Yoshikawa, M., S. Yamaguchi, T. Murakami, H. Matsuda, J. Yamahara and N. Murakami, 1993. Absolute stereo-structures of trifoliones A, B, C and D ~ a new biologically active diterpenes from the tuber of *Sagittaria trifolia* L. *Chem. Pharm. Bull.* (Tokyo), **41**(9): 1677-9.
36. Yoshikawa, M., S. Yoshizumi, T. Murakami, H. Matsuda, J. Yamahara and N. Murakami, 1996. Medicinal foodstuffs. ll. On the bioactive constituents of the tuber of *Sagittaria trifolia* L. (Kuwai, Alismataceae): absolute stereostructures of trifoliones A, B, C and D, sagittariosides a and b, and arabinothalictoside. *Chem. Pharm. Bull.* (Tokyo), **44**(3): 492-9.
37. Yuan, J.L., R.S. Jiang, Y.W. Lin and W.P. Ding, 1993. Chemical constituents of *Sagittaria sagittifolia* L. *Zhongguo Zhong Yao Za Zhi.*, **18**(2): 100-01, 126.

38. Zepeda G., Carmen and Antonio Lot, 2005. Distribucion y uso tradicional de *Sagittaria macrophylla* Zucc. y *S. latifolia* Willd. en el Estado de México. Ciencia Ergo Sum [online] 2005, 12 (noviembre-febrero) : [Date of reference: 18/abril/2014] Available in: <http: //estudiosterritoriales.org/articulo.oa?id=10412308> ISSN 1405-0269.

39. Zheng X.W., X.D. Wei, P. Nan, Y. Zhong and J.K. Chen, 2006. Chemical composition and antimicrobial activity of the essential oil of *Sagittaria trifolia*. *Chem. Nat. Comp.*, **42**(5): 520-22.

4

Arrowroot (*Maranta arundinacea* L.)

Family: Marantaceae.

Other common names: Araru, Ararao, Araruta, Bermuda arrowroot, Hulankeeriya, Indian arrowroot, Maranta, Obedience plant, St Vincent Arrowroot and West Indian Arrowroot; in **Hindi**: Araaruta, Ararot/Arroroot or Sitalpati; in **Sanskrit**: Sita tavakshira.

Synonyms: *Maranta indica* Tussac, *M. minor* Chantrier ex Andre, *M. ramosissima* Wall., *M. sylvatica* Roscoe ex Sm., *M. tessellata* var. *kegeljanii* E.Morren, and *Phrynium variegatum* N.E.Br., nom. illeg.

Introduction and Origin

Botanically it was named *Maranta* to honour Bartolomea Maranta, a 16th century Italian physician and botanist. Specific name *arundinacea* refers to the 'reed or cane-like' form of the stems. Its common name 'arrowroot' has derived probably from its use as a poultice to remove poison from arrow wounds or from Arawak word *aru-aru* which means 'meal of meals'. Arrowroot is a lesser known minor edible tuber crop. It is found in rainforest habitats. Arrowroot is a plant of moist areas that grows in partial shade near or in the margins of water as it can be found growing on moist stream banks, moist hammocks, shaded ditches, irrigation channels, on the soil of moist to wet forests, along roadsides, disturbed sites, thickets and wasteland at low to lower middle elevations (up to 1,000m). It is common in clearings where light levels are relatively high (Acevedo-Rodríguez and Strong, 2005). In India, it is cultivated mainly in NE States, West Bengal, Asom, and in South India, especially in Kerala in the homesteads and as intercrops in coconut, arecanut and rubber plantations (Ramesh, 2019).

Arrowroot is a starchy root vegetable similar to yam, cassava, sweet potato and taro, having easily digestible starch in its rhizomes. Though often harvested from the wild, it is also widely cultivated in tropical regions for its starch-rich root (Uphof, 1959; Whistler, 2000). It is often grown as an ornamental pot plant and ground cover in tropical gardens and as a pot plant in cooler climates (Whistler, 2000). It has escaped from cultivation and can be found naturalized in a wide variety of substrates and habitats ranging from open areas with full sunlight to deep-shaded sites. The pieces of rhizome left in the soil after harvest are difficult to eradicate as they re-sprout easily and often become a nuisance invasive weed.

Maranta arundinacea, native from Mexico to Brazil, probably originated in the Amazon rainforest of NW Brazil and neighbouring countries, especially Central America, the West Indies (Cuba, Hispaniola, Puerto Rico, Trinidad, Lesser Antilles) and South America (Brazil, Peru, Ecuador, Colombia, Venezuela, Suriname, Guyana and French Guiana) (Anon., 2018c). It is widely cultivated in many of the warm countries and is considered naturalized in Jamaica, Bahamas, Bermuda, the Netherlands, Antilles, India, Sri Lanka, China, Taiwan, Volcano Islands, Mauritius, Reunion, Equatorial Guinea, Gabon, Florida-USA, Cambodia, Indonesia and the Philippines (Anon., 2018a). It is one of the earliest plants to be domesticated for food in northern South America and is now widely grown commercially for producing arrowroot flour mostly in the Caribbean island nation of St. Vincent and the Grenadines ~ the world's largest grower of arrowroot and producer of arrowroot flour. In Kerala (India), where arrowroot is locally called 'bilathi koova' is cultivated to produce an easily digestible starch (Koova, 2016).

Botany

Its diploid chromosome number is 2n=48. Arrowroot is an erect, herbaceous and 0.3-1.8m tall dichotomously branched perennial with thin stem that is usually much forked towards the apex. It bears subterranean rhizomes which are usually long, fleshy, cylindrical and obovoid. It bears alternate, mid-green and lanceolate or ovate leaves having entire margin. The petiole often absent in upper leaves is terete and base-sheathing with a thickened, glabrous to sparsely hairy and pulvinus at the apex. The blade is ovate-oblong, 10-30×3-10cm, rounded to truncate at base, acute-acuminate at top, largest in basal leaves, glabrous or more usually hirsute, green or sometimes variously streaked white or brownish-purple with prominent midrib and numerous pinnately arranged, fine and closely spaced parallel veins.

Its paniculate inflorescence is terminal and often branched, each branch subtended by a deciduous bract and ending in a stalked flower pair. At the end of the long and thin stems, twin clusters of bisexual, zygomorphic, tubular, small and white flowers with asymmetrical petals appear about 90 days after planting, usually in spring and summer. Thin peduncle of flower pair is up to 4cm long. The pedicel of one flower of the pair is 0.7-1.5cm long and other one 0-3mm. About 2cm long flowers have 3 green, free, persistent and lanceolate sepals 1.2-1.8cm long; 3-lobed early caduceus white corolla is tubular, androecium appearing in 2 whorls but attached to corolla, outer whorl consists of 2 petaloid staminodes some 1cm long, inner whorl about half as long, consisting of 1 fertile stamen, a large fleshy staminode and a smaller hooded staminode, fertile stamen with a 1-celled half anther joined to a petaloid appendage, pistil 1 with a 1-celled *cum* 1-ovuled inferior ovary, style adnate to corolla tube, stigma 3-lobed and enclosed by hooded staminode. The oblongoid berry-like fruit is about 7mm long, leathery, brown and glabrous to hairy. The 3-sided seed is scabrous, pinkish with yellow and 2-lobed with basal aril.

Its fleshy, edible and white to dull white to creamy or reddish rhizomes are cylindrical or spindle-shaped, 5-45×2-5cm and covered with overlapping, persistent or deciduous brownish-white scale leaves. The plant rarely produces seed and reproduction is typically by planting part of a rhizome with a bud. It has a shallow root system, with rhizomes that penetrate more deeply into the soil (ecocrop.fao.org). Fresh rhizome is prominently marked with nodal rings and scale leaves which completely encircle the nodes.

Cultivars

Out of many cultivars, 22 were reported to be growing in a Philippine Germplasm Nursery. Two main cultivars recognised in St. Vincent are:

1. 'Creole': It has long thin rhizomes spreading more widely and penetrating more deeply into the soil.
2. 'Banana': It has shorter, thicker but less fibrous rhizomes produced near soil surface which are more easily adapted to mechanical harvesting.

Some varieties being grown in Kozhikode (Kerela, India) are locally known as 'Bilathi Koova', 'Neela Koova' and 'Manja Koova'. Cultivars with blue rhizomes give higher yield of starch though yellow-coloured local cultivars are usually grown.

Nutritional Value and Health Benefits

Arrowroot flour contains several nutrients and is high in protein even greater than that of cassava flour (Aprianita *et al.*, 2014). Since it contains low glycemic index value and high fibre content, therefore, food products made from arrowroot flour are easier to digest compared with those made from other types of flour (Lestari *et al.*, 2017), making it ideal for children and older people who may need gentler food. Its easily digested starch is of high quality composed of simple ovoid or ellipsoid grains of 15-70 microns in length. Food products made by arrowroot flour is also recommended for people with digestive disorder (Silva *et al.*, 2000; Heredia-Zarate and Vieira, 2005; Mason 2009; Silveira *et al.*, 2013). The rhizome contains starch 25-30 per cent, moisture 64 per cent, crude protein 1.6 per cent, fat 0.2 per cent, dextrin *cum* sugars 2.1 per cent and crude fibre 3.9 per cent (https://www.bimbima.com).

Ayurveda describes the properties and action of arrowroot as *rasa* (sweet in taste), *madhur vipaka* (sweet after digestion) and *virya* (cool in effect) (https://www.bimbima.com) as it is *sheet virya. Sheet virya* or 'cool potency' herb subdues *pitta* (bile) and *vata* (wind), gives nourishment and steadiness to the body and supports the building of the body fluids. Such herbs have cooling *cum* anabolic effects on the body and reduce *pitta cum* swelling.

Arrowroot, Raw

Nutritional Value per 100g

Energy	271 kJ (65 kcal)
Carbohydrates	13.39g
Dietary fiber	1.3g
Fat	0.2g
Protein	4.24g
Vitamins	***Quantity per cent DV†***
Thiamine (B1)	12 per cent 0.143mg
Riboflavin (B2)	5 per cent 0.059mg
Niacin (B3)	11 per cent 1.693mg
Pantothenic acid (B5)	6 per cent 0.292mg
Vitamin B6	20 per cent 0.266mg
Folate (B9)	85 per cent 338μg
Minerals	***Quantity per cent DV†***
Iron	17 per cent 2.22mg
Magnesium	7 per cent 25mg
Manganese	8 per cent 0.174mg
Phosphorus	14 per cent 98mg
Potassium	10 per cent 454mg
Zinc	7 per cent 0.63mg

Link to USDA Database entry

Units ~ μg = micrograms, mg = milligrams, IU = International units

†Percentages are roughly approximated using US recommendations for adults. *Source*: USDA Food Data Central

Source: Wikipedia.com

Uses

Arrowroot is a starchy vegetable with a significant amount of protein and several minerals. It is used as a thickening agent and gluten-free flour. It provides over 100 per cent of the DV for folate. Although whole rhizomes can be eaten raw or cooked (boiled/roasted) when still they are not fibrous. Arrowroot is often cultivated for production of highly digestible rhizome-starch that is currently used in food preparations and confectioneries, and for industrial applications. The starch is often used as a thickener in all kinds of foods, dressings, soups, sauces, candies, cookies, and desserts such as puddings and icecreams. Unlike flour, it does not add a mealy taste.

Due to its high starch content, arrowroot may be suitable for gluten-free diets, to promote weight loss, to help treat diarrhoea and to strengthen the immune system. The resistant starch content of arrowroot may stimulate immune system. Arrowroot tuber is a potential source of prebiotics that are a type of fibre that feed beneficial gut bacteria which produce multiple vitamins and absorb key minerals that the immune system needs to function properly. The pulp of fresh rhizomes can be used in poultices to cure wounds and ulcers. It is usually prepared with water or milk to cure stomach problems as in case of poisoning and diarrhoea or to prepare easily digestible food. It helps to relieve acidity, indigestion and colic. The root has also been used traditionally as an antidote for poison and also to treat urinary infections.

Arrowroot starch has several industrial uses, *i.e.* in manufacturing of paper, cardboard, cushions and wallboards. The starch is basic ingredient of powders, glues and soap. Its starch is also used as a base for face powders and in certain types of glues. The rhizomes are also a good substitute for maize in broiler rations and the fibrous residue after starch extraction can be fed to livestock (Heuze and Tran, 2017). Arrowroot cultivars with strikingly coloured leaves especially the brown-purple ones are also popular as ornamentals and can be grown as groundcover. Leaves are used for wrapping.

Arrowroot (Ararot) Starch

Like other pure starches, pure arrowroot is a light, white powder that is odourless when dry but emits a faint peculiar odour when mixed with boiling water. It gets swelled into a perfect jelly on cooking which can be used to make a food that is very smooth in consistency ~ unlike adulterated articles, mixed with potato flour and other starches of lower value which contain larger particles. The pure arrowroot powder feels firm to the finger and cracks like newly fallen snow when rubbed or pressed.

Although the gluten-free arrowroot flour can be used as a replacement for wheat flour for those having gluten intolerance, it does not provide a complete substitute for wheat flour in bread-making because it is relatively high in carbohydrates and low in protein (approximately 7.7 per cent) (Aprianita *et al.*, 2013).

As compared to flour or cornstarch, arrowroot thickens at a lower temperature and is not weakened by acidic ingredients. Moreover arrowroot has a more neutral taste and is not affected by freezing. It does not mix well with dairy where it forms a slimy mixture. It is recommended that arrowroot be mixed with a cool liquid before adding to a hot fluid. The mixture should be heated only until it thickens and removed immediately to prevent the mixture from thinning. Overheating tends to break down arrowroot's thickening property. Two teaspoons of arrowroot can be substituted for one tablespoon of cornstarch or one teaspoon of arrowroot for one tablespoon of wheat flour (wikipedia.com).

Other Arrowroot Plants with almost Similar Uses

Arrowroot is also the name of a starch obtained from the rhizomes (rootstock) of several tropical plants, traditionally from arrowroot plant (*Maranta arundinacea*), but also Florida arrowroot or Coontie palm [from *Zamia integrifolia*], and tapioca from Brazilian arrowroot [or cassava (*Manihot esculenta*)], which is often labelled as arrowroot. East Indian arrowroot [Polynesian arrowroot or pia (*Tacca leontopetaloides*)], Japanese arrowroot [or kudzu (*Pueraria lobata*)], and Queensland arrowroot [or Indian shot/African arrowroot/edible canna/purple arrowroot/Sierra Leone arrowroot (*Canna indica*)] are used in similar ways; their rhizomes can be eaten raw, but usually cooked; rhizomes can be ground into a flour to be used in several ways; they can be used as femine food; the easily digestible starch is used in various food preparations as well as have industrial uses. Another arrowroot plant with nutritional value as a source of starch for Indian foods is Indian arrowroot [or East Indian arrowroot/narrow-leaved turmeric/*tikhur* (*Curcuma angustifolia*)]. Its rhizomes are typically ground into flour which can then be mixed together with milk or water to form a nutritious meal.

Some other species of *Maranta* such as *M. amplifolia*, *M. linearis* and *M. incrassata* also develop starch-storing rhizomes. *M. amplifolia* is very closely related to *M. arundinacea* and is sometimes even united with it. It is a more robust plant with glabrous leaves, having 1.6-2cm long sepals, obligate allogamous and fruiting very rarely. *M. linearis* differs from *M. arundinacea* especially in its linear or very narrowly oblong leaf blades, smaller inflorescences and shorter sepals. *M. incrassata* is a small plant up to 1m tall with ovate leaf blades and simple inflorescences.

Extraction of Starch

Mature arrowroot rhizomes contain about 23 per cent starch. They are first washed then the paper-like scales that have a disagreeable flavour are carefylly cleaned (Rines and Edwin, 1920). The roots are washed again, drained and finally reduced to a pulp. The milky liquid thus obtained is passed through a coarse cloth or hair-sieve and the pure starch which is insoluble is allowed to settle at the bottom. The wet starch is dried in the sun or in a drying house to obtain the arrowroot powder that is quickly packed for market in air-tight cans, packages or cases. In modern factories most of this starch can be extracted by steam grinding. Microscopically the arrowroot starch is oval in shape and with hilum at the proximal end.

Arrowroot Substitutes

Cornstarch can be used as a substitute for arrowroot starch for both culinary as well as cosmetic purposes. About 8g of cornstarch is used for every 5g of arrowroot, about 8g of tapioca flour is used for every 2.5g of arrowroot, some 2.5g of potato starch is used for every 5g of arrowroot and about 2.5g of rice flour is used for every 5g of arrowroot. **Tapioca flour** is a gluten-free flour similar in flavour to arrowroot. **Potato starch** and arrowroot contain similar amounts of amylase, a starch compound that acts as a thickener. **Rice flour** and arrowroot have very similar carb contents but rice flour is higher in starch.

Propagation

Arrowroot can reproduce sexually by seeds and vegetatively by rhizomes and suckers though in cultivation it is rarely propagated by seed. In cultivation, arrowroot can be propagated through suckers and rootstock or rhizomes with two or more nodes each. It is usually propagated from 4-7cm long rhizome apices ('bits') not too thin, preferably with 2-4 nodes, each weighing around 25g. Planting depth is kept usually 5-7.5cm and covered with soil. CTCRI recommended that a distance of 30×15cm planting during early June *i.e.* just before the onset of the monsoon will accommodate 2,20,000 plants/ha which is oiptimum for obtaining high rhizome dry biomass yields to the tune of 13.9t/ha. Bits can be stored for some months before being planted in the field. In parts of Asia the 'bits' are sometimes treated with smoke to aid germination.

In India, it is usually propagated through rhizome bits as per hectare requirement is about 3,000-5,000kg. Suckers are occasionally used for planting when some 30cm tall. Thin rhizome or cigar roots are planted for the shoots to come up within 15 days. Suckers are separated from the clump at harvest and immediately planted at 30-45cm apart. These suckers give rise to new plants which are then uprooted and canopy removed by retaining only 10cm length of the shoot intact with roots.

Growth and Development

Shoot sprouts in 1-3 weeks after planting the tip of a rhizome. The new plant gets established within 6-7 days by forming adventitious roots after which more leaves and a stem develop above ground, and more roots and new rhizome material develop underground. Some new rhizomes get developed as specialized storage rhizomes while others as suckers. The storage rhizomes may not be found throughout the growing season; they are usually initiated only after the plant has reached a certain stage of development. In the deciduous growth habit the plants die back to the rhizome in the dry season but grows continuously under optimum conditions producing suckers and rhizomes; old leaves wither; fleshy rhizomes deteriorate or develop new plants when not harvested. Flowering starts 3-6 months after planting and flowers open in the evening. Some authors claim that arrowroot is predominantly autogamous though certain others say it being obligately allogamous. The flowers seem devised for cross-pollination by insects. Fruit development is rather abundant with about one fruit per 4 flowers but seed viability is poor.

Environmental Factors

In the wild, *M. arundinacea* grows mainly in the understorey of tropical deciduous or semi-deciduous forest near temporary pools and brooklets though sometimes also in dry pine forest. It grows best under warm humid conditions preferring annual daytime temperatures of 23-30°C but can tolerate 17-34°C. Dormant rhizomes can withstand temperatures as low as 5°C (Ecocrop). It requires an average annual rainfall of 1,500-2,000mm or more but with 1-2 dry months though can tolerate between 700-4,000mm (Ecocrop). Arrowroot can tolerate up to 50 per cent shading without notable yield reduction. It can survive waterlogging and saturated soil conditions, however, though under such conditions these do not produce good rhizomes. The weather in Kerala (India) is suitable for arrowroot farming. It can be planted by the onset of monsoon in June and is usually grown as rainfed crop though harvesting season extends from October to May.

Soil

Arrowroot can be grown under sunny or partially shaded position (Whistler, 2000) on various soil types though thrives best on well-drained moist and loose sandy loam fertile soils having pH range of 5-8 (preferable one being 5.5-6.5) (Ecocrop). Friable soils are preferred for better rhizome development and easy harvesting (Ecocrop). Clay soils should always be avoided because there rhizomes are poorly developed, deformed and at harvesting these are broken.

Land Preparation and Planting Method

Field preparation for arrowroot planting starts one month before planting, *i.e.* in May just by 2-3 deep ploughings, followed by planking after every ploughing and collection of polythene shreds, wood pieces, bricks and stones, *via-a-vis* weed rootstocks. Ploughting should be carried out when soil is still moist. When soil has reached to a fine tilth, the beds of convenient sizes are prepared making provision for irrigation channels and bunds. In the beds, the 'bits' weighing about 25g are planted 5-7.5cm deep and about 30-90cm apart preferably at the beginning of the rainy season. Planting distance depends on the mode of cropping, *e.g.* when intercropped with annual crops row distance is wider and plant distance closer, *i.e.* 75×15-20cm and in square planting patterns, however, in sole cropping the appropriate distance is 20-50cm where it permits hand-weeding especially in poor soil conditions. In a studies conducted at CTCRI (Kerala), it was revealed that planting arrowroot bits at 30×15cm spacings (plant density being 2,20,000 plants/ha) was ideal for obtaining high rhizome dry biomass yields to the tune of 13.9 tonnes per hectare. Since arrowroot tolerates shade quite well, it can be grown successfully under perennial tree crops such as coconut and other fruit trees. Many farmers, especially from Catanauan and Quezon (Philippines) traditionally practice 'Bugsok-tani' method of planting where shoots or stump (arrowroot plant remnants after harvest) are replanted at the same time when the rhizomes are harvested, without ploughing and without applying fertilizers and no planting distance is maintained. Detopping in case of arrowroot is the cutting of the top leaves of the plant and excess of the roots before planting with the aim of getting good photosynthetic efficiency which can be increased for advancement of crop maturity.

Manure and Fertilizers

Arrowroots respond well to both organic and commercial fertilizers. Application of cowdung manure or FYM (farmyard manure) at 10 tonnes per hectare at the time of land preparation increases soil organic matter so rhizome yield increases significantly. For its proper growth and yield, usually 350-650kg of ammonium sulphate and 300kg each of single superphosphate and KCl per hectare are used. The entire quantity of P and K can be applied at planting time whereas N in 2-3 split doses.

Irrigation

Arrowroots require plenty of water for its growing. If a dry spell occurs during the first 3-4 months, supplementary irrigation at weekly intervals should be provided as optimum soil moisture throughout the growing period is essential for good yield though drying out of soil between two watering is beneficial.

Interculture

If planting has been done during the dry season it is advisable to **mulch** the beds to prevent evaporation. Mulching also helps controlling the weeds. Young plantings are especially susceptible to weed competition. Arrowroot field usually needs 3-4 **weedings** that can be done manually or mechanically. The field should be kept clean and free from weeds during the first 3-4 months. The crop is also **earthed up**/hilled up along with weeding to facilitate better rhizome development. To save energy for rhizome development, emerging inflorescences should immediately be **sheared off** so that there remains a maximum nutrient flow to the rhizomes (Matthews, 2008). Pieces of rhizome left in the soil after harvest act as ground keepers and often pose a weed problem in subsequent crops because they are then very difficult to eradicate (Anon., 2018e).

Crop Rotation and Intercropping

Successive crops are usually grown on the same land for 5-6 years (Ecocrop). Crop rotation is recommended after at least every 5-6 years for better nutrient and pest management. Arrowroot does very well under coconut plantings provided competing vegetation is kept in check. Thus inside copra plantations, it would be a quite suitable intercrop. It can also be intercropped with papaya, banana, breadfruit, rubber, screwpine and arecanut.

Harvest

Depending mainly on environmental factors and varieties, rhizomes are ready for harvesting after 8–12 months of planting when leaves of the plant begin to yellow and wilt. Harvesting should be done before the next rainy season otherwise the rhizomes will start sprouting. Starch content reaches a maximum in the rhizomes about 12 months after planting but then they are more fibrous and the starch is more difficult to extract. When storage rhizomes are left in the soil for more than 12 months, the starch is gradually converted into sugar. Harvesting can be done manually using a fork or by ploughing carefully. Weeds and aboveground parts are usually buried to provide organic matter, and the 'bits' are separated from the harvested rhizomes to serve as new planting material.

On the larger estates, the harvesting of rhizome usually proceeds from the base of a hill towards the top. Harvesting involves breaking off the rhizome from the shoot. Planting and harvesting may be inter-related; when the rhizomes are harvested the stumps (arrowroot plant remnants after harvest) are replanted at the same time. Mechanical harvesters have recently been introduced allowing faster arrowroot harvesting. Under favourable conditions, the rhizome yield of 30-35t/ha can be obtained where amount of starch obtained is to a tune of 16-18 per cent of the fresh weight of the rhizome. Average per hectare starch yield is 3-3.5 tonnes. CTCRI has developed a mobile starch extraction unit with 20kg/h of crushing capacity compared with 1kg/h through conventional method. Likewise, the starch recovery using the extraction unit is 14-15 per cent as compared to 9-10 per cent through conventional method.

Marketable rhizomes are mature rhizomes as indicated by a rounded tip, mainly preferred by rhizome buyers for its high starch content. Non-marketable rhizomes are mature rhizomes indicated by a pointed tip, not preferred by buyers due to its high fibre content.

Storage

There are two options for storing arrowroot rhizomes ~ one is to store it in processed form, the other to store it unprocessed in a cool dark place such as in pits along the beach. Unprocessed tubers can be stored for up to six months after which they begin to sprout. Processed starch is kept dry and away from weevils, ants, cockroaches and alikes and then this will keep indefinitely as a suitable trade item. Sometimes the rhizomes are stored in dark embedded in dry sand layers.

Yield

Rhizome yield varies considerably ranging from 7-47t/ha. The amount of starch obtained is 16-18 per cent of the fresh weight of the rhizomes. The highest rhizome yield can be obtained with a spacing of 50×30cm and through application of NPK fertilizer @60:60:60.

Pests and Diseases

Sometimes the arrowroot leaf roller (*Calpodes ethlius*) causes defoliation and consequently starch losses which requires to be controlled by using arsenicum-based sprays. Arrowroot does not have any serious diseases or pests though under poor drainage conditions burning disease caused by *Rosellinia bunodes* may cause havoc. In India, sometimes the banded leaf blight disease is caused by *Pellicularia filamentosa* and *Rhizoctonia solani* fungi so in such cases the crop requires to be sprayed with Bordeaux mixture. Bacterial wilt is caused by a pathogen *Xanthomonas marantae* which infects the vascular system of the roots. It can be managed by avoiding planting in areas where disease has been previously detected. Immediately after spotting the infected plants, they should immediately be burnt. Leaf mosaic disease is caused by vector insects such as aphids. The most common symptom of leaf mosaic disease is the presence of small whitish dots on the leaves. Sometimes in South America the crop is attacked by *Ascia monuste, Neocurtilla hexedactyla* and *Scapteriscus vicinus* (Kay and Gooding, 1987; Anon., 2007; Anon., 2018e).

Physiological Disorders

On poor soils, arrowroot may produce so-called 'cigar roots', *i.e.* long, thin and fibrous rhizomes containing little starch. This can be prevented by applying more fertilizer.

References

1. Acevedo-Rodríguez, P. and M.T. Strong, 2005. *Maranta arundinacea* GRIN 2018. Contributions from the United States National Herbarium. Department of Systematic Biology - Botany. Smithsonian Institution 52, National Museum of Natural History, Washington, USA, 415 pp.
2. Aprianita, A., T. Vasiljevic, A. Bannikova and S. Kasapis, 2014. Physicochemical properties of flours and starches derived from traditional Indonesian tubers and roots. *J. Food Sci. Technol.*, **51**(12): 3669-79.
3. Ecocrop; http: //ecocrop.fao.org/ecocrop/srv/en/home.
4. Anon. 2007. Crop Ecological Requirements Database, FAO.
5. Anon. 2018a. Flora of China. Missouri Botanical Garden and Harvard University Herbaria. St. Louis, Missouri and Cambridge, Massachusetts, USA. http: //www.efloras.org/flora_page.aspx?flora_id=2.
6. Anon. 2018b. Flora of North America North of Mexico. Missouri Botanical Garden and Harvard University Herbaria. St. Louis, Missouri and Cambridge, Massachusetts, USA. http: //www.efloras.org/flora_page.aspx?flora_id=1.
7. Anon. 2018c. *Maranta arundinacea.* Germplasm Resources Information Network *(GRIN).* Agricultural Research Service *(ARS),* United States Department of Agriculture (USDA). *Retrieved* 12 January 2018.
8. Anon. 2016d. Koova (arrowroot), http: //papanasini.blogspot.com/2014/01/koova.html, accessed 21 Feb 2016.
9. Anon. 2018e. Plant Resources of South-East Asia. Bogor, Indonesia. PROSEA Foundation.http: //proseanet.org/prosea/e-prosea.php.
10. Heredia-Zarate, N.A. and M.C. Vieira, 2005. Produção da araruta 'comum' proveniente de três tipos de propágulos [Production of the 'comum' arrowroot from three types of propagules]. *Ciência e Agrotecnologia*, **29**(5): 995-1000.
11. Heuze V. and G. Tran, 2017. Arrowroot (*Maranta arundinacea*). Feedipedia, a programme by INRA, CIRAD, AFZ and FAO. https: //www.feedipedia.org/node/545. Last updated on July 17, 2017, 14: 10.
12. https: //en.wikipedia.org/wiki/Arrowroot
13. Kay, D. E. and E.G.B. Gooding, 1987. Root crops. Tropical Products Institute, London, UK.
14. Lestari, L.A., E. Huriyati and Y. Marsono, 2017. The development of low glycemic index cookie bars from foxtail millet (*Setaria italica*), arrowroot

(*Maranta arundinacea*) flour, and kidney beans (*Phaseolus vulgaris*). *J. Food Sci. Technol.*, **54**: 1406-13.

15. Mason, W. 2009. Starch use in foods. In: Starch: Chemistry and Technology, 3rd ed (eds Whistler, R.L., J.N. BeMiller and E.F. Paschall), 879p. Academic Press, New York.
16. Mathews, J. 2008. Underutilized and Underexploited Horticultural Crops, pp. 21-29. New India Publishing Agency, Kerala, India.
17. Ramesh, V. 2019. Arrow Root. In: Handbook of Horticulture, vol. 1, pp. 653-4. ICAR, New Delhi.
18. Rines, George Edwin (ed.), 1920. Arrow-Root. Encyclopedia Americana.
19. Silva, J.R., R. Assumpção and C.L.R. Vegro, 2000. A inserção da fécula de mandioca no mercado de amido [Insertion of cassava starch in the starch market]. *Informações Econômicas*, **30**(7): 31-41.
20. Silveira, J.R.S., C.M.F.T. Santos, J.B. Silva, A.J. Batista and J.A. Costa, 2013. Resgate da cultura da araruta junto aos agricultores familiares no Território do Recôncavo da Bahia [Arrowroot crop rescue from family farmers in the Reconcavo da Bahia Territory]. *Cadernos de Agroecologia*, **8**(1): 1-5.
21. Uphof, J.C. Th. 1959. Dictionary of Economic Plants. Weinheim Publisher.
22. Whistler, W.A. 2000. Tropical Ornamentals; A Guide. Timber Press Inc., Oregon, USA. https: //www.bimbima.com/herbs/arrowroot/2923/#important_medicinal_properties.

5

Bengal arum (*Typhonium trilobatum* (L.) Schott)

Family: Araceae. **Subfamily**: Aroideae.

Common names: Kharkan (Tropical Plants Database, 2023), Bengal arum, Lobed-Leaf Typhonium; Bengali: Ghatkanchu, Ghatkol, Ghet kachu, Kharkol; Tamil: Karunai-k-kilanku, Karunai, Pitikarunai; and Assamese: Chema kachu (www.flowersofindia.net).

Synonyms: *Arisaema pumilum* Blume; *Arum auriculatum* Sims; *A. foetidum* Salisb.; *A. orixense* Roxb. ex Andrews; *A. pumilum* Lam.; *A. trilobatum* L.; *Desmesia orixensis* (Roxb. ex Andrews) Raf.; *Dracunculus trilobatus* (L.) Raf.; *Typhonium orixense* (Roxb. ex Andrews) Schott; *T. siamense* Engl.; *T. trilobatum* var. *fulvus* H.Ara and M.A.Hassan; *T. triste* Griff. (POWO, 2023; Tropical Plants Database, 2023).

Introduction and Origin

Typhonium trilobatum is an aroid with native range from India to S. China and Malaya. This tuberous geophyte grows primarily in the wet tropical biome (POWO, 2023) in wooden areas across India, Bangladesh, China, Thailand, Vietnam, Malaysia, and Sri Lanka (Anandhi *et al.*, 2020; WHO, 1990). The plant is harvested from the wild for local use as a medicine and as a food (Tropical Plants Database, 2023).

Typhonium trilobatum has a huge potential as an alternative source of antioxidant since the price is cheaper and has high antioxidant contents compared to traditional source spices and herbs. Therefore, root extract of *Typhonium trilobatum* can be a very important contributor in new drug discovery including antioxidant and antidepressant drug (Shafa, 2015).

Typhonium trilobatum plant with tubers

Source: Anandhi et al. (2020)

Chromosome Number

Chromosome number: 2n = 18 (Sharma and Mukhopadhyay, 1965).

Number of chromosome (2n) presence in different morphological form of *Typhonium trilobatum*.

Morphological Form	***Number of Chromosomes (2n)***	***References***
Typical form	18	Huq *et al.*, 2007
Tall form	18	Huq *et al.*, 2007
Slender form	19	Huq *et al.*, 2007
Typhonium having green colour petiole form	18	Warasy and Alam, 2009
Typhonium having dark purple colour petiole form	17	Warasy and Alam, 2009
Typhonium having light purple colour petiole form	18	Warasy and Alam, 2009

In various studies, it was found that *Typhonium trilobatum* has tiny chromosomal variation in different morphological forms. Establishment of a standard karyotype of this plant has not been accomplished yet due to lack of distinguishing features and polyploidy. Further Sousa *et al.* (2013) reported the accurate chromosome number (2n=19) using the combination techniques of C- banding and Fluorescent *in situ* Hybridization (FISH).

Botany

Bengal arum is an erect, small to moderate-sized and up to 0.85m tall perennial herb having subterranean stem ~ a short 3-5 cm × 1.5-2cm subglobose or subcylindric tuberous rhizome that is light brown outside. The rhizome produces few (2-4) annual offsets or splitting up. The 25–40cm long petiole is green or variously flushed with purple. The 10-15cm long and 6-11cm wide leaf blade is cordate-ovate in outline, usually deeply 3-lobed and rarely 5-lobed; central lobe ovate, 10–15 × 6–11cm, acuminate, sometimes mucronate and lateral lobes 8–13cm (www.flowersofindia.net; Flora of China, 1994; GBIF Secretariat, 2022; Manna *et al.*, 2016).

Araceae family comprises of monocotyledonous flowering aroids producing spadix as inflorescence (Anandhi *et al.*, 2020). Flowering of Bengal arum usually occurs in August. The plant has very narrow 90cm long flowerheads emerging before leaves in spring. Like most arums, when flowering it emits a distinctive stinky odour for a few hours to entice insects for pollination (www.flowersofindia.net). The spathe of the flower is larger than the spadix. The ovoid or ellipsoid flowers measuring 10-20cm are dark to reddish-purple inside and green outside. Spathe convolutes (rolling longitudinally inwards) at the base. It has a greenish-white barrel-shaped chamber below the spathe with occasional red streaks (National Parks, 2023). Ellipsoid and 1- or 2-seeded fruits are berries that are initially green with purple spots and white when mature (GBIF Secretariat, 2022).

Toxicity

All parts of *Typhonium trilobatum* contain calcium oxalate crystals that are toxic fresh. It is safe to eat the plant after either thorough cooking or after drying it fully. People with a tendency to rheumatism, arthritis, gout, kidney stones and hyperacidity should cautiously include this plant in their diet (Bown, 1995; Tropical Plants Database, 2023).

Nutritional Value

Leaf Composition

It is an important source of vitamin such as thiamine, niacin, carotene and folic acid (Paul *et al*, 2011, WHO, 1990). On fresh weight basis the nutrient contents of *Typhonium trilobatum* leaves are:

Proximate Composition		*Minerals*	
Parameters	*Percentage*	*Parameters*	*Percentage*
Moisture per cent	88.84±0.43	Sodium mg per cent	0.03±0.01
Carbohydrate gm per cent	2.94±0.20	Potassium mg per cent	5.23±0.02
Lipid gm per cent	0.66±0.03	Calcium mg per cent	22.59±0.06
Total Ash gm per cent	1.30±0.20	Iron mg per cent	58.40±0.50

Proximate Composition		*Minerals*	
Parameters	***Percentage***	***Parameters***	***Percentage***
Protein gm per cent	2.81±0.60	**Antinutritionals**	
Crude Fibre gm per cent	2.44±0.25	Phytate µg per cent	0.35±0.01
Vitamin C mg per cent	108.08±1.70	Oxalate mg per cent	7.19±0.58

Values are presented as mean ± SD, n = 3.

Source: Banerjee *et al.* (2015).

Tuber Composition

Of the 150-200g/plant tuber yield, the approximate dry matter is 33 per cent, starch 22.50, sugar 0.71 per cent and lipid 0.92 per cent (Mukherjee *et al.*, 2015).

Analysis of Indian material shows the edible portion of the rhizome to consist of water 69.9 per cent; protein 1.4 per cent; fat 0.1 per cent; fibre 1.0 per cent; other carbohydrates 26.0 per cent; and minerals 1.6 per cent. Alcoholic extracts have given β-sitosterol, two undetermined sterols and a crystalline compound (JSTOR, 2023).

Phytochemicals

Phytochemical screening of the roots revealed the presence of phenols and flavonoids as well as carbohydrate in the form of methanol, ethanol and chloroform extract. No glycoside, tannin, steroid and saponin were detected in any of these extracts. Most abundant ascorbic acid obtained from the ethanol extract was 51.13 mg/0.25g dry extract. Methanol extract also contained 48.16mg whereas chloroform extract contained 44.65mg of ascorbic acid per 0.25g of total extract (Shafa, 2015). Ascorbic acid, total phenolic and flavonoid contents were determined which have antioxidant activity. After pharmacological studies with root extracts of *Typhonium trilobatum*, it revealed mild anti-depressant activity. The result clearly indicates that the extracts of *Typhonium trilobatum* may be a very important contributor in different drug discovery including antioxidant and anti-depressant drugs (Shafa, 2015).

Uses

Leaves and roots of *Typhonium trilobatum* can be eaten after thorough cooking. Tubers are eaten in some tribal societies and the plant also has various medicinal uses (www.flowersofindia.net; Tropical Plants Database, 2023). The plant is widely used in folk medicine for the treatment of various diseases. The herb possesses anti-diarrhoeal, antiseptic and analgesic qualities (Ali *et al.*, 2012). The tuberous rhizome is used in the treatment of vomiting, cough, asthma, excessive expectoration, pyogenic sore throat, headache, gastric ulcer, abscess and snake-bite. The rhizomes may be eaten to induce relaxation of bowels in patients suffering from haemorrhoids. The boiled rhizomes are also eaten with bananas to treat stomach complaints. A poultice of the crushed fresh rhizome is used as a treatment on abscesses and snake-bite. This plant must be used very cautiously in treating pregnant women (Tropical Plants Database, 2023).

Cultivation

Bengal arum is propagated by division of offsets in the dormant season. Like its aroid cousins, *Typhonium* too prefers partial shade and well-drained, loose and humus-rich amended soil that is neutral to slightly acidic site. It may be grown in a woodland rock garden or at the front of a shaded perennial bed. Protection with thick, dry mulch may be needed in cold areas. If the plant is being grown in a container, it can be overwintered indoors. Tuberous rhizomes or offsets are planted 15cm deep. Moderate watering is required during growth but can be kept dry in winter.

The tuberous rhizomes of the plant are harvested when the leaves die in autumn. They can be taken out of the soil and stored dry until they grow back in spring (cycadales.eu).

References

1. Ali, K., A. Ashraf and N.N. Biswas, 2012. Analgesic, anti-inflammatory and anti-diarrheal activities of ethanolic leaf extract of *Typhonium trilobatum* L. Schott. *Asian Pacific J. Trop. Biomed.*, **2**(9): 722-6. (https://core.ac.uk/download/pdf/61806325.pdf).
2. Anandhi, D., K.N.J. Raju Prakash, M.P. Vivekanandan and V. Kumaresan, 2020. The First Case Series Report of *Typhonium trilobatum* Tuber Poisoning in Humans. *Indian J. Crit. Care Med.*, **24**(7):581–4.
3. Banerjee, S., Kamala Adak, Mohini Mohan Adak, Sumana Ghosh and Amitava Chatterjee, 2015. Effect of Some Antinutritional Factors on the Bioavailability of Minerals along with the Study of Chemical Constituents and Antioxidant Property in *Typhonium trilobatum* and *Spinacia oleracea*. *Chem. Sci. Rev. Lett.*, **4**(14): 429-39. (Article CS31204603. ISSN 2278-6783).
4. Bown, D. 1995. Encyclopaedia of Herbs and their Uses. Dorling Kindersley, London.

5. cycadales.eu/produit/typhonium-trilobatum-a/?lang=en
6. Flora of China, 1994. Missouri Botanical Garden Press, St. Louis, USA. (http://flora.huh.harvard.edu/china/).
7. GBIF Secretariat, 2022. *Typhonium trilobatum* (L.) Schott in GBIF Secretariat, GBIF Backbone Taxonomy. Checklist dataset https://doi.org/10.15468/39omei accessed via GBIF.org on 2023-05-16.
8. Huq, S., H. Ara, A. Hassan and S.S. Alam, 2007. Comparative karyotype analysis of *Typhonium trilobatum* L. and its two morphological forms. *Cytologia*, **72**(2): 233–8.
9. JSTOR ~ Global Plants. Retrieved on 16 May, 2023 plants.jstor.org/stable/10.5555/al.ap.upwta.1_452.
10. Manna, K., B. Debnath, M. Das and S.A. Marwein, 2016. Comprehensive Review on Pharmacognostical Investigation and Pharmacology of *Typhonium trilobatum. The Nat. Products J.*, **6**, 172-8. (https://www.researchgate.net/publication/306240164).
11. Mukherjee, A., B. Vimala, B. Nambisan, S.K. Chakrabarti, J. George and H. Gowda, 2015. Underutilized Tropical Tuber Crops with Hidden Treasure of Food, Nutrition and Medicine. *Intern. J. Trop. Agric.*, **33**(4): 3812. (https://serialsjournals.com/abstract/37417_56.pdf).
12. National Parks, 2023. Flora and Fauna Web. Retrieved May 16, 2023. (https://www.nparks.gov.sg/florafaunaweb/flora/8/3/8385).
13. Paul, A.K., H.A. Arif, S. Seraj, A. Nahar, D. Nasrin, M.H. Chowdhury, F. Islam, R. Jahan, A.A.B.M. Bashar, R. Freedman, and M. Rahmahatullah, 2011. A survey of plant items eaten by the low income groups of the rural population of Talbunia village in Bagerhat district, Bangladesh with an account of their folk medicinal applications. *Adv. Natural and Appl. Sci.*, **5**: 41-54.
14. POWO (Plants of the World Online), 2023. Kew Royal Botanical gardens. (Retrieved: 2023-05-15. https://powo.science.kew.org/taxon/urn:lsid:ipni.org:names:89268-1).
15. Shafa, F. 2015. Characterization of Phytoconstituents, In vitro Antioxidant Activity and Pharmacological Investigation of the Root Extract of *Typhonium trilobatum*. (https://core.ac.uk/download/pdf/61806325.pdf).
16. Sharma, A. and S. Mukhopadhyay, 1965. Chromosome studies in *Typhonium* and *Arisaema* with a view to find out the mode of origin and affinity of the two. *Cytol.*, **30**(1):58–66. (doi: 10.1508/cytologia.30.58).

17. Sousa, A., N. Cusimano and S.S. Renner, 2013. Combining FISH and model-based predictions to understand chromosome evolution in *Typhonium* (Araceae). *Ann. Bot.*, pp. 1-12. (doi: 10.1093/aob/mct302).

18. Tropical Plants Database, 2023. <tropical.theferns.info/viewtropical.php?id=Typhonium+trilobatum>.

19. Warasy, A.A. and S.S. Alam, 2009. Comparison of Orcein and CMA Stained Karyotypes in Three Morphological Forms of *Typhonium trilobatum* L.(Araceae). *Cytologia*, **74**(3): 311–6.

20. WHO, 1990. World Health Organization, Medicinal plants in Viet Nam. WHO Regional Publications, WHO Regional Office for the Western Pacific, Manila., 1990, 3.

21. World Health Organization, 1990. Medicinal plants in Vietnam. WHO Regional Publications, Western Pacific Series No. 3, WHO Regional Office for the Western Pacific, Manila. ISBN 92 9061 101 4.

22. www.flowersofindia.net/catalog/slides/Bengal per cent 20Arum.html

6

Bitter Potato (*Solanum* x *curtilobum* Juz. and Bukasov and *Solanum* x *juzepczukii* Bukasov)

Family: Solanaceae

Other common names: In Aymara: Luki; in Quechua: Ruku; in Spanish: Choquepito or Ococuri.

Introduction and Origin

The specific epithet *juzepczukii* honours Russian botanist Sergei Vasilievich Juzepczuk. The specific epithet *curtilobum* means 'short nightshade', which might refer to the whole plant but probably refers only to the often short and broad leaves. The domestication of bitter potatoes probably began some 8,000 years ago and that as cultivated species extensively for at least 3,000 years (Hernando and Leon, 1994). Two species of bitter potatoes are *S.* x *juzepczukii* (triploids) and *S.* x *curtilobum* (pentaploids). Because of their ploidy levels caused by a high degree of sterility, it is difficult to use the characteristics of bitter potatoes in improvement programmes. Origin of bitter potatoes seem to have appeared due to crossings of various wild species such as *S. acaule* (Hernando and Leon, 1994). *S. juzepczukii* is high in glycoalkaloids and is only consumed after processing into the freeze-dried form ~ known as chuno.

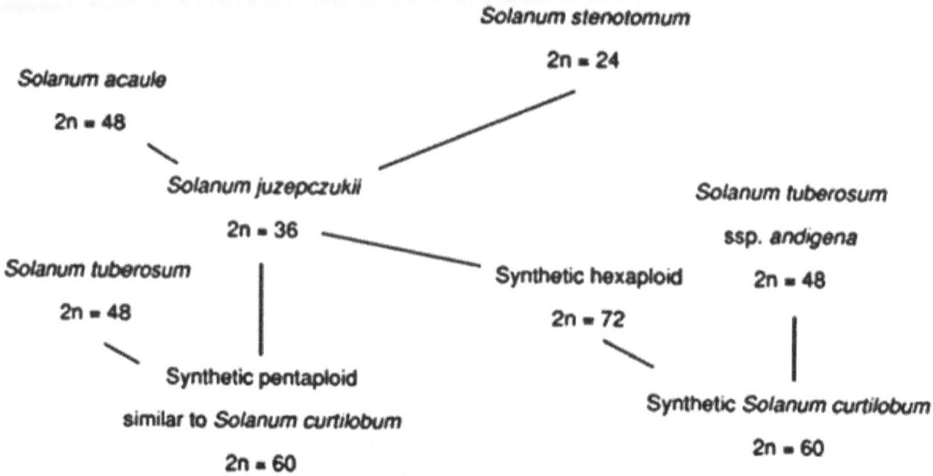

The Origin of Bitter Potatoes
***Source*: Hernando and Leon (1994)**

Varieties

Over centuries, peasants of the southern region of Peru and the Bolivian high plateau have bred a great number of varieties that are suited to various ecological conditions in the highest region of the Andes. There is an extensive collection of bitter potatoes in Cuzco and Puno in southern Peru while a collection from the Bolivian high plateau is maintained at the experimental station of Patacamaya (Hernando and Leon, 1994). The Aymara as per palatability recognizes only two groups of bitter potatoes: i) Choque potatoes which are not bitter or moderately bitter and can be eaten fresh; and ii) Luki potatoes which are bitter so must be processed before eating. Most, if not all, varieties of *S. juzepczukii* are in the Luki group.

There is greater variability in the species *S.* x *juzepczukii* with the main cultivated varieties being 'Ruki', 'Luki', 'Pinaza', 'Parina', 'Locka', 'Parko', 'Keta' and 'Kaisallu' having white or purple tubers. In the species *S.* x *curtilobum*, we may distinguish those of the Choquepito group and the so-called Ococuri which are characterized by a lower glycoalkaloid content than *S.* x *juzepczukii* (Hernando and Leon, 1994).

Botany

Bitter potatoes are similar in form to domesticated diploids although with more delicate foliage. *Solanum* x *juzepczukii* grows to about 30-50cm long having a semi-rosette growth habit, long straight leaves, short petioles and a small blue corolla. *S.* x *curtilobum* is distinguished by its more coriaceous leaves, and bigger *cum* purple corolla with very short lobules and a pointed end (Hernando and Leon, 1994).

S. x *curtilobum*

Solanum x *juzepczukii*

The tubers of bitter potatoes vary in size and shape ranging from rounded ('Pinaza') to elliptical, oblong or elongated-oblong ('Luki'). Clones of Ococuri have purple and white tubers. Most varieties have blue skin and some have blue flesh, less commonly white skin and yellow flesh (Hernando and Leon, 1994).

Glykoalkaloid Content

Total glycoalkaloid levels for *S.* x *curtilobum* ranges from 3.8 to 29mg/100g and these levels are not particularly high but that do span the safety limit of 20mg/100g (Osman, 1978). Most of its varieties can be eaten fresh in moderate quantity.

Total glycoalkaloids levels for *S.* x *juzepczukii* ranges 11.7 to 46.8mg/kg so this species straddles the generally accepted safety limit of 20mg/kg. Some varieties are edible fresh while others must be processed to reduce glycoalkaloid content (Osman, 1978).

Uses and Nutritional Value

If bitter potatoes are to be eaten they must first undergo processing to remove the glycoalkaloids although some varieties are edible fresh. Traditional processes in the upper Andean area consist of freezing by exposing the tubers to several night-frosts and drying them in strong sunlight at altitudes of 4,000m to obtain black *chuno*. Larger bitter potatoes are preferably used to prepare white *chuno*, also called *tunta* (Aymara) or *moraya* (Quechua). Freezing is followed by the peeling, hydrating for up to 30 days and then drying. Both white *chuno* and black *chuno* are very rich in energy (Hernando and Leon, 1994). It has been found that, between August and March, black *chuno* can account for 70 per cent of the food of rural populations of the Peruvian and Bolivian high plateau (Hernando and Leon, 1994).

Growth and Development

There is great variation in the growing cycle which is usually 5-8 months, the clone 'Pinaza' being one of the earliest taking 150 days, while clone 'Ruki' the latest taking up to 195 days (Hernando and Leon, 1994). Between the two the start of tuberization coincides with the start of flowering, approximately 7-9 weeks after emergence, and lasts for about 4 weeks during which time the absence of humidity and severe frosts is vitally important. In this respect there is a differentiation between early, intermediate and late ecotypes which may mature between 4 to 6 months hence a wide range is available for improvement programmes (Hernando and Leon, 1994). Bitter potatoes are usually propagated from tubers. The tubers are generally planted 5-7.5cm deep or left as such in the prepared beds.

Environmental Factors

Bitter potatoes are cultivated at altitudes between 3,000 and 4,300m which are characterized by mean growth period temperatures between 6 and 14°C with precipitations which vary according to the region and year between 400 and 1,400mm, and these are spread over 5 to 6 months, coinciding with the summer period in the Southern Hemisphere (October-May) (Hernando and Leon, 1994). Frosts may occur during the growth stage when temperature mostly drops to -5°C. A greater incidence of low temperatures is observed in the dry period and these affect production heavily with damage varying according to the species. Recently, in an area of Peru with frosts and temperatures of -5°C there was 5 per cent harvest reduction in case of *S.* x *juzepczukii*, 30 per cent in *S.* x *curtilobum* and 40 per cent in common potato (Hernando and Leon, 1994). Planting time of bitter potatoes is quite conditioned by the presence of rain since the crop is grown under rainfed conditions. The period extends from September to November depending on whether the rains begin early or late, so the tradition is to stagger planting on two or three dates as a means of reducing the climatic risk (Hernando and Leon, 1994).

Soil and Planting

The cultivation of bitter potatoes requires soils which have sufficient organic matter, *i.e.* 3-5 per cent and which have had a period of fallow or adequate rotation. The best yields are obtained on soils in which enough manure, *i.e.* around 2 to 3 tonnes per hectare has been applied (Hernando and Leon, 1994). The soil is generally prepared with local implements such as the *chakitaklla* or foot-plough with the entire peasant family taking part in the operation, since it involves plots located in isolated places. The tubers are usually small enough and hilling is generally not required. The varieties of *S.* x *juzepczukii* are better suited to shallow soils than *S.* x *curtilobum*, which has deeper roots (Hernando and Leon, 1994). If lots of satellite plants are seen, the stolons are running just under the surface and a little more soil should be added. The crop needs to be earthed up once or twice when the plants reach a height of 30cm (Hernando and Leon, 1994).

Manure and Fertilization

In Andean region the manuring is limited to sheep manure. However, there have been positive responses to the addition of chemical fertilizers in intermediate doses. The var. 'Pinaza' responds better to fertilization than the var. 'Ruki', the latter having higher dry matter content (Hernando and Leon, 1994).

Crop Rotation

Bitter potatoes predominate on land where the main production is livestock and where there is natural pasture and thus little pressure on the land. Because of this, land can be put under a rotational system of crops with canihua (*Chenopodium pallidicaule*) or fodder plants such as barley or oats, including a prolonged period (up to six years) of fallow during which the natural vegetation covers the soil again. In areas which have a quite broken topography and where the puna zone is very close to the suni or quechua (valley) zones, rotation includes other crops suited to these conditions with tubers such as oca (*Oxalis tuberosa*) and ullucu (*Ullucus tuberosus*) or mixtures of these species (Hernando and Leon, 1994).

Pests and Diseases

Bitter potatoes are susceptible to nematodes (*Naccobus aberrans*), to the Andean weevil (*Premnotrypes* spp.) and also to wart fungus (*Synchytrium endobioticum*) (Hernando and Leon, 1994) but further investigation is to be carried out to ascertain their exact damage and control measures.

References

1. Hernando Bermejo, J.E. and J. Leon (eds), 1994. Andean tubers (by Arbizu, C. and M. Tapia). In: Plant Production and Protection Series No. 26, pp. 149-63. FAO, Rome, Italy (CIP. Lima, Peru). https: //hort.purdue.edu/newcrop/1492/ tubers.html.
2. Osman, S. F., S.F. Herb, T.J. Fitzpatrick and P. Schmiediche, 1978. Glycoalkaloid composition of wild and cultivated tuber-bearing *Solanum* species of potential value in potato breeding programs. *J. agric. Food Chem.*, **26**(5): 1246-8.

7

Bugleweed (*Lycopus* L.)

Family: Lamiaceae (mint family). **Subfamily**: Nepetoideae.

Synonym: *Phytosalpinx* Lunell, *Euhemus* Raf.

Other common names: Water Bugle (keys2liberty), Water horehound or Gypsywort (en.wikipedia.org 'a').

Introduction and Origin

The genus name is from the Greek *lycos* 'a wolf' and *pous* 'foot', referring to the likeness of leaves of some species resembling to a wolf's footprint. *Lycopus* is a genus of species often found in wetlands, damp meadows and stream banks. Its native range is temperate Northern Hemisphere to NW Africa and E and SE Australia. *Lycopus* species are hardy to USDA zone 5 and 4. About 10 species of *Lycopus* occur in eastern North America; most are very similar so making identification quite difficult.

Botany

Lycopus includes perennial dicot herbs usually with enlarged rhizomes. Some species produce swollen rhizome tips while others have slender rhizomes. The species vary in size but generally grow to about 0.91–1.22m. The plants spread through seeds and rooting stems as they grow along the ground (en.wikipedia.org 'a'). Dentate or pinnatifid leaves are bright green, pointed, lobed and in opposite pairs. Some species start with curled purple leaves that unfurl to a bright green colouration. Small white flowers appear in late summer from leaf axils (en.wikipedia.org 'a'). Though *Lycopus* species are not or faintly aromatic, they look somewhat like species of *Mentha* plants that have strong mint or fruit scent. Moreover, *Lycopus* plants usually have 2 stamens while *Mentha* plants have 4. Fruit is flattened at the top in *Lycopus* which is sometimes with a crest or tubercles but rounded in *Mentha*. All six species can become difficult to control in almost any climate since even tiny pieces of rhizome can generate a new plant. It is best to grow these plants in pots or in a hard sided bed. Although they are normally shallow rooted, they may set tubers as deep as 45cm (cultivariable.com).

Uses

The rhizomes can be eaten raw or cooked. Leaves of most species are also edible. The leaves, flowers and roots of most species are used medicinally for treatment of hyperthyroidism, premenstrual syndrome (PMS), breast pain, nervousness, insomnia and bleeding, especially nosebleeds and heavy bleeding during menses though there is insufficient reliable information available about its medicinal effectiveness.

Species

The genus *Lycopus* comprises 20-23 species including two hybrids. Some authorities consider *L. laurentianus* as the subspecies of *L. americanus* (Roll.-Germ.) B. Boivin. The genus *Lycopus* includes *Lycopus alissoviae* Prob. (Primorye region of Russia), *L. americanus* Muhl. ex W.P.C.Barton, *L. amplectens* Raf., *L. angustifolius* Elliott, *L. asper* Greene, *L. australis* R.Br., *L. cavaleriei* H.Lev., *L. charkeviczii* Prob., *L. cokeri* H.E.Ahles ex Sorrie, *L. europaeus* L., *L. exaltatus* L.f., *L. hirtellus* Kom., *L.* × *intermedius* Hausskn., *L. kurilensis* Prob. (en.wikipedia.org 'a'), *L. laurentianus* Roll. (en.wikipedia.org 'a'), *L. lucidus* Turcz. ex Benth., *L. rubellus* Moench, *L.* × *sherardii* Steele, *L. sichotensis* Prob., *L. uniflorus* Michx. and *L. virginicus* L. (en.wikipedia.org 'a').

Rough bugleweed (*Lycopus asper* Greene 1898)

Other common names: Western water-horehound (gobotany).

Synonyms: *Lycopus lucidus* is now merged with *L. asper* so now it is recognized as its synonym (en.wikipedia.org 'd').

Introduction and Origin

The species name *asper* means rough, probably referring to the pubescence on the stems of some individuals. ***Lycopus asper*** is a temperate species native to North America (western Canada and W and C United States) (en.wikipedia.org 'b') but is now found as an introduced species in new areas including New England where it occurs in marshes and shorelines in Massachusetts. It is harvested from the wild for local use as a food and medicine. It is often found in damp areas such as near the lakes (en.wikipedia.org 'b'). It is typically found at elevation ranging from sea level to 300m a.s.l. In N. America it can be found from Michigan to Kansas, west to Manitoba, British Columbia, California and Arizona (keys2liberty).

Lycopus asper ~ By Robert Tatina-http://plants.usda.gov/java/largeImage?imageID=lyas_002_avp.tif, Public.Domain, https://commons.wikimedia.org/w/index.php?curid=7762255

Botany

Rough bugleweed (*Lycopus asper*) is reported having 2n=22 chromosomes. It is an erect perennial that usually grows up to 1m high (en.wikipedia.org 'b'). Hollow, unbranched, stout and green stems have a single vertical groove on each side of the square (Glen, 2012). Non-aromatic, sessile and unlobed leaves (Valo, 2012; Glen, 2012) are narrowly oblong to inversely lanceolate, 1.5-4cm long and 0.25-1.37cm wide though these becoming gradually smaller above (Cao and Berent, 2022). Each pair of opposite leaves is at right angle to the leaf pairs above and below it (Valo, 2012). Each leaf tapers to a point at the tip and expands a broad base at the point of attachment with the stem (Cao and Berent, 2022). The lower surface is hairless while there may be minute but stiff hairs on the upper surface; however, it can even be hairless (Cao and Berent, 2022). The margins are serrate, coarsely *cum* evenly toothed with sharp forward-pointing teeth (Cao and Berent, 2022). The leaves do not smell of mint when crushed.

Lycopus asper flowers appear from July to September with seeds dispersing shortly afterward (Godfread, 2004). The inflorescence consists of a tight cluster of small, white, stalkless (Glen, 2012) and non-fragrant (Valo, 2012) flowers. Pairs of clusters in opposite leaf axils form false whorls on the upper 2/3 of the stem (Valo, 2012). Corolla is slightly longer than calyx (Cao and Berent, 2022). The 5 white corollas (petals) unite at the base into a tube and then separate at the tip into 4 lobes (Cao and Berent, 2022). There are 2 fertile stamens (longer than corolla tube) with brown anthers and 2 club-like sterile stamens (staminodes) shorter than the corolla tube (Cao and Berent, 2022). Style is also longer than corolla tube (Cao and Berent, 2022). The species is hermaphrodite and is pollinated by bees, flies (pfaf.org1), *etc.* Calyx lobes consist of 5 green, narrow and acutely pointed sepals that remain united for most of their 4mm length, and bell-shaped tube with 5 teeth (Glen, 2012). The calyx teeth are firm and awl-shaped (Valo,

2012). The fruit is a set of 4 single seeded, brown and ridged nutlets that are egg-shaped, hairless and depressed at the centre (Cao and Berent, 2022). Corky ridges on the nutlets aid in their dispersal by allowing the nut to float on water (Valo, 2012). After maturity of the fruits, the calyx lobes distinctly surpass the mature nutlets (Cao and Berent, 2022). The seeds are buoyant and can be transported *via* waterways and irrigation canals (Valo, 2012). Approximately 50 per cent of the seeds are capable of germination (Kelley and Bruns, 1975).

Below ground the plant has long, slender and horizontal rhizomes with thick, knobby tips and fibrous roots while the next year stem rises from the rhizome tip (nas.er.usgs.gov). It does not produce stolons (Valo, 2012). Rhizomes are intermediate in size as compared with *L. uniflorum* which is smaller and *L. lucidus* (syn. *L. asper*) of East Asia which is larger (cultivariable.com).

Uses

Rough bugleweed has slightly bitter after-taste and can be eaten boiled or steamed, and can also be dried for later use. The Iroquoi used it as a laxative so it is probably a good idea to take it slowly when introducing this plant to the diet. The part used for the laxative preparation is not specified so it could have been the leaves rather than the rhizomes (cultivariable.com).

Lycopus lucidus Turcz. ex Benth.

Unabridged misapplied name: *Lycopus lucidus* Turcz. ex Benth., misappl.

Common name: Bugleweed, Rough bugleweed or Chinese bugleweed

Introduction

This E. Asian (China, Japan and Korea) plant is now recognized as a synonym of *Lycopus asper* (native to North America) and has been merged into that taxon (en. wikipedia.org 'd') and *Lycopus lucidus* is also now recognized as a synonym of *L. asper*. We have found three reasons for discussing this species as a separate topic: i) its native range and uses have been mentioned in most of the

Source: cultivariable.com

Source: commons. wikimedia.org

Chinese Bugleweed Plant and Tubers

literatures of E. Asia; ii) its rhizomes are somewhat larger as compared to the plants found in North America (cultivariable.com); and iii) the medicinal uses of these plants as reported in several literature sources are quite different from North American plants.

It is found growing in wet places by the edges of streams or in other damp areas (pfaf.org2). It makes some of the largest rhizomes among the tuberous mints. The plants are also very attractive with woody stems and glossy leaves, looking more ornamental (cultivariable.com).

Uses

It is eaten boiled or preserved in salt in China and Japan (Zhang *et al.*, 1977). The aerial parts and the dried rhizomes are used in traditional Chinese medicine and have a reputation for promoting blood circulation and energy, *vis-a-vis* preventing water retention, and are said to have a 'warming' property and have a mild bittersweet flavour (en.wikipedia.org 'd').

Northern Bugleweed (*Lycopus uniflorus* Michx.)

Introduction and Origin

Lycopus uniflorus is native to much of North America (Canada, United States) and east Asia (China, Japan, Korea and Russian Far East) (en.wikipedia.org 'c'). It is found growing in low, wet or boggy ground in the north of its range, and wet woodland in the south (keys2liberty). In N. America it grows from Newfoundland to British Columbia, North Carolina, Nebraska and Oregon. It is hardy to USDA zone 5 (keys2liberty) and is generally found at elevation below 2,000m.

Botany

Northern bugleweed is a perennial herb growing from a slender rhizome with abruptly thicker tuber-like tips. The plant grows upright, 10-50cm tall where its puberulent to finely strigose stem is lined with pairs of generally short-petioled and 2–8cm long toothed leaves with heads of flowers in their axils (en.wikipedia.org 'c'). Leaf blade is elliptic to lanceolate, generally serrate and glabrous to sparsely puberulent. A few millilitre long white flowers can be seen from July to September. Flower has ovate, obtuse to acute calyx lobes and corolla is 2.5–4mm long and larger than calyx. Fruit is a 1–2mm long nutlet with truncate tip and may or may not be finely toothed.

Uses

The rhizome of the plant was used as a staple food by several native American groups (en.wikipedia.org 'c'). The rhizomes can be peeled and eaten raw, cooked or pickled (Elias and Dykeman, 2009). The crisp white rhizomes can be eaten raw in salads or cooked in soups, as vegetables, or fried like French fries, *etc.* When boiled for a short time they are said to make an agreeable vegetable somewhat like Chinese artichokes (keys2liberty). Whole plant is antitussive and sedative (keys2liberty).

Propagation

Similar to other tuberous mints such as *Stachys*, the *Lycopus* plants can also be propagated through seeds as well as by rhizomes. Seeds can be sown in spring or autumn in a cold frame and when they are large enough to handle the seedlings should be pricked out into individual pots to grow them on in the greenhouse for their first winter. They are then planted out into their permanent positions in late spring or early summer after the last expected frosts. Unlike certain *Stachys* species, *Lycopus* seeds are not reported to require stratification, but the site cultivariable.com mentions it to have germinated a lot better with 30 days of cold and wet stratification.

Division is done in spring or autumn. Larger clumps can be replanted direct into their permanent positions, though it is best to pot up smaller clumps and grow them on in a cold frame until they are rooting well. They are planted out in the spring.

Factors Affecting Growth

Climate

The American *Lycopus* species grow in some of the coldest parts of the country so they are probably good down to zone 4. The *Lycopus* species are tolerant of chilly conditions and can survive the coldest weather (around -4 °C). The foliage will die at temperature below -4 °C but the tubers survive shallow freezing pretty well. These plants are tolerant to wide temperature range (cultivariable.com). It can grow under full sun or semi-shaded (light woodland) conditions but preferring moist or wet soil (pfaf.org1). Plants seem to withstand even waterlogged conditions in the winter. The yield will suffer if the soil is allowed to dry out too frequently. These plants will happily grow in shade but they need full sun to get the energy required to produce good rhizome yield. None of these plants have any photoperiod dependencies. Presumably, increased growth in long day conditions is simply due to the increased amount of sunlight (cultivariable.com).

Soil and Field Preparation

Their soil requirement is akin to other members of mint family. *Lycopus* species can be grown even in most soil types so long as they are wet and even in very alkaline soils. These can easily be grown in moisture-holding but free-draining light (sandy-loam), medium (loamy) and heavy (clay) soils (pfaf.org1) enriched with organic matter and having a pH of 6.6-7.5 (iplantz.com; pfaf.org1). It succeeds in damp meadows or in wet places by ponds or streams (pfaf.org1).

For field preparation, a sunny or semi-shaded location is chosen where soil remains wet throughout. Soil is fortified with 100-150 quintals of compost or farmyard manure. Soil is pulverized by 2-3 cultivation, levelled and then beds of convenient sizes are prepared. Its cultivation is akin to other members of mint family. Its rhizomes are planted preferably during spring by burying the rhizomes about 7cm deep and some 25-30cm apart in the soil. It takes about one week for sprouting.

Irrigation and Mulching

Similar to other tuberous mints, *Lycopus* species are not drought-tolerant, these too need a consistent supply of water for harvesting good rhizome yield. Plants seem to withstand even water-logged conditions in the winter. The yield will suffer if the soil is allowed to dry out too frequently.

Until the canopy covers the soil it should be kept weed-free or covered by mulch or black polyethylene.

Thinning, Trimming and Intercropping

It is virtually impossible to collect all the rhizomes securely while harvesting so a lot of tubers escape from collection which grow the following season. If these crops have been grown previously in the same field, they may require thinning in spring to maintain the preferred spacing (cultivariable.com). Moreover, failure to thinning out will result in crowding and poor yield of small rhizomes. Apart from thinning, these also require trimming to prevent flowering and loss of energy that the plant will divert to seed formation.

These species can make good understorey plants for shrubs or trees. The only problem is that they will probably escape from control in such an environment and become very weedy (cultivariable.com).

Harvest

The tubers require 5-7 months for full development for harvesting (cultivariable.com). They are ready for harvest any time after the foliage dies back in the fall. In cool summer climates, rhizomes are generally harvested in the fall and in frost-free climates they may be best harvested in spring.

Storage

Rhizomes must be stored in cool, dark and under high humid conditions. These have thin skin and can be stored only for a few days, *i.e.* for a week in crisper drawer of a refrigerator without much decline in quality (cultivariable.com). Rhizomes tend to discolour and keep quite poorly out of the ground. Due to this reason, they are best left in the ground and harvested as needed in climates where the ground doesn't freeze. In very cold climate or where animals eat the tubers, they can be kept lightly wetted and sealed in a plastic bag to be stored through the winter for replanting in the spring (cultivariable.com).

References

1. Cao, L. and L. Berent, 2022. *Lycopus asper* Greene. Nonindigenous Aquatic Species Database, U.S. Geological Survey. Gainesville, FL, and NOAA Great Lakes Aquatic Nonindigenous Species Information System. Ann Arbor, MI. https: //nas.er.usgs.gov/queries/GreatLakes/FactSheet.aspx?Species_ID=2688. Revision Date: 9/23/2012, Access Date: 12/19/2022.
2. cultivariable.com: .www.cultivariable.com/instructions/root-crops/how-to-grow-chinese-artichoke-and-tuberous-mints/
3. Elias T.S. and A.P. Dykeman, 2009. Edible Wild Plants: A North American Field Guide to Over 200 Natural Foods. Sterling, New York, 89pp.
4. en.wikipedia.org a: https: //en.wikipedia.org/wiki/Lycopus
5. en.wikipedia.org 'b': https: //en.wikipedia.org/wiki/Lycopus_asper
6. en.wikipedia.org c: https: //en.wikipedia.org/wiki/Lycopus_uniflorus
7. en.wikipedia.org 'd': https: //en.wikipedia.org/wiki/Lycopus_lucidus
8. Glen,M.2012.*Lycopus.asper*.Saskatchewan's.Wildflowers.EM.ca.Available. http: //www.em.ca/garden/native/nat_lycopus_asper.html. Accessed 6 September 2012.
9. gobotany: https: //gobotany.nativeplanttrust.org/species/lycopus/asper/
10. Godfread, C. 2004. Plant Inventory at Fort Union Trading Post National Historic Site 2003-2004, 37pp. Final Report. Bismarck, North Dakota, USA.
11. iplantz.com: https: //www.iplantz.com/plant/1442/stachys-affinis/
12. Kelley, A.D. and V.F. Bruns, 1975. Dissemination of weed seeds by irrigation water. *Weed Sc.*, **23**(6): 486-93.
13. keys2liberty: .https: //keys2liberty.wordpress.com/tag/lycopus-asper/#: ~: text=Aug%20to%20October.-,Edible%20Uses%3ARoot%20

%E2%80%93%20raw%20or%20cooked.,of%20hyperthyroidism%20and%20related%20disorders.

14. nas.er.usgs.gov: https: //nas.er.usgs.gov/queries/GreatLakes/FactSheet.aspx?Species_ID=2688
15. pfaf.org1: https: //pfaf.org/user/Plant.aspx?LatinName=Lycopus+asper
16. pfaf.org2: https: //pfaf.org/user/Plant.aspx?LatinName=Lycopus+lucidus
17. Valo, J.F. 2012. Rough bugleweed, *Lycopus asper*. Nature Tourism in Minnesota. Available.http: //minnesotaseasons.com/Plants/rough_bugleweed.html. Accessed 6 September 2012. https: //nas.er.usgs.gov/queries/GreatLakes/FactSheet.aspx?Species_ID=2688
18. Zhang, Z.Y., A.M. Lu and W.G. D'Arcy, 1977. Verbenaceae through Solanaceae. Flora of China. Volume 17.

8

Cassava (*Manihot esculenta* Crantz)

Family: Euphorbiaceae.

Other common names: Brazilian arrowroot, Manioc; in South America: Tapioca and Yuca. Indian languages: Kuchi Kizhangu or Maravallli Kizhangu in Tamil; Kappa in Malayalam; Kavva pendalam in Telugu; Mara Genasu in Kannada and Shimla Aaloo in Hindi.

Synonyms: *Janipha aipi* (Pohl) J.Presl, *J. manihot* (L.) Kunth, *Jatropha aipi* (Pohl) Gopp., *J. diffusa* (Pohl) Steud., *J. digitiformis* (Pohl) Steud., *J. dulcis* J.F.Gmel., *J. flabellifolia* (Pohl) Steud., *J. loureiroi* (Pohl) Steud., *J. manihot* L., *J. mitis* Rottb., *J. paniculata* Ruiz and Pav. ex Pax, *J. silvestris* Vell., *J. stipulata* Vell., *Mandioca aipi* (Pohl) Link, *Mandioca dulcis* (J.F.Gmel.) D.Parodi, *M. utilissima* (Pohl) Link, *Manihot aipi* Pohl, *M. aypi* Spruce, *M. cannabina* Sweet, *M. diffusa* Pohl, *M. digitiformis* Pohl, *M. dulcis* (J.F.Gmel.) Baill., *M. edule* A.Rich., *M. edulis* A.Rich., *M. flabellifolia* Pohl, *M. flexuosa* Pax and K.Hoffm., *M. loureiroi* Pohl, *M. melanobasis* Müll. Arg., *M. sprucei* Pax, and *M. utilissima* Pohl. (*Manihot esculenta*. Anon., 2017).

Introduction and Origin

Cassava is a widely cultivated crop in tropical and subtropical regions. In India it was introduced to Kerala in 17th century by the Portuguese. It is one of the most drought-tolerant crops capable of growing on marginal soils where during prolonged drought, its fibrous root system grows deeper into the soil to absorb water while leaves stop growing and the older leaves drop off. Cassava, grown for its edible starchy tuberous root ~ a major source of carbohydrates, is recognized as the second most important tuber crop after potato. Providing a basic diet for over half a billion people, it is a major staple food in the developing world (Anon., 1995). Nigeria is the world's largest producer of cassava, while Thailand is the largest exporter of cassava starch known as tapioca. Cassava is the third-largest source of food-carbohydrates in the tropics after rice and maize (Claude and Denis, 1990; Afedraru and Lominda, 2019). Total area under cassava cultivation in India is estimated to be 0.21 million hectare with a production of 7.24 million tonnes of tubers. Predominantly, it is cultivated in southern states such as Kerala and Tamil Nadu which are responsible for 46 and 43 per cent of area, *vis-à-vis* 38 and 59 per cent of tuber production, respectively. Primarily for industrial purpose it is being grown in Tamil Nadu, Andhra Pradesh, Karnataka and Maharashtra though as a subsidiary food crop in Kerala, Asom, Nagaland, Mizoram and Andaman and Nicobar Islands. There are many cassava processing factories alongside National Highway 68 between Thalaivasal and Attur in Tamil Nadu (India). In the Indian states of Andhra Pradesh and Kerala, cassava is widely cultivated and eaten as a staple food. Its adventitious roots are an important source of carbohydrates in Asom, especially for natives of hilly areas. The cassava plant gives the third-highest yield of carbohydrates per cultivated area among crop plants, after sugarcane and sugar beets. Though its tuber is predominantly consumed in boiled form as vegetables but substantial quantities are used to extract cassava starch. The cassava starch in powdered form is called tapioca but when made into small pearl-like balls it is known as *sago, sabudana* or *Jawwarisi.* Cassava tubers are grated, pressed to remove moisture, and then dried and roasted to obtain edible coarse flour called *farinha* in Brazil and *garri* in West Africa.

Manihot esculenta, native to tropical regions of South America, is not known in the wild state and may have arisen by mutation or hybridization. Probable centre may be southern Mexico/Guatemala or NE Brazil or both. Botanically it is a tropical perennial shrub whose centre of origin is the Amazon Basin (Olsen and Schaal, 1999). Walker's manihott (*Manihot walkerae*), a close relative of cassava, has been studied for its role in introducing valuable traits into the latter (Barrett Cindy, 2010). The tubers of *M. walkerae* exhibit delayed postharvest physiological deterioration (PPD) and this trait can be passed to *M. esculenta* × *M. walkerae* hybrids allowing the roots to remain intact one month after harvest (Lokko *et al.*, 2007).

Botany

Typically cassava is a diploid species (2n = 36) (Raji *et al.*, 2009; Sakurai *et al.*, 2013). It is highly heterozygous and vegetative propagation is through

Cassava

Leaf Tuber Seeds

stakes in agriculture. Although cassava is a woody perennial shrub, it is mainly cultivated as an annual crop for its enlarged starchy tuberous roots. The plant has many cultivars, some are adapted to dry areas of alkaline soil and others to acid mud banks along rivers. Cultivars also vary greatly in size and morphology, ranging from low herbs to branching shrubs and slender unbranched trees. Almost all cultivars have latex in all their parts. The plant grows to about 1-3m or more in height having usually slender and glabrous stems with leaves being borne near the apex. Some spreading cultivars branch out near the base whereas others may be erect with branching nearer the apex. Stems vary in colour, being grey or silvery, green, greenish-yellow, reddish-brown or streaked with purple. The lower parts of the stems have nodes formed when the older leaves are shed leaving the prominent leaf scars. The green to reddish leaves are almost palmate (fan-shaped) and deeply parted into 5-9 lobes. The 5-30cm long petioles and the midrib may be green, pale-green or deep red. The monoecious flowers borne in axillary racemes near the ends of branches are pale yellow or red, and 1-1.5cm in diameter. The fruit is a six-winged capsule with three being about 12mm long ellipsoidal seeds. Root tubers develop by a process of secondary thickening as swellings on adventitious roots a short distance from the stem.

The long, cylindrical and tapered fleshy roots that develop in clusters of 5-10 at the base of the stem are reminiscent of dahlia tubers and vary greatly in the number, shape and size of the tubers and the angle at which they penetrate the ground. The tubers are 2.5-15cm in diameter and 0.15-1m long, occasionally even longer. Commercial cultivars are generally 5-10cm in diameter at the top and around 15-30cm long. The firm and homogeneous flesh of chalk-white or yellowish colour is encased in about 1mm thick rind that is brown on the outside and rough.

Varieties

Over the last 50 years, several varieties with special traits such as short duration, long duration, edible types, starchy types and processing types were developed and the most popular ones are as under (George, 2019):

'CO (TP) 4' is a non-branching hybrid with erect stem where petioles are purple with green tinge and the root tuber is conical with brown skin, white rind and white flesh. It matures in 270 days with an average yield of 45t/ha of root tubers having 40 per cent starch. It was released by TNAU, Coimbatore through All India Coordinated Research Project on Tuber Crops.

'Indira Cassava 2' was developed and released by AICRPTC centre at IGKV, Raipur (Chhattisgarh) which is a spreading and branching type of cassava bearing green stem, green petiole and narrow leaves. Its cylindrical tubers have light brown skin, creamy rind and light yellow flesh. Its average yield is 38t/ha with 30 per cent starch content.

'Mulluvadi 1' was developed by State Department of Agriculture, Tamil Nadu which matures in 270 days with an average tuber yield of 34.5t/ha and is good for industrial use as well as for home consumption. It is a non-branching type spreading tall variety bearing purple petiole where tuber skin is brown, rind light cream and flesh white. It is tolerant to CMD.

'M 4' is a non-branching type tall variety introduced by KAU, Kerala from Malaysia. It matures in 300-320 days with an average yield of 20-25t/ha. It is widely grown in Kerala for edible purpose as tuber's cooking quality is excellent.

'Yethapur 1' (maturing in 270-300 days with tuber yield of 49.51t/ha) is a non-branching, erect and tall selection by TNAU, Coimbatore released through AICRPTC, having very starch rich tubers. Its leaves are shiny dark green with greenish-purple petiole. Its tubers are long cylindrical, skin brown, rind light cream and flesh white.

Varieties Developed by CTCRI

1. **'H-97'** (maturity period: 10 months; average yield: 25-35t/ha) is a medium size (1.5-2m tall), high yielding hybrid variety of tapioca. Its tubers contain 27-31 per cent starch. This drought tolerant variety is also tolerant to Cassava Mosaic Disease (CMD) as well as field resistant to leaf spot, spider mite and scale insect.
2. **'H-165'** (maturity period: 220-240 days; average yield: 33-38t/ha) is a quite popular, 1.5-2m tall and high yielding industrial hybrid variety of tapioca having starch content of 23-25 per cent on fresh weight basis. This was released by CTCRI which is grown in the hilly tracts of Tamil Nadu for starch industry. Its conical tubers are relatively short with golden-brown skin, cream-coloured rind and golden flesh. This easily harvestable hybrid has good tuber shape and is field tolerant to CMD, spider mite and scale insect. This variety is suitable for rotation cropping.

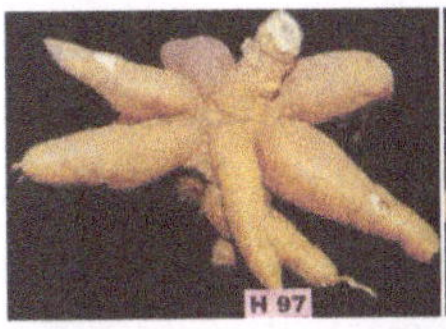

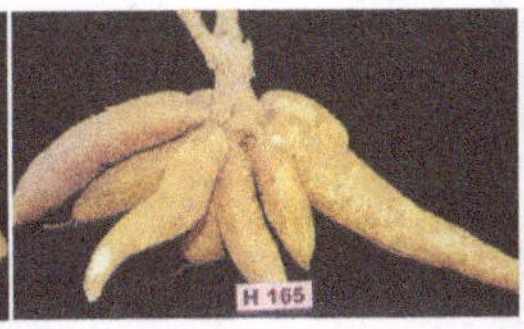

Sree Sahya

Sree Visakham

Sree Prakash

Sree Jaya

Sree Vijaya

Sree Harsha

Sree Athulya

Sree Apoorva

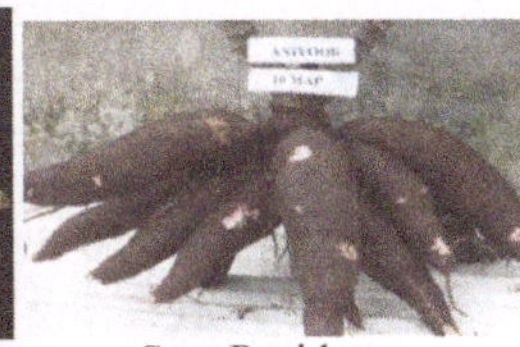
Sree Pavithra

3. **'H-226'** (maturity period: 260-300 days; average yield: 30-35t/ha), released by CTCRI, is a quite popular hybrid variety in starch industry belt of Tamil Nadu. It is 1.5-2m tall variety where tubers are smooth with light purple and white rind, containing 28-30 per cent starch and is field tolerant to spider mite and scale insect.
4. **'Sree Athulya'** (average yield: 38.7t/ha with starch content of 34.8 per cent) is an erect, greyish-brown triploid hybrid bearing branching type erect and stout stem, thick and broad dark purple leaves overlaid with green tinges, and long cylindrical tubers with brown skin, cream rind and white flesh. This cultivar is suitable for starch extraction as well as cattle feed. It has been recommended for cultivation in the states of Tamil Nadu and Andhra Pradesh. It takes 300 days for maturity
5. **'Sree Apoorva'** (average yield: 37-38t/ha) is a high yielding triploid variety having long cylindrical tubers with brown skin, white flesh and high starch content (33.3 per cent). It matures in 300 days. It is suitable for starch extraction and for culinary uses. It has non- or top-branching

habit. It has been recommended for cultivation in Tamil Nadu and Kerala.

6. **'Sree Harsha'** (average yield: 35-40t/ha), the first triploid variety, is an erect non-branching, drought-tolerant and high yielding variety where tubers have high starch content (38-41 per cent). It is suitable for industrial purposes. This is also resistant to leaf spot, spider mite and scale insect. Its tubers mature for harvesting in 270-300 days.
7. **'Sree Jaya'** (maturity period: 180-210 days) is a selection which is early maturing variety suitable for low land cultivation as a rotation crop in paddy-based cropping system. Its brown-skinned conical tubers with purplish rind and white flesh have good cooking quality. Its tuber yield is 26-30t/ha with starch content of 24-27 per cent.
8. **'Sree Kaveri-8S501'** (average yield: 61-62t/ha) is a very high yielding variety for industrial use. It is drought tolerant *cum* CMD resistant variety suitable for cultivation in Kerala, Tamil Nadu and Andhra Pradesh.
9. **'Sree Padmanabha'** (maturity period: 9-10 months; average yield: 38t/ha). It is the first CMD resistant variety. It is suitable for irrigated plains of Tamil Nadu and rainfed areas of Kerala.
10. **'Sree Pavithra'** is non-branching, tall and high yielding variety that grows well at low to marginal levels of potassium in soil. Its broad leaves bear green petiole with purple tinge. Its conical tuber has brown skin, pink rind and white flesh. It yields 35-45t/ha of tubers. It has good starch content (29 per cent) having excellent cooking quality and low cynogenic (bitterness) glucoside (25.8ppm). This variety is suitable for cultivation in the soils of Kerala which are inherently low to marginal in soil exchangeable K.
11. **'Sree Prabha'** is a semi-spreading medium-tall variety bearing light brown stem, intermediate leaf size, purple petiole, conical root tubers with brown skin, light yellow rind and light yellow flesh. It matures in 240-300 days with tuber yield of 40-45t/ha and starch content of 26-29 per cent, *vis-à-vis* having excellent cooking quality. It is suitable for upland and lowland cultivation. It is also field-tolerant to leaf spot and spider mite.
12. **'Sree Prakash'** (maturity period: 210 days; average yield: 30-35t/ha) is a non-branching, short (1-1.5m) and high yielding variety selected from indigenous germplasm collection of cassava. The starch content in fresh tubers is 29-31 per cent. This variety is suitable for lowland cultivation.

13. **'Sree Rekha'** (maturity period: 240-300 days; average yield: 45-48t/ha) grows erect with branched stem where leaves are broad *cum* brownish and petioles purple. Its conical tuber is long with light brown skin whereas rind and flesh are cream with excellent cooking quality. It yields starch to a tune of 28-30 per cent. It is suitable for upland and lowland cultivation. It is field tolerant to leaf spot and spider mite.
14. **'Sree Sahya'** (maturity period: 300-320 days; average yield: 35-40t/ha and starch recovery 29-31 per cent) is a multi-cross hybrid variety 2-2.5m tall with erect branching and sometimes even tubers are also branched. Its stems are dark grey with purple petiole. Long cylindrical tubers are long-necked. Skin of the tuber is light brown, rind cream and the flesh is white. It is a high-yielding multiple cross hybrid. This highly drought tolerant variety is moderatly resistant to leaf spot.
15. **'Sree Swarna'** (maturity period: 210 days; average yield: 40t/ha and starch recovery of 25 per cent) is a medium-tall variety with top branching. Stem is reddish brown with broad leaf and purple petiole. This produces cylindrical tubers having brown skin, pink rind, light yellow flesh and good culinary quality. It is tolerant to cassava mosaic disease (CMD).
16. **'Sree Vijaya'** (maturity period: 180-210 days; average yield: 25-28t/ha with 27-30 per cent starch) is a variety with erect branching habit where stem is greenish brown and the leaves are broad having light green petiole. Tubers are conical with brown skin, creamy rind, light yellow flesh and good cooking quality which has been selected from indigenous germplasm collection of cassava. After cooking its tubers become light yellow. The starch content in fresh tubers is 27-30 per cent.
17. **'Sree Visakham'** (maturity period: 270-300 days; average yield: 35-38t/ha; starch recovery 25-27 per cent) is a non-branching 2-2.5m tall variety where carotene content is to a tune of 466 IU/100g. Its conical tubers are densely set. Skin of the tuber is brown, rind cream and flesh light yellow. It is less susceptible to spider mite and scale insect.

Nutritional Value and Health Benefits

The edible portion of the tubers contains 60-65 per cent water and is very rich in starch, with small amounts of calcium, phosphorus and vitamin C (Anon., 2016a). However, tuber is poor in protein (0.7-2.6 per cent) and other nutrients, and has negligible fat (0.2-0.5 per cent) (Tewe, 2004). In contrast, cassava leaves are a good source of protein for animal and human nutrition (Latif and Muller, 2015) but deficient in amino acid methionine (Ravindran, 1992). In a 100g reference serving, raw cassava provides 670 kilojoules (160 kilocalories) of food energy and 25 per cent of the daily value (DV) of vitamin C, but otherwise has no micronutrients in significant amount (*i.e.* above 10 per cent of the relevant DV). Cooked cassava starch has a digestibility of over 75 per cent (Tewe, 2004). Arginine, histidine, isoleucine, leucine and Iysine are the principal amino acids present in the protein. Because of the low protein content, the disease Kwashiorkor is prevalent in areas where cassava is the main staple item of diet. Cassava flour paste can be inoculated with *Rhizopus* or other suitable fungi to increase the protein content to about 3.25 per cent.

Cassava, Raw

Nutritional Value per 100g

Energy	160 kcal (670 kJ)
Carbohydrates	38.1g
Sugars	1.7g
Dietary fiber	1.8g
Fat	0.3g
Protein	1.4g
Vitamins	***Quantity per cent DV†***
Thiamine (B1)	8 per cent 0.087mg
Riboflavin (B2)	4 per cent 0.048mg
Niacin (B3)	6 per cent 0.854mg
Vitamin B6	7 per cent 0.088mg
Folate (B9)	7 per cent 27μg
Vitamin C	25 per cent 20.6mg
Minerals	***Quantity per cent DV†***
Calcium	2 per cent 16mg
Iron	2 per cent 0.27mg
Magnesium	6 per cent 21mg
Phosphorus	4 per cent 27mg
Potassium	6 per cent 271mg
Sodium	1 per cent 14mg
Zinc	4 per cent 0.34mg
Other Constituents	***Quantity***
Water	60g

Units: μg = micrograms; mg = milligrams; IU = International units.

†Percentages are roughly approximated using US recommendations for adults.

Source: Wikipedia.com

Cassava Starch

Cassava starch granules of various shapes (round, truncated, *etc.*) and sizes from 5 to 35microns are available in the markets. The amylose content is about 17 per cent compared with 22 per cent for potato starch and 27 per cent for maize starch. The approximate composition of commercial samples is: moisture 9-18 per cent; protein 0.31-1 per cent; fat 0.1-0.4 per cent; starch (and a little fibre, *etc.*) 81-89 per cent; and ash 0.1-0.8 per cent. Good quality starch is pure white in colour, absolutely free from specks, with a pH of 4.7-5.3, the moisture content of 10-13.5 per cent and an ash content of less than 0.2 per cent (Humanity development library 2.0).

Cassava is being substantially used in the manufacture of starch for use in the foodstuff, textile and paper industries, for manufacturing of plywood and veneer, adhesives, dextrin and glucose. Minor industrial applications include use in the manufacture of explosives, dyes, drugs, chemicals, carpets, linoleum, for production of alcohol and for coagulation of rubber latex (Humanity development library 2.0).

Tapioca, made from cassava starch is used for preparation of puddings and invalid foods. Glucose is also produced from cassava in Kerala. *Sabudana* ~ small pearl-like balls made by powdered form of cassava starch is used to make *sabudana vada, upma, payasam, etc.* Sabudana papad, sabudana khichdi or sabudana kheer (porridge: usually prepared by Hindus during fasting) are the common Indian preparations of *sabudana.*

Toxicity

Like other roots and tubers, cassava too has varieties that are either sweet or bitter. Both bitter and sweet cassava varieties contain antinutritional factors and toxins, with the bitter varieties containing much larger amounts. Improper preparation of cassava can leave enough residual cyanide to cause acute cyanide intoxication, goitres, and even ataxia, partial paralysis or death (Soto-Blanco and Silvana Lima Gorniak, 2010; Sri Suharti *et al.*, 2021).

Cassava tubers, peels and leaves contain cyanogenic glucosides (linamarin and lotaustralin) that are decomposed by the enzyme linamarase ~ a naturally occurring enzyme in cassava, liberating hydrogen cyanide (HCN) (Cereda and Mattos, 1996). The antinutritional and unsafe glycoside's concentration varies significantly due to varieties, climatic variations and cultural conditions. Varieties grown during drought are especially high in these toxins (Aregheore and Agunbiade, 1991; White *et al.*, 1998). While some smaller-rooted sweet varieties can be used after boiling, the larger-rooted bitter varieties used for production of flour or starch must be treated and prepared properly prior to consumption.

More toxic varieties of cassava are a fall-back resource ('a food security crop') in times of famine or food insecurity in some places. Farmers often prefer the bitter varieties because they deter pests, animals and thieves (Chiwona *et al.*, 2002).

HCN contents in tubers can vary from 10 to 490mg/kg, being highest in roots grown on soils of low fertility, particularly if there is a potassium deficiency, and also in the first year's growth, *vis-a-vis* in dry season. It is usually highest in the rind and in the fibrous core at the centre but considerable variation can occur between various parts of the same tuber. For edible purposes, cultivars with a high starch and protein content and a low HCN content are preferred. For starch manufacture, the cultivars with high starch content are favoured where HCN content is of less importance.

Detoxification

Cassava is used for food, as animal feed and for industrial purposes but before consumption the tuber should be detoxified by proper cooking. The tuber of the sweet variety has a delicate flavour and can replace potatoes. Peeling and cooking give a partial detoxification but soaking the roots for long periods, repeated boiling with changed water, soaking after comminution and fermentation or fermentation followed by heat treatment all are used. However, there appears to be no evidence that animals develop toxicity symptoms from continuous intake of cassava or cassava forage.

Culinary Uses

Fresh peeled tubers can be eaten as a vegetable after boiling or roasting. The boiled and mashed tubers can be added to soups and stews. The tubers are often preserved in the form of sun-dried chips and consumed after cooking or ground into flour. Cassava flour is used in the preparation of bread, biscuits and confectionery and in products such as macaroni, spaghetti and rice substitutes, also as an adulterant of cereal flour and in the production of adhesives. In West Africa it is eaten as a fermented meal known as *garri*, while in C and S America a product (*farinha* de manioc) similar to *garri* except that much less fermentation occurs during its preparation, is very popular. In the Philippines 'landang' or 'cassava rice' is prepared which retains much of the original small quantity of protein. The detoxified tubers can be made into flour that can be used in breads, cakes and cookies.

The juice obtained from the tubers during starch production is sometimes concentrated, and the spices added to obtain a sauce known as 'cassari po' or 'cassareep' in the West Indies and 'tucupi' in Brazil (Humanity development library 2.0). Alcoholic beverages, *e.g. tarul ko* (Darjeeling and Sikkim, India) can be made from cassava. On a small scale, in some countries fermented beers are prepared.

Cassava leaves are a good source of protein for animal and human (Latif and Muller, 2015) nutrition. Young leaves can be eaten as a vegetable, whereas the mature leaves can be used for animal feeding and are sometimes dried and ground into a meal. They have high lysine content.

Biofuel

Cassava (tapioca) chips have gradually become a major source of ethanol production. On 22 December 2007, the largest cassava ethanol fuel production facility was completed in Beihai, with annual output of 200 thousand tonnes, which would need an average of 1.5 million tonnes of cassava.

Other Uses

Increasing quantities of dried cassava roots are being used for livestock feeding. The pellets made by grinding the dried chips and compressing the powder are used as a carbohydrate source in animal feed rations. About 10-20 per cent residue that is left after the extraction of starch and tapioca, can be used as a livestock feed or as a raw material for the production of adhesives. Spaying of water-diluted cassava starch over fabrics before ironing helps stiffen the collars.

Other Plants with almost Similar Uses

The starch, often labelled as arrowroot, is obtained from the rhizomes (rootstock) of several tropical plants, traditionally from arrowroot plant (*Maranta arundinacea*), but also Florida arrowroot [from *Zamia integrifolia*] and tapioca from cassava (or Brazilian arrowroot). East Indian arrowroot [Polynesian arrowroot or pia (*Tacca leontopetaloides*)], Japanese arrowroot [or kudzu (*Pueraria lobata*)], and Queensland arrowroot [or Indian shot/African arrowroot/ edible canna/purple arrowroot/Sierra Leone arrowroot (*Canna indica*)] are used in similar ways; their rhizomes can be eaten raw, but usually cooked; rhizomes can be ground into a flour to be used in several ways; they can be used as femine food; the easily digestible starch is used in various food preparations as well as have industrial uses. Another plant with nutritional value as a source of starch for Indian foods is Indian arrowroot [or East Indian arrowroot/narrow-leaved turmeric/*tikhur* (*Curcuma angustifolia*)]; its rhizomes are typically ground into flour which can then be mixed together with milk or water to form a nutritious meal.

Cassava Substitutes

Cornstarch can be used as a substitute for tapioca starch for both culinary as well as cosmetic purposes. Arrowroot flour is the gluten-free flour similar in flavour to **tapioca flour**. About 8 grammes of tapioca flour is used for every 2.5g of arrowroot. **Potato, arrowroot** and **rice flour** can be used as a substitute to cassava to obtain flour and starch.

Propagation

Cassava can be either propagated from seed or vegetatively from stems, but the latter is the most common practice for commercial production. Sexual reproduction is the key requirement for conventional breeding (Kawano, 1980; Alves, 2002) and several breeding programmes are currently active in generating new genetic variation through crossing (Hershey, 2017).

Cassava is usually vegetatively propagated through 15-30cm long and 2.5-3.75cm thick cuttings with 1-5 buds taken from 8-18 months old stems. For hand planting, about 20-30cm long sticks with 3-5 buds are recommended. For mechanical planting, shorter sticks (15-20cm long) with 2-3 buds are used. The sticks may be cut obliquely or at right angles to the axis. Disease-free planting material should be selected and it would be better to treat the cuttings with a mixture of fungicides and insecticides (maneb + propineb + copper oxychloride + malathion). The sticks can be stored for up to 8 weeks in cool and well-ventilated conditions except when harvested under rainy days when storage is normally reduced to 7-10 days. Depending on weather conditions and planting method (machine or hand), the cuttings are buried vertically, flat or tilted. Sprouting usually takes place in about a week with only about a 5 per cent failure rate. New cuttings are substituted within a month.

Pre-sprouting

It is a good idea to dip the stem cuttings in an appropriate fungicide prior to planting to help prevent the development of diseases. Though the cuttings can directly be planted in the main field, some farmers presprout the cuttings in partial shade in good quality soil, either in a nursery bed or in trays/polyethylene bags. The stems can be planted horizontally in a nursery bed to encourage the growth of multiple stems. In trays or polyethylene bags, the cuttings are generally planted vertically or inclined. Stem cutting should be watered immediately after planting and on a regular basis thereafter and the soil must be kept moist but not wet. Stems should begin to sprout 7-10 days after planting. Plants propagated from stem cuttings are ready to be transplanted after approximately 4-6 weeks.

Growth and Development

According to the 'University of Florida IFAS Extension Bulletin', cassava plants have two distinct growth stages:

1. First stage lasts for roughly eight weeks and includes ~ planting, sprouting and formation of leaves, stems and roots. Once the aerial part and supporting root system of the plant get formed, the formation of starchy, edible tuber begins during this stage.
2. During the second growth stage, the plant and tuber grow rapidly until the tuber reaches a mature and harvestable size.

Technically, there is no 'mature stage' for cassava because harvesting occurs at variable times depending on the varieties, climatic conditions and soil conditions. The maturity period is generally 6-24 months. Though most of the varieties give good yield after 9-12 months, a few quick-growing cultivars developed by CTCRI such as 'Sree Jaya', 'Sree Prakash', 'Sree Swarna', 'Sree Vijaya', *etc.* can be harvested in 6-7 months. For food usage, the cassava is usually harvested within 12 months when the root has achieved the desirable size. The cassava can spend more than one season underground in tropical regions but as the tuber grows longer, a portion may become inedible and then used for their starch content only.

Factors Influencing Growth and Quality

To be disease-free, the planting area should be carefully selected. Environmental factors such as light, temperature, moisture and soil type influence the growth and quality of cassava. Other factors such as variety of cassava, stem cutting, planter operation, irrigation, nutrition, pest management, planting date and harvest date, *etc.* can be controlled by the grower. Also, a good ground preparation, harrowing and ploughing, and planting method are always needed. Only when all factors are at optimum levels the most profitable yields of quality storage roots can be attained.

Environmental Factors

Cassava can grow from sea level up to about 1,000m in equatorial regions but the growth is slow and yields are reduced at the highest altitudes. Being a short-day (photo-insensitive) crop, it produces fewer tubers in regions with day lengths greater than 10-12 hours with exception to some cultivars that tolerate longer days. Although cassava can tolerate temperature range between 12 to 40°C, the optimum temperature range is 25-29°C (Willy, 2010). The yield gets significantly reduced at temperature below 18°C and above 30°C. It cannot withstand freezing and must have a minimum of 300 frost-free days. Also, a well-distributed annual rainfall of 100-150cm is regarded as ideal, but the crop can be successfully grown in areas with rainfall ranging from 50 to 250cm. Occasionally, with rainfall of less than 75cm, the crop is irrigated as in Kerala (Humanity development library 2.0). Except at planting, cassava can withstand prolonged periods of drought and is therefore a useful crop in areas of low or uncertain rainfall.

Soil

Cassava gows best in light sandy-loam soil of medium fertility with a pH range of 5.5 to 7.5, but cultivars can be grown successfully on soils ranging from stiff marine clays with a pH of 8-9, to sands or loose laterites with a pH of 5-5.5. If required, soil acidity can be corrected by addition of limestone, between 3 months and up to two weeks prior to planting. Many cultivars give poor yield when grown on clay soils where the plant produces stem and leaf growth at the expense of the roots. Saline and swampy soils are unsuitable. Although it responds well to fertilization, cassava can tolerate soils of low fertility, especially if the feeder roots can penetrate to depths of 40-60 cm; deep cultivation before planting is therefore recommended.

Land Preparation, Planting Season and Planting Method

Soil preparations and fertilizer use for cassava are almost same as for sweet potatoes. To bring the soil to a fine tilth before planting land must be deep ploughed 2-3 times followed by harrowing and planking. At second ploughing if FYM or compost is added in the soil @ 150 quintals per hectare, that will certainly influence increasing tuber yield significantly. Any existent weed should be removed. Planting is done preferably at the beginning of rainy season often in the flat beds though in wet regions 30cm high ridges are desirable where stakes may be planted about 5cm deep. Deeper planting of stakes results in swelling of the mother stem with consequent reduction in root size and yield, especially in compact soil. Both the interplant and interrow spacings are usually maintained to about 0.9-1.2m though the preferred plant density may vary greatly depending on the region *cum* local customs, the cultivar and soil condition. Rows with some 1.2m apart and about 0.9m spacing along the row will accommodate nearly 9,250 plants pere hectare. Though more usual range is 7,000 to 10,000 plants per hectare but for local consumption close planting is preferred, *i.e.* 3,000-20,000 plants per hectare especially on quite infertile soils. Closer spacing gives higher yield but with smaller tubers for the purpose of processing so in this case mechanical harvesting of tubers is favoured. With certain cultivars, in parts of Brazil some 30,000 plants/ha are also used though high densities are however more vulnerable to certain fungal diseases (Humanity development library 2.0).

Though most growers usually prefer to plant the stem cuttings directly in the main field, some growers presprout the cuttings in nursery and then transplant these in the main field. The sticks are planted vertically or at an angle with half their length in the soil or flat underground at a depth of about 10cm. When sticks are planted vertically or inclined, tubers form only at the extreme end of the cut forming a slanted tuber cluster often of irregular size but when a stick is cut at right angle and planted vertically the roots are evenly distributed around the circumference and are more uniform in size. Horizontally-planted sticks produce tubers at each node but lodging of the aerial parts is increased so yield is reduced. With vertical or inclined planting the roots penetrate more deeply and tubers may be formed at intervals along the planted portion but in low rainfall areas desiccation of the cuttings may occur.

Although hand planting is still practiced in majority of areas, several types of machine for planting the cuttings vertically, at right angle or in a trench have been developed or modified by vegetable planters that are being used on a large scale in Brazil, Africa and elsewhere.

Farmyard Manure and Fertilizer

Although cassava can grow on relatively infertile soils which are not suitable for other crops, it responds well to fertilization. Farmyard manure and fertilizers recommended by CTCRI are given below:

- ✰ For high yielding varieties of 10-11 months duration yielding 30-35t/ha
  - ❖ Farmyard manure: 12.5t/ha.
  - ❖ NPK: 75:50:75kg/ha.
- ✰ For high yielding varieties of 6-7 months duration yielding 25-30t/ha of tubers
  - ❖ Farmyard manure: 12.5t/ha.
  - ❖ NPK: 100:50:100kg/ha.
- ✰ For local varieties yielding 20-25t/ha of tubers
  - ❖ Farmyard manure: 12.5t/ha.
  - ❖ NPK: 50:25:50kg/ha.

Cassava crop giving a 30t/ha tuber yield, removes 187kg N, 33kg P and 233kg K, therefore the soil should be fortified with 12t/ha of FYM along with a fertilizer dose of 50kg each of nitrogen, phosphorus and potassium at the time of land preparation. Again at 1st intercultural operation *i.e.* after 45-60 days after transplanting, George (2019) has suggested a basal dose of 75, 50 and 75 kg N, P_2O_5 and K_2O, respectively for short duration *cum* local type cassavas, and 35kg each of N and K_2O as topdressing.

Irrigation

While cassava can tolerate drought or water scarcity to a greater extent but it is important that the crop has adequate moisture at crop establishment and for the first 3 months for root and shoot growth, *vis-a-vis* tuber formation. Supplementary irrigation at IW:CPE (Irrigation water/cumulative pan evaporation) ratio of 0.70 during drought spell is beneficial in maximising productivity. Most cost-efficient irrigation system for cassava production is drip irrigation or fountain tubing. Overhead irrigation system is the poor choice because the thick canopy of the plant can lead to high evaporation before water gets to the soil. Mulching and intercropping strategies can also be used to conserve soil moisture. George (2019) says that under drip irrigation at 100 per cent CPE provided water use efficiency of 61.01kg/ha/mm with water saving of 47.9 per cent and enhanced tuber yield by 44 per cent.

Interculture

Water availability to the cassava plant can be improved by organic mulch such as hay or live mulch such as peanuts. **Mulching** can also regulate the solar radiation penetrating the soil thereby lowering soil temperature, reducing wind velocity *cum* evaporation from the soil and increasing relative humidity (Anon., 2021).

Weed control is extremely important especially for young plantings that are susceptible to weed competition. After planting, the first weeding is usually done 2-3 weeks after sprouting followed by 3-4 weedings during the following 4-6 months (Humanity development library 2.0). Chemical control gives excellent results; pre-emergence with subsequent application of the latest weedicides such as pendimethalin will give satisfactory results (Humanity development library 2.0). After 8-12 weeks the plants are usually earthed up to encourage tuber formation (Humanity development library 2.0). CTCRI recommends that along with other intercultural operations only two shoots facing opposite directions should be retained by removing the excess ones at 30 days after transplanting.

Intercropping

Small farmers usually grow cassava in mixed cultivations with vegetables, bananas, yams or sweet potatoes, *vis-a-vis* with rubber and coconuts. When interplanted with coconuts a ground cover of the legume *Stylosanthes* sp. (stylo) leads to substantial increase in yield (Humanity development library 2.0). Short duration legumes such as bunchy varieties of groundnut and vegetable cowpea are found ideal. Intercrop is dibbled while cassava is planted (CTCRI). When there is no any assured source of irrigation, rice as rainfed crop during May-August period should be taken first and then short duration cassava crop should be taken for harvesting in April.

Harvest

The harvest time depends greatly on the cultivar and growth conditions.

Harvesting season extends from October to May. Delays in harvesting do not seriously affect tuber quality or yield. The plants are usually topped, *i.e.* the aboveground parts removed by hand using a machete or through machine, *e.g.* pushed down by a heavy screen mounted on the front of a tractor. The tubers may then be dug by hand or through machine harvester. Damage to tubers is still a problem with mechanical harvesting but smaller tubers that are less liable to damage are more easily lifted. For this reason mechanical harvesting is particularly suitable for tubers to be used for processing where small tubers are satisfactory and this way a greater degree of damage can be tolerated.

For food usage, the cassava is usually harvested by hand within 12 months when the tubers have achieved the desirable size. This is done by raising the lower part of the stem, pulling the roots out of the ground, and removing them from the base of the plant. The leaves and upper part of the plant are removed before harvesting begins as this enables the farmers to grab the visible stem and pull the tubers from the ground. The young branches are cut and discarded and before harvesting the roots the main stems (0.5–2m in length, depending on the cultivar and growing conditions) are cut and tied together in bundles of approximately 50 stems (Ceballos and Hershey, 2017).

Postharvest Physiological Deterioration (PPD)

Cassava roots cannot be kept in the fresh state for more than a few days after harvest. Moreover, once the tubers are separated from the main plant these start deteriorating rapidly due to inevitable physical damage of the roots during harvesting. Deterioration can be either post-harvest physiological deterioration (PPD) or due to entry of microorganisms into the mechanically damaged roots thus causing rapidly-spreading internal rotting. PPD reduces starch quality and renders the roots unpalatable and unmarketable. After about 15 minutes of damage, the tubers start responding to the wound with a healing mechanism which involves coumaric acids. PPD occurs when this healing mechanism fails to switch off in harvested tubers and continues until the entire tuber is oxidized and blackened within 2-3 days after harvest. PPD is related to the accumulation of reactive oxygen species (ROS) initiated by cyanide release during mechanical harvesting.

Shelf-life of tuber may be increased up to three weeks by overexpressing a cyanide insensitive alternative oxidase which can suppress ROS by 10-fold (Zidenga, 2012). As to those of potatoes fresh tubers can be preserved using thiabendazole or bleach as a fungicide, then wrapping in plastic, coating in wax or freezing (Anon., 2016b). Some measure of control for PPD has been obtained by holding the roots at low temperature or storing in air with low oxygen levels or in carbon dioxide.

Strategies to Delay Postharvest Physiological Deterioration (PPD)

Plant breeding has resulted in different strategies for cassava tolerance to PPD (Venturini *et al.*, 2016; Morante *et al.*, 2010). Cassava genotypes with delayed PPD trait have also been generated through plant breeding and transgenic approaches. One was induced through gamma ray mutagenesis which putatively silenced one of the genes involved in PPD genesis while other was a group of high-carotene clones in which the antioxidant properties of carotenoids are postulated to protect the roots from PPD (Venturini *et al.*, 2016).

Farmers usually follow one or more of the pre- or post-harvest methods given below to delay PPD:

1. Pre-harvest Methods

To delay PPD includes extended ground storage and pre-harvest pruning.

a. **Extended ground storage**: Harvest is delayed until the storage roots are required for consumption. This ties up the land that could otherwise be used for farming. Extended ground storage also decreases starch content and quality, *vis-a-vis* storage roots become more fibrous with the result their cooking time also increases (Wenham, 1995; Salcedo and Siritunga, 2011).

b. **Pre-harvest pruning**: Removal of the aerial parts before harvest may reduce the impact of PPD by about 25 per cent as it increases the sugar/starch ratio content in the storage roots which results into delayed onset of PPD (Wheatley and Schawabe, 1985; van Oirschot *et al.*, 2000) and limits the accumulation of scopoletin (Wheatley, 1980). However, the time of pruning is critical and requires a level of expertise because it can affect texture, flavour as well as general consumer acceptability (van Oirschot *et al.*, 2000). Since pruning can negatively impact cassava yield (Ayoola and Agboola, 2004), its adoption by farmers is also dependent on the establishment of pruning practices performed at the right physiological stage to preserve yield and starch content (Curcelli *et al.*, 2014). Pre-pruning of aerial portions of the plant 2-3 weeks before harvesting has shown to minimise physiological deterioration in the tubers (Humanity development library 2.0).

c. **Post-harvest methods** to reduce losses due to wounding of the storage roots during harvest include ~ proper storage of harvested cassava roots using antioxidant molecules, chemical application and avoidance strategies.

d. **Storage**: Cassava storage roots and roots of other tropical crops such as yam transpire and loose moisture which reduce their quality during storage (Uchechukwu-Agua *et al.*, 2015). Storing unwounded roots under high humidity and limited oxygen can reduce water loss and oxidative stress (Marriott *et al.*, 1978; Noon and Booth, 1977). Storing at high humidity also helps wounding the periderm formation (Booth, 1975; Marriott *et al.*, 1978; Rickard, 1985). Using polyethylene bags, storage boxes and coating cassava storage roots with paraffin wax also maintain high humidity and exclude the air. Other approaches to reduce PPD include storage of cassava roots under low temperature (0–4°C) (Ravi *et al.*, 1996) and hot-water treatment at 54–56°C for 10 minutes combined with modified atmosphere packaging (Acedo and Acedo Jr., 2012). However, these approaches are costly and therefore only suitable for high premium cassava products (Ravi *et al.*, 1996).

e. **Chemical treatment**: Post-harvest chemical treatment can extend the shelf-life of cassava tubers (Ravi *et al.*, 1996), however, their potential toxicities could restrain their broad use. Applications of ethanol (>20 per cent), sodium sulphite (10 per cent), sodium dithiocarbamate (10 per cent), saturated sodium chloride, benomyl (500 ppm) and dicloran (1,000 ppm) can delay the onset of PPD for several days (Booth, 1976, 1977).

f. Antioxidant molecules can also be used for treatment of storage roots to delay PPD as suggested by the longer shelf-life of cassava genotypes with high antioxidant contents (Sanchez *et al.*, 2006). Application of melatonin (500ppm) to cassava storage roots could attenuate vascular discolouration of root slices. Exogenous melatonin leads to a moderate but significant reduction of PPD symptoms (Ma *et al.*, 2016). Experimental data showed that exogenous melatonin acts by reducing H_2O_2 content and improving activities of catalase and peroxidase (Hu *et al.*, 2016). Reduction of vascular discolouration might also be due to the antioxidant properties of melatonin (Reiter *et al.*, 2002).

g. **Immediate processing (Avoidance strategy)**: After harvest, immediate processing of root tubers into more stable traditional food or industrial products avoids the PPD problem (Wenham, 1995; Reilly *et al.*, 2004).

Pests and Diseases

In some areas baboons, monkeys, wild boars and rodents are reported to cause severe damage, therefore, to discourage these predators sometimes smallholders plant bitter cassava around the areas of sweet cassava plantation.

Insects and Pests

Infestation by insects is a major cause of losses during cassava storage (Osipitan *et al.*, 2015). A wide range of insect species infestation is reported to cause losses in the field. Several species that feed directly on dried cassava chips have been reported as a major factor in spoiling stored cassava with losses ranging between 19-30 per cent of the harvested produce (Osipitan *et al.*, 2015). The cassava mealybug (*Phenacoccus manihoti*) and cassava green mite (*Mononychellus tanajoa*) were found rampant during 1970s and 1980s in Africa causing up to 80 per cent crop loss. Two South American natural enemies, *Anagyrus lopezi* (a parasitoid wasp) and *Typhlodromalus aripo* (a predatory mite) were found effective in controlling cassava mealybug and cassava green mite, respectively. Except for severe cases of insect infestation the pesticides should be avoided as this also reduces population of beneficial insects such as natural enemies.

1. **Variegated grasshopper** (*Zonocerus variegatus*) and **Elegant grasshopper** (*Z. elegans*): These brightly coloured large grasshoppers have a wide host range and mainly attack seedlings. Cooperation between neighbouring farms can help controlling the grasshoppers as the females lay their eggs with the plantation with nymphs migrating nearby to the crop to feed.

Variegated grasshopper

Cassava whitefly

Scales on cassava stem and tuber

Cassava green mite infestation

Tetranychus urticae

Two females with an egg

Male

Symptoms: Plants get defoliated and the bark removed from stems.

Control: Handpicking and killing of grasshoppers; destroying the egg-laying sites; spraying with fenitrothion pesticide; and use of 'green muscle' and other similar biopesticides as available in S and W Africa are the means to control these pests. Products containing neem oil make good control of variegated grasshoppers.

2. **African root and tuber scale** (*Strictococcus vayssierrei*): Indigenous to Africa, this insect is particularly prevalent in forest areas of Ghana and Congo.

   Symptoms: Oval-shaped scales on stems, roots and/or tubers are seen. Early stage infestation kills the plants so root tuber formation and growth is hampered, plants shrivel and show discolouration at feeding sites.

   Control: Soil is improved and made fertile through addition of organic matter, scale infested stems are removed and destroyed, and care should be taken that only non-infested stems are planted.

3. **Cassava scale**: Scale insects are reported to cause damage especially in areas experiencing drought. These include the **cassava stem mussel scale** (*Aonidomytilus albus*), the **white peach scale** (*Pseudaulacaspis pentagona*) and the **black scale** (*Parasaissetia nigra*). The insect is a flattened oval scale with an elongated white cover. Using pesticides to control scale insects reduces even beneficial insect population.

   Symptoms: Stem surfaces become covered with white waxy substance where leaves wilt and drop from the plant. Severe infestations may result in stunted growth of plants and poor tuber yield. Cuttings taken from infested plants normally do not sprout.

   Control: Only healthy plants should be selected for taking cuttings. Infested stems from existing plantation should be removed and destroyed. To improve fertility in the soil, organic matter should be incorporated. Oil emulsion along with Malathion should be sprayed to control its infestation on the plants.

4. **Cassava hornworm** (*Erinnyis ello*): It attacks the crop in the Americas and Caribbean.

   Control: This is controlled biologically by the use of *Bacillus thuringiensis* sprays, egg parasite (*Trichogromma fasciatum*) or through use of predatory wasp (*Polistes canadensis*).

5. **Leaf cutting ants** (*Atta* sp. and *Acromyrmex* sp.): It can cause severe defoliation.

   Control: Recommended control is poison bait.

6. **Cassava lacebug** (*Vatiga manihotae*): It attacks the crop in South America and may cause early defoliation.

   Control: In severe cases, fenitrothion will control the pest.

7. **Cassava green mite** (*Mononychellus tanajoa* and *M. progresivus*): Green spider mites are very common 'arachnid' pests in most African growing regions where these become problematic during dry season. Their infestation causes significant tuber losses.

   Symptoms: Leaves due to green mite attack start showing 'yellow stipping of leaves', 'chlorotic spots on leaves'; 'chlorosis of entire leaves'; in case of heavy infestation leaves may be stunted and deformed; and terminal leaves may die and drop from the plant.

   Control: Plant mite-tolerant cassava varieties where possible. The crop should be planted at beginning of the rainy season to encourage vigorous growth which allows plant to tolerate attack to a great extent. Intercropping with crops such as cowpea reduces its incidence to a great extent. Predaceous mites such as *Typhlodromalus aripo* as well as staphylinid beetles should be introduced in the crop for successful control of green spider mite as is being done in many regions of Africa. In the last, one can try acaricides to control such pests.

8. **Two-spotted spider mite** (*Tetranychus* spp.): Spider mites thrive in dusty conditions. Water-stressed plants are more susceptible to attack by these mites, especially the *Tetranychus cinnabarinus*.

   Symptoms: Its infestation causes leaves becoming stippled with yellow. Its attack may cause webbing, yellowing *cum* bronzing and dropping of leaves. Mites may be visible as tiny moving dots on the webs or underside of leaves, best viewed using a hand lens. These are usually not spotted until there are visible symptoms on the plant.

   Control: These are attacked by predaceous mites and staphylinid beetles. Spraying plants with a strong jet of water can help reduce build up of spider mite populations. If mites become problematic, apply insecticidal soap to plants. Certain chemical insecticides may actually increase mite populations by killing off natural enemies and promoting mite reproduction.

9. **Shootflies** (*Silba pendula* and *Carpolonchaea chalybea*) are borers causing damage in South America. They burrow into the growing point at the start of the rainy season.

   Control: Resistant cultivars are recommended. Its larvae can be controlled with diazinon, dimethoate or certain other chemicals.

10. **Weevil borers** (*Coelosternus* spp.): Stems may be attacked by the weevil borers.

    Control: Cutting off the infected branches is recommended.

Nematodes

A wide range of parasitic nematodes are associated with cassava worldwide, including *Pratylenchus brachyurus*, *Rotylenchulus reniformis*, *Helicotylenchus erythrinae*, *Scutellonema* spp. and *Meloidogyne* spp. Nematodes prefer sandy soils and damage is most likely in the field where such soils exist. Use of nematicides results in lowering the number of galls per feeder root coupled with a lower number of rots in the storage roots (Coyne *et al.*, 2006). When the organophosphorus nematicide (femaniphos) is used, crop growth and yield parameter variables are not affected at harvest. The use of tolerant and resistant cultivars is the most practical management method (Coyne *et al.*, 2018; Coyne, 1994).

Root-knot Nematodes (*Meloidogyne* spp.)

Root-knot nematodes are common but seldom of economic importance except when cassava is intercropped with good hosts for the nematode, *e.g. Solanum melongena* or *Hibiscus sabdariffa*. Most widely reported and economically important are 'root knot nematodes' (*Meloidogyne incognita* and *M. javanica*) (Mc-Sorley *et al.*, 1983). *Meloidogyne* spp. feeding produces physically damaging galls with eggs inside them. The galls later merge as the females grow and enlarge, and they interfere with water and nutrient supply of plants (Gapasin, 1980). Galls can appear as quickly as a month prior to planting. Other pests and diseases can gain entry through the physical damage caused by gall formation which leads to rotting of tubers.

Symptoms: Yellowing of plants occurs with reduction in plant vigour due to galling of root-knot nematodes on the roots (galls are usually smaller but up to 3.3cm in diameter) and such plants in severe attack may wilt in hot weather.

Control: Planting of resistant varieties if nematodes are known to be present in the soil. If symptoms indicate nematode infestation, immediately check the roots of plants. Before planting, soil should thoroughly be solarized which will kill insect-pests and spores of other diseases too.

Fungal Diseases

Of the fungal diseases attacking this crop, most important ones are leaf spots, root rots, rusts and bud necrosis as discussed below:

Brown Leaf Spot

Anthracnose canker on cassava stem

1. **Anthracnose** (*Colletotrichum gloeosporioides*): This disease emerges at the beginning of wet season as has been observed in Africa. Its spores spread by wind.

   Symptoms: Cankers appear on stems and leaf petioles. Its infection causes downward drooping *cum* wilting, followed by falling of leaves so plants defoliate, shoots die and soft plant parts become twisted and distorted

   Control: Anthracnose usually does not cause large-scale economic damage to cassava crop. To avoid this disease, canker infected cuttings should be avoided from planting. Proper sanitation in the field should be maintained and after harvest plant debris should be removed.

2. **Brown leaf spot** is caused by *Cercospora henningsii* and **white leaf spot** by *C. caribaea*. These diseases are prevalent in wetter and warmer regions where cassava is usually regularly grown, *e.g.* Brazil and Colombia and in West Africa.

   Symptoms of brown leaf spot: Circular or irregular brown spots appear with darker margin between leaf veins on older leaves. Centres of lesions may drop out giving leaves a shot-hole appearance. In severe infection leaves may turn yellow, dry out and drop from the plant.

   Symptoms of white leaf spot: Large diffused white spots on upper surface of leaves develop during humid weather, with grey centre on underside of leaves.

   Control: Leaves, crop debris and weeds around plants should be removed to prevent spread of disease. Wider spacing provides little chance for building up of excess humidity. This disease may be controlled by use of copper fungicides, thiophanate or benomyl but chemical methods are unlikely to be economic, therefore, planting of resistant cultivars is recommended.

3. ***Phyllosticta*** sp. also causes leaf-spotting so for this only resistant cultivars should be used.
4. **Root rots** include those caused by *Phytophthora* spp., *Rigidoporus lignosus* (white rot, white thread), *Rosellinia necatrix* (black rot), *Corticium rolfsii*, and a number of others of lesser importance. Control is by cultural practices including proper drainage, early harvest, crop rotation, removal of crop debris and planting of healthy planting material.
5. ***Uromyces*** spp. though causes leaf and stem rusts but not of serious economic importance.
6. ***Glomerella*** sp., ***Botryodiplodia theobromae*** and some members from Basidiomycetes and Ascomycetes are also causative agents. Control involves careful selection and handling to avoid wounding, and storage not under excessive humid conditions. 'Seed treatment' fungicides, *e.g.* captan+carbendazim, mancozeb and chloroneb are among those recommended.
7. **Bud necrosis**: This fungal disease is most commonly found on cassava grown in humid regions (e.g. humid forest zones) and is quite less common in drier savannah areas. Fungal spores are carried by wind to new plants and farms.

   Symptoms: Dark brown to gray patches of fungal growth appear on stems with necrotic areas covering buds on the stem.

   Control: Cuttings for planting should be collected only from healthy plants. Plants should be widely spaced to allow good air circulation in between so that disease incidence is quite reduced. Weeds around plants should be removed. All the necrotic stems and crop debris must immediately be burnt after harvest to prevent the spread.

Virus Diseases

Viruses limit cassava production in the tropics. Recommended control measures include planting of healthy planting material, removal and destruction of diseased plants, use of resistant cultivars and the control of vectors and host plants where these are known.

1. **Cassava mosaic disease (CMD)**: CMD is caused by Gemini virus which is widespread in Africa. Disease is spread by infected cuttings and through whitefly (*Bemisia tabaci*). Wild cassava (kisamvu cha mpira in Kiswahili) also hosts the disease. The disease was first observed in the late 19th century in Tanzania. **African cassava mosaic** (curly leaf disease,

Severe CMD symptoms on leaf

Whiteflies on cassava leaf

CBSD symptoms on the lower leaves

Rotting tuber caused by CBSD

Witche's broom symptom

manihot virus) is transmitted by a white fly (*Bemisia* sp.) and through transplanting of diseased plants into new fields. No host plant is known for this virus. This virus limits the root growth along with withering of leaves. **Asian mosaic** occurs in India producing effects similar to African mosaic and cucumber acts as its host plant. In northern South America another virus causing mosaic, chlorosis and leaf curl has been reported with several hosts such as other species of *Manihot*, and some members of Chenopodiaceae as well as Malvaceae.

Symptoms: CMD causes discoloured pale green, yellow or white mottled and distorted leaves reduced in size. In highly susceptible cassava cultivars, plant growth stunts which results in poor root yield and low quality stem cuttings. The infected plants can express a range of symptoms such as leaf distortion and chlorotic discolouration from yellow to green, *vis-à-vis* folding and shrivelling of leaf blades.

Control: New crop cuttings should be collected only from disease-free healthy plants. Plants should be regularly inspected for symptoms of disease and the infected plants should be removed and destroyed. Its vector whitefly should be regularly controlled. CMD resistant varieties should be collected from various countries and planted. *M. esculenta* cultivars most widely used on African continent have *M. carthaginensis* subsp. *glaziovii* genes where some have resistance to CMD (Lebot, 2020). CMD resistant/tolerant varieties from CTCRI, India are 'H-97', 'H-165', 'Sree Kaveri', 'Sree Padmanabha', 'Sree Swarna', *etc.* Other resistant varieties include 'Cyizere', 'Garukunsubire', 'Mavoka', 'Mbakungahaze',

'Ndamirabana', 'Rwizihiza', 'Seruruseke', *etc.* Harvest can be done early to avoid severe losses due to necrosis of tubers. Proper plant quarantine practices should be used to avoid spread of virus to new region. Whiteflies can be controlled by introducing beneficial insects in the field like spiders, ladybird beetles, *etc.* Yellow sticky traps can be used to monitor infestation of whiteflies. Insecticidal soaps can be sprayed on undersurface of leaves to kill flies.

2. **Cassava brown streak disease (CBSD)**: CBSD is caused by two distinct species of single-stranded RNA (ssRNA) viruses ~ Cassava brown streak virus (CBSV) and Ugandan cassava brown streak virus (UCBSV), belonging to the genus Ipomovirus of the family Potyviridae. Cassava brown streak virus has been identified as a major threat to cassava cultivation worldwide. Prevalent in East Africa (Tanzania, Uganda and Kenya), the virus is transmitted through whiteflies and stem cuttings. Though the economic loss from brown streak virus depends on region, cultivars and environmental conditions but in general the loss is estimated up to 70 per cent.

   Symptoms: About 2.5-3 months after planting chlorotic or necrotic vein banding can be seen in mature leaves which may later on merge to form large yellow patches. Dark brown elongated necrotic lesions can be seen on upper portion of green stem (young stems) after about 5-6 months of planting. About 9-10 months after planting the disease causes tuber malformation and root constriction (roots developing knots and tubers necrosis) where internal tissues of roots and tubers are stained brown and may rot due to secondary fungus infection. Severe cases cause drying of leaves and shoot die-back. Varying symptoms of cassava brown streak disease makes its field-identification quite difficult. The matter becomes utterly complicated if 'cassava brown streak' and 'cassava mosaic disease' occur together.

   Control: Cuttings from disease-free plants, *vis-à-vis* resistant varieties should be used for planting the fresh crops. Plants should regularly be inspected for symptoms of disease and the infected plants should be removed and destroyed. Harvest can be done early to avoid severe losses due to necrosis of tubers. Proper plant quarantine practices should be used to avoid spread of virus to new region. Whiteflies can be controlled by encouraging beneficial insects in the field like spiders, ladybird beetles, *etc.* Yellow sticky traps can be used to monitor infestation of whiteflies. Insecticidal soaps can be sprayed under leaf surface to kill flies.

3. **Witche's broom**: Cassava witche's-broom caused by phytoplasma has been detected in Vietnam, Thailand, Cambodia, Laos, Indonesia and the Philippines with yield loss from 30 to 80 per cent.

   Symptoms: Plants are stunted with an excessive proliferation of branches where shoots have shortened internodes with small leaves. Infected plants produce quite weak shoots but with no visible symptoms.

   Control: plants that are suspected of being infected must be removed and destroyed. All cassava debris should be removed from field after harvest. All tools and equipment used for cuttings should be disinfected before cutting.

Bacterial Diseases

Most important bacterial disease is **cassava bacterial blight** caused by *Xanthosoma manihotis*. This disease is more prevalent in South America and Africa. It becomes quite severe under wet conditions, spreading through infected cuttings, water splash and infected tools. Bacterial stem rot (*Erwinia cassavae*) and bacterial stem gall (*Agrobacterium* sp.) have also been found infecting cassava plantation.

Symptoms: It appears as leaf spotting and blight, wilting, die-back of shoots, brown gum exudation and vascular necrosis throughout the plant. Small, angular, brown and water-soaked lesions appear between leaf veins on lower surface of leaves. Leaf blades turn brown as lesion expands. Lesions may have a yellow halo. Lesions coalesce to form large necrotic patches. Defoliation occurs with leaf petioles remaining in horizontal position, the shoots in severe infection may die back and the stems, petioles and leaves may present gumming.

Control: Resistant cultivars should be planted. Cassava crop rotation with non-host is effective. Crop debris after harvest should be removed and destroyed. Infected parts from plant should be pruned off. Planting material should be taken only from healthy plants. Intercropping of cassava with corn (maize) and melon is also effective.

Integrated Pest Control

Integrated pest control recommended for cassava to ensure healthy and resistant plants include:

- ☆ Good soil preparation with proper drainage.
- ☆ Robust and disease-free planting sticks from well-lignified stems should be taken, dipped in fungicide (captan + benomyl) before planting. Planting should be properly done with wide spacing, proper fertilizing and good weed control measures should be adopted.

- Disease- and pest-resistant cultivars should be used where possible.
- In monoculture practice, different cultivars should be planted together or in adjacent plots to provide different degrees of susceptibility (or resistance) within the planted areas.
- Regular inspections should be made so that appropriate pest control may be applied at an early stage when necessary.
- Natural biological control is important and chemicals should be used only when very necessary.
- After harvest all plant residues should be removed.
- Continuous planting of adjacent areas should be avoided as this gives the opportunity for pests to have uninterrupted access to plants at their most susceptible age.

References

1. Acedo, J. and A. Acedo Jr. 2012. Controlling postharvest physiological deterioration and surface browning in cassava (*Manihot esculenta* Crantz) roots with hot water treatment. *Southeast Asia Symposium on Quality Management in Postharvest Systems and Asia Pacific Symposium on Postharvest Quality*, pp. 357-61.
2. Afedraru and Lominda, 2019. Uganda to launch innovative gene-edited cassava research. *Alliance for Science*. Retrieved 15 August 2021.
3. Alves, A.A.C. 2002. Cassava botany and physiology. In: Cassava: biology, production and utilization (eds Hillocks, R.J., J.M. Tresh and A.C. Bellotti), pp.67–89. CABI Publishing, Wallingford (UK).
4. Anon. 2016a. Cassava, raw. National Nutrient Database for Standard Reference Release 28. ARS, USDA. May 2016. Retrieved 7 December 2016.
5. Anon. 2016b. Storage and processing of roots and tubers in the tropics. U.N. Food and Agriculture Organization. Retrieved 4 May 2016.
6. Anon. 1995. Dimensions of Need: An Atlas of Food and Agriculture. *Food and Agriculture Organization of the United Nations. Rome, Italy.*
7. Anon. 2021. Growing cassava - Soil and Water Management. FACTSHEET No: 04 (May 2021). https: //www.cardi.org.
8. Anon. 2017. *Manihot esculenta*. World Checklist of Selected Plant Families (WCSP). Royal Botanic Gardens, Kew. Retrieved 21 March 2017.
9. Aregheore E.M. and O.O. Agunbiade, 1991. The toxic effects of cassava (*Manihot esculenta* Crantz) diets on humans: a review. *Vet. Hum. Toxicol.*, **33**(3): 274–5.

10. Ayoola, O.T. and A.A. Agboola, 2004. Influence of cassava planting patterns and pruning methods on crop yield in a cassava-based cropping system. *Afr. Crop Sci. J.*, **12** (2): 115-22.

11. Barrett, Cindy, 2010. *Manihot walkerae.* CPC National Collection Plant Profiles. Center for Plant Conservation. Archived from the original on 2010-12-15. Retrieved 2011-05-07.

12. Booth, R.H. 1975. Cassava Storage: Post-harvest Deterioration and Storage of Fresh Cassava Roots. CIAT, Cali, Colombia.

13. Booth, R.H. 1976. Storage of fresh cassava (*Manihot esculenta*). I. post-Harvest deterioration and its control. *Exp. Agric.*, **12** (02): 103.

14. Booth, R.H. 1977. Storage of fresh cassava (*Manihot esculenta*): II. simple storage techniques. *Exp. Agric.*, **13**: 119-28.

15. Ceballos, H. and C.H. Hershey, 2017. Cassava. In: Genetic improvement of Tropical Species (eds Campos, H. and P.D.S. Caligari). Springer, Cham (Switzerland), p.129–80.

16. Cereda, M.P. and M.C.Y. Mattos, 1996. Linamarin: the Toxic Compound of Cassava. *J. Venomous Animals and Toxins*, **2**: 6–12.

17. Chiwona-Karltun Linley, Chrissie Katundu, James Ngoma, Felistus Chipungu, Jonathan Mkumbira, Sidney Simukoko and Janice Jiggins, 2002. Bitter cassava and women: an intriguing response to food security. *LEISA Mag.*, **18** (4): 22-6.

18. Claude, F. and F. Denis, 1990. African Cassava Mosaic Virus: Etiology, Epidemiology and Control. *Pl. Dis.*, **74**(6): 404–11.

19. Coyne, D.L., F. Kagoda, E. Wambugu and P. Ragama, 2006. Response of cassava to nematicide application and plant-parasitic nematode infection in East Africa, with emphasis on root-knot nematode. *Intern. J. Pest Manag.*, **52**(3): 215–23.

20. Coyne, D. L. 1994. Nematode pests of cassava. *Afr.Crop Sci. J.*, **2**(4): 355–9.

21. Coyne, D.L., Laura Cortada, Jonathan J. Dalzell, Abiodun O. Claudius-Cole, Solveig Haukeland, Nessie Luambano and Herbert Talwana, 2018. Plant-Parasitic Nematodes and Food Security in Sub-Saharan Africa. *Ann. Rev. Phytopath.*, **56**(1): 381–403.

22. Curcelli, F., S.J. Bicudo, E.B. Aguiar and M.I.N. Valdivie, 2014. Pruning management of cassava for animal feeding: parameters of the root. *Afr. J. agric. Res.*, **9**(16): 1238-43.

23. Gapasin, R.M. 1980. Reaction of golden yellow cassava to *Meloidogyne* spp. inoculation. *Ann. Trop. Res.*, **2:** *: 49–53.*

24. George, J. 2019. Cassava. In: Handbook of Horticulture, vol. 1 (eds Verma Shashi A., Som Dutt and Sudhir Pradhan), pp. 655-8. ICAR, New Delhi.

25. Hershey, C.H. 2017. Drivers of change for cassava's multiple roles in sustainable development. In: Achieving Sustainable Cultivation of Cassava. Vol. 1 (ed.Hershey, C.). Burleigh Dodds Series on Agricultural Science, Philadelphia (PA), No. 21, pp.175–98.

26. Hu, W., H. Kong, Y. Guo, Y. Zhang, Z. Ding, W. Tie, Y. Yan, Q. Huang, M. Peng, H. Shi and A. Guo, 2016. Comparative physiological and transcriptomic analyses reveal the actions of melatonin in the delay of postharvest physiological deterioration of cassava. *Front. Plant Sci.*, **7**: 736.

27. Humanity development library 2.0. Root crops: Cassava (*Manihot esculenta*). nzdl.org.

28. Kawano, K. 1980. Cassava. In: Hybridization of Crop Plants (eds Fehr, W.R. and H.H. Hadley), pp. 225–233. Madison (WI), USA.

29. Latif, S. and J. Muller, 2015. Potential of cassava leaves in human nutrition: a review (PDF). Trends in Food Science and Technology.

30. Lebot, V. 2020. Tropical Root and Tuber Crops: Cassava, Sweet Potato, Yams and Aroids. Wallingford, Oxfordshire, Boston, UK and MA, USA. CABI (Centre for Agriculture and Bioscience International), 541p.

31. Lokko, Y., E. Okogbenin, C. Mba, A. Dixon, A. Raji and M. Fregene, 2007. Cassava (Chapter 14). Pulses, Sugar and Tuber Crops. *Genome Mapping and Molecular Breeding in Plants* (vol. 3). Springer, p.263.

32. Ma, Q., T. Zhang, P. Zhang and Z.Y. Wang, 2016. Melatonin attenuates postharvest physiological deterioration of cassava storage roots. *J. Pineal Res.*, **60**: 424-34

33. Marriott, J., B.O. Been and C. Perkins, 1978. The aetiology of vascular discolouration in cassava roots after harvesting: association with water loss from wounds. *Pl. Physiol.*, **44**: 38-42.

34. Mc-Sorley, R., S.K. Ohair and J.L. Parrado, 1983. Nematodes of Cassava, *Manihot esculenta* Crantz. *Nematropica*, **13**: 261–87. Archived from the original on 3 June 2016.

35. Morante, N., T. Sanchez, H. Ceballos, F. Calle, J.C. Perez, C. Egesi, C.E. Cuambe, A.F. Escobar, D. Ortiz, A.L. Chávez and M. Fregene, 2010. Tolerance to postharvest physiological deterioration in cassava roots. *Crop Sci.*, **50** (4): 1333–8.

36. Noon, R. and R.H. Booth, 1977. Nature of post-harvest deterioration of cassava roots. *Trans. Br. Mycol. Soc.*, **69** (2): 287-90.

37. Olsen, K.M. and B.A. Schaal, 1999. Evidence on the origin of cassava phylogeography of *Manihot esculenta. Proc. Natl. Acad. Sci.*, **96** (10): 5586-91.

38. Osipitan, A. A., V.T. Sangowusi, O.I. Lawal and K.O. Popoola, 2015. Correlation of chemical compositions of cassava varieties to their resistance to *Prostephanus truncatus* Horn (Coleoptera: Bostrichidae). *J. Insect Sci.*, **15** (1): 13.

39. Raji, A.A., J.V. Anderson, O.A. Kolade, C.D. Ugwu, A.G. Dixon and I.L. Ingelbrecht, 2009. Gene-based microsatellites for cassava (*Manihot esculenta* Crantz): prevalence, polymorphisms and cross-taxa utility. *BMC Plant Biol.*, **9**: 118.

40. Ravi, V., J. Aked and C. Balagopalan, 1996. Review on tropical root and tuber crops: I. Storage methods and quality changes. *Crit. Rev. Food Sci. Nutr.*, **36:** 47-59.

41. Reilly, K., R. Gomez-Vasquez, H. Buschmann, J. Tohme and J.R. Beeching, 2004. Oxidative stress responses during cassava post-harvest physiological deterioration. *Plant Mol. Biol.*, **56** (4) 625-41.

42. Reiter, R.J., D. Tan and S. Burkhardt, 2002. Reactive oxygen and nitrogen species and cellular and organismal decline: amelioration with melatonin. *Mech. Ageing Dev.*, **123**: 1007-1019.

43. Rickard, J.E. 1985. Physiological deterioration of cassava roots. *J. Sci. Food Agric.*, **36**: 167-76.

44. Sakurai, T., K. Mochida, T. Yoshida, K. Akiyama, M. Ishitani and M. Seki, 2013. Genome-wide discovery and information resource development of DNA polymorphisms in cassava. *PLoS One*, **8**(9): e74056.

45. Salcedo, A. and D. Siritunga, 2011. Insights into the physiological, biochemical and molecular basis of postharvest deterioration in cassava (*Manihot esculenta*) roots. *Am. J. Exp. Agric.*, **1**(4): 414-31.

46. Sanchez, T., A.L. Chavez, H. Ceballos, D.B. Rodriguez-Amaya, P. Nestel and M. Ishitani, 2006. Reduction or delay of post-harvest physiological deterioration in cassava roots with higher carotenoid content. *J. Sci. Food Agric.*, **86**(4): 634-9.

47. Soto-Blanco, B. and Silvana Lima Gorniak, 2010. Toxic effects of prolonged administration of leaves of cassava (*Manihot esculenta* Crantz) to goats. *Exper. and Toxicol. Path.*, **62**(4): 361–6.

48. Sri Suharti, Hafini Oktafiani, Asep Sudarman, Myunggi Baik and Komang Gede Wiryawan, 2021. Effect of cyanide-degrading bacteria inoculation on performance, rumen fermentation characteristics of sheep fed bitter cassava (*Manihot esculenta* Crantz) leaf meal. *Ann agr. Sci.*, **66**(2): 131–6.

49. Tewe, O.O. 2004. The Global Cassava Development Strategy. U.N. Food and Agriculture Organization, Rome, Italy.

50. Uchechukwu-Agua, A.D., O.J. Caleb and U.L. Opara, 2015. Postharvest handling and storage of fresh cassava root and products: a review. *Food Biopr. Technol.*, **8:** 729-48.

51. van Oirschot Q.E.A., G.M. O'Brien, D. Dufour, M.A. El-Sharkawy and E. Mesa, 2000. The effect of pre-harvest pruning of cassava upon root deterioration and quality characteristics. *J. Sci. Food Agric.*, **80**(13): 1866-73.

52. Venturini, M.T, L.R. Santos, C.I. Vildoso, V.S. Santos and E.J. Oliveira, 2016. Variation in cassava germplasm for tolerance to post-harvest physiological deterioration. *Genet. Mol. Res.*, **15**(2): 71-74.

53. Wenham, J.E. 1995. Post-harvest deterioration of cassava: a biotechnological perspective. FAO, Rome.

54. Wheatley, C. 1980. Studies related with the nature of post-harvest physiological deterioration in cassava roots. CIAT, Cali, Colombia.

55. Wheatley, C.C. and W.W. Schawabe, 1985. Scopoletin involvement in post-harvest physiological deterioration of cassava root (*Manihot esculenta* crantz). *J. exp. Bot.*, **36** (166): 783-91.

56. White, W.L.B., D.I. Arias-Garzon, J.M. McMahon and R.T. Sayre, 1998. Cyanogenesis in cassava: the role of hydroxynitrile lyase in root cyanide production. *Plant Physiol.*, **116** (4): 1219–25. doi: 10.1104/pp.116.4.1219.

57. Willy, H.V., ed. 2010. Tropical Root and Tuber Crops. Soils, Plant Growth and Crop Production (vol. II). EOLSS Publishers, USA, 273p.

58. Zidenga, T. 2012. Extending cassava root shelf life via reduction of reactive oxygen species production. *Pl. Physiol.*, **159** (4): 1396–1407.

9

Chinese Potato (*Coleus* Lour.)

Family: Lamiaceae

Synonyms: *Ascocarydion* G.Taylor, *Briquetastrum* Robyns and Lebrun, *Burnatastrum* Briq., *Calchas* P.V.Heath, *Capitanya* Schweinf. ex Gürke, *Englerastrum* Briq., *Holostylon* Robyns and Lebrun, *Isodictyophorus* Briq., *Leocus* A.Chev., *Majana* Rumph. ex Kuntze, *Mitsa* Chapel. ex Benth., *Neohyptis* J.K.Morton, *Neomuellera* Briq., *Rabdosiella* Codd, *Saccostoma* Wall. ex Voigt, *Solenostemon* Thonn., *Symphostemon* Hiern (Anon., 2022).

Cultivar of *Coleus scutellarioides*

The name *Coleus* is derived from the Greek *koleos*, meaning 'a sheath', referring to the fused stamens of the type species for the genus. *Coleus*, found in the Old World tropics and subtropics, is a genus of around 300 species of annual or perennial herbs or shrubs that can be sometimes succulent or with a fleshy or tuberous rootstock. Earlier, *Coleus* and *Solenostemon* were sunk into *Plectranthus*, but recent phylogenetic analysis found *Plectranthus* to be paraphyletic with respect to other related genera in the subtribe Plectranthinae (https://en.wikipedia.org). Most recent taxonomic treatment of the genus resurrected *Coleus*, and 212 names were changed from combinations in *Plectranthus*, *Pycnostachys* and *Anisochilus*. *Equilabium* was segregated from *Plectranthus*, after phylogenetic studies supported its recognition as a phylogenetically distinct genus (Yeshitila and Jens, 2012).

Coleus plants are usually cultivated as ornamentals for their brightly coloured foliage, particularly *Coleus scutellarioides* (syn. *Coleus blumei, Plectranthus scutellarioides*) but some species such as *Coleus esculentus, C. rotundifolius* and *C. maculosus* subsp. *edulis* produce edible root tubers and are cultivated for food (https://en.wikipedia.org). The edible coleus is a minor tuber crop being grown seasonally in India on a small scale in the south including Tirunelveli (North Kerala) and Karnataka. To a limited scale it is also grown in NE States, M.P., and in tribal settlements throughout India.

As of October 2022, Plants of the World Online listed 301 species in the genus (Anon., 2022). Some of these species are *Coleus amboinicus* (syn. *Plectranthus amboinicus*), *C. argentatus, C. australis, C. barbatus* (syn. *Plectranthus barbatus*), *C. caninus* (syn. *Plectranthus caninus*), *C. cataractarum, C. comosus* (syn. *Plectranthus ornatus*; but is sold as *C. canina* or as cat plants), *C. cremnus, C. dissitiflorus, C. esculentus* (syn. *Plectranthus esculentus*), *C. fredericii* (syn. *Plectranthus welwitschii*), *C. graveolens, C. maculosus, C. neochilus, C. rotundifolius* (syn. *Plectranthus rotundifolius*), *C. scutellarioides* (syn. *Coleus blumei, Plectranthus scutellarioides*), *C. socotranus* (syn. *Plectranthus socotranus*), *C. unguentarius* (syn. *Plectranthus unguentarius*) (Paton *et al.*, 2019).

Coleus rotundifolius (Poir.) A.Chev. and Perrot

Synonyms: *Calchas parviflorus* (Benth.) P.V.Heath, *Coleus dysentericus* Baker, *C. pallidiflorus* A.Chev., *C. parviflorus* Benth., *C. rehmannii* Briq., *C. rotundifolius* var. *nigra* A.Chev., *C. rugosus* Benth., *C. salagensis* Gürke, *C. ternatus* (Sims) A.Chev., *C. tuberosus* (Blume) Benth., *Germanea rotundifolia* Poir., *Majana tuberosa* (Blume) Kuntze, *Nepeta madagascariensis* Lam., *Plectranthus coppinii* Heckel, *P. ternatus* Sims, *P. rotundifolius* (Poir.) Spreng., *P. tuberosus* Blume and *Solenostemon rotundifolius* (Poir.) J.K.Morton.

Common names: Chinese/Country/Coleus/Frafra/Hausa/Native/Sudan/Zulu potato; in **India**: Chinese potato (an English name), Sambrani (Kannada), Jhade Kanaga or Kooka/Kuka (Konkani), Koorka/Koorkka (Malayalam), Innala (Sinhala), and Ciru kilanku/Siru kizhangu (Tamil).

Introduction and Origin

The specific epithet *rotundifolius* is derived from New Latin words *rotundus* (round) and *folium* (leaf). Chinese potato, native to central or east Africa, was formerly of considerable importance as a staple foodstuff in tropical Africa but its production has declined drastically and has been largely replaced by higher-yielding starch crops such as cassava, potatoes and sweet potato (Nkansah, 2004; https://en.wikipedia.org). Now it is a minor tuber crop grown in homesteads for its edible tubers. Chinese potato occurs wild in grassland in East Africa, especially Kenya, up to 2,200m altitude. It is suited for cultivation in marginal areas in dry savannah with poor soils. It can still be found in cultivation in the Kita region in Mali, the northern districts of Ghana, the Jos Plateau in Nigeria and N and E South Africa. International trade has been reported to take place between northern Ghana and Burkina Faso (Nkansah, 2004). More recently, its cultivation has spread in tropical Asia including India, Sri Lanka, Malaysia and Indonesia where it is cultivated on a small scale (Anon., 2006; Humanity development library 2.0). It is one of 3 African *Coleus* species with same vernacular names grown for edible tubers. *C. esculentus* from southern Africa and *C. edulis* from Ethiopia have not spread beyond Africa. *C. rotundifolius* is closely related to widely cultivated ornamental *Coleus* and is now again placed in the genus *Coleus* after being placed in the defunct genus *Solenostemon* and in *Plectranthus* (Paton *et al.*, 2019).

Coleus rotundifolius

Botany

It is a branched semi-succulent small herbaceous perennial 15-60cm tall having aroma resembling to that of mint. The quadrangular stem is decumbent or ascending with densely pubescent on the angles and root nodes. Simple and opposite leaves are rather thick, juicy and faintly aromatic when bruised. The leaf-stalk is 1-3cm long and puberulous. The blade is ovate to nearly round, measuring 2-6×1.5-4cm, basally wedge-shaped and coarsely toothed at margins. Nkansah (2004) reported its diploid chromosome number as 2n=64 and 84.

It flowers in September-October. The inflorescence is a terminal and slender false spike measuring 5-15cm long with distant whorls of 4-6 flowers. The zygomorphic flowers are bisexual having pedicel up to 1–2mm long; calyx campanulate, 1.5–3mm long, glandular-hispid, upper tooth ovate, 2 lateral teeth small, 2 lower teeth almost completely fused; corolla 4–8mm long, pubescent and gland-dotted with curved tube, 2-lipped, upper lip erect, 4-lobed, whitish, lower lip boat-shaped, bluish purple; stamens 4, shortly united at base, curved within the lower corolla lip; ovary superior, 4-celled and style 2-fid. Fruit consists of 4 nutlets though develop rarely. Though it flowers profusely but developing only a few seeds.

The tips of the adventitious roots modify into small round tubers. Unlike other tuber crops this is a transplanted vegetable. The small dark-brown tubers being produced in clusters at the base of the stem are ovoid and upto 2.5-8cm long in India and Sri Lanks (Nkansah, 2004). The sweet-flavoured tubers resemble the potato but smaller in size. In Sri Lanka, two main types are recognised, the small-tuber type favoured for its delicate flavour, and the larger type that produces heavier crops which are easier to harvest. In West Africa, there are three recognised types: 'Nigra' which is widespread in Mali with small tubers having blackish skin; 'Rubra' with small reddish-gray or reddish-yellow tubers; and 'Alba' which is whitish (Nkansah, 2004).

Nutritional Value

The composition of edible portion of tubers has been quoted as containing water (75-75.6 per cent), energy [394 kJ (94 kcal)], protein (1.3-1.4 per cent), fat (0.2-0.5 per cent), carbohydrate (21-21.9 per cent), fibre (0.7-1.1 per cent), ash (1 per cent), Ca (17 mg/100g), Fe (6 mg/100g), thiamine (0.05mg/100g), riboflavin (0.02mg/100g), niacin (1mg/100g) and ascorbic acid (1mg/100g). The principal amino acids in the protein are arginine, aspartic and glutamic acids (Leung *et al.*, 1968; Nkansah, 2004). Suja (2019) reported that its tubers contain 31-33 per cent dry matter, 18-20 per cent starch and 1-1.5 per cent protein and further mentions that analysis of protein suggests a chemical score of 72-78 with lysine, leucine, cysteine and methionine as the marginally limiting amino acids. On cutting or peeling, its tubers start rapid darkening due to phenol content (0.13-0.22 per cent).

Uses

The tubers can be used as a potato substitute; usually cooked with spices either alone or in various combinations with other foods such as beans and cooked vegetables or curries. These can also be boiled, baked or fried similar to potato. Non-peeled roasted tubers are eaten whole along with skin as a snack. The tubers can also be made into alcoholic drinks. The leaves are occasionally used as a potherb but more often in traditional medicine, *e.g.* for treatment of dysentery and certain eye disorders in Nigeria.

Coleus esculentus (N.E.Br.) G.Taylor

Synonyms: *Coleus coppinii* Heckel; *C. dazo* A.Chev.; *C. floribundus* (N.E.Br.) Robyns and Lebrun, nom. illeg., non *C. floribundus* Baker; *C. floribundus* var. *longipes* (N.E.Br.) Robyns and Lebrun; *C. langouassiensis* A.Chev.; *Englerastrum floribundum* (N.E.Br.) T.C.E.Fr.; *Plectranthus esculentus* N.E.Br.; *P. floribundus* N.E.Br.; *P. floribundus* var. *longipes* N.E.Br.

Common names: African potato, Kaffir potato, Livingstone potato, Veld potato and Wild potato.

Introduction and Origin

The specific epithet *esculentus* is a Latin word meaning 'edible' and this refers to its edible tubers. Previously known as *Plectranthus esculentus*, this plant has recently been re-classified as being part of the *Coleus* genus. Wild potato, indigenous to South Africa (Ben-Erik, 2005) is a bit more difficult to cultivate as compared to *Coleus rotundifolius*, though gives greater yield. It is a pleasant aromatic herb with beautiful yellow flowers that are seen in winter and spring. It is able to grow in a variety of climatic and soil conditions.

The crop is locally grown for its edible and nutritious tubers, but has not spread beyond Africa. It was also first cultivated in the Upper Niger Valley of the Hausaland in Nigeria and in the Central African Republic (Kyesmu, 1994). The cultivation of *C. esculentus*, *C. rotundifolius* and *C. edulis* has been largely displaced by the spread of cassava which was introduced by the Portuguese to Africa from South America about 500 years ago (Blench, 2006).

Botany

Coleus esculentus is an erect to trailing, semi-succulent, deciduous and perennial herb or shrublet growing upto 2m tall with hairy, long arching, sparsely unbranched to slightly branched stems that are woody at base, arising from tuberous roots that are fleshy and swollen similar to the ordinary potato. Like the ordinary potato it has true tubers (underground stems that are modified and where new shoots are able to arise).

The subsessile leaves are deciduous, 5–8×1–3 cm, oblong-elliptic to oblanceolate, roughly hairy on both surfaces, thick-textured with brown gland-dots on the undersurface and with toothed leaf margins. The leaf base is obtuse or cuneate and the apex is obtuse.

Inflorescences are many-flowered in axillary racemes and 5–8cm long. The boat-shaped individual flowers are pedicellate and solitary; the corolla is 1.3–1.5cm long, yellow and expanding to the throat. The flowers appear during winter or spring after the plants have shed their leaves. The calyx is 4–5mm which elongates up to 1cm after flowering has taken place. The fruits are small and ovoid with brown nutlets that are 1mm long. It is the only species within the genus to flower in this season.

Nutritional Values

In addition to its high starch content, it contains 13 per cent essential amino acids which are essential to the body for constructing proteins as well as other important vitamins and mineral elements such as calcium, iron and vitamin A. Its 100g edible portion of tuber contains 80g carbohydrates, 13-14g crude protein level, 0.6g total lipids, 140mg calcium, 50mg iron and 0.17mg vitamin A (Ben-Erik, 2005).

Uses

The edible tubers are a great source of nutrition and can be used to make a variety of foods. They are often used as a substitute for a potato or sweet potato. Directly after lifting it can be boiled or roasted. The stems have been used to sweeten gruel (porridge). The leaves can be cooked in sauces as well. The plant is believed to help with digestive problems and used to treat stomachache and abdominal pain (https://en.wikipedia.org/wiki/Plectranthus_esculentus - cite_note-lukhoba1-10). It has also been used as anthelmintics.

Coleus maculosus subsp. *edulis* (Vatke) A.J.Paton

Synonyms: *Coleus aquaticus* Gurke; *C. clivicola* S.Moore; *C. edulis* Vatke; *C. fimbriatus* Lebrun and L.Touss.; *C. palustris* Vatke; *C. rivularis* Vatke; *C. tuberosus* A.Rich., nom. illeg., non *C. tuberosus* (Blume) Benth.; *Majana richardiana* Kuntze, nom. illeg.; *Plectranthus edulis* (Vatke) Agnew; *P. fimbriatus* (Lebrun and L.Touss.) Troupin and Ayob.; and *P. punctatus* subsp. *edulis* (Vatke) A.J.Paton.

Common names: Ethiopian potato (Yeshitila and Jens, 2012) and Welayta dinich (in Amharic).

Ethiopian potato, a perennial plant indigenous to Ethiopia, is grown for its edible tubers. The tubers are cooked before they are eaten (Zemede, 2012).

Propagation

Chinese potato is normally propagated vegetatively through tubers, suckers or soft-wood stem cuttings (Nkansah, 2004). Nursery should be raised about 45 days before planting. An area of 500 to 600m^2 is sufficient to produce cuttings required for one hectare of land. Healthy tubers weighing about 15-20g from the previous harvest are usually planted in raised and well-manured nursery beds in rows 5×15cm distance and about 4cm deep which requires 75-100kg tubers. It would be good to apply 125 to 150kg FYM in the nursery area. This is then top-dressed with urea @ 5kg/500m^2 at about 3 weeks after planting to encourage good growth of vine. At about 45 days of planting the terminal portion of vines devoid of roots is clipped off to a length of 10-15cm. The soil should be kept moist but must not be waterlogged. The tubers sprout in 10-15 days and give rise to a cluster of sprouts which are ready for transplanting in the field after about 3 months.

In some areas such as Tamil Nadu, the seed tubers are planted in a corner of the field where about a month after sprouting the top suckers with 4-5 leaves are ripped off and planted 15-20cm apart in another part of the field. These quickly become established and in about another fortnight a further set of suckers is available for planting out. In this way in about 2 months a hectare of planting material is obtained from an initial 0.2 ha, *vis-a-vis* out of relatively some tubers (Humanity development library 2.0). *In-vitro* multiplication through tissue culture is successful using stem meristems, apices and nodes.

Varieties

CTCRI, Thiruvananthapuram (Kerala) has evolved a high potential variety 'Sree Dhara' which yields 20-25t/ha over a 5-month period of cropping. KAU Regional Agricultural Research Station, Pattambi (Kerala) released 'Nidhi' which is also a 5-month crop. KAU released 'Suphala' ~ a tissue culture mutant of a local cultivar which matures within 5 months (120-140 days) and can be cultivated throughout the year in Kerala. In different parts of India, local cultivars are being grown which are of 2 types, *viz.* i) small but uniform size of tubers having good flavour, and ii) with large tubers having higher yield (Suja, 2019).

Growth and Development

Unlike other tuber crops, this is a transplanted vegetable. The crop normally reaches maturity in 5-8 months depending on the climatic conditions, planting methods and cultural practices adopted. It has been reported that tuber initiation may be accelerated by about 2 weeks by treatment of the young plants with chlormequat or ethephon (ethrel) (Humanity development library 2.0).

Tubers occur in clusters of 3–7, either at the base of the stem or at the nodes below the soil surface. They are ready for harvesting 150–200 days after planting by which time the plant has flowered and aerial parts have become senescent. Most tubers found in Africa are 2.5–4×1–1.5cm though some attaining even up to 8cm. Such larger tubers are the norm in India and Sri Lanka where yields are also higher than those in Africa's semi-arid zones (Nkansah, 2004).

Factors Influencing Growth and Quality

Coleus is easy to grow and thrives in many climates and soil conditions with exception of winter-rainfall region, and is best suited to warm, subtropical conditions. The plant is suited to evenly-distributed high rainfall areas coupled with low night temperatures as such conditions favour development of tubers though can not tolerate drought. It is grown in India as a monsoon crop and is sometimes grown under irrigation in West Africa (Humanity development library 2.0). Tubers may not be formed when rain is insufficient or when there is too much of rain and under such conditions tubers tend to branch out which is disliked by consumers because they are then difficult to peel. Rainfall of 700–1,100mm or with watering to make up a shortfall is preferred. The crop requires full sunlight though shading in any form reduces the yield (Nkansah, 2004). Optimum temperatures for growth and development range from 25-30°C.

Soil

A fertile well-drained sandy loam to alluvial soil with a pH range of 5.5-7 (best being 6.6-7) is most suitable for its cultivation. Heavy clay soils are not suitable for this crop at all. Tubers are formed with relative ease in well-drained sandy soil but will not develop well in compacted heavy soil. Waterlogging deforms the tubers with considerable reduction in yield. It is usually grown on ridges except in very well-drained soils (https://en.wikipedia.org; Nkansah, 2004).

Season

Planting in the main field is done at the beginning of rainy season. Irrigation may be required during dry season for crops planted towards the end of the rainy season. Cultivation throughout the year is possible under tropical conditions.

Land Preparation and Planting Methods

Before one month of planting, *i.e.* May-June when usually it takes about a month to arrive the monsoon, the field should be ploughed or dug thoroughly to a depth of 15-20cm, cleaned of weeds and mixed with plenty of manure and compost. Planting is usually done at the beginning of the rainy season *i.e.* July either in mounds or ridges or in well-drained loose soil on flat ground. Planting materials, methods, field spacing and planting depth may differ depending on the region but usually these are planted on 15-20cm raised beds stretched 45cm apart. The vines are usually planted at a spacing of 30cm apart on the ridges either in vertical form or horizontally. Horizontal planting to a depth of 4-5cm with exposed terminal bud ensures quick establishment with higher tuber yield.

- ✰ Ghana: Sprouted tubers are planted with the growing end placing at the soil surface uncovered by soil. Ridges are spaced at 90cm and the plants at 15–20cm in the ridges, resulting in a plant density of about 50,000 plants/ha (Nkansah, 2004).
- ✰ South Africa: Tubers are planted at a depth of 5–10cm with a spacing of 25-30cm on ridges spaced 75cm. Cuttings are used occasionally and planted in pairs facing opposite directions. They are placed at a depth of 5cm but with the growing point clearly above the soil surface. Farmers in Ghana apply wood ash and diluted cattle urine prior to planting to promote growth and development of the crop (Nkansah, 2004).
- ✰ Sri Lanka: Three methods of planting are usually used (Humanity development library 2.0):
  - ❖ Ordinary planting - Vine cuttings about 15cm long having 3-4 leaves at the top-end are planted 7cm deep at about 22cm spacing on ridges 90cm apart. Around 20,000 tubers are required for planting in nursery to provide the necessary number of cuttings to plant one hectare.
  - ❖ Coiled planting – Vine cuttings about 22cm long are used and about 12cm of the more mature portion is coiled and planted in holes about 7cm wide and 5cm deep. Around 40,000 tubers are planted in the nursery to provide necessary number of cuttings to plant one hectare. This method is reported to give the best results.

- ❖ Horizontal planting - Cuttings about 30cm long are placed horizontally across the ridge, two at a time, in opposite directions and almost touching each other. About 22cm of the cuttings remain on the ridge and 7cm outside and there is about 7cm between each pair of cuttings. This way some 60,000 tubers are required for planting in the nursery to provide necessary number of cuttings to plant one hectare.

- ☆ India: A slightly closer spacing of 15-20cm along the ridge is sometimes used. The top suckers are taken at intervals from a nursery bed. In this method some 2,500 tubers will produce enough suckers to plant one hectare of land.

Manure and Fertilizers

People hardly apply manure and fertilizer in this crop as the crop is not so economic, *vis-à-vis* there is no commensurate tuber yield even after applying the nutrients. The inclusion of 10-25t/ha of FYM at planting is recommended and subsequently a 16:8:8 complete NPK fertiliser at the rate of 125kg/ha is applied (Nkansah, 2004). In parts of Africa, it is customary to add some wood ash and even diluted cattle urine.

Interculture and Intercropping

Once planted, the crop normally requires two weedings, once about 3 weeks after planting, and second one again after another 3 weeks together with earthing up along with top dressing of fertilizers. When the crop is well established, earthing up with loose soil is necessary for good tuber development (Nkansah, 2004). In case there is soil erosion from the plant base, one more earthing up is given 1 month later to promote tuber formation (Suja, 2019).

It is grown either as a sole crop or intercropped with 'bambara groundnut', yam, okra, millet, maize or sorghum (Nkansah, 2004).

Harvest

This is a crop of 4-6 months duration, depending on the variety. The tubers reach maturity when leaves and stems begin to wither. Harvesting cannot be delayed as the mature tubers if left in the soil deteriorate rapidly and start rotting but can be stored successfully in dry sand or in a cool and well-ventilated shed. The plants are harvested with their tubers still attached by gently excavating them. The tubers are then removed and packed in baskets or sacks (Nkansah, 2004). Under congenial conditions tuber yield ranges 5-15t/ha, though under very favourable conditions they may reach 18-20t/ha but become considerably lower when soil fertility or rainfall is poor (Nkansah, 2004; Humanity development library 2.0). Experimental work carried out at Roodeplaat (South Africa) has indicated that the potential yield could be up to 45t/ha when adequate irrigation and plant nutrients are provided together with good agronomic practices (Nkansah, 2004).

Postharvest

Traditionally the tubers are stored in the ground under a tree where it is cooler than in the open. When stored in this way under hot conditions the special taste of tubers usually lasts only for two months. The tubers can also be packed in bags or baskets stuffed with straw but if these are kept under warm conditions the tubers will soon shrivel and become inedible. To keep the tubers longer, especially the small ones that are needed for the next planting season, people put them in pots sealed with cow dung. In cooler conditions such as in highland regions or in South Africa storage is easier (Nkansah, 2004).

Pests and Diseases

Coleus species are largely unaffected by insects, pests or diseases. Caterpillars of various moth and butterfly species have been reported to feed on the leaves but these can be removed manually by hand. Termites, centipedes and potato weevils also damage the tubers. Millipede damage is observed as the tubers are bored by these so these become unfit for consumptiojn. Grasshoppers feed on the leaves. Stem borers may also be found attacking the standing crop. In South Africa it is believed that the crop suppresses nematode populations but Crop Research Institute in Kumasi (Ghana) has found that infestation by nematodes may lead to large yield losses so this demands further studies (Nkansah, 2004). In case of root-knot nematode infestation on tubers the gall formation or root swelling occurs which results into stunted growth and plant wilting. These enter the roots of the seedlings so at the time of planting only the healthy tubers should be selected. However, solarization kills even the nematodes. Suja (2019) suggests growing of 'Sree Bhadra' sweet potato variety as preceding crop to trap these nematodes in the soil. Moreover, deep ploughing of field in summer, proper crop rotation and burning of plant residues including the roots will keep away the problem of root-knot nematodes. To get rid of leaf-folding caterpillars and vine borers the vines at planting should be dipped in dimethoate, *i.e.* Rogor 30EC @ 1.7ml in 1 litre of water for 10 minutes. In case of severe infestation the crop is sprayed with chlorpyriphos 20EC 2ml/l or imidacloprid 0.01 per cent. Usually, in case of tuber crops chemicals are not used due to environmental and food safety reasons. Therefore, one should restrict for planting of resistant varieties, use of biological agents for control of insect-pests and through cultural methods with a need-based application of chemical pesticide. In case of virus infected plants, such plants should be uprooted and burnt. Proper cultural treatment will protect the tubers even from rotting.

References

1. Anon. 2006. Lost Crops of Africa, Vol. II: Vegetables, pp. 269-85. National Research Council. National Academic Press, Washington D.C.
2. Anon. 2022. *Coleus* Lour. Plants of the World Online. Royal Botanic Gardens, Kew (Retrieved 7/10/2022).
3. Ben-Erik, van Wyk, 2005. Food Plants of the World. Timber Press, Oregon, 302p.
4. Blench, R. 2006. Archaeology, Language, and the African Past. Altamira Press.
5. https: //en.wikipedia.org/wiki/Coleus_rotundifolius.
6. Humanity Development Library 2.0. Root crops: Hausa potato (*Solenostemon rotundifolius*). nzdl.org.
7. Kyesmu, P.M. 1994. *Plectranthus esculentus* N.E.Br. - a minor tuber crop in dire need of rescue from extinction. *Lamiales N.L.*, pp. 3–4.
8. Leung, W.T.W., F. Busson and C. Jardin, 1968. Food Composition Table for Use in Africa, 306p. FAO, Rome.
9. Nkansah, G.O., 2004. *Solenostemon rotundifolius* (Poir.). Record from PROTA4U. Ressources végétales de l'Afrique tropicale (Plant Resources of Tropical Africa), edited by Grubben, G.J.H. and O.A. Denton. Wageningen, Netherlands. <http: //www.prota4u.org/search.asp>. Accessed 14 October 2022.
10. Paton, A.J., Montfort Mwanyambo, R.H.A. Govaerts, Kokkaraniyil Smitha, Somran Suddee, Peter B. Phillipson, Trevor C. Wilson, Paul I. Forster and Alstair Culham, 2019. Nomenclatural changes in *Coleus* and *Plectranthus* (Lamiaceae): a tale of more than two genera. *PhytoKeys*, No. 129, pp. 1–158.
11. Suja, G. 2019. Chinese Potato. Handbook of Horticulture, vol. 1, pp. 659-60. ICAR, New Delhi.
12. Yeshitila, M. and Jens Weibull, 2012. Local Customary Use and Management of Ethiopian Potato, *Plectranthus edulis* (Vatke) Agnew in Sodo Zuria District, South Ethiopia. *Ethnobot. Res. Applic.*, v. 10: p. 381-7. (http: //journals.sfu.ca/era/index.php/era/article/view/793).
13. Zemede, A. 1995. Conservation and use of traditional vegetables in Ethiopia. *Proceedings of the IPGRI Intern. Works. Gen. Res. of Tradit. Veg. in Africa* (Nairobi, 29–31 Aug. 1995). Archived 2012-07-07 at the Wayback Machine.

10

Chinese Water Chestnut [*Eleocharis dulcis* (Burm.f.) Trin. ex Hensch.]

Family: Cyperaceae.

Other common names: Matai, Water chestnut, Waternut; in Hindi: Chini Singhara.

Homotypic synonym: *Andropogon dulcis* Burm.f. in Fl. Indica: 219 (1768).

Heterotypic synonyms: *Carex tuberosa* (Schult.) Blanco in Fl. Filip.: 35 (1837), nom. illeg.; *Eleocharis austrocaledonica* Vieill. in Ann. Sci. Nat., Bot., sér. 4, 16: 38 (1862); *Eleocharis dulcis* var. *tuberosa* (Schult.) T.Koyama in J. Fac. Sci. Univ. Tokyo, Sect. 3, Bot. 8: 97 (1961); *E. equisetina* J.Presl and C.Presl in Reliq. Haenk. 1: 195 (1828); *E. esculenta* Vieill. in Ann. Sci. Nat., Bot., sér. 4, 16: 37 (1862); *E. indica* (Lour.) Druce in Bot. Soc. Exch. Club Brit. Isles 1916: 621 (1917); *E. plantaginea* (Retz.) Roem. and Schult. in Syst. Veg. ed. 15bis. 2: 150 (1817); *E. plantaginea* var. *stolonifera* Boeckeler in Beitr. Cyper. 2: 14 (1890); *E. plantagineiformis* Tang and F.T.Wang in Fl. Reipubl. Popularis Sin. 11: 225 (1961); *E. plantaginoidea* W.Wight in Contr. U.S. Natl. Herb. 9: 267 (1905), nom. superfl.; *E. plantaginoides* (Rottb.) Domin in Biblioth. Bot. 20(85): 445 (1915); *E. tuberosa* Schult. in Mant. 2: 86 (1824); *E. tumida* (Roxb.) Schult. in Mant. 2: 86 (1824); *Limnochloa plantaginea* (Retz.) Nees in R.Wight, Contr. Bot. India: 114 (1834); *L. tumida* (Roxb.) Nees in R.Wight, Contr. Bot. India: 114 (1834); *Scirpus dubius* Roxb. in Fl. Ind. 1: 215 (1820); *S. plantagineus* Retz. in Observ. Bot. 5: 14 (1788); *S. plantaginoides* Rottb. in Descr. Icon. Rar. Pl.: 45 (1773); *S. spiralis* Willd. ex

Kunth in Enum. Pl. 2: 153 (1837); *S. tuberosus* Roxb. in Pl. Coromandel 3: 25 (1815), nom. illeg.; and *S. tumidus* Roxb. in Fl. Ind. 1: 212 (1820).

Introduction and Origin

Water Chestnut Growing in a Bog Garden

(Photo credit: flickr/techieoldfox; https://awkwardbotany.com/tag/eleocharis-dulcis/

Chinese water chestnut is a perennial aquatic plant that grows in marshes, bogs and the margins of other wetland and riparian areas in tropical and subtropical climates (awkwardbotany.com) though it may also thrive in warm temperate wetlands (pondinformer.com). It is native to tropical and subtropical Old World ~ Andaman Island and elsewhere (India), Angola, Bangladesh, Botswana, Burkina, Cambodia, Caroline Island, Chad, SC and SE China, Congo, Fiji, Gabon, Gambia, Ghana, Guinea, Guinea-Bissau, Hainan, Ivory Coast, Japan, Jawa, Kenya, Korea, Lesser Sunda Island, Liberia, Madagascar, Malaya, Mali, Maluku, Marianas, Mauritius, Mozambique, Myanmar, Namibia, Nansei-shoto, New Caledonia, New Guinea, New South Wales, Nigeria, Northern Territory, Pakistan, Philippines, Queensland, Reunion, Samoa, Senegal, Seychelles, Sierra Leone, Sri Lanka, Sulawesi, Sumatra, Taiwan, Tanzania, Thailand, Togo, Tonga, Uganda, Vietnam, Wallis-Futuna Island, Western Australia, Zambia, Zaire and Zimbabwe (Anon., 1833a).

Water chestnut is grown in many countries for its edible corms (Anon., 1833b) and is commercially cultivated in China and Japan. Water chestnut is usually grown in paddy fields, often as a substitute crop of rice (where rice is overproduced) (sciencedirect.com). It grows wild in many parts of India, SE Asia and Polynesia (Humanity development library 2.0). It was first cultivated in SE China in humid monsoon areas and is now also grown commercially in Japan, Hong Kong, the Philippines, Hawaii and other Pacific islands, India and the southern USA. It is rather widely cultivated in both mainland China and Taiwan and some parts of Thailand not only for local consumption but also as a source of preserved (canned) vegetables for export (sciencedirect.com).

Raw water chestnuts are slightly sweet and crunchy. Boiled water chestnuts have a firm and slightly crunchy texture with a mild and slightly nutty flavour that may easily be overpowered through cooking or seasonings or through sauces with which the water chestnut is served (en.wikipedia.org). Some varieties are not sweet and are grown for starch or as pig feed though Australian variety is small but quite sweet and Chinese varieties are superior in size and sweetness (Alanna Moore, 2008).

Two very distinct forms of *E. dulcis* are recognized. One is a wild form which generally grows in stagnant water and produces very small *cum* quite dark-skinned (almost black) secondary corms and is sometimes referred to in the literature as *E. plantaginea* or *E. plantaginoides*. The second form occurring only under cultivation and producing large and sweet secondary corms was originally described as a separate species, *E. tuberosa* (Humanity development library 2.0).

Botany

Chinese water chestnut is a grass-like but tuberous marginal aquatic perennial sedge having a dense clump of 50-100, erect, tubular, hollow, branchless, about 1.5m tall (en.wikipedia.org) green stems; many of which are topped by a straw-brown and cylindrical flower spikelet up to 5cm long (missouribotanicalgarden.org). Leaves of a 'normal' kind disappear, being reduced to pale or purple-brown sheaths surrounding the base of the stems (sciencedirect.com). Stem leaves are bladeless sheaths with the stems actually providing the photosynthesis for the plant. Inflorescences at the top contain about 50 insignificant flowers; female (pistillate) flowers appearing when the stem tips reach about 15cm above the water, later followed by male (staminate) flowers (Humanity development library 2.0). In temperate regions, flowers are seldom produced (Kleinhenz *et al.*, 2001). Tiny 'seeds' (achenes) are produced but are of no economic importance. Its chromosome number is n=38, 2*n*=212 (Hideyuki and Uchiyama, 2009).

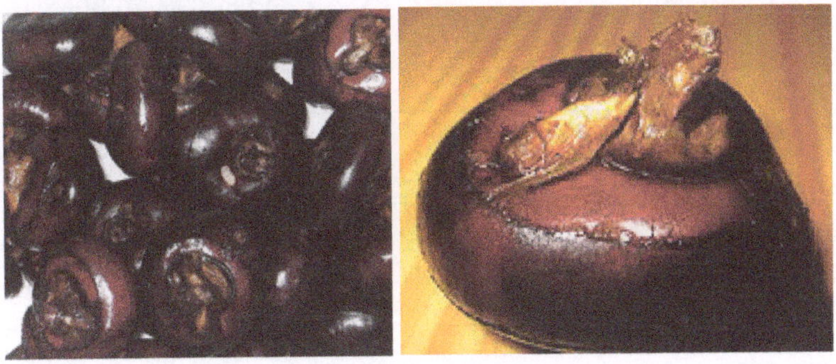

Corms of Water Chestnut

Source: Public Domain, https://commons.wikimedia.org

The plant produces two types of subterranean rhizomes: i) under long-daylength conditions and 6-8 weeks after planting, rhizomes grow horizontally and then turn upwards forming daughter plants, and ii) somewhat later and under short-daylength conditions the rhizomes grow downward and produce an edible rounded or depressed-globose corm at the tip (missouribotanicalgarden.org). The corms are small (2-5cm in diameter), white when young, clothed with thin and purple–brown to black–brown scales at maturity and their white flesh is crisp (pondinformer.com; sciencedirect.com).

Nutritional Values

The corms show considerable variation in composition. Edible portion of fresh Chinese corms has given approximately 77.29 per cent moisture, 1.53 per cent protein, 0.15 per cent fat, 18.9 per cent nitrogen-free extract, 1.94 per cent reducing sugars, 6.35 per cent sucrose, 7.34 per cent starch, 0.94 per cent fibre, 1.19 per cent ash, 2-10mg/100g calcium, 0.43-0.6mg/100g iron, 52.2-65mg/100g phosphorus, 0.24mg/100g thiamine, 0.007mg/100g riboflavin, 0.94mg/100g niacin and 9.2mg/100g ascorbic acid (Humanity development library 2.0).

Approximate Nutritional Value per 100g

Energy	**406 kJ (97 kcal)**
Carbohydrates	23.94g
Sugars	4.8g
Dietary fiber	3g
Fat	0.1g
Protein	1.4g
Vitamins	***Quantity (per cent DV†)***
Thiamine (B1)	12 per cent (0.14mg)
Riboflavin (B2)	17 per cent (0.2mg)
Niacin (B3)	7 per cent (1mg)
Pantothenic acid (B5)	10 per cent (0.479mg)
Vitamin B6	25 per cent (0.328mg)
Folate (B9)	4 per cent (16μg)
Vitamin C	5 per cent (4mg)
Vitamin E	8 per cent (1.2mg)
Minerals	***Quantity (per cent DV†)***
Calcium	1 per cent (11mg)
Iron	0 per cent (0.06mg)
Magnesium	6 per cent (22mg)
Manganese	16 per cent (0.331mg)

Phosphorus	9 per cent (63mg)
Potassium	12 per cent (584mg)
Zinc	5 per cent (0.5mg)
Other constituents	Quantity
Water	73.5g

Units: μg = micrograms; mg = milligrams; IU = International units

†Percentages are roughly approximated using US recommendations for adults. Source: USDA FoodData Central.

Table source: https://en.wikipedia.org/wiki/Eleocharis_dulcis

Phytochemicals

Water chestnuts are unusual among vegetables for retaining their crisp texture even after being cooked or canned. The reason for this textural stability is that their cell walls are cross-linked and strengthened by certain phenolic compounds such as oligomers of ferulic acid (Terri Grassby, 2008). The corms contain the antibiotic agent puchiin which remains stable to high temperature (Hodge, 1956).

Bhaisare (2014) reported presence of medicinally active constituents like alkaloids, steroids, flavonoids and terpenes in the corms of this plant. These constituents are known to show medical activity as well as exhibiting physiological activity and also important to pharmacy.

Uses

Chinese water chestnut is usually grown for its dark brown edible corms that are commonly used in Asian cooking. They are noted for their sweetness and distinctive nut-like crunchy texture. The small rounded corms have a crispy white flesh that may be eaten raw, slightly boiled or grilled and often pickled or tinned. These may also be ground into flour and used for making water chestnut cake (en.wikipedia.org). Water chestnut may be combined with bamboo shoots, coriander, ginger, sesame oil and snow peas (en.wikipedia.org). These can be added to steamed or sauteed vegetable dishes such as noodle or rice recipes (Green Aliza, 2004). These may also be added to minced-meat fillings and dishes such as cantonese dim sum-style steamed meatballs to add a crunchy texture (Anon., 2019). Thai dessert *thapthim krop* with water chestnuts as its main ingredient was named one of the world's best 50 desserts in 2019 by CNN Travel (Smith Jen Rose, 2019). Apart from the edible corms the leaves are used as cattlefeed, mulch or compost (Morton *et al.*, 1988). Certain cultivars are sometimes used for preparation of starch while only a small portion of corms is useful for poultry feed (Humanity development library 2.0).

Water chestnuts are used as an infusion in traditional folk medicine by the tribal people called *Vaidu* for treatment of stomach disorder, diarrhoea, dysentery, fever, pain, other inflammatory problems and dysmenorrheal irregularities (Bhaisare, 2014). The juice extracted from the corms has been shown to contain an antibiotic principle designated as puchiin (Humanity development library 2.0).

The starch similar to that obtained from sweet potatoes or cassava has large grains up to 27 microns in length which may be rounded, have regular geometric shapes or be completely irregular (Humanity development library 2.0).

Processing

Canning

The washed and peeled corms are processed in a manner similar to potatoes. The recommended processing times are 30 minutes for No. 2 cans, 35 minutes for No. 2.5 and 45 minutes for No. 10 at 115°C temperature after an initial heating before sealing to 60°C (Humanity development library 2.0).

Quick Freezing

The washed and peeled corms are blanched for 4 minutes in steam at 99-100°C in single layers on wire mesh trays, they are then cooled immediately in an air-blast, packed into cans and frozen in a blast freezer at -32°C then held at -18°C for periods up to 12 months (Humanity development library 2.0).

Starch

In China, sometimes starch is extracted from the corms by a very primitive method. The corms are cleaned by thorough washing and then crushed. The resultant starchy mass is put in a fine bamboo basket which is set in a filter cloth and hung over a wood fire. The basket is then placed in a pan and after adding water the contents are thoroughly stirred for 15 minutes. Three parts of starch milk are collected which contains the largest proportion of starch and is set aside to allow the starch to separate out and the rest 1 part of the starch milk is re-used to wash more pulp. After about 5 hours the starch gets separated out and is collected and dried in the sun on bamboo trays (Humanity development library 2.0).

Adverse Effects

If eaten uncooked the surface of the plants may transmit fasciolopsiasis (Bhatti *et al.*, 2000) ~ an infection caused by the *Trematode fasciolopsis* buski (Lankester and Kuchenmeister, 1857), the largest human intestinal fluke (up to 7.5cm in length).

Propagation

Chinese water chestnut is always propagated asexually using corms, thus producing clones of its parent. The crop can be established by re-growth of corms from the previous year's planting ('ratooning'), direct planting of corms and transplanting of nursery-raised 'seedlings'. Since the planting material is actually a corm and not a seed, the young plants are not real 'seedlings' ~ they are the vegetative parts of the parent plant ~ a type of clone (Kleinhenz *et al.*, 2001).

Factors Influencing Growth

Soil

A rich clay or peaty soil with a pH of 6.9-7.3 is required for optimum yields. Slightly more acidic soils can be neutralized with limestone (Humanity development library 2.0).

Climate, Field Preparation and Planting

The crop of water chestnut develops quite slowly in the temperate than in the sub-tropical conditions (Kleinhenz *et al.*, 2001). It requires a long warm growing season with at least 220 frost-free days though soil temperature of 14-15.5°C is necessary for corm sprouting (Humanity development library 2.0). The photoperiod significantly influences the rate of corm growth as the corms begin to develop quite slowly if photoperiod exceeds 12h (Li *et al.*, 2015). The crop may be grown at altitudes ranging from sea level to 1,200m (Humanity development library 2.0). This can be cultivated in paddy fields or in a hydroponic culture. Paddy design depends on the adopted harvesting technique as any paddy bed width is possible for suction-type harvesting equipment though some growers prefer a narrow width (2-3m) in which harvesting can be done in one sweep. A paddy field width of less than 5m has good access from either side. The suction machine requires at least 50cm water depth to become buoyant so paddy banks need to be approximately 1m high. If trench type paddy fields are adopted then the field will need to be excavated sufficiently to accommodate this water depth. If tractor is to be used for harvesting, provision for access of harvester into the field as well as sufficient turning area at each end will be required (Kleinhenz *et al.*, 2001).

Due to faster crop growth under sub-tropical conditions, planting date for corms is less critical and can occur from early September to end of November. In cooler temperate regions, its corms are planted as early as possible in spring. Typically, corms are planted in a greenhouse in mid-August and transplanted in mid-October. Alternatively, the growers may plant corms directly from late September to October into flooded paddy fields. These direct-planted corms are prone to minor frost damage resulting in 'burnt' stem tips if water depth is not sufficiently high (20 cm) (Kleinhenz *et al.*, 2001).

If the paddy is designed for mechanical harvest the width is kept largely on the size of the harvesting equipment. Bigger the equipment, the space requirement is also more. Paddy bed width depends on the reach of the excavator. Reach is typically 7.5m and this should be the paddy bed width if there is access only along one side (length-wise) and twice that width if there is access on both the sides (Kleinhenz *et al.*, 2001).

Approximately 5 quintals of corms are required to plant one hectare of land. Although the corms are directly planted by hand into the prepared paddy field at 2-5 corms per square metre at a soil depth of 2–3cm (Kleinhenz *et al.*, 2001) though in little temperate climate it is usually started in protected nursery beds followed by hand-transplanting when top growth is 20-30cm high (Humanity development library 2.0). However, direct planting of seed corms into the paddy field has labour-saving advantage over the more intensive method of using seedling transplants (Kleinhenz *et al.*, 2001). Small gardeners usually plant the corms manually with a hand trowel but in larger plantings the furrows are opened with a plough or courter and the corms are dropped in at intervals of 75cm and then covered with a covering plough or hiller. Field spacing of 75×75cm is recommended in USA (Humanity development library 2.0) and Australia. In China, a triangular spacing of 45-60cm and 45cm between plants is a common practice. As with the effects of 'ratooning', higher planting densities tend to increase total yields through greater numbers of corms but corm-size becomes smaller. Lower planting densities have the potential to produce bigger corms though paddy fields are more prone to weed infestation due to canopy closure (*i.e.* cover over the soil by green stems) (Kleinhenz *et al.*, 2001).

Weeding

Although weeds are not usually a problem provided the soil has been well-tilled just before planting. Aquatic- and waterlogging-tolerant weeds become more prevalent with each successive season as a weed seed bank builds up within the paddy field so leaving the paddy field fallow once every 4-5 seasons will help contain weed populations, *vis-a-vis* breaking the life cycle of possible crop pests and diseases (Kleinhenz *et al.*, 2001). In USA, use of pre-emergence herbicides has been tried with success as 2,4-D amine at 1.9 kg/ha gave good weed control and was effective for 3 months onward (Humanity development library 2.0).

Mulching

Weeds can be suppressed effectively by applying a dense layer of mulch about 2cm thick early in the growing season when water chestnut canopy is insufficiently developed to shade the paddy floor. Mulch can be applied to moist soil before planting corms or seedlings. It will also conserve soil moisture if the paddy is not kept flooded. Organic mulches contain nutrients that can become available to the crop during growing season. Although these mulches do not supply significant quantities of nitrogen to Chinese water chestnut, there is release of potassium and presumably also micronutrients during anaerobic decomposition of the organic materials. Water chestnut shows good response to mushroom compost (Kleinhenz *et al.*, 2001).

Irrigation

Being an aquatic plant, it thrives in areas where there are well-controlled irrigation facilities giving a continuous supply of water throughout the year. After planting, the fields are flooded for 24 hours and then allowed to drain naturally. When the top growth reaches 20-30cm the fields are again flooded with water level at least 10-12.5cm throughout the growing season (Humanity development library 2.0).

Manure and Fertilizers

Mushroom compost is quite helpful in its cultivation. Where N:P:K uptake ratio is 1:0.5:1.75, potassium and phosphorus mineral fertilizers are applied before planting whereas nitrogen in two split doses ~ once applying just before planting and the other when corms begin to enlarge though too much nitrogen application before planting may result in production of excessive stem biomass at the expense of yield, hence, not more than 50 per cent of the total nitrogen rate should be applied before planting (Kleinhenz *et al.*, 2001).

Harvesting and Handling

The corms usually reach maturity in about 7-8 months, and harvested after the culms have turned brown or have been yellowed or killed off by frost (Humanity development library 2.0). It is recommended to harvest the corms once they turn dark brown because if left in the soil at this stage the corms will get sweeter together with decreased shelf life. In temperate climates, the quality of harvested corms deteriorates rapidly but under tropical conditions corms can be harvested through to October with a notable rise in sugar content over that period (Kleinhenz *et al.*, 2001). However, storability of corms harvested after July is low due to greater susceptibility to fungal infection and internal rots which results in reduced shelf life. Since the corms get sweetened in cool storage so harvesting even in tropical climates should probably not be extended beyond July (Kleinhenz *et al.*, 2001).

In most areas irrigation is stopped at least 3-4 weeks before harvest so that the ground dries out and the corms are carefully dug out by hand to avoid bruising or can be harvested using a 'modified gladiolus corm or potato harvester'. Alternatively, a 'water-suction harvester' can be used without any need to drain the paddy field. In the harvesting of small plots in USA, the soil is carefully lifted on to a 0.9cm wire mesh screen and worked over with rubber pads or paddles when about 98 per cent of the corms are left on the screen (Humanity development library 2.0). These are picked up and dropped into water for perfect cleaning. In larger plantings a small plough is used which turns a furrow to a depth of approximately 15cm. The furrow is then raked with a potato rake having rubber-covered prongs. The corms are carefully picked out by hand, washed thoroughly, and all damaged ones removed before they are air-dried preferably in the shade.

The dried corms can be packed in 64-litre moisture-proof containers which are sealed but not airtight. These can be kept satisfactorily at -1 to 4°C for up to 6 months though at temperatures above 13.6°C the corms will begin to sprout (Humanity development library 2.0). At 4°C, transpiration and weight loss are minimized so sprouting is delayed *vis-a-vis* deterioration due to minor injuries is minimised (Kleinhenz *et al.*, 2001). If the corms need to be stored longer they can also be kept in a bleach solution of 1,000ppm (Kanes and Vines, 1977).

Yield

Per plant yield is about 2.3kg per season. In China, 20–40t/ha average yields are reported (Humanity development library 2.0) while in USA it is in a range of 47 to 85t/ha (Morton *et al.*, 1988).

Pests and Diseases

Chinese water chestnuts are usually not very prone to serious attacks from pests and diseases though when grown on acid soils with a pH of 5.5 the plants may be attacked by a stem fungus (*Cylindrosporium* spp.) (Humanity development library 2.0) which causes **stem blight**. **Rust** (*Uromyces* species) attacks the crop in warmer climates and can be controlled in its early stages by sulphur sprays. Chemicals such as tebuconazole (Folicur) are effective but not registered for use (Kleinhenz *et al.*, 2001).

Caddisfly larvae (*Trichoptera* spp.) and other leaf-eating **caterpillars** also feed on the stems (en.wikipedia.org). **Plant hoppers** (Fulgoridae) and **scale insects** feed on plant sap and may cause subsequent death (Kleinhenz *et al.*, 2001). Currently there are no insecticides registered for use on Chinese water chestnut so control options are limited. Stocking paddies with insect-eating fish or creating appropriate habitat such as rank grass along the bank verge for other predatory insects such as 'mud-eye' may reduce insect attack (Kleinhenz *et al.*, 2001). In the Philippines, the crop is sometimes attacked by a **grasshopper** (*Aiolopus thalassinus*) which is easily controlled through use of insecticides (Humanity development library 2.0). In Florida, the most serious insect-pest is **billbug** (*Calendra cariosa*) and a **stem nematode** (*Ditylenchus* spp.) as well as an **awl nematode** (*Dolichodorus heterocephalus*) (Humanity development library 2.0).

Caddisflies

By Bruce Marlin - Own work http://www.cirrusimage.com/Trichoptera_caddisfly.htm, CC BY-SA 2.5. https://commons.wikimedia.org/w/index.php?curid=664241

Rodents, especially rats can cause considerable crop losses at harvest (Humanity development library 2.0). **Water fowl** may damage the stems and corms especially when plants are young. These can effectively be controlled using flashing lights, scare guns or water fowl deterrent devices such as nets or nylon lines (triangular shaped) across the paddy field at 1-5m intervals. Allowing paddy fields to dry out after plant senescence will prevent damage by ducks as it is difficult to rummage through semi-hardened soil. If water fowls are frequently visiting the paddy fields in large numbers before draining then it is a good indication that the crop is mature and is ready for harvest (Kleinhenz *et al.*, 2001).

References

1. Alanna Moore, 2008. Water Chestnuts. https: //www.permaculturenews.org/2008/11/29/water-chestnuts/
2. Anon. 1833a. *Eleocharis dulcis* (Burm.f.) Trin. ex Hensch. Plants of the World Online.Royal Botanic Garden, Kew. First published in Vita Rumphii: 186 (1833).
3. Anon. 1833b. *Eleocharis dulcis* (N. L. Burman) Trinius ex Henschel, Vita Rumphii. Flora of China, Vol. 23 Page 191, biqi, 186. 1833. efloras.org
4. Anon., 2019. Chinese Cooking Demystified. Cantonese Meatballs, Dim Sum-style. YouTube. Archived from the original on 2021-12-13. Retrieved 13 February 2020. https: //www.youtube.com/watch?v=NsEPOKveJFQ
5. awkwardbotany.com/tag/eleocharis-dulcis/

6. Bhaisare, M.S. 2014. Phytochemical constituent in *Eleocharis dulci*, ethnomedicinal plant from Bhandara district of Maharashtra. *Intern. J. Res. Biosci., Agric. and Techn.* (Online) 2347-517X.
7. Bhatti, H.S., N. Malla, R.C. Mahajan and R. Sehgal, 2000. Fasciolopslasis - a re-emerging infection in Azamgarh (Uttar Pradesh). *Ind. J. Path. and Microbiol.*, **43** (1): 73–6. PMID 12583425.
8. en.wikipedia.org/wiki/Eleocharis_dulcis
9. Green Aliza, 2004. Field Guide to Produce. Quirk Productions. p. 284. ISBN 978-1-931686-80-8.
10. Hideyuki, M. and H. Uchiyama, 2009. Chromosome counts of aquatic and wetland plants growing in acid sulfate soil from the central plain, Thailand. *J. Jap. Bot.*, **84**: 313–20. www.jjbotany.com/pdf/JJB_084_313_320.pdf
11. Hodge, W.H. 1956. Chinese water chestnut or matai—a paddy crop of China. *Econ. Bot.*, **10** (1): 49–65. doi: 10.1007/BF02985317. S2CID 23026535.
12. Humanity development library 2.0. Root crops: Chinese water chestnut (*Eleocharis dulcis*). nzdl.org.
13. Kanes, C.A. and H.M. Vines, 1977. Storage conditions for Chinese water chestnuts, *Eleocharis dulcis* (Burm F.) Trin. *Acta Hort.*, No.62, pp. 151–60. doi: 10.17660/ActaHortic.1977.62.15.
14. Kleinhenz, V., Geoff Lodge and David Midmore, 2001. A Grower's Guide to Cultivating Chinese Waterchestnut in Australia. *Rural Industries Research and Development Corporation*. P. 7-30. Archived from the original on 2 March 2011. Retrieved 15 November 2016.
15. Lankester, E. and F. Kuchenmeister, 1857. Appendix B: On the occurrence of species of *Distoma* in the human body. On animal and vegetable parasites of the human body: a manual of their natural history, diagnosis, and treatment. Vol. 1. Sydenham society, pp. 433–7.
16. missouribotanicalgarden.org/PlantFinder/PlantFinderDetails.aspx?kempercode=b811
17. Morton, J.F., C.A. Sanchez and G.H. Snyder, 1988. Chinese waterchestnut in Florida - past, present, and future (PDF). *Proc. Fla. St. hort. Soc.*, **101**: 139–44. Archived from the original (PDF) on 20 December 2016. Retrieved 15 November 2016.
18. pondinformer.com/water-chestnut-eleocharis-dulcis/
19. https: //powo.science.kew.org/taxon/urn: lsid: ipni.org: names: 306797-1

20. sciencedirect.com/topics/agricultural-and-biological-sciences/eleocharis-dulcis
21. Smith Jen Rose, 2019. 50 of the world's best desserts. CNN Travel. Retrieved 28 May 2020. https: //edition.cnn.com/travel/article/world-50-best-desserts/index.html
22. Terri Grassby, 2008. Phenolics and Phenolic-polysaccharide Linkages in Chinese Water Chestnut (*Eleocharis dulcis*) Cell Walls. Ph.D. Thesis. University of East Anglia. ueaeprints.uea.ac.uk/id/eprint/33057/1/Thesis_-_TGrassby.pdf.

11

Coontie Palm (*Zamia integrifolia* L.f.)

Family: Zamiaceae.

Synonyms: *Encephalartos prunifer* Sweet; *Palmifolium floridanum* (A.DC.) Kuntze; *P. integrifolium* (L.f.) Kuntze; *P. medium* (Jacq.) Kuntze; *P. tenue* (Willd.) Kuntze; *Zamia angustifolia* var. *floridana* (A.DC.) Regel; *Z. floridana* A.DC.; *Z. floridana* var. *purshiana* J.Schust.; *Z. floridana* f. *silvicola* (Small) J.Schust.; *Z. floridana* var. *umbrosa* (Small) D.B.Ward; *Z. integrifolia* var. *broomei* D.B.Ward; *Z. integrifolia* var. *floridana* (A.DC.) D.B.Ward; *Z. integrifolia* var. *silvicola* (Small) D.B.Ward; *Z. integrifolia* var. *umbrosa* (Small) D.B.Ward; *Z. media* Jacq.; *Z. media* f. *brevipinnata* J.Schust.; *Z. media* f. *calciola* J.Schust.; *Z. media* var. *commeliniana* J.Schust.; *Z. silvicola* Small; *Z. subcoriacea* H.L.Wendl. ex J.Schust.; *Z. tenuis* Willd.; *Z. umbrosa* Small. (Anon., 2023a).

Other common names: Coontie, Comfort root, Florida arrowroot, Florida zamia, Sago zamia, Seminole bread, Smooth zamia and Wild sago (Aiton, 1789).

Introduction and Origin

Native range of *Zamia integrifolia* is SE. Georgia to Florida, Bahamas, Cuba and Cayman Islands but now it is considered as an endangered species in Florida (Anon., 2023a). Its common name 'Coontie' or 'Koonti' is derived from the Seminole Native American language *conti hateka*. It prefers filtered sunlight to partial shade and well-drained sandy or sandy loam soils (en.wikipedia.org/

Male *Z. integrifolia* with Multiple Cones of Various Ages

By Donald Albury - Own work, CC BY-SA 4.0, https://commons.wikimedia.org/w/index.php?curid=124794442

Female *Z. integrifolia* Releasing Seeds from Cone

By Donald Albury - Own work, CC BY-SA 4.0, https://commons.wikimedia.org/w/index.php?curid=128662255

wiki/Zamia_integrifolia). The names 'Florida arrowroot' and 'wild sago' refer to the former commercial use of this species as the source of an edible starch.

Botany

Zamia integrifolia

CC BY-SA 2.5, https://commons.wikimedia.org/w/index.php?curid=571927

Zamia integrifolia is a tough, woody and small clumping dioecious cycad producing reddish seed cones with a distinct acuminate tip and a stem that is initially solitary, but over time it gets divided dichotomously or branches out from the side forming a multibranched cluster with a larger tuberous root system which is actually an extension of the aboveground stems which grow to a height of 3-25cm but are often subterranean and tuberous bearing 2-15 leaves arranged in terminal rosettes up to 2m in diameter (Aiton, 1789). Leaves are 20-100cm long with 5-30 pairs of leaflets (pinnae) that are linear to lanceolate or oblong-obovate, 8–25 cm long and 0.5–2 cm broad, entire or with indistinct teeth at the tip. They are often revolute, with prickly petioles (en.wikipedia.org/wiki/Zamia_integrifolia). Unlike *Zamia pumila, Z. integrifolia* does not have distinct apical teeth (Aiton, 1789).

Cylindrical to ovoid-cylindrical and often 1-30-clustered dark reddish-brown and pedunculate male cones grow to 3-15cm long having 8-20mm diameter with acute apex. Female cones are solitary or up to 5-clustered, pedunculate, dark reddish brown, elongate-ovoid with a distinct acuminate tip 5–19 cm long and 4–6 cm in diameter. The seeds are red to orange-red, ovoid and 1-2cm long (Aiton, 1789). Its diploid chromosome number is 2n=16 (Anon., 2023).

Uses

Seminoles, the indigenous people of Florida and European settlers grind its root and after soaking it overnight they rinse it with running water for several hours to remove the rest of the water-soluble toxin cycasin and then resulting paste is fermented, dried, powdered and then used for preparing bread, biscuits and arrowroot puddings (Anon., 2023b). However, it requires to be thoroughly cooked in order to destroy the poisonous principle. In World War I Florida mills processed coontie for military use, however, afterwards in 1980s in Miami area several mills started functioning to produce Florida arrowroot (Anon., 2013).

Toxicity

The whole plant contains a toxin called cycasin which can cause liver failure and ultimate death. Cycasin is water-soluble and can be leached with water through proper processing (Wag, 2023). The seeds contain a toxic glycoside Beta-methylamino-alanine which causes headaches, vomiting, stomach pains and diarrhoea as well as central nervous system failure if ingested (en.wikipedia.org/wiki/Zamia_integrifolia).

Propagation

This plant can be propagated through seeds or by using the suckers around the base of the trunk of old plants.

The germination process of seeds is quite slow so difficult to achieve in cultivation. Seeds usually need several months of storage before the inner embryo is ready to germinate. Therefore, it is best to clean the seeds of external fruit and setting these aside before attempting to propagate these. Seeds should be sown in shallow container lightly covered with sand but after germination these are potted on in small pots filled with moderately rich but light soil. The growing plants do best in partial shade (Aiton, 1789).

The old plants frequently send up suckers around the base of the trunk. When dormant, these suckers may be taken out, leaves removed to check excessive transpiration and rooted separately (Aiton, 1789).

Factors Effecting Growth

Zamia integrifolia is a popular drought-tolerant plant for tropical and subtropical regions which is easy to grow with low maintenance. This plant may be cultivated as an indoor plant in temperate countries. It prefers filtered sunlight to partial shade with some protection from mid-day and afternoon heat but not in constant deep shade. It should also be protected from extreme cold conditions (Aiton, 1789).

It prefers a deep and well-drained sandy or sandy-loam soil which is neutral or basic (tolerates limestone). Mulching with organic materials (bark or leaf mould) is recommended. It is salt resistant and is even happy living out on seaside condominium balconies (Aiton, 1789).

Though it is drought-tolerant but prefers to be watered when dry. It should be fertilized only when terminal bud begins to swell, indicating the start of the annual growth cycle (Aiton, 1789).

Pests and Diseases

In coontie, no disease is of major concern (Edward, 1999), however, sooty mould has been observed though it initiates initially due to feeding by certain insects such as aphids or mealy bugs. Florida red scales as well as mealy bugs have sometimes been observed feeding on this plant where Florida red scale sometimes becomes very serious which may be controlled through applying tar oil wash in early to mid-winter when the plants are in dormant state, however, in summers when newly hatched nymphs are present these may be killed by spraying malathion. Alata caterpillar feeds only on the coontie by devouring foliage at a rapid rate for about two weeks and then disappears but then plants appear normal in a fortnight with appearance of the new foliage (Edward, 1999).

References

1. Aiton, W., 1789. *Zamia integrifolia* L.f. ex Aiton. The Encyclopedia of Palms and Cycads. Encyclopedia of Living Forms. *Hort. Kew.*, **3**: 478. www.llifle.com/Encyclopedia/PALMS_AND_CYCADS /Family/Zamiaceae/23863/Zamia_integrifolia. Retrieved 25 February, 2023.
2. Anon. 2013. Do You Know Coontie? Florida's Native Indians and Settlers Did. Florida Department of Agriculture and Consumer Services – Division of Plant Industry. https: //fdacsdpi.wordpress.com/2013/01/23/do-you-know-coontie-floridas-native-indians-and-settlers-did. Retrieved 25 February, 2023.
3. Anon., 2023a. Coontie (Florida Arrowroot) - Arca del Gusto. Slow Food Foundation (www.fondazioneslowfood.com/en/ark-of-taste-slow-food/coontie-florida-arrowroot). Retrieved 2023-02-26.
4. Anon., 2023b. *Zamia integrifolia* L.f. Plants of the World Online. Royal Botanic Gardens, Kew. https: //powo.science.kew.org/taxon/urn: lsid: ipni.org: names: 270527-2. Retrieved 25 February, 2023.
5. Anon., 2023c. *Zamia integrifolia* Linnaeus f. Flora of North America. FNA Vol. 2. http: //www.efloras.org/florataxon.aspx?flora_id=1 and taxon_id=233501359. Retrieved 25 February, 2023.
6. Edward, F.G. 1999. *Zamia floridana* - Coontie. FPS-617, one of a series of the Environmental Horticulture Department, UF/IFAS Extension; University of Florida. Original publication date October 1999. Reviewed February 2014. https: //edis.ifas.ufl.edu/publication/FP617.
7. en.wikipedia.org/wiki/Zamia_integrifolia.
8. Wag, 2023. Coontie Palm Poisoning in Dogs. https: //wagwalking.com/condition/coontie-palm-poisoning. Retrieved 2023-02-26

12

Curcuma (*Curcuma* L.)

Family: Zingiberaceae. **Subfamily**: Zingiberoideae. **Type species:** *Curcuma longa* L.

Synonyms: *Kua* Rheede ex Medic., *Erndlia* Giseke, *Stissera* Giseke 1792, illegitimate homonym not Heist. ex Fabr. 1759, *Hitcheniopsis* (Baker) Ridl. and *Paracautleya* R.M.Sm.

Introduction and Origin

Curcuma takes its name from the Sanskrit *kunkuma*, referring to turmeric. *Curcuma*, a native to SE Asia, Indian subcontinent, southern China, New Guinea and northern Australia is a genus of about 90 species distributed widely throughout tropical and subtropical regions of the world. Some species are naturalized in other warm parts of the world such as tropical Africa, Central America, Florida and various islands of the Pacific, Indian and Atlantic Oceans. The greatest diversity of *Curcuma* species by number alone is 40-45 in India alone and Thailand 30-40 species. Other countries in tropical Asia also have numerous wild *Curcuma* species (https://en.wikipedia.org/wiki/Curcuma). Most of the species grow well in shaded areas in loose and sandy soil.

Nutritive Value

Almost all the edible species of *Curcuma* are a good source of carbohydrates, proteins and dietary fibres along with vitamins and minerals. Studies on the nutritional values of species such as *C. amada, C. leucorrhiza, C. longa, C. phaeocaulis* and *C. pseudomontana* have shown that these are nutritionally rich containing sugars, starch, fibres, essential oils, proteins, amino acids, minerals such as iron, zinc, copper, molybdenum, calcium, chromium, manganese, magnesium, phosphorous, nitrogen, potassium, sodium and sulphur as well as vitamins such as thiamine, riboflavin, vitamin A, niacin, pyridoxine, vitamin C and vitamin E (Jagtap, 2015; Sharma *et al.*, 2011; Sajitha and Sasikumar, 2014; Ibrahim and Jansen, 1996; Bharathi, 2014; Singh *et al.*, 2014a, b; Hiremath and Kaliwal, 2014; Wang and Wang, 2001). Due to presence of minerals such as Ca, Mg, Fe, K, Na, Cu, Mn and Zn, the rhizomes of *C. phaeocaulis* are also used in Chinese Pharmacopoeia as health food, in addition to being used as food supplement (Wang and Wang, 2001).

The plants have been found containing bioactive molecules that possess pharmacological properties like anti-inflammatory, antimicrobial, hypocholestraemic, antirheumatic, antiviral, antifibrotic, antivenomous, antihepatotoxic, antidiabetic, antinociceptive, anticancerous and gastroprotective properties (Padalia *et al.*, 2013; Policegoudra *et al.*, 2010; Srivastava *et al.*, 2006).

Uses

Various parts of these plant species are reportedly eaten either raw or cooked as vegetables in many Asian countries. They are also considered as nutritionally rich food products since the plants are a rich source of starch, carbohydrates, proteins, fats, vitamins and minerals (Ahamefula *et al.*, 2014; Jose and Thomas, 2014; Anjusha and Gangaprasad, 2014; Jagtap, 2015; Suphrom *et al.*, 2012). The rhizomes, inflorescence, tuberous roots and rootstocks of *C. angustifolia*, *C. australasica*, *C. caulina*, *C. pierreana* and *C. amada* have been reported to be a rich source of proteins and carbohydrates and are used in culinary preparations as vegetables and food appetizers (Sharma *et al.*, 2011; Watt, 1908; Pukhrambam, 2002; Uphof, 1959; Huxley, 1992; Ibrahim and Jansen, 1996; Patiri and Borah, 2007; Cribb and Cribb, 1987; Bruce, 2006; Parthasarathy and Sasikumar, 2006).

The rhizome of *Curcuma angustifolia* as well as several other species including *C. caulina*, *C. leucorrhiza* and *C. xanthorrhiza* can be used as nutritional food supplement and substitute for true arrowroot powder (Ravindran *et al.*, 2007; Uphof, 1959; Prakash, 1972; Rajashekhara and Sharma, 2010; Kokate, 1994; Devi *et al.*, 2015; Kay, 1987; Azam *et al.*, 2014). The rhizome is a good source of starch so it is a better supplement for new-born babies. The fresh tips of young rhizomes and shoots of *C. manga* are used in making pickles and its aromatic leaves are used for flavouring steamed and baked fish (Wrong *et al.*, 1999). In addition, the rhizomes of *C. pseudomontana* are crushed, boiled and eaten in Deccan region (India) whereas rhizomes of *C. xanthorrhiza* are either eaten raw or used as spice (Pukhrambam, 2002; Ochse *et al.*, 1980; Facciola, 1990; Burkill, 1966).

Rhizomes of *C. aeruginosa, C. amada, C. aromatica, C. longa, C. pierreana, C. pseudomontana, C. purpurascens, C. xanthorrhiza* and *C. zedoaria* are also employed as dye, spice and food flavouring and colouring agents in food preparations mainly due to their exotic aroma (Jose and Thomas, 2014; Anjusha and Gangaprasad, 2014; Ravindran *et al.*, 2007; Azam *et al.*, 2014; Sajitha and Sasikumar, 2014; Ibrahim and Jansen, 1996; Govindarajan, 1980; Thaikert and Paisooksantiatana, 2009; Koller, 2009; Benzie and Wachtel- Galor, 2011).

Genetics

The basic chromosome number in majority of Indian taxa (belonging to subgenus *Curcuma*) is x=7; published counts correspond to 6x, 9x, 11x, 12x and 15x ploidy levels. Only a few species-specific C-values were found, but karyological and/or flow cytometric data may support taxonomic decisions in some species alliances with morphological similarities.

A Synopsis of Published Chromosome Counts and Genome Sizes in the Genus *Curcuma*

Species	*No. of Chromosomes*		*Genome Size (pg)**	*Origin of Plant Material*	*Reference*
	n	*2n*			
subgen. *Curcuma* K.Schum.					
C. aeruginosa Roxb.		63		Indonesia	Prana (1977, 1978)
		63		Thailand	Apavatjrut *et al.* (1996)
		63		India	Joseph *et al.* (1999)
		63		Thailand	Paisooksantivatana and Thepsen (2001)
		63		Indonesia	Ardiyani (2002)
	28–35	63		Thailand	Sirisawad *et al.* (2003)
		63, 84	3·203–5·302	Bangladesh	Islam (2004)
C. amada Roxb		42		India	Raghavan and Venkat. (1943)
		42		India	Chakravorti (1948)
		42		India	Raghavan and Arora (1958)
		42		India	Sharma and Bhattacharya (1959)
		42		India	Ramachandran (1961, 1969)
		40	4·234/4C	India	Das *et al.* (1999)
		42	2·132	Bangladesh	Islam (2004)

Species	No. of Chromosomes		Genome Size (pg)*	Origin of Plant Material	Reference
	n	2n			
C. amarissima Roscoe		63	3·289	Bangladesh	Islam (2004)
C. angustifolia Roxb.		42		India	Chakravorti (1948)
		42	2·121–2·141	Bangladesh	Islam (2004)
C. aromatica Salisb.		42		India	Raghavan and Venkat (1943)
		42		India	Chakravorti (1948)
		63, 86		India	Ramachandran (1961, 1969)
		84		India	Nambiar *et al.* (1982)
		63		China	Chen and Chen (1984)
		42		India	Sarkar (1990)
		63		Thailand	Paisooksantivatana and Thepsen (2001)
		63	3·184	Bangladesh	Islam (2004)
C. attenuata Wall. ex Baker	42	84		Thailand	Apavatjrut *et al.* (1996)
	42	84		Thailand	Sirisawad *et al.* (2003)
C. australasica Hook.f			2·153–1·181	Bangladesh	Islam (2004)
C. aurantiaca Zijp		42		Indonesia	Prana (1977, 1978)
	21			Pen. Malaysia	Beltran and Kam (1984)
	21	42		Thailand	Sirisawad *et al.* (2003)
C. brog Valeton		63, 64, 70		Indonesia	Prana (1977)
		63, 64		Indonesia	Prana (1978)
C. caesia Roxb.		22	3·12/4C	India	Das *et al.* (1999)
		63		India	Joseph *et al.* (1999)
		63	3·333	Bangladesh	Islam (2004)
C. colourata Valeton		62, 63		Indonesia	Prana (1977, 1978)
C. comosa Craib		42		India	Joseph *et al.* (1999)
C. cf. *comosa* Roxb.		63		Thailand	Paisooksantivatana and Thepsen (2001)

Species	No. of Chromosomes		Genome Size (pg)*	Origin of Plant Material	Reference
	n	2n			
C. decipiens Dalzell	21	42		India	Ramachandran (1961, 1969)
C. elata Roxb.		63		Thailand	Apavatjrut *et al.* (1996)
	28–35	63		Thailand	Sirisawad *et al.* (2003)
		63	3·181	Bangladesh	Islam (2004)
C. haritha Mangaly and M. Sabu		42		India	Joseph *et al.* (1999)
C. heyneana Valeton and Zijp		63		Indonesia	Prana (1977, 1978)
		63		Indonesia	Ardyiani (2002)
C. kwangsiensis S.G.Lee and C.F.Liang	42	84		China	Chen *et al.* (1988)
C. latifolia Roscoe		63	3·435	Bangladesh	Islam (2004)
C. longa L.		64		Unknown	Sugiura (1931, 1936)
		62		India	Raghavan and Venkat (1943)
		62, 63, 64		India	Chakravorti (1948)
		32		Unknown	Sato (1948)
		62, 93		India	Sharma and Bhattacharya (1959)
		63		India	Ramachandran (1961, 1969)
		63		Indonesia	Prana (1977, 1978)
		48	5·1–5·26/4C	India	Das *et al.* (1999)
		63		Thailand	Paisooksantivatana and Thepsen (2001)
		63	3·275	Bangladesh	Islam (2004)
		48	4·30–8·84/4C	India	Nayak *et al.* (2006)
C. malabarica Velay., Mural. and Amalraj		42		India	Joseph *et al.* (1999)

Species	*No. of Chromosomes*		*Genome Size (pg)**	*Origin of Plant Material*	*Reference*
	n	*2n*			
C. mangga Valeton and Zijp		42		Indonesia	Prana (1977, 1978)
		63		Indonesia	Ardyiani (2002)
C. neilgherrensis Wight		42		India	Chakravorti (1948)
		42		India	Ramachandran (1961, 1969)
C. aff. *oligantha* Trimen		42		Thailand	Eksomtramage *et al.* (2002)
		40		Thailand	Saensouk and Chantaranothai (2003)
C. petiolata Roxb.		64		India	Venkatasubban (1946)
		42		Indonesia	Prana (1977, 1978)
		42		Thailand	Apavatjrut *et al.* (1996)
	21	42		Thailand	Sirisawad *et al.* (2003)
		42	2·142	Bangladesh	Islam (2004)
C. cf. *petiolata* Roxb.		42		Thailand	Paisooksantivatana and Thepsen (2001)
C. phaeocaulis Valeton		62, 63, 64		Indonesia	Prana (1977, 1978)
C. purpurascens Blume		63		Indonesia	Prana (1977, 1978)
C. raktakanta Mangaly and M.Sabu		63		India	Joseph *et al.* (1999)
	21	42		Thailand	Sirisawad *et al.* (2003)
C. roscoeana Wall.	21	42		Thailand	Apavatjrut *et al.* (1996)
		42		Thailand	Eksomtramage *et al.* (1996*a*,*b*)
		42		Thailand	Paisooksantivatana and Thepsen (2001)
C. rubescens Roxb.	28–35	63		Thailand	Sirisawad *et al.* (2003)
		63	2·204	Bangladesh	Islam (2004)
		42		Bangladesh	Islam (2004)
C. sessilis Gage		84		Thailand	Saensouk *et al.* (1998)
		46, 92		Thailand	Saensouk and Chantaranothai (2003)
C. soloensis Valeton		63		Indonesia	Prana (1977, 1978)

Species	*No. of Chromosomes*		*Genome Size (pg)**	*Origin of Plant Material*	*Reference*
	n	*2n*			
C. viridiflora Roxb.		42	2·164–2·173	Bangladesh	Islam (2004)
C. wenyujin Y.H.Chen and C.Ling		63		China	Chen and Chen (1984)
C. zanthorrhiza Roxb.		63		Indonesia	Prana (1977, 1978)
		63		China	Chen and Chen (1984)
		63		Thailand	Apavatjrut *et al.* (1996)
		63		Indonesia	Ardiyani (2002)
	28–35	63		Thailand	Sirisawad *et al.* (2003)
		63	3·285	Bangladesh	Islam (2004)
			1·30/1C	Origin unknown	Bharathan *et al.* (1994)
C. zedoaria (Christm.) Roscoe		64		India	Venkatasubban (1946)
		63, 64		India	Chakravorti (1948)
		63		India	Ramachandran (1961, 1969)
		63, 64, 66		Indonesia	Prana (1977, 1978)
		66		India	Chatterjee *et al.* (1989)
		63		Thailand	Apavatjrut *et al.* (1996)
		42		Thailand	Paisooksantivatana and Thepsen (2001)
		63		Indonesia	Ardiyani (2002)
	28–35	63		Thailand	Sirisawad *et al.* (2003)
		63	3·321	Bangladesh	Islam (2004)
subgen. *Hitcheniopsis* (Baker) K.Schum.					
C. alismatifolia Gagnep.	16	32		Thailand	Apavatjrut *et al.* (1996)
		32		Thailand	Saensouk *et al.* (1998)
		32		Thailand	Paisooksantivatana and Thepsen (2001)
		32		Thailand	Saensouk and Chantaranothai (2003)
	16	32		Thailand	Sirisawad *et al.* (2003)

Species	No. of Chromosomes		Genome Size (pg)*	Origin of Plant Material	Reference
	n	2n			
C. gracillima Gagnep.		24		Thailand	Saensouk *et al.* (1998)
		24		Thailand	Saensouk and Chantaranothai (2003)
	16	32		Thailand	Sirisawad *et al.* (2003)
C. cf. *gracillima* Gagnep.		40		Thailand	Paisooksantivatana and Thepsen (2001)
C. harmandii Gagnep.	10	20		Thailand	Eksomtramage *et al.* (1996*a*,*b*)
		20		Thailand	Paisooksantivatana and Thepsen (2001)
	10	20		Thailand	Sirisawad *et al.* (2003)
C. parviflora Wall.	14, 17, 18, 28	28, 34, 36		Thailand	Apavatjrut *et al.* (1996)
		32		Thailand	Weerapakdee and Krasaechai (1997)
		30		Thailand	Saensouk *et al.* (1998)
		42		Thailand	Paisooksantivatana and Thepsen (2001)
		32		Thailand	Eksomtramage *et al.* (2002)
		30		Thaialnd	Saensouk and Chantaranothai (2003)
	16	32		Thailand	Sirisawad *et al.* (2003)
	12, 14, 17, 18, 28	24, 28, 34, 36, 56		Thailand	Sirisawad *et al.* (2003)
C. rhabdota Sirirugsa and M.F.Newman		24		Thailand	Eksomtramage *et al.* (2002)
	12	24		Thailand	Sirisawad *et al.* (2003)

Species	*No. of Chromosomes*		*Genome Size (pg)**	*Origin of Plant Material*	*Reference*
	n	*2n*			
C. thorelii Gagnep.		36		Thailand	Eksomtramage and Boontum (1995)
	17	34		Thailand	Apavajrut *et al.* (1996)
		36		Thailand	Paisooksantivatana and Thepsen (2001)
		38?		Thailand	Ardiyani (2002)
		34		Thailand	Saensouk and Chantaranothai (2003)
	17	34		Thailand	Sirisawad *et al.* (2003)

* Genome sizes expressed in 2C-values unless indicated otherwise.

Source: https://www.ncbi.nlm.nih.gov/pmc/articles/PMC2533610/

Edible or Medicinal Species

Curcuma aeruginosa Roxb.

Common names: Blue ginger and Pink ginger.

It is native to Myanmar though is also found in West Bengal, Bihar, Coromandel Coast, South Karnataka and Kerala (Khan *et al.*, 2005). It is a perennial tropical plant with unbranched leafy stems up to 2m tall with mildly aromatic greenish-blue rhizome (Ahmed, 2011; Newman *et al.*, 2004).

Curcuma amada Roxb.

Synonyms: *Curcuma mangga* Valeton and van Zijp.

Common names: Mango ginger; in Hindi: Amba haldi.

Curcuma amada

Curcuma amada is native to India, commonly distributed in many parts of NE India and in the hills of south India (Sharma, 2012). The taxonomy of the species is a subject of some confusion as some authorities have considered the name *C. mangga* as identical while others describe it a distinct species. *C. mangga* is found in southern India while *C. amada* is of East Indian origin. Major chemical components include starch, phenolic acids, volatile oils, curcuminoids and terpenoids like difurocumenonol, amadannulen and amadaldehyde.

These perennial rhizomatous plants grow to about 80cm in height and are morphologically similar to *Zingiber officinale* and *Curcuma longa* but due to presence of car-3-ene and cis-ocimene these emit a raw mango aroma but no pungency (Rao *et al.*, 1989; Ravindran *et al.*, 2007; Sharma, 2012). The flavour is sweet with subtle earthy floral and pepper overtones as to that of raw mango. This is a popular spice and vegetable used especially in South Asian and SE Asian as well as Far East Asian cuisines. In north India it is used for chutney making though in south India for pickling. It is a delicious addition in salad and stir fries.

Curcuma mangga extracts have shown cytotoxic activities on the human cancer cell lines MCF-7 (a hormone-dependent breast cell line), KB (a nasopharyngeal epidermoid cell line), A549 (a lung cell line), Ca Ski (a cervical cell line), and HT-29 (a colon cell line). The extracts showed no cytotoxicity against the non-cancerous human fibroblast cell line MRC-5. It works as an antibacterial, antifungal, anti-inflammatory, platelet aggregation inhibitory, antioxidant, antiallergic, hypotriglyceridemic, brine-shrimp lethal, enterokinase inhibitory, CNS depressant and analgesic. Ayurveda medicines have been using *Curcuma amada* as a part of its herbal remedies for centuries as a starter, diuretic, laxative, expectorant, aphrodisiac and more. It is locally used to relieve cold *cum* cough, and bronchitis in some cases.

Ethanolic extracts of mango ginger have shown to have an antiallergic effect while the volatile oil didn't have any effect on allergic symptoms.

Curcuma angustifolia Roxb.

Common names: East Indian arrowroot and Narrow-leaved turmeric. **Regional names**: Aipah (in Thadou-Kuki), Gaddalu (in Telugu), Kaadu arrowroot (in Kannad), Koova (in Malayalam), Kutupah (in Poula), Tavakshira (in Sanskrit), Tikhur (in Hindi/Gujarati), Tikkur (in Bengali) and Yaipan (in Manipuri).

Curcuma angustifolia is endemic to northwestern, central and south India where plants are utilized for its medicinal properties by local herbalists (Watt, 1872). Most commonly it is found growing wild in India, especially in the NE and W coastal regions (plains and hills both) such as Maharashtra, Madhya Pradesh, Andhra Pradesh, Himachal Pradesh, Odisha, Chhattisgarh, Tamil Nadu and Kerala. This species is found in the eastern Himalays and inhabits bright open hillsides and woods. East Indian Arrowroot is found in the Himalayas from Kumaon to NE India and SE Asia, at altitudes of 900-1,210m.

Botany

Curcuma angustifolia prefers shady areas and grows best in moist soil. This perennial rhizomatous herbaceous ginger grows up to 1–1.5m in height in its

entirety, bearing stout underground rhizomes which lie dormant during winter. It is characterized by the presence of green, glabrous and lanceolate leaves with entire margins, *vis-a-vis* small inflorescences having yellow flowers bearing pink coma bracts (Sharma, 2012). Deciduous leaves appearing in an opposite arrangement may grow to about 36–37cm long and 8–10cm wide. Upper surface of leaves is usually a darker shade of green than the lower ones. They smell and taste similar to turmeric.

The flowers are produced before the leaves and are usually seen at the beginning of the monsoon from July to August and the inflorescence lasts in full bloom on the plants for about or more than three weeks. The bract shape is very variable and colourful which make this a showy species. Small spiked-inflorescences have 3-4 funnel-shaped yellow flowers within tufts of pink terminal coma bracts. Typical of other members of this genus, the flowers of *C. angustifolia* have double anthers, a slender style and a globular stigma. The hairy calyx of the flower has 3 lobes that may appear to be triangular or obtuse. Its cut flowers last for about 10 days. Seeds are reddish-brown. The strong rhizome is the primary source of its nutritive and medicinal properties. The rhizomes can grow up to 1.5m in length and are means of vegetative propagation.

Uses

The starch extracted from the rhizome of *C. angustifolia* has nutritional value as a source of starch for Indian foods and is a good substitute for West Indian arrowroot (*Maranta arundinacea*) (Uphof, 1959; Facciola, 1998). Rhizomes are typically ground into flour which can then be mixed with milk or water to form a nutritious meal. It is also used as an ingredient in cakes, fruit preserves, biscuits and puddings or utilized in mixture with the starch of other species such as sweet potato (*Ipomoea batatas*), rice (*Oryza sativa*) and manioc (*Manihot esculenta*). It can be used as a source of nutrition and as a non-irritating diet for patients suffering from specific chronic ailments, recovering from fevers or experiencing irritations of the gastrointestinal tract, the lungs or the excretory system. A drink with *C. angustifolia* as an ingredient is also used as a replacement of breast-milk or as a nutritional supplement for babies. In Manipur, *pakodas* made using its flowers are considered a delicacy.

The rhizomes are used to soothe coughs and to treat bronchitis. A paste of the root is applied in cases of dislocated bones (Manandhar, 2002). Its essential oil is used in antifungal medications. Compounds in the leaves of this plant have also shown to have potential as antibacterial agents. *Amlapitta*, a disease caused by increase of *Amla Guna of Pitta*, is a commonly occurring psychosomatic disease which can be correlated with diseases like gastritis, hyperacidity, non-ulcer dyspepsia, *etc.* Starch obtained from the rhizomes of *Curcuma angustifolia* and *Maranta arundinacea* are used as *Tugaksheeree*, which has been mentioned as an ingredient in many Ayurvedic formulations for the treatment of *Amlapitta* and other diseases.

Cultivation

Like other members of its genus, the vegetative propagation is by division of the rhizome when the plant is dormant. It sprouts best at temperatures around 20°C (Huxley, 1992) and is well adapted to growing in areas of seasonal drought in monsoonal forests (Huxley, 1992), preferring shady areas and moist soils that is sandy, pebbly or loamy. It is often found at the edges or in the clearings of forests (Babu, 1977).

Before planting, the field should be well ploughed, weeds should be removed and beds levelled. It does best in well-drained humus-rich soil (Facciola, 1998; Ahmed, 2011) that is rich in organic matter.

Curcuma aromatica Salisb

Common names: Aromatic turmeric, Wild turmeric or Yellow zedoary; in Hindi: Jangli/Kasthuri haldi.

Curcuma aromatica is native to Tamil Nadu (India) and is naturalized in the south Asian region, predominantly in eastern Himalayas and especially in South India. It is distributed in India, China and Sri Lanka (Sharma *et al.*, 2011) and is the most widely used *Curcuma* species next to the common turmeric (*Curcuma longa*) (Brown, 1995).

Botany

The plant grows vigorously for weeks during summer and monsoon season. The stalk grows to about 20–30cm tall and is crowned with enlarged coloured bracts with pink tips. Leaves are broad and very decorative, elliptic with a leaf stem running as long to the tip of the blade. Leaves often appear even after the flowers. When in full growth the plants can reach a height of about 40cm. Its foliage dies down in late autumn and rhizomes remain dormant in winter. Fresh rhizomes of the plants are yellowish or creamish with camphoraceous smell and produce clumps of about 1m tall, erect and unbranched leaf stems. The inflorescence appears in early spring from the base of the rhizomes and flowers are pinkish-white in colour with an orange lip (Sharma, 2012; Sharma *et al.*, 2011).

Uses

A fresh cut stalk with flowers and leaves can be enjoyed for vase arrangement indoors for up to 10 days. The rhizomes are used as a culinary ingredient in limited quantities as a food flavour, *vis-a-vis* often in cosmetic herbal medicines.

Curcuma australasica J. D. Hook. f.

Common names: Cape York lily and Native turmeric.

Curcuma australasica Rare white native turmeric

Curcuma australasica, the only Australian native (Sharma, 2012; Sharma *et al.*, 2011) is usually distributed in shady rainforest margins of the Cape York Peninsula in northern Queensland, along the coast of the Gulf of Carpentaria and New Guinea. (Watt, 1908). It is usually found growing in moist pockets in sandstone areas, and in wet woodland and forest areas. The Aboriginal Gugu Yimithirr language name for this plant is '*kumbigi*'.

It grows well in containers and can be propagated by division. It requires regular watering during summer though dry soil during winter, in correspondence with monsoonal patterns.

Though the native form is usually less than 50cm tall but its cultivated variety 'Aussie Plume' which is grown in US grows up to 2m tall. Often it is cultivated for its long-lasting beautiful flowers (1.5–2cm) and long green leaves (14-45×6–19cm). Its yellow flowers bearing pink, red, mauve or green bracts appear from November to March in Queensland though in India during summer. Occasionally a white form is also available in the wild. Aboriginal people in the Cape York Peninsula roast and relish its rhizomes.

Curcuma caesia Roxb.

Synonyms: *Curcuma kuchoor* Royle.

Common names: Black turmeric/Black zedoary; in Hindi Kali haldi, Nar Kachura, Krishna Kedar, *etc.*

Curcuma caesia is native to NE India, Java and Myanmar (Sharma *et al.*, 2011). The rhizome possesses pungent smell and is bitter *cum* hot in taste. The cultivation and harvest practices are similar to that of common turmeric.

It is a perennial herb characterized by deep bluish-black or grayish-black rhizome having pungent smell and hot bitter taste, and the leaves with a deep red-violet patch running through the length of lamina. Usually, the upper side of the leaf is rough and velvety but this characteristic is variable. Floral bracts are green and the petals may be deep pink or red.

It is claimed to be useful in treating piles, leprosy, bronchitis, asthma, cancer, epilepsy, fever, wounds, impotency, infertility, menstrual disorders, toothache, vomiting, *etc.*

Curcuma caulina J. Graham.

Common names: Arrowroot lily or Indian arrowroot.

Synonym: *Hitchenia caulina* (J. Graham) Baker.

Curcuma caulina, native to India, grows in the wild mainly in the Mahabaleshwar plateau and neighbouring forest areas with high annual rainfall (Pukhrambam, 2002). It is a herbaceous perennial emerging from underground rhizome, producing erect, unbranched and leafy stems up to 1.2m tall. It bears inflorescence with prominent greenish white or pinkish-white bracts and yellow or white flowers (Sharma, 2012).

Curcuma comosa Roxb.

Common names: Wan-chuk-mot-luk in Thai.

Curcuma comosa is native to much of Asia including Thailand, Indonesia and Malaysia. It is cultivated in Thailand, especially in the Northern Province, including Petchaboon, and the NE Province including Loei.

Curcuma comosa is widely used as a traditional herbal remedy. The rhizome of *C. comosa* is traditionally used for its anti-inflammatory properties in the treatment of postpartum uterine bleeding and uterine inflammation. Various studies have been conducted on the effects of *C. comosa* showing that it can reduce blood cholesterol, increases the thickness of epithelial cells lining the vagina and decreases uterine smooth muscle contraction. The plant is also used to repair vaginal wall prolapse, tightens vaginal muscles and relieves premenstrual syndrome and cramping.

It has been used for centuries in Thailand as a remedy for many other gynaecological problems including vaginal dryness, hot flashes, excessive white *cum* yellow vaginal discharge, bad odour, and irregular *cum* excessive or absent menstrual cycles. (https://en.wikipedia.org/wiki/Wikipedia:Citation_needed). It appears to increase the effect of ovarian and testicular hormones, and may have an estrogenic effect. It is used in Thai luxury spas to lift sagging muscles, promoting circulation and providing good health and glowing skin.

Curcuma leucorrhiza

Common name: East Indian arrowroot.

Curcuma leucorrhiza is grown widely in Manipur (India) for its rhizomes which are a good source of an edible starch known as East Indian arrowroot (Uphof, 1959; Huxley, 1959). The plant grows to a height of up to 60cm.

Curcuma longa L. (*C. domestica* Valeton)

Common names: Turmeric; in Hindi: Haldi.

Inflorescence Turmeric rhizome and powder

Name turmeric derives possibly from Middle English or Early Modern English as *turmeryte* or *tarmaret*. It may be of Latin origin, *terra merita* (meritorious earth). Recent studies have shown that taxonomy of *Curcuma longa* is problematic with only the specimens from South India being identifiable as *C. longa*. The phylogeny, relationships, intraspecific and interspecific variation and even identity of other species and cultivars in other parts of the world still need to be established and validated. Various species currently utilized and sold as turmeric in other parts of Asia belong to several physically similar taxa with overlapping local names. Species similar to *Curcuma longa* are *Alpinia galangal, Etlingera fulgens, Etlingera maingayi, Kaempferia galangal* and certain others.

Curcuma longa, native to Indian subcontinent and SE Asia is cultivated in the tropics of entire India and elsewhere (Sharma *et al.*, 2011), growing well at 20-30°C temperature with considerable amount of annual rainfall to thrive. Each year the rhizomes are gathered from the field either for consumption or for planting the following season. In India, it is grown in every tropical and subtropical region though Andhra Pradesh, Tamil Nadu and Odisha account for 75 per cent of total production in the country whereas Kerala, Maharashtra, Bihar, Karnataka and Uttar Pradesh account only for 25 per cent of total production. In India, turmeric is being grown roughly in 1,30,000ha area with a total production of about 4,10,000 metric tonnes of rhizomes.

Botany

Curcuma longa is a rhizomatous perennial growing to 1m tall with highly branched, elliptic to cylindrical, deep yellow to orange and aromatic underground rhizomes. Characteristically the plants have white to green and sometimes tinged reddish-purple stem bracts with tapered upper ends at the top of the inflorescence (Kaewkroek *et al.*, 2009). Alternate leaves arranged in two rows are divided into leaf sheath, petiole and leaf blade. From the leaf sheaths, a false stem is formed. The petiole is 0.5-1.15m long and simple though leaf blades are oblong to elliptical and usually 0.76-1.15m long (rarely up to 2.3m) and 38-45cm wide with narrowed tips, and white to green but rarely tinged reddish-purple stem bracts with tapered upper ends are present at the top of 20 cm long, cylindrical and terminal inflorescence where no flower occurs. Floral bracts are pale-green with white tips while upper bracts are rosy, the corolla is yellow and the lips are deep yellow. In East Asia it flowers usually in August where hermaphrodite flowers are zygomorphic and threefold, three white sepals being 8-12mm long and fused with fluffy hairs, the three calyx teeth are unequal, the three bright-yellow petals are fused into a corolla tube, three corolla lobes are triangular with soft-spiny upper ends while the average corolla lobe is larger than the two laterals, and only the median stamen of the inner circle is fertile with the dust bag being spurred at its base while all other stamens are converted to staminodes. Outer staminodes some 1.2-2cm long are shorter than the obovate labellum which is yellowish with a yellow ribbon in its centre. Three carpels are under a constant trilobed adherent ovary which is sparsely hairy. The fruit capsule opens with three compartments. Usually turmeric does not set inflorescence at north Indian climatic conditions.

Improved Varieties

i) **'Co-1'** released by TNAU, Coimbtore in 1982 is a late maturing variety taking 285 days, with 5.85t/ha of rhizome yield. It possesses 3.2 per cent of curcumin.

ii) **'Duggirala Red'** is a high-yielding variety released for Telangana region which matures in 240-270 days with 2.5t/ha of rhizome yield. It contains 4.1 per cent curcumin. Its dry recovery is to a tune of 23.5 per cent.

iii) **'Krishna'** developed in 1983 by M.A.U. Maharashtra is a medium maturing variety requiring 235 days to mature. Rhizome yield is 4t/ha.

iv) **'BSR-1'** developed by Bhawanisagar (TNAU) in 1984 is a late maturing (285 days) variety with 6t/ha of rhizome yield. It yields 4.2 per cent curcumin.

v) **'BSR-2'** is a high yielding short duration variety maturing in 245 days with an average rhizome yield of 3.27q/ha.

vi) **'IISR Alleppey Supreme'** matures in 210 days with rhizome yield of 3.54t/ha and curcumin 5.5 per cent.

vii) **'IISR Kedaram'** matures in 210 days with an average rhizome yield of 3.45t/ha and curcumin 5.5 per cent.

viii) **'IISR Prabha'** matures in 205 days with an average rhizome yield of 3.7q/ha and curcumin *cum* essential oil 6.5 per cent.

ix) **'IISR Pratibha'** matures in 225 days with an average rhizome yield of 3.9q/ha and curcumin *cum* essential oil 6.2 per cent.

x) **'Kanthi'** matures in 240-270 days with average rhizome yield of 3.7q/ha, curcumin 7.2 per cent and essential oil 5.1 per cent.

xi) **'Krishna'** matures in 240 days with rhizome yield of 0.92t/ha, curcumin 2.8 per cent and essential oil 2 per cent.

xii) **'Megha Turmeric**-1' is a quite late variety maturing in 300-315 days with 2.3q/ha of rhizome yield.

xiii) **'Narendra Haldi-1'** has high yield potential with high amount of curcumin and essential oil.

xiv) **'Narendra Haldi-2'** matures in 240-270 days having high yield potential.

xv) **'Narendra Haldi-3'** matures in 240-270 days with high rhizome yield.

xvi) **'Pant Peetabh'** is a late maturing variety with average rhizome yield of 2.9q/ha, 7.5 per cent curcumin and only 1 per cent of essential oil.

xvii) **'Rajendra Sonia'** matures in 225 days. This yields 4.2q/ha of rhizomes, 8.4 per cent curcumin and 5 per cent essential oil.

xviii) '**Ranga**' matures in 250 days and is suitable for growing in low-lying areas as well as for late-season sowing. It yields 2.9q/ha of rhizomes, 6.3 per cent curcumin and 4.4 per cent essential oil.

xix) '**Rashmi**' matures in 240 days with average rhizome yield of 3.2q/ha, 6.4 per cent curcumin and 4.4 per cent essential oil. It is suited to early as well as late planting, *vis-à-vis* for rainfed as well as irrigated conditions.

xx) '**Roma**', released in 1988 by O.U.A.T., Odisha with average rhizome yield of 6.43t/ha. It is a mid-season variety maturing in 253 days. Its curcumin content is 9.3 per cent and oil 4.2 per cent.

xxi) '**Sobha**' matures in 240-270 days with average rhizome yield of 3.59q/ha. Its curcumin recovery is 7.4 per cent and essential oil 4.2 per cent.

xxii) '**Sona**' matures in 240-270 days with 2.113q/ha rhizome yield. It is most suitable for growing in central Kerala. Its curcumin recovery is 7.1 per cent and essential oil 4.4 per cent.

xxiii) '**Sonia**' is a mid-season variety maturing in 230 days. It was released in 1989 by RAU, Bihar for cultivation in Bihar region. It yields 4.5-5t/ha of rhizomes which contains 8.4 per cent of curcumin.

xxiv) '**Sudarsana**' is 190 days maturing early variety released by NRCS, Calicut in 1989 for cultivation in Kerala and Andhra Pradesh regions. It contains 7.9 per cent curcumin and 7 per cent oil.

xxv) '**Sugandham**' is a variety released in 1984 by CPCRI, Kasargod for growing in Gujarat region. Fresh rhizome yield is 15t/ha.

xxvi) '**Suguna (PTC-13)**' is a mid-season variety released by NRCS, Calicut in 1988 for cultivation in Kerala and Andhra Pradesh regions. It is an early maturing (190 days) variety yielding 7.2t/ha of rhizomes. Its curcumin yield is 4.9 per cent and oil 6 per cent.

xxvii) '**Surangi**' is suitable for growing on the hills as well as in the plains. It matures in 180-200 days with average rhizome yield of 2.34q/ha. It contains curcumin content 4.5-6.5 per cent and essential oil 4.6 per cent.

xxviii) '**Suranjana (TCP-2)**' is suitable for open and shaded condition. Rhizomes mature in 239 days and is suitable for growing in West Bengal. Rhizome yield is 2.9t/ha with 5.7 per cent curcumin and 4.1 per cent essential oil.

xxix) '**Suroma**' is a mid-season variety released by OUAT, Odisha in 1989 for cultivation in Kerala which matures in 210 days. This yields 9.19t/ha of rhizomes which contain 9.3 per cent curcumin and 4.4 per cent oil.

xxx) '**Swarna**' was developed by N.R.C.S., Calicut in 1986 for Kerala, Karnataka and Andhra Pradesh regions. It matures in 210 days (mid-season) and has 4.6t/ha of rhizome yield. It contains 8.7 per cent curcumin and 7 per cent oil.

xxxi) '**Varna**' takes 240-270 days for rhizome maturity with average rhizome yield of 2.19t/ha. It yields 7.9 per cent curcumin and 4.6 per cent essential oil.

Uses

Turmeric has been used in the animistic rituals of Austronesian peoples. It was first used as a dye and then later for its supposed properties in folk medicine and in cuisines. Curcumin, a bright yellow chemical produced by the turmeric rhizome is approved as a food additive by the World Health Organization, European Parliament and United States Food and Drug Administration. It also contains an orange-coloured volatile oil. The essential oil produced from the dried rhizomes is used in face creams, body lotions, perfumery and soap fabrication. Turmeric as colouring is commonly used in Indian clothing such as saris and Buddhist monks' robes.

The rhizomes are used fresh, or dried ones after water-boiling and then again drying, after which they are ground into a deep orange-yellow powder commonly used as a spice for colouring and flavouring many Asian cuisines, especially the curries, gravies, sauces and cooked vegetables. Turmeric powder has a warm, bitter and black pepper–like flavour *cum* earthy and mustard-like aroma. It is used mostly in savory dishes, *vis-à-vis* in some of the sweet dishes such as the cake *sfouf*. In India, turmeric leaf is used to prepare special sweet dishes, *patoleo*, by layering rice flour and coconut-jaggery mixture on the leaf, then closing and steaming it in a special utensil (*chondro*). Most turmeric is used in the form of rhizome powder to impart a golden yellow colouration. In combination with annatto, turmeric has been used to colour numerous food products. It is used to impart a yellow colouration to some prepared mustards, canned chicken broths and certain other foods ~ often as a much cheaper replacement for saffron.

It is used in many products such as canned beverages, baked products, dairy products, icecream, yogurt, yellow cakes, orange juice, biscuits, popcorn, cereals, sauces and gelatin. It is a principal ingredient in curry powders. Although typically used in its dried and powdered form, turmeric is also used fresh like ginger. It has numerous uses in East Asian recipes such as a pickle that contains large chunks of fresh soft turmeric. Hot milk mixed with turmeric powder is a traditional Indian recipe used as an anti-oxidant and for increasing immunity, *vis-a-vis* roasted and then powdered rhizomes are eaten after rubbing it over Indian breads. In USA and UK, the drink known as 'golden milk' is prepared using non-dairy milk and sweetener, and sometimes black pepper after the traditional recipe.

Turmeric has been used in Asia for centuries and is a major part of Ayurveda, Siddha medicine, traditional Chinese medicine and Unani. The rhizome is the part that is most widely used. People use rhizome to stimulate digestive, respiratory and circulatory systems, *vis-a-vis* uterus. It has anti-inflammatory and antibiotic effects. It is also used to lower blood-cholesterol levels and inhibiting blood clotting. It is taken internally in the treatment of a wide range of disorders including digestive and skin complaints, gastric and duodenal ulcers, poor circulation, uterine tumours, menstrual problems, liver disease, rheumatism and to relieve cough and tuberculosis. Externally, the rhizome is used to treat a range of skin conditions including itch, insect bites, small injuries, scores and ringworm. The powdered rhizome is used in a treatment for sores and rashes in infants. A solution is used to treat eye diseases and open wounds.

Curcuma mangga Val. and Zijp.

Common name: Mango turmeric.

Though the species was first described from Java, it is a common growing crop in Thailand also (Liu and Wu, 1999). *Curcuma mangga* is characterized by the mango-like smell of rhizome as in *C. amada* (Sharma, 2012). This is an erect herbaceous perennial producing clumps of about 0.3–1.1m tall pseudostems from a branched underground rhizome (Perry, 1980).

Curcuma phaeocaulis Valeton

Common name: Rhizoma curcumae.

Curcuma phaeocaulis is widely distributed in southern regions of the People's Republic of China including Sichuan, Yunnan, Guangdong and Fujian Provinces (Sharma, 2012). The plant grows up to 1m tall. Its rhizomes are pale blue, green, yellowish green or yellow; roots are tuberous at tip and inflorescences are borne on separate shoots arising from rhizomes. Sirirugsa *et al.* (2007) reported that this plant was earlier misidentified as *Curcuma zedoaria, C. caesia* or *C. aeruginosa* in China.

Curcuma pierreana Gagnep

Common name: False arrowroot.

Curcuma pierreana has its origin in Cambodia and Thailand (Tyagi, 2005). *Curcuma pierreana* is an herbaceous perennial producing clumps of leafy stems about 60cm tall from a rhizomatous rootstock. Plants belonging to this species are distinguished by the presence of sessile inflorescence and white staminodes with large purple blotched apices.

Curcuma prakasha S.Tripathi

Curcuma prakasha is somewhat new to science. It was collected from open grassland (altitude 400m) in the Garo Hills of Meghalaya (its only known location) by its binomial author, Sunil Tripathi. He named it in Honor of Dr. Ved Prakash, an economic botanist, ethnobotanist and taxonomist. *C. prakasha*, closely related to *C. neilgherrensis* Wight, flowers and fruits in May and is pollinated by ants. The rhizomes of *C. prakasha* are used locally to treat bruises and swollen throats.

Curcuma pseudomontana J. Graham.

Common name: Hill turmeric.

Curcuma pseudomontana is endemic to the Western and Eastern Ghats of peninsular India and is grown as a potential ornamental species in Karnataka, Maharashtra and Andhra Pradesh. It is characterized by its beautiful well-developed coma, deep yellow flowers, and broadly ovate *cum* prominently sulcate leaves with bright green colour. This plant grows abundantly on rocky, shrubby slopes and flowers richly from June to October (Sharma, 2012).

Curcuma purpurascens Blume

Curcuma purpurascens is one of the lesser known species of *Curcuma* which is considered of minor importance where a limited phytochemical and biological investigations have been done. It is a perennial herb with up to 1m tall pseudostems arising from a branched rhizome and each pseudostem is comprised of several leaves having 55–70cm long and 19–23cm wide leaf blades (Anon., 1950).

Curcuma zanthorrhiza Roxb.

Common names: Java ginger, Javanese ginger, Javanese turmeric and Temulawak.

Synonyms: *Curcuma xanthorrhiza* Roxb. orth. var.

Curcuma zanthorrhiza originates from Indonesia, more specifically from Java Island. It is known in Javanese as *temulawak*, in Sudanese as *koneng gede* and in Madurese as *temu labak*. *Curcuma xanthorrhiza* is its orthographical variant though sometimes is also referred as its synonym. It flourishes up to 1,500m above sea level in a tropical climate. Its rhizomes develop well in loose soil. Currently, most of the temulawak is cultivated in Indonesia, Malaysia, Thailand and the Philippines. Outside of South East Asia, cultivars may also be found in China, Indochina, Barbados, India, Japan, Korea, the United States and some countries in Europe. *Curcuma zanthorrhiza* is characterized by its deeply coloured rhizome. The plant bears a cluster of erect pseudostems up to 2m tall from an underground rhizome and each pseudostem is made up of about eight leaves with blades that can be 40–90cm long and 15–21cm wide.

Curcuma zanthorrhiza is used as spice and as a medicine. It is used in dyspepsia. Earlier it was used as a dye but now it is often used as a substitute for *C. aromatica* in cosmetics (Ravindran *et al.*, 2007; Sharma, 2012; Liu and Wu, 1999).

Curcuma zedoaria (Christm.) Roscoe

Common names: Zedoary, White turmeric, Temu putih.

Synonyms: *Amomum latifolium* Lam., *A. latifolium* Salisb., *A. zedoaria* Christm., *Costus luteus* Blanco, *Curcuma malabarica* Velay., Amalraj and Mural., *C. pallida* Lour., *C. raktakanta* Mangaly and M.Sabu, *C. speciosa* Link, *Erndlia zerumbet* Giseke, *Roscoea lutea* (Blanco) Hassk. and *R. nigrociliata* Hassk.

Curcuma zedoaria is indigenous to Bangladesh, Sri Lanka and India (Facciola, 1990; Burkill, 1966) but is now naturalized in other places including Florida (USA). This perennial herb grows in tropical and subtropical wet forest regions. Its use as a spice in the West today is extremely rare, having been replaced by ginger, and to a lesser extent, yellow turmeric.

The fragrant plant bears yellow flowers with red and green bracts, and the underground stem section ~ a rhizome is large and tuberous with numerous branches. The plant is a rather pretty fragrant species about 1m in height which flowers in the beginning of the rainy season when the leaves start to develop. The plant bears yellow flowers with red and green bracts, and the rhizomes are white inside with strong and bitter taste, and a flavour reminiscent of mango (Sharma, 2012).

Indonesians ground it to a powder and make white curry pastes whereas in India it is usually used fresh or in pickling and also to flavour cooked *dals* and chutneys. In Thai cuisine it is used raw as thin strips in certain Thai salads. It can also be served cut into thin slices together with other herbs and vegetables with certain types of *nam phrik* (Thai chilli pastes). Traditionally, the plant is used to treat inflammation, pain and a variety of skin ailments including wounds as well as menstrual irregularities and ulcers. The essential oil produced from the dried rhizomes of *Curcuma zedoaria* is used in perfumery and soap fabrication, as well as an ingredient in bitter tonics.

Cultural Practices with Special Reference to Turmeric

Curcuma requires hot and moist climate for its cultivation. It does well in well-drained, loose, friable and fertile loamy or alluvial soils under rainfed or irrigated conditions, ranging from sea level to an altitude of 1,220m m.s.l. It likes slightly acidic soil for its cultivation with annual rainfall of 2,250-2,500mm

These are propagated through disease-free healthy rhizomes having 1-2 sprouting buds and are planted 30×25cm on well-prepared flat beds or 45cm apart in case of ridges and furrows at a depth of 4-5cm. At the time of field preparation, the soil should have been mixed thoroughly with about 20 tonnes per hectare of farmyard manure. Usually it requires 2,500kg of rhizomes per hectare with 1-2 eyes each for planting. After planting, it would be better to mulch it thickly (200 quintals per hectare of *Sesbania,* sunhemp or with leaves of any other green plant) with green leaves just after planting. These may sprout within one fortnight in case of irrigated land otherwise in a month when planted on raised beds. To get optimum yield, the crop should be given 30-100kg N, 30-80kg P and 60-100kg K per hectare. Full dose of phosphorus and half of potassium should be mixed in the soil before planting whereas full dose of nitrogen and half of remaining dose of potassium should be applied in the soil after one month of planting. Remaining 1/4th dose of potassium should be applied in the soil two months after planting. In Kerala, where 30:30:60kg/ha NPK is recommended, full phosphorus and 2/3rd of nitrogen is applied after one month of planting though remaining nitrogen and potassium is applied after two months of planting (Sharma *et al.*, 2001).

The turmeric crop requires regular watering which depends much on the soil type, the weather condition and whether the crop was mulched or not. Looking into climatic conditions the crop is irrigated at an interval of 1-3 weeks. At the time of rhizome formation it requires plenty of soil moisture during autumn. Though water requirement is commensurately much less in green-leaf mulched field. In any case, there should be proper and sufficient drainage facilities.

From time to time when weeds are seen, these are weeded out immediately though weed problem may not be so serious due to use of leaf-mulching. When weeding operation is being followed, care should be taken that soil is also little dug up for proper soil aeration. This is the time when earthing up is also followed.

The crop matures for harvesting in 8-9 month period, *i.e.* February to April when planted in April-May. Yellowing *cum* drying of leaves indicate that crop is mature for harvesting. While harvesting, through forks, shovels or spades the clumps are lifted out and the leafy tops are twisted or cut off, the roots removed swiftly, rhizomes are washed of adhering soil and dried in shade at a well-aerated place. The fingers from the mother rhizomes are conveniently removed and shade-dried for 2-3 days.

Turmeric Rhizome Harvesting, Yield and Storage

Turmeric as a crop matures for rhizome harvesting in 210 days in case of early crop, 240 days in case of mid-season and 300-315 days in case of late varieties. When mature, the leaves start yellowing and drying as by this time the rhizomes are fully developed. **Harvesting** of the rhizomes should be done carefully so that these are neither injured nor any one is left in the field. From the field these can be lifted through loosening of clumps by shovels, spades or forks as this time field should have sufficient moisture to facilitate the harvesting. While harvesting, the plant stalks along with the leaves should be cut away, the rhizomes are cleaned of their soil and finally are further washed to remove all the adhering soils any way and then these are shade-dried for 4-5 days in a well-aerated room. This is the time when fingers are separated from the mother rhizomes and then again shade-dried for 2-3 days for healing of the exposed portions. For subsequent planting the fingers are **stored** but mother rhizomes are sent for sale in the market if there is demand otherwise these are boiled along with little of fresh cow dung and turmeric leaves for better colouration as a traditional system, however, scientists have recommended treating the rhizomes not with fresh dung and turmeric leaves but with light lime water, sodium bicarbonate (0.1 per cent) or sodium carbonate. While boiling when rhizomes have started becoming soft as felt so through finger-press the boiling is stopped as this time the vapours start giving turmeric smell in the atmosphere. It usually takes 45-60 minutes. The boiled turmeric rhizomes are now sun-dried for 10-15 days in direct sun to around 5 per cent moisture content. After thorough drying the rhizomes are securely rubbed with some rough-surfaced object to clean these of their peels. Now the machines for taking out the peels are developed, and if there such facility is available that should be availed. While rubbing, the rhizomes are sprinkled with turmeric powder so that these are properly smoothened with shining surface. Now these are ready for making turmeric powder or for direct sale to the market as such. The cured and perfectly dried rhizomes can be stored for up to 3 years at a cool and dry place.

Through *barani* cultivation the **yield** is about 1/3rd to ½ as compared to irrigated ones, and so per hectare yield of raw turmeric is 150-200 quintals from the irrigated areas. Through curing and drying there is 15-25 per cent yield loss.

The rhizomes as planting stocks (mother rhizomes and fingers having 1-2 developed eyes) are not boiled but **stored** as such after dipping in 0.25 per cent Ceresan fungicide followed by shade-drying. These are traditionally stored in a pit dug up for the purpose. At the bottom, in the side and between the layers of rhizomes the pit should have either a thick layer (8-10cm) of saw dust or dried grass or leaves of *Glycosmis pentaphylla* or *Strychmos nux-vomica* as cushioning material to check moisture coming from either side of the pit. The pits are partitioned in several cells through perforated wooden planks so that the whole stock gets ample aeration throughout the storage period. These rhizomes remain in this position there until planting time approaches, *i.e.* 2-4 months. Some farmers do not lift the rhizomes from the field as these remain as ground-keepers until next planting season. This area is covered with a thick layer of dried grass or leaves until the next planting.

Insect Pests and Diseases

Major insect-pests attacking *Curcuma* are **leaf roller** (*Udaspes folus*) larvae feeding voraciously on turmeric leaves. Apart from this there are certain other leaf-eating caterpillars such as *Colasposoma splendieum, Epilachna sparsa* (beetle), *Lema lacordairei* and *Myllocerus viridanus.* **Thrips** (*Pachaetothrips indicus*) also infest on leaves and other tender parts by rasping their sap. Spraying of Dimethoate 30EC at 2ml/l or Fipronil 5 SC at 2ml/l will control all these pests. **Shoot borer** larvae (*Conogethes punctiferalis*) bore into the stem which causes plant wilting. These are controlled by applying 1kg/ha of Carbofuran or neem cake 5q/ha at plant base while planting is going on. **Rhizome scale** (*Aspidiella hartii*) adults and nymphs suck the sap of the rhizomes with the result rhizomes are disfigured and become sickly. If quite essential, the plants should be sprayed with Buprofezin 5EC @ 1ml/l. **Leaf roller** (*Udaspes folus*) larvae feed on turmeric leaves along with other insects and live therein by inrolling the leaves longitudinally through silky threads these produce. They can easily be killed through fingers or by destroying such leaves. These can also be controlled through use of 0.1 per cent Dichlorvos spraying.

Root-knot nematode (*Meloidogyne arenaria* and other species) infestation causes root-galling, plant stunting and leaf yellowing. This is controlled through fallowing and soil treatment with Carbofuran. Soil application of *Pochonia chlamydosporia* grown on sorghum seeds @ 50g/3-metre row will also control root-knot nematodes. **Burrowing nematode** (*Radopholus similis*) infestation causes dark lesions on rhizomes, topmost leaves becoming chlorotic with scorched tips, *vis-à-vis* reduced tillering *cum* plant vigour along with stunting of plants. Though nematode-free planting material should be used and the rhizomes should be given hot-water treatment for 20 minutes at 40°C before planting. **Root lesion nematode** (*Pratylenchus* spp.) infestation causes light brown to dark lesions on rhizomes and roots along with rotting of cortical portion, therefore plants are stunted, growth is reduced and the rhizomes become black and initiate rotting. At the time of planting in infested soil, neem and mustard cakes @1.5t/ha should be applied. Application of phenaphos @ 5.6kg/ha, ethoprophos @6kg/ha or oxamyl @ 1kg/ha will also kill lesion nematodes. **Grey rhizome rot** of turmeric is caused through *Pratylenchus Fusarium* complex where rhizomes turn grey and rot. To control this, the nematicide Carbofuran at 1g/m^2 area along with 0.1 per cent Bavistin drenching to the rhizomes will control this problem.

Leaf spot (*Colletotrichum capsici*). Its infection causes 4-5cm long and 3cm wide spots on leaves and their peduncles which after coalescing become quite large so affecting adversely the rhizome formation. To control this malady, before start of the symptom the crop should be sprayed with 1 per cent Bordeaux mixture, 0.2 per cent Captan or 0.2 per cent Dithane Z-78 starting from August to December at monthly intervals. **Leaf blotch** (*Taphrina maculans*) deforms the leaves by forming reddish-grey spots which is followed by yellowing of entire leaf by which leaves are not able to function properly and ultimately rhizome yield is badly affected. This disease is significantly minimized by frequent spraying of Bordeaux mixture, Aureofungin (2.5 microgram in 1ml of water), zineb 0.1 per cent or Dithane Z-78 at 0.2-0.3 per cent fortnightly. **Rhizome rot** (*Pythium graminicolum, P. aphanidermatum*) causes leaf drying starting from the margins but afterwards engulfing entire leaf. After drying of the leaves, basally watery spots as rotting of rhizomes start appearing and together even roots start rotting. Now the disease symptoms start appearing on the rhizomes where their colour changes from light orange to grey, followed by flaccid rotting. This disease is sometimes highly devastating in Andhra Pradesh and Tamil Nadu regions. Since there is no any sure remedy for this problem, so such plants should immediately be taken out from the field and the field treated with chestnut compound or Ceresan wet (1 per cent) drenching. At the time of planting only healthy materials should be used and that too after treating with some effective fungicide. Apart from these diseases, turmeric also suffers from **blight** (*Corticium sasaki*) in the field and **rhizome rot** (*Sclerotium rolfsii*) in the storage. Proper sanitation of the field and proper curing of the rhizomes followed by chemical treatment especiaaly with 0.2 per cent Captan dipping for 30 minutes, followed by thorough shade drying are quite useful.

References

1. Ahamefula, I., G.L. Onwuka and N. Chibuzo, 2014. Nutritional Composition of Tumeric (*Curcuma longa*) and its Antimicrobial Properties. *Int. J. Scientific Eng. Res.*, 5(10): 1085–89.

2. Ahmed, F. 2011. Encyclopedia of Tropical Plants - Identification and Cultivation of over 3,000 Tropical Plants. Firefly Books Ltd., New Zealand.

3. Anjusha, S. and A. Gangaprasad, 2014. Phytochemical and Antibacterial Analysis of Two Important *Curcuma* Species, *Curcuma aromatica* Salisb. and *Curcuma xanthorrhiza* Roxb. (Zingiberaceae). *J. Pharmacogn. Phytochem.*, 3(3): 50–53.

4. Anon. 1950. The Gardens' Bulletin Singapore. Singapore, pp. 13.
5. Apavajrut, P., T. Sirisawad, P. Sirirugsa, P. Voraurai and C. Suwanthada, 1996. Studies on chromosome number of seventeen Thai *Curcuma* species. *Proc. 2nd Nat. Conf. on Flower and Ornam. Plants*, India, pp. 86–99.
6. Ardiyani, M. 2002. Systematic Study of *Curcuma* L.: Turmeric and Its Allies (Ph.D. Thesis). Scotland University of Edinburgh, England.
7. Azam, M. G., M.S. Noman and M.M. Al-Amin, 2014. Phytochemical Screening and Antipyretic Effect of *Curcuma zedoaria* Rosc. (Zingiberaceae) Rhizome. *Brazilian J. Pharmacol. Res.*, **4**(5): 569–75.
8. Babu, C.R. 1977. Herbaceous Flora of Dehra Dun. Publications and Information Directorate, New Delhi, 501p.
9. Beltran, I.C. and Y.K. Kam, 1984. Cytotaxonomic studies in the Zingiberaceae. *Notes from the Royal Botanic Garden Edinburgh,* 41: 541–59.
10. Benzie, I. F. and S. Wachtel- Galor, 2011. Herbal Medicine: Biomolecular and Clinical Aspects. CRC Press, Boca Raton, London and New York.
11. Bharathi, S. V. 2014. *In Silico* Analysis of Proteins of *Curcuma caesia* Roxb. *Int. J. Pharm. Sci.*, **6**(2): 2–5.
12. Bharathan, G., G. Lambert and D.W. Galbraith, 1994. Nuclear DNA content of monocotyledons and related taxa. *Amer. J. Bot.*, **81**: 381–6.
13. Bown, D. 1995. Encyclopedia of Herbs and Their Uses. Dorling Kindersley, London.
14. Bruce, R. F. 2006. Food Composition Tables for Food Plants in Papua New Guinea, Tasmania. 38 West St., Burnie.
15. Burkill, I. H. 1966. A Dictionary of the Economic Products of the Malay Peninsular, 2nd ed. Ministry of Agriculture and Cooperatives. Kuala Lumpur, Malaysia, pp. 1–2.
16. Chakravorti, A.K. 1948. Multiplication of chromosome numbers in relation to speciation in Zingiberaceae. *Sci. and Cult.*,**14**: 137–40.
17. Chen, Z.Y.and S.J. Chen, 1984. A report on chromosome numbers of Chinese Zingiberaceae. *Guihaia*, **4**: 13–18.
18. Chen, Z.Y., S.J. Chen, X.X. Huang and S.F. Huang, 1988. A report on chromosome numbers on Chinese Zingiberaceae (5). *Guihaia*, **8**: 143–7.
19. Cribb, A. B. and J.W. Cribb, 1987. Wild Food in Australia (2nd ed.). Fontana Collins, Sydney, Australia, pp. 240.
22. Das, A.B., S. Rai and P. Das, 1999. Karyotype analysis and cytophotometric estimation of nuclear DNA content in some members of the Zingiberaceae. *Cytobios*, **97**: 23–33.

20. Devi, A. D., T.C. Singh, O.L. Devi, S.S. Singh, A.R. Singh and E.J. Singh, 2015. Phytochemical Analysis of Some Traditional Aromatic Plant Species of Thoubal District, Manipur. *Asian J. Pharm. Sci. Technol.*, **5**(1): 50–53.

21. Eksomtramage, L. and K. Boontum, 1995. Chromosome counts of Zingiberaceae. *Songklanakarin J. Sc. Techn.*, **17**: 291–7.

22. Eksomtramage, L., P. Sirirugsa, P. Jivanit and C. Maknoi, 2002. Chromosome counts of some zingiberaceous species from Thailand. *Songklanakarin J. Sci. Techn.*, **24**: 311–9.

23. Eksomtramage, L., P. Sirirugsa and S. Mayakul, 1996a. Chromosome numbers of some Thai Zingiberaceae. *Songklanakarin J. Sci. Techn.*, **18**: 153–9.

24. Eksomtramage, L., P. Sirirugsa and S. Mayakul, 1996b. Chromosome counts of Thai Zingiberaceae. *Proc. 2nd Symp. fam. Zingiberaceae* (eds Wu, T.L., Q.G. Wu and Z.Y. Chen). Zhongshan University Press, Guangzhou, pp. 107–11.

25. Facciola, S. 1998. Cornucopia II. Kampong Publications, California.

26. Facciola, S. (ed.) 1990. Cornucopia. In: A Source Book of Edible Plants. Kampong publication, California, 677p.

27. Govindarajan, V.S. 1980. Turmeric- Chemistry, Technology and Quality. *CRC Crit. Rev. Food Sci. Nutr.*, **12:** 199–301.

28. Hiremath, G.B. and B.B. Kaliwal, 2014. Pharmacognostic Evaluation of Rhizome of *Curcuma pseudomontana Int. J. Pharm. biol. Sci.*, **5**(2): 242–50.

29. https: //en.wikipedia.org/wiki/Curcuma

30. Huxley, A. 1992. The New RHS Dictionary of Gardening. MacMillan Press, Great Britain.

31. Ibrahim, H. and P.C.M. Jansen, 1996. *Curcuma* Roxburg. In: Plant Resources of South East Asia (No.9) ~ Plants Yielding Non-Seed Carbohydrates (eds Flach, M. and F. Rumawas). Prosca Foundation, Boogor, pp. 72–78.

32. Islam, M.A. 2004. Genetic diversity of the genus *Curcuma* in Bangladesh and further biotechnological approaches for *in vitro* regeneration and long-term conservation of *C. longa* germplasm (Ph.D. Thesis). University of Hannover, Germany.

33. Jagtap, S. 2015. Phytochemical screening, anti-oxidant activity, multi-elemental analysis by ICP-spectroscopy and antimicrobial activity of rhizome extracts of *Curcuma pseudomontana. J. Grahm, Eur. J. Biomed. Pharm. Sci.*, **2**(4): 1152–69.

34. Jose, S. and T.D. Thomas, 2014. Comparative phytochemical and anti-bacterial studies of two indigenous medicinal plants *Curcuma caesia* Roxb. and *Curcuma aeruginosa* Roxb. *Int. J. Green Pharm.*, **8:** 65–71.
35. Joseph, R., T. Joseph and J. Joseph, 1999. Karyomorphological studies in the genus *Curcuma* Linn. *Cytologia*. **64**: 313–7.
36. Kaewkroek, K., C. Wattanapiromsakul and S. Tewtrakul, 2009. Nitric oxide inhibitory substances from *Curcuma mangga* rhizomes. *Songklanakarin J. Sci. Technol.*, **31**(3): 293–7.
37. Kay, D. E 1987. Root Crops. In: Tropical Development and Crop Research (2nd ed. rev. by Gooding). London.
38. Khan, S. K., N.B.M. Karnat and D. Shankar, 2005. India's foundation for the revitalization of local health traditions pioneering *in situ* conservation strategies for medicinal plants and local cultures. *Herbal Gram*, **68**(34): 48.
39. Kokate, C. K. 1994. Practical Pharmacognosy (4th ed.). Vallabh Prakashan, New Delhi, 112p.
40. Koller, E. 2009. Javanese Medical Plants Used in Rural Communities (Ph.D. Thesis). Faculty of Life Sciences University of Vienna, Kubelka (Wolfgang), Austria.
41. Liu, N. and T.L. Wu, 1999. Notes on *Curcuma* in China. *J. Trop. Subtropical Bot.*, **7:** 146–50.
42. Manandhar, N.P. 2002. Plants and People of Nepal. Timber Press. Oregon, USA.
43. Nambiar, M.C., P.K.T. Pillai and Y.N. Sharma, 1982. Seedling propagation in turmeric (*Curcuma aromatica* Salisb.). *J. Plantation Crops*, **10**: 81–5.
44. Nayak, S., P.K. Naik, L.K. Acharya and A.K. Pattnaik, 2006. Detection and evaluation of genetic variation in 17 promising cultivars of turmeric (*Curcuma longa* L.) using 4C nuclear DNA content and RAPD markers. *Cytologia*, **71**: 49–55.
45. Newman, M. F., A. Lhuillier and A.D.A. Poulsen, 2004. Checklist of the Zingiberaceae of Malaysia. *Blume*, **16**: 63.
46. Ochse, J.J. and R.C.B. van der Brink, 1980. Vegetables of the Dutch Indies (3rd ed.). Asgher and Co., Amsterdam, 1016p.
47. Padalia, R.C., R.S. Verma, V. Sundaresan, A. Chauhan, C.S. Chanotiya and A. Yadav, 2013. Volatile terpenoid compositions of leaf and rhizome of *Curcuma amada* Roxb. from northern India. *J. Essential Oil Res.*, **25**(1): 17–22.

48. Paisookoantivatana, Y. and O. Thepsen, 2001. Phenetic relationship of some Thai *Curcuma* species (Zingiberaceae) based on morphological, palynological and cytological evidence. *Thai J. agr. Sci.*, **34**: 47–57.

49. Parthasarathy, V.A. and B. Sasikumar, 2006. Biotechnology of *Curcuma*. In: *Perspect. Agriculture, Vet. Sci., Nutr. Nat. Resources, CAB Rev.*, **1**: 20.

50. Patiri, B. and A. Borah, 2007. Wild Edible Plants of Asom. Geetakhi Printers and Publishers, Guwahati.

51. Perry, L.M.1980. Medicinal Plants of East and Southeast Asia: Attributed Properties and Uses. MIT Press, Cambridge, United States and London, 436p.

52. Policegoudra, R.S., K. Rehna, L.J. Rao and S.M. Aradhya, 2010. Antimicrobial, antioxidant, cytotoxicity and platelet aggregation inhibitory activity of a novel molecule isolated and characterized from mango ginger (*Curcuma amada* Roxb.) rhizome. *J. Biosci.*, **35**(2): 231–40.

53. Prakash, S. 1972. Bharat Ki Sampada. CSIR, New Delhi.

54. Prana, M.S. 1977. Studies on some Indonesian *Curcuma* species (Ph.D. Thesis). University of Birmingham, England.

55. Prana, M.S., S. Sastrapradja, J.G. Hawkes and I. Lubis, 1978. A cytological study of some Indonesian *Curcuma* species. *J. Root Crops*, **4**: 31–5.

56. Pukhrambam, C. 2002. Evaluation of Antioxidant Properties in Some Members of Zingiberales and Micropropagation of Elites. Ph.D Thesis, Manipur University, Imphal.

57. Raghavan, R.S. and C.M. Arora, 1958. Chromosome numbers in Indian medicinal plants. II. *Proc Indian Acad. Sci.* (Ser. B), pp. 352–8.

58. Raghavan, T.S. and K.R. Venkatasubban, 1943. Cytological studies in the family Zingiberaceae with special reference to chromosome number and cyto-taxonomy. *Proc. Indian Acad. Sci.* (Ser. B), **17**: 118–32.

59. Rajashekhara, N. and P.P. Sharma, 2010. A comparative study of efficacy of tugaksheeree (*Curcuma anglustifolia* Roxb. and *Maranta arundinacea* Linn. In: *Management of Amlapitta, Ayurveda*, **31**(4): 482–6.

60. Ramachandran, K. 1961. Chromosome numbers in the genus *Curcuma Linn. Cur. Sci.*, **30**: 194–6.

61. Ramachandran, K. 1969. Chromosome numbers in Zingiberaceae. *Cytologia*, **34**: 213–21.

62. Rao, A.S., R. Bandaru and S. Ramachandran, 1989. Volatile aroma components of *Curcuma amada* Roxb. *J. Agric. Food Chem.* **37**: 740–43.

63. Ravindran, P.N., K. Nirmal Babu and K. Sivaraman, 2007. Turmeric: the genus *Curcuma.* In: Medicinal and Aromatic Plants. CRC press (Taylor and Francis group), Boca Raton, London and New York, 504p.
64. Saensouk, S., B. Sumonthip and A. Luangpirom, 1998. Chromosome numbers of some Zingiberaceae in Phu Phan National Park. 24th Congress on Science and Technology of Thailand (eds Saensouk, S. and P. Chantaranothai, pub. in 2003), held on 19–21 October. Queen Sirikit Botanical Garden, Khon Kaen.
65. Saensouk, S. and P. Chantaranothai, 2003. The family Zingiberaceae in Phu Phan National Park. *Proc. 3rd Symp. family Zingiberaceae* (eds Chantaranothai, P., K. Larsen, P. Sirirugsa and D. Simpson). Applied Taxonomic Research Centre, Khon Kaen University, Khon Kaen, pp. 16–25.
66. Sajitha, P.K. and B. Sasikumar, 2014. Qualitative and quantitative variation in starch from four species of *Curcuma. Cytologia*, **80**(1): 45–50.
67. Sarkar, A.K. 1990. Cytological investigation of certain members of Zingiberaceae Lindl. to ascertain their taxonomic affinities. *Proc. Indian Sci. Congr.*, **77**: 150.
68. Sato, D. 1948. Karyotype and systematics of Zingiberales. *Japanese J. Genet.*, **23**: 44–5.
69. Sharma, A. 2012. Traditional processing of shotti (*Curcuma angustifolia* Roxb.)-a rhizome based ethnic weaning food. *Indian J. Trad. Know.*, **11**(1)154–5.
70. Sharma, A.K. and N.K. Bhattacharya, 1959. Cytology of several members of Zingiberaceae. *La Cellule*, **59**: 297–346.
71. Sharma, G.J., P. Chirangini and R. Kishor, 2011. Gingers of Manipur: diversity and potentials as bioresources. *Genet. Resour. Crop Evol.*, **58:** 753–67.
72. Sharma, R.K., D.S. Bhati and B.N. Bhatt, 2001. Masalon ki Kheti (in Hindi), 353p. ICAR, New Delhi.
73. Singh, H., B.K. Sarkar, P. Kirar and Y.A. Chowdary, 2014a. Nutritional value of some dietary herbs commonly used in Ayurveda for their medicinal values. *Int. J. Nutr. Agric. Res.*, **1**(1): 68–71.
74. Singh, T.S., T.B. Singh and T. Phucho, 2014b. Preliminary phytochemical screening and inorganic element analysis of rhizomes of *Curcuma leucorhiza* Roxb. *Am. J. Pharmatechnology Res.*, **4**(4): 1–9.
75. Sirirugsa, P., K. Larsen and C. Maknoi, 2007. The Genus *Curcuma* L. (Zingiberaceae): Distribution and Classification with Reference to Species Diversity in Thailand. *Gardens' Bull. Singapore*, No. 59;1–2.

76. Sirisawad, T., P. Sirirugsa, C. Suwanthada and P. Apavtjrut, 2003. Investigation of chromosome numbers in 20 taxa of *Curcuma. Proc. 3rd Symp. family Zingiberaceae* (eds Chantaranothai, P., K. Larsen, P. Sirirugsa and D. Simpson). Applied Taxonomic Research Centre, Khon Kaen University, Khon Kaen, South Korea, pp. 54–62.

77. Srivastava, S., N. Chitranshi, S. Srivastava, M. Dan, A. Rawat and P. Pushpangadan, 2006. Pharmacognostic evaluation of *Curcuma aeruginosa* Roxb. *Nat. Prod. Sci.*, **12**: 162.

78. Sugiura, T. 1931. A list of chromosome numbers in angiospermous plants. *Bot. Mag. Tokyo*, **45**: 353–5.

79. Sugiura, T. 1936. Studies on the chromosome numbers in higher plants, with special reference to cytokenesis, I. *Cytologia*, **7**: 544–95.

80. Suphrom, N., G. Pumthong, N. Khorana, N. Waranuch, N. Limpeancho and K. Ingkaninana, 2012. Antiandrogenic effect of sesquiterpenes isolated from the rhizomes of *Curcuma aeruginosa* Roxb, *Fitoterapia*, **83**(5): 864–71.

81. Thaikert, R. and Y. Paisooksantiatana, 2009. Variation of total curcuminoids contents, antioxidant activity and genetic diversity in turmeric (*Curcuma longa* L.) collections. *Kasetsart J. Nat. Sci.*, **43**: 507–18.

82. Tyagi, D.K. 2005. Pharma Forestry: A Field Guide to Medicinal Plants. Atlantic Publishers and Distributors, New Delhi.

83. Uphof, J.C. Th., 1959. Dictionary of Economic Plants. Weinheim, Germany.

84. Venkatasubban, K.R. 1946. A preliminary survey of chromosome numbers in Scitamineae of Bentham and Hooker. *Proc. Indian Acad. Sci.* (Ser. B), pp. 281–300. Wang, Y. and M.Z.

85. Wang, 2001. The Research Progress of *Curcuma. Chinese Pharm. J.*, **36**: 80.

86. Watt, G. 1872. A Dictionary of the Economic Products of India. Today and Tomorrows Pub., New Delhi.

87. Watt, G. 1908. The Commercial Products of India: Being an Abridgement of The Dictionary of the Economic Products of India. J. Murray Pub., London, 1189p.

88. Weerapakdee, W. and A. Krasaechai, 1997. Collection and development studies of certain *Curcuma* spp. *J. Agric.*, **13**: 127–36.

89. Wong, K.C., T.C. Chong and S.G. Chee, 1999. Essential oil of *Curcuma mangga* Val. and Zijp. rhizomes. *J. Essential Oil Res.*, **11**: 349–51.

13

Elephant-Foot Yam (*Amorphophallus* Blume ex Decne.)

Family: Araceae

Introduction and Origin

Titan arum

Amorphophallus is derived from ancient Greek word *amorphous* meaning 'without form' or 'misshapen', and *phallus* refers to the shape of the prominent spadix. Of about more than 170 species of tropical and subtropical cormous herbaceous plants, only a few species are edible as famine foods, that too, only after careful processing to remove irritating chemicals. *Amorphophallus* is native to Asia, Africa, Australia and various oceanic islands, and about 10 species are endemic to India. Many species disappeared due to habitat destruction, especially in the tropical forests of Western Ghats (India), Java, Madura, Bali, Sumatra and Indonesia. In general, the genus produces starch and glucomannan. Since *Amorphophallus* corms are rich in glucomannan they have been widely used in food, medicine, the chemical industry and so on.

Botany and Genetics

Members of Araceae are monocotyledonous monoecious flowering plants where flowers appear from inflorescence spadix. They are small to very large perennial plants growing from the subterranean corms that vary greatly in shape and size from species to species ~ from nearly uniformly globose (*A. konjac*) to elongated (*A. longituberosus* and *A. macrorhizus*) to bizarre clustered rootstock (*A. coaetaneus*). Atop a trunk-like petiole, a single leaf which can be several metres across in some species is produced from the top of corm. The leaf that lasts one growing season has a vertical leaf stalk and a horizontal blade which may consist of a number of small leaflets. The peduncle (primary floral stalk) can be short or large. The inflorescence consists of an elongated to ovate spathe as typical to Araceae family which usually envelops the spadix. Spadix possesses tiny corollaless flowers with female flowers being no more than a pistil at the base, then upward the male flowers each with one stamen and then further upward a blank sterile area (the appendix) that consists of usually large sterile (staminode) flowers. Spathes are mostly brownish-purple or whitish-green though may have different colours. Inside these are ridges or warts which function as insect-traps. Inflorescence in several species emits a decaying-flesh odour that attracts insects though certain species air pleasant odour. Mature female flowers remain receptive usually only for a day. When female flowers become receptive, these trap pollinating insects inside for about the receptive duration and at the same time male flowers also mature and together release the pollen falling over these pollinating insects. While exiting the spathe, these pollen-loaded insects can pollinate flowers in other spathes. Usually each of these pollinated flowers develops into red, orange-red, blue, yellow or white globose berries (fruits).

Basic chromosome numbers in *Amorphophalus* are most frequently found as x=13, though certain species also show x=14. The genome size of *Amorphophallus* is quite large, approximately 20 times larger than the rice genome, together with large variation existing in the genomic sequencing. Therefore, sequencing the whole genome of *Amorphophallus* species is very difficult. Several authors reported chromosome numbers as 2n=26, 2n=28 and 2n (3x)=39 for most of the species as given below in the table:

Species	***Chromosome Numbers***	***References***
A. abyssinicus (A. Rich.) N.E. Br.	2n=26	Chauhan and Brandhamt, 1985
A. ankarana Hett., Ittenb. and Bogner	2n=26	Hetterscheid *et al.*, 1999
A. albus P.Y.Liu and J.F.Chen	2n=26	Liu *et al.*, 1985
A. asterostigmatus Bogner and Hett.	2n=26	Petersen, 1992
A. bannanensis H. Li	2n=26	Gu *et al.*, 1992
A. bulbifer (Roxb.) Blume	2n (3x)=392n=26	Chauhan and Brandhamt, 1985; Ishida, 2004;Kuruvilla *et al.*, 1989

Species	*Chromosome Numbers*	*References*
A. commutatus (Schott) Engl.	2n=26	Chauhan and Brandhamt, 1985
A. dracontioides (Engl.) N.E. Br.	2n=26	Chauhan and Brandhamt, 1985
A. dubius Blume	2n=28	Chauhan and Brandhamt, 1985
A. dunnii Tutcher	2n=26	Zheng and Liu, 1989
A. goetzei (Engl.) N.E. Br.	2n=26	Chauhan and Brandhamt, 1985
A. henryi N.E. Br.	2n=26	Ishida, 2004
A. hildebrandtii (Engl.) Engl. and Gehrm.	2n=26	Chauhan and Brandhamt, 1985
A. hirtus N.E. Br.	2n=26	Ishida, 2004
A. johnsonii N.E. Br.	2n=26	Chauhan and Brandhamt, 1985
A. kerrii N.E. Br.	2n=26	Chauhan and Brandhamt, 1985
A. kiusianus (Makino) Makino	2n=26	Ishida, 2004
A. konjac K. Koch	2n=26	Chauhan and Brandhamt, 1985
A. krausei Engl.	2n=26	Ishida, 2004
A. lambii Mayo and Widjaja	2n=26	Chauhan and Brandhamt, 1985
A. laxiflorus N.E. Br.	2n=26	Chauhan and Brandhamt, 1985
A. longiconnectivus	2n=3x=39	Lekhak and Yadav, 2011
A. macrorhizus	2n=26 (several cells with 56)	
A. mairei H. Lev.	2n=26	Zheng and Liu, 1989
A. margaritifer (Roxb.) Kunth	2n=262n=3x=39	Sarker and Hetterscheid, 1997Lekhak and Yadav, 2011
A. muelleri Blume	2n (3x)=39	Ishida, 2004
A. oncophyllus Prain ex Hook. f.	2n(3x)=39	Chauhan and Brandhamt, 1985
A. paeoniifolius (Dennst.) Nicolson	2n=28	Chauhan and Brandhamt, 1985; Ishida, 2004
A. palawanensis Bogner and Hett.	2n=26	Petersen 1992
A. pingbianensis H. Li and C.L. Long	2n = 26	Ishida, 2004
A. prainii Hook. f.	2n=28	Chauhan and Brandhamt, 1985
A. rivieri Durieu ex Carriere	2n=26	Liu *et al.*, 1985
A. sinensis Belval	2n=26	Sun, 1999
A. sutepensis Gagnep	2n=26	Chauhan and Brandhamt, 1985
A. sylvaticus	2n=28	Chauhan and Brandhamt, 1985
A. taurostigma Ittenb., Hett. and Bogner	2n=26	Hetterscheid *et al.*, 1999

Species	*Chromosome Numbers*	*References*
A. titanum (Becc.) Becc.	2n=26	Ishida, 2004
A. variabilis Blume	2n=26	Chauhan and Brandhamt, 1985
A. yunnanensis Engl.	2n=26	Ishida, 2004

Sources: (as of 31 August 2022): <http://www.tropicos.org/Name/40012273?projectid=9>; and Tropicos.org. Missouri Botanical Garden. 31 Aug 2022 <http://www.tropicos.org/Name/40012273>

Studies are being conducted towards development of suitable *in vitro* culture protocols as to raise the '*In vitro* Active Gene Bank' (IVAG) of the crop and its wild relatives. Germplasm collection not only safeguards the species from future extinction but also enables variety improvement through breeding programmes and large scale propagation by *in vitro* techniques, which is albeit naturally hindered.

Nutrititive Values and Uses

Amorphophallus is a highly resourceful crop with wide adaptability. It is a good source of energy, sugar, starch, proteins as well as minerals. These convert solar energy efficiently and have high dry matter productivity as compared to other conventional energy sources like rice, wheat and maize. The corms are the prime edible part which has a high protein content of 3.8-14 per cent. Besides, the whole plant has immense medicinal and pharmaceutical values.

The composition of mature corms varies with species, origin and growing conditions. Out of 9 *Amorphophallus* species cultivated in China, *A. konjac* and *A. albus* have glucomannan (konjac) as their main storage carbohydrate (Liu, 2004). The konjac occurs in the rhizomes of East Asian *Amorphophallus* species where it is the principal carbohydrate reserve. It is a water-soluble polysaccharide that is considered a dietary fibre (Nishinari and Yoshimura, 1999). This soluble fibre which can not be metabolized by human body but by intestinal bacteria is formed from mannan *cum* glucose (Nishinari and Yoshimura, 1999). Glucomannan obtained from the rhizome has a wide range of industrial uses. It is a hemicellulose component in the cell walls of some plant species. It has gelling and film-forming properties and is widely used as an emulsifier and stabilizer in foods, drinks, cosmetic and the pharmaceutical industries. Inside the digestive system, glucomannan absorbs water up to 200 times its weight without any negative action upon the digestive system. It gives a feeling of fullness so helps in weight loss, relieves constipation *cum* irritable bowel syndrome as well as reduces serum cholesterol levels. It is a major source of mannan oligosaccharide

(MOS) found in nature, the other being galactomannan which is insoluble. It is also used as a non-clay drilling fluid in the diamond and oil industries.

Like most other members of the family, the plant contains an irritant which causes intense discomfort to the lips, mouth and throat. This acridity is caused in part by microscopic needle-like raphides of calcium oxalate monohydrate. Calcium oxalate is broken down by thoroughly cooking the plant or by fully drying it, rendering it safe to eat. People with a tendency to rheumatism, arthritis, gout, kidney stones and hyperacidity should take special precaution if including this plant in their diet.

Subgenera and Species

Based on phylogenetic analysis in 2017, the genus was divided into 4 subgenera: i) ***Amorphophallus*** (around 70 species including *A. paeoniifolius, A. variabilis* and *A. titanium*); ii) ***Scutrandrium*** (around 36 species including *A. konjac*); iii) ***Metandrium*** (around 62 species including *A. margaritifer* and *A. muelleri*); and iv) ***Afrophallus*** (>39 species including *A. abyssinicus*) (Claudel *et al.*, 2017) along with a number of SE Asian genera currently unplaced. The most popular edible species are *A. paeoniifolius* (Dennst.) Nicolson, and *A. konjac* K. Koch. Some other edible species include *A. abyssinicus* (A.Rich.) N.E.Br., *A. muelleri* Blume and *A. variabilis* Blume (Bailey, 1950). Other species are either weeds or are valued for their medicinal properties or as ornamentals. The genus also includes one of the world's largest flowers 'titan arum' (*Arum titanum*).

Out of more than 170 species, 10 are endemic to India, *viz. A. carnosus, A. longistylus* and *A. oncophyllus* to Andaman Islands; *A. arunachalensis* and *A. bognerianus* to Arunachal Pradesh; *A. mysorensis* to Karnataka; *A. longiconnectivus* to Madhya Pradesh; and *A. bonaccordensis, A. nicolsonianus* and *A. smithsonianus* to Kerala. *A. paeoniifolius, A.hirsutus, A. brooksi, A. jijas* and *A. prainii* are species that occur in Sumatra. *A. konjac* is seen in Korea and China. The tall type with largest inflorescence (up to 3m tall) is *Amorphophallus titanium* occurring in Indonesia. *A. konjac* and *A. jaijas* (*A. brooksi*) have tall spadices ranging from 1.2 to 3m. *A. bulbifer, A. kachinensis, A.hirsutus, A. nepalansis, A. paeoniifolius* and *A. sylvaticus* are the species with wide distribution in India, Myanmar, Thailand, Laos, China, Sumatra, Bhutan, Nepal, Sri Lanka and Pacific Islands.

Amorphophallus konjac K. Koch

Synonyms: *Amorphophallus mairei* H.Lev., *A. nanus* H.Li and C.L.Long, *A. rivierei* Durand ex Carriere, *Brachyspatha konjac* (K.Koch) K.Koch, *Conophallus konjak* Schott, *Hydrosme rivierei* (Durand ex Carriere) Engl., *Proteinophallus rivierei* (Durand ex Carrière) Hook.f., *Tapeinophallus rivierei* (Durand ex Carrière) Baill.

Common names: Devil's tongue, Elephant yam (also for *A. paeoniifolius*), Konjac/Konjak, Konjaku, Konnyaku potato, Snake palm, or Voodoo lily.

Introduction and Origin

The specific epithet konjac comes from the colloquial name for both the plant as well as the starchy food made from rhizomes. **Konjac** is prized for its large starchy corms, used to create a flour and jelly of the same name. Native to the forest margins and open thickets in Yunnan province of China, konjac is cultivated as an ornamental as well as a food crop, especially in warm subtropical to tropical East and SE Asia, from China and Japan south to Indonesia and Vietnam. Wild forms grow naturally in China and SE Asia. Konjac has been known in Japan since the 6th century as a medicinal food.

Botany

This perennial plant grows from a large, round and flattened corm up to 25cm in diameter. The corm produces a single, bipinnate and highly divided leaf with a pale pink petiole (leaf stem) mottled with olive-green splotches. The petiole can reach 1.2-1.5m tall and the leaf blade up to 1.2m across. The leaf is divided into numerous leaflets. Before the leaf emerges, the mature plants produce aroid-type inflorescence that is made up of a dense, spike-like spadix which bears numerous small male and female flowers and a leafy dark maroon to purple-brown spathe with ruffled margins. The bloom is 0.9-1.2m tall which attracts the pollinators by emitting a strong odour of rotten flesh.

Nutrition and Chemical Constituents

By weight, dormant corms of *A. konjac* contain 49–60 per cent glucomannan, 10–30 per cent starch, 2.6–7 inorganic elements (aluminum, calcium, chromium, cobalt, iron, magnesium, manganese, phosphorus, potassium, selenium, silicon, sodium, tin and zinc), 5–14 per cent crude protein, 3–5 per cent soluble sugars, 3.4–5.3 per cent ash and a small amount of alkaloids (trigonelline) and saponin at their stem base (Li *et al.*, 2005). Organic compounds such as β-carotene, choline, niacin, riboflavin and thiamine as well as serotonin and its derivatives, *viz.* cis-N-(p-coumaroyl) serotonin and trans-N-(p- coumaroyl) serotonin have also been identified in the fresh corm tissue (Niwa *et al.*, 2000).

Konjac glucomannan (KGM) is extracted from the rhizome of *A. konjac.* Primary components of the corm contain mainly KGM. KGM is a high molecular weight polysaccharide consisting primarily of mannose and glucose sugars and is a source of soluble dietary fibre (Nishinari and Yoshimura, 1999). KGM has also been reported to have the ability to control blood cholesterol and sugar levels so helping with weight loss and promoting intestinal activity.

Uses and Acridity

The corm is used as a food source and traditional medicine in Asia. In Japan, the corm is known as a great source of '*konnyaku*' ~ a type of flour used in many dietary products and jelly. The corms are starchy and edible when cooked as potato. Starch extracted from the cooked corms is edible and is used to make noodles, jellies, fruit candies, vegan seafood and as a thickening or stabilizing agent in food and drink. Cooked corms also have various traditional medicinal uses. The food prepared from its corms is universally known as *konnyaku* which in Japanese means 'yam cake' that is being cooked and consumed primarily in Japan and Korea. There are two basic types of cakes ~ white and black. Noodles made from *konnyaku* are called *shirataki.*

Like several other species in the genus, this plant too contains calcium oxalate crystals which before processing if swallowed causes piercing itching in the mouth, tongue and throat as well as those body parts where it was persistently touched. Calcium oxalate breaks down to become safe to eat through cooking or drying the edible parts thoroughly.

Amorphophallus paeoniifolius (Dennst.) Nicolson, 1977

Synonyms: *Amorphophallus campanulatus* (Roxb.) Blume ex Decne, *A. chatty* Andrews, *A. decurrens* (Blanco) Kunth., *A. dixenii* K.Larsen and S.S.Larsen, *A. dubius* Blume, *A. gigantiflorus* Hayata, *A. malaccensis* Ridl., *A. microappendiculatus* Engl., *A. rex* Prain, *A. sativus* Blume, *A. virosus* N.E.Br., *Arum decurrens* Blanco, *A. phalliferum* Oken, *A. rumphii* Oken, *Conophallus sativus* (Blume) Schott, *Dracontium paeoniifolium* Dennst., *D. polyphyllum* G.Forst., *Hydrosme gigantiflora* (Hayata) S.S.Ying, *Plesmonium nobile* Schott, *Pythion campanulatum* Mart.

Common names: Elephant foot yam, Elephant yam, Sweet yam, Telinga potato, White/Whitespot giant arum; in **Hindi**: Gandira, Kanda, Jangli suran,

Jimikand/Zimikand, Madana masta, Oal, Suran, Suranakanda, *etc.* In West Bengal, Asom and in neighbouring country Bangladesh it is called 'Ol'; in Bihar 'Oal' and in U.P. and Gujarat 'Suran'.

Amorphophallus paeoniifolius Inflorescence Fruits Leaves Corms

Introduction and Origin

Species epithet *paeoniifolius* means 'leaves being paeony-like', named after ancient Greek physician Paeon, who was supposed to be the first to use plants medicinally. **Elephant foot yam** is a profitable cormous tuber crop, gaining popularity due to its shade tolerance, easiness in cultivation, high productivity, fewer incidences of pests and diseases, steady demand and reasonably good price. It is an important cash crop of tropical and sub-tropical countries because of its high yield potential and culinary properties. Although it is available year-round, the peak season is late fall. Elephant foot yam is a local staple food in several countries and is cultivated as an intercrop with turmeric and bananas. Not only the corms but the pseudo-stem (culm) and inflorescence are also edible if harvested at an earlier stage. It is grown primarily in Africa, S and SE Asia and the tropical Pacific Islands. This perennial cormous crop is native to the hot and humid tropical regions of SE Asia, including the islands of Indonesia, Malaysia and the Philippines, spreading westwards into Thailand and India. In Indonesia, after rice and corn, it is the third most important source of carbohydrate. In India, elephant foot yam is found almost throughout the tropical and subtropical parts of the country and has gained the status of a cash crop due to its high production potential, market acceptability and lucrative economic returns with a good production potential (50-80t/ha) (Misra *et al.*, 2002). In India, it is cultivated in several states including West Bengal, Uttar Pradesh, Madhya Pradesh, Tamil Nadu, Andhra Pradesh, Bihar, Chhattisgarh, Karnataka, Kerala, Bihar, NE States, Odisha, Gujarat, Maharashtra and Jharkhand. This crop also offers excellent export potential from India as generally it is not cultivated commercially in other countries.

Botany

It is a stout indigenous herbaceous perennial growing 1-1.5m tall. Full-grown depressed, globose to hemispherical corms are approximately 20-25cm in diameter, dark brown outside though flesh is dull pale-yellow to white with numerous long roots. Magnificent deeply lobed leaves are tripartite and 30-90cm broad. Occasionally, the plant blooms annually around the beginning of the rainy season (April–May), bearing large, purple-red to greenish-white but with usually unpleasantly-scented spathes. The pistillate (female) and staminate (male) flowers are on the same plant and are crowded in cylindrical masses as an inflorescence; the top part is responsible for secreting mucous that gives off a putrid, pungent smell attracting pollinating insects, the middle part of the inflorescence contains staminate, and the base of the inflorescence pistillate. The stigmas of the female flowers become receptive on the first day of the bloom when foul smell attracts pollinating insects inside, and the inflorescence closes, trapping them for a night to allow the pollen deposited on the insect to be transferred to the stigmas. Later on the second day, the female flower is no longer receptive to pollens, the male flowers start to bloom, and the inflorescence opens again. This allows the pollen to be deposited on the emerging insects to pollinate other flowers while preventing the pollen from the same inflorescence fertilising itself, preventing inbreeding. In 24–36h after the first bloom of the inflorescence, the inflorescence's female flowers start developing into berries, *i.e.* bright red fruiting bodies, and other parts of the inflorescence start wilting away. The subglobose or ovoid berries are red when ripe.

Varieties

In India, some high yielding vigorous varieties with better adaptability mainly of *A. paeonifolius* are developed by selective breeding programme for wide cultivation in the country, as given below:

- ☆ '**Bidhan Kusum**' (BCKV, Nadia, W.B.): It is a non-acrid short-duration (180-210 days) variety with excellent cooking quality, having yield potential and other features similar to 'Gajendra'. This is recommended for cultivation in West Bengal region.
- ☆ '**Gajendra**' (APAU, Hyderabad): It is a short duration (180-210 days) local selection from Kovvur area of Andhra Pradesh yielding 50-60t/ha corms. The corms are non-acrid, well-shapen and generally devoid

of cormels or propagules. This is quite suitable for growing in eastern and northern India though is recommended for growing throughout the country.

- '**NDA-5**': This early maturing clonal selection recommended for cultivation in U.P. produces non-acrid corms with good cooking quality and taste. This variety is recommended for cultivation in Uttar Pradesh.
- '**NDA-9**': The corms of this early maturing clonal selection are non-acrid with good cooking quality. This is recommended for cultivation in U.P.
- '**Santragachi**': This variety is recommended for cultivation in W.B. It is a medium yielder producing several rough daughter corms. Its flesh is light cream with slight acridity.
- '**Sree Athira**' (CTCRI): It is the first genetically improved elephant foot yam developed by crossing 'AM 15' × 'AM 45' which matures in 270-300 days with an average yield of 40.5t/ha. Its corms are brown with pink flesh colour and has excellent cooking quality. This variety is recommended for cultivation in Kerala.
- '**Sree Padma**' (CTCRI): The first variety from Kerala, it is a selection from indigenous germplasm collection of elephant foot yam from Wayanad district of Kerala. This variety has a maturity period of 240-270 days with an average corm yield of 42t/ha with starch content of 12-13 per cent. Its non-acrid corms generally bear one mother corm with a few cormels or propagules. These are field-tolerant to collar rot and fairly tolerant to mosaic virus. This is recommended for cultivation in Kerala.

Chemical Constituents

Elephant foot yam is a good source of energy, proteins, fat, fibre, carbohydrates, starch, oxalic acid, vitamin A and minerals (Ca, P and Fe). Besides these, glucose, galactose and xylose are also present. The presence of an active diastatic enzyme is also reported. The corm contains betulinic acid, beta sitosterol, stigmasterol and beta sitosterol palmitate. The corms are irritant due to presence of calcium oxalate. Average nutritional profile contains starch (11-28 per cent), sugar (0.7-1.7 per cent), protein (0.8- 2.60 per cent), fat (0.07-0.40 per cent) and mean energy value (236- 566.70KJ/100g). Macro mineral and soluble oxalate between different yam varieties range are K 230-417mg/100g, P 120-247 mg/100g, Ca 131-247 mg/100g, Fe 1.97-5.56 mg/100g, Mn 0.19-.65mg/100g, Zn 12-1.92mg/100g and soluble oxalate 6.65-18.50mg/100g (Misra and Sriram, 2001). The mean soluble oxalate content (13.53mg/100g) is safe from the viewpoint of accumulation of urinary oxalate leading to kidney stones (Chattopadhyay *et al.*, 2009).

The petroleum ether extract contains alkaloids, steroids, fats and fixed oil. The chloroform extract contains alkaloids. The methanol extract contains alkaloids, steroids, flavonoids and carbohydrates. The aqueous extract contains flavonoids, tannins, proteins and carbohydrates (De *et al.*, 2010). Ethyl acetate and hexane extracts contain alkaloid, flavones, carbohydrate and saponins (Firdouse and Alam, 2011). Methanolic extract (ME) and 70 per cent hydroalcoholic extract (AE) of *Amorphophallus paeoniifolius* were analysed for flavonoid content (FC) in terms of rutin and total phenolic content (TPC), and measured in terms of catechol equivalent. Thin layer chromatography (TLC) study of methanolic extract was conducted. The flavonoid content of ME and AE were found to be 46.33mg/g and 36.88mg/g, respectively. Similarly TPC of study extracts (ME and AE) were found to be 12.67mg/g and 6.25mg/g. ME showed higher contents of flavonoid and phenolics. Upon TLC of the ME, it was observed that there were seven spots at different Rf values (Firdouse and Alam, 2011). A flavonoid (quercetin) from the ethylacetate fraction of tuber of *Amorphophallus paeoniifolius* was isolated by column chromatography using gradient elution method (Nataraj *et al.*, 2009a,b).

Uses and Acridity

Amorphophallus paeoniifolius is usually eaten as vegetable; mashed, fried or added to curries; maceration of boiled one added with raw mustard oil, raw chopped onion *cum* garlic, green chilli, and salt to prepare a popular Indian dish *chokha/bharta/bhurta*; and quite rarely used in pickling and for chutney making; or to make chips. In some households, young green leaves and stems are also cooked as green vegetables. 'Oal chutney', or 'Barabar chutney', known so as usually it contains equal quantity of peeled raw mango fruit slices, ginger and boiled oal flesh + chilli powder and salt as per taste. *Suran* is rich in nutrients and minerals. The stems can also be used as cattle feed.

The elephant-foot yam is widely used in Indian medicine and is recommended as a remedy in all three of the major Indian medicinal systems ~ Ayurveda, Siddha and Unani. The corm is prescribed in these systems for a variety of ailments. The fresh ones are used as stimulant and expectorant, and to increase appetite *cum* taste. They are carminative, aperient and expectorant. They are applied to treat acute rheumatism. They are also used in dysentery, piles and haemorrhoids as recommended by Ayurveda.

The corm of this plant is associated with acridity (itchy sensation in mouth and throat) upon oral consumption due to presence of calcium oxalate which is broken down by thoroughly cooking the plant or by fully drying it, rendering it safe to eat.

Some other Edible Species

Amorphophallus abyssinicus (A.Rich.) N.E.Br

Common name: Bagana (Sidamo).

Amorphophallus abyssinicus is native to S. Ethiopia where it is grown in gardens, hence its specific epithet *abyssinicus* which is derived from Latin meaning from Ethiopia (former name Abyssinia) (Eggli and Newton, 2004). It is cultivated in Gojjam, and the Semien Omo Zone and Konso special woreda of the Southern Nations, Nationalities and Peoples Region, on lands between 1,300 and 2,000m a.s.l. This herbaceous perennial grows up to 30cm tall from a subterranean flattened corm up to 7cm in diameter, bearing fleshy leaves. It includes *A.a.* subsp. *abyssinicus*; *A.a.* subsp. *akeassii* Ittenbach, 1997; and *A.a.* subsp. *unyikae* (Engl. and Gehrm.) Ittenb. ex Govaerts and Frodin, 2002.

Its corms as well as leaves are harvested and cooked for a long time before eating, as raw corm is sharply acrid (Zemede, 1995). Corms are sometimes gathered from the wild as local food source mainly in times of shortage. Corm is used in Nigeria to induce delivery for 'delayed birth'.

Amorphophallus albus P.Y.Liu and J.F.Chen

Amorphophallus albus is native to Yunnan Province (China) where it is cultivated as well as harvested from the wild for food, medicine and glucomannan. This easy to grow species is more tolerant of over-watering and rotting than many others. It grows up to 1m tall from corms that are brown, subglobose, 7-10cm in diameter and 5-6cm thick. The stalks are either solid green or green with grey spots, topped with a glossy green leaf. It is a rapidly offsetting species that forms a nice 60cm wide patch (in 5 years) in the woodland garden. In spring, just as the leaf emerges, it is joined by a 30cm tall floral stalk with a 15-20cm tall flower perched atop. Each flower resembles a large ivory-shell showcasing a white, cigar-like appendage, green at the tip and white at the base. Edible and medicinal uses, as well as acridity features are almost similar to *A. konjac*, *A. muelleri* or other similar East Asian species. The corm is thoroughly cooked before eating.

Amorphophallus muelleri Blume

A. muelleri is native to E. Asia ~ the Andaman Islands, Myanmar, Thailand, Malaysia and Indonesia. Though, the plant can be cultivated, it is usually harvested from the wild for local use as a food. It is being increasingly harvested on a commercial basis for its corm which has a wide range of applications in the food and medicine. The corm must be thoroughly boiled or baked before cooking due to presence of acrid calcium oxalate crystals whose effect is nullified only after thorough cooking. Fresh corms are chopped and dried or turned into tasteless curd that absorbs flavours well and so can be used as a carbohydrate element in a wide range of other foods. The corms, especially just before dormancy when foliage starts declining, are the rich source of mannan ~ a polysaccharide of mannose and glucose. The mannose is extracted and widely used in foods for diabetics. Glucomannan, a soluble fibre is formed from mannan and glucose.

Propagation

Amorphophallus has an irregular flower bearing nature, coupled with very poor seed setting, and the seeds those formed also are usually non-viable causing a genetic barrier to the conventional breeding works. Propagation in nature is thus mainly through vegetative means which leads to a very poor variability, narrow genetic base and perpetuation of systemic infection from season to season. Cultivators therefore perpetuate *Amorphophallus* vegetatively through small corms, offsets of mother corms, cuttings of large corms weighing 0.5-1kg, aerial cormules (bulbils) or apomictic seeds.

Amorphophallus species are usually propagated through small corms weighing up tro 100g or its cuttings (seed setts) of the previously harvested dormant corms (usually in November in India) that are stored in well-ventilated rooms (for Group-1 species) or are reburied in the soil (for Group-2 species). Before planting during February, the corm is cut into setts of 75-200g, each bearing a portion of central bud. Cut corms are smeared with cow dung slurry or wood ash and allowed to dry in partial shade. Rapid seed-corm production technique suggests use of cormels and mini-sett transplants of around 100g size for planting at a closer spacing of about 45×30cm (eagri.org). The seed sett weight and spacing for the production of planting material, especially for elephant foot yam, though may not show any appreciable difference for plant height but wider spacing (60×45cm) coupled with larger seed sett size (170-200g) favours canopy spread and substantially higher seed-corm yield per plant. Most economical way to produce preferred size of seed-corms (500–600g) in elephant foot yam is by planting seed setts of 75g at 60×45 cm apart.

Offsets

Every year of vegetative growth, most of the Group-1 species readily produce offsets, which may still be attached to the main corm which should be detached when mother corm is replanted for a new season. In other species the offsets are already loose when the plant is dug up after a season's growth. These offsets produce the second generation when planted. Not all of the offsets grow into new plants as there are always some inert offsets that may linger on for years without sprouting and finally die. Slim cormous offsets should be kept in soil during resting as in case of *A. hohenackeri* and *A. krausei* (Hetterscheid and Ittenbach, 1996). Offsets left on the main corm usually do not develop into new plants and though often but not always devour when the corm is utilized while nourishing the developing leaf (Hetterscheid and Ittenbach, 1996).

Bulbils

Some species produce cormules (bulbils) on the leaves which depending on species may be of three types: i) bulbils are produced entirely on the surface of branching points of the leaf (epiphyllar bulbil, *e.g.* in *A. bulbifer* and *A. muelleri*); ii) bulbils are formed when entire branching point of a leaf transforms into a bulbil and is dislodged when the leaf rots away (intercalary bulbil, *e.g.* in *A. angulatus, A. manta* and *A. sparsiflorus*); iii) third type has been found only in *A. yuloensis*, which is half epiphyllar and half intercalary (Hetterscheid and Ittenbach, 1996).

Apomictic Seeds

Some species produce apomictic seeds (vegetative seeds without pollination) that can be used to produce clonal plants. These are *A. bulbifer*, *A. muelleri* and *A. kiusianus*. The first two species exhibit a genome with 39 chromosomes and can thus not form normal haploid gametes for sexual reproduction (Hetterscheid and Ittenbach, 1996).

Tissue Culture

Vegetative propagation through tissue culture has succeeded in several occasions as witnessed by the literature. Here cell suspensions are prepared from tuber as well as from leaf parts, *e.g.* in *A. titanium* for tissue culturing (Hetterscheid and Ittenbach, 1996).

Sexual Propagation

In cultivation, different clones can be successfully pollinated by placing the pollen on the stigmas on the first day of flowering as early as possible. Since the stigmas are very sticky so no pollen will fall off though on the day the pollen is released the stigmas are not receptive (Hetterscheid and Ittenbach, 1996).

Pollination within one and the same clone has been successful in a limited number of cases (*e.g.* with *A. atroviridis*) but usually leads to only partial seed-set (two selfings in *A. curvistylis* yielded only three viable seeds) (Hetterscheid and Ittenbach, 1996).

Due to the naturally hindered sexual propagation, the seed setting is rarely observed, although an account of the propagation from seeds is presented and can be considered as an alternative approach. True seeds are produced by artificial pollination and appropriate breeding methods in some species. Fresh seeds of *Amorphophallus* usually germinate quickly (7-21 days). The seed coat of *Amorphophallus* is rather thin and there is no endospermic layer to protect the embryo from desiccation. The seeds taken from the flesh of the berries must not be stored dry. Once dry, *Amorphophallus* seeds cannot be rehydrated successfully. The first leaf is already 3- or 5-parted (in *A. prainii* often 7-parted) and soon new leaves will emerge. This may go on for more than a year before the first real resting period starts. The young corms are best left in the soil until they are bigger and more resistant to drought (only Group I tubers may be stored dry). Seeds are sown in a typical sowing soil (poor in nutrients and slightly acidic). Difficult seeds may be forced to germinate in sphagnum (Hetterscheid and Ittenbach, 1996).

Growth and Development

Majority of species are not very difficult to grow provided minimum favourable conditions are met. For about 80 per cent of the species, these favourable conditions are roughly a rich fertile soil, shading against direct sunlight, a minimum of 22°C temperature during daytime and 19°C during the night, a well defined resting period, and with a proper use of plant protection measures. In the end, the health of an *Amorphophallus* is measured by the increase of the size of the corm every season, especially in youngs to submature plants. Corm may increase its weight threefold per season (Hetterscheid and Ittenbach, 1996).

Although environmental factors such as temperature, moisture, light and soil types influence the growth and quality, other factors such as species and varieties, weight of seed setts, planter operation, irrigation, nutrition, pest management, planting date and harvest date, *etc.* are equally responsible and can be controlled by the grower. Also, a good ground preparation, harrowing and ploughing, and a proper planting method are always needed. The most profitable yields can be attained only when all factors are at optimum levels.

Amorphophallus is a tropical to subtropical crop and hence thrives well under warm humid climate with a mean annual temperature of 20-25°C (Hetterscheid and Ittenbach, 1996) and a well-distributed rainfall of 1,000-1,500mm over a period of 6-8 months. It grows well on a variety of well-drained soils with a preferred pH range of 5.5-7. *Amorphophallus* requires hot and humid climate with a temperature range of 25-35°C during vegetative growth, especially for elephant foot yam (Maniyam, 2019). Humid climate supports at the initial stages of crop growth whereas dry climates facilitate corm bulking.

Temperature requirement varies considerably among different species. *A. kiusianus* from southern Japan does not appreciate high temperature as vegetative parts die quickly if temperature rises above 25°C. Similarly, *A. konjac* thrives optimally at 20-25°C temperature range although it is more resistant than *A. kiusianus. A. bulbifer* tolerates a rather large range of temperature and thrives well at 15-30°C.

Soil Type

Well-drained and fertile sandy loam, sandy clay-loam or black soil with near neutral soil reaction is ideal for its cropping. Soil should be rich in organic matter with adequate amount of available plant nutrients. A pH range of 5.5-7 is preferred. *Amorphophallus* can be divided into 2 groups that should be grown in different types of soils, relating to the climatic conditions that the plants experience.

- **Group or Class-1** includes species such as *A. paeonifolius* and *A. variabilis* whose corms should be stored dry when dormant. Plants that often dries out completely towards the end of the dry season with the surrounding vegetation thinning creating virtually no overhead canopy causing the corms to have to survive in a layer of hard dry soil. *A. paeonifolius* and *A. variabilis* corms that have adapted to these conditions should be suited to forage out of the pot without any soil. Corms can only be harvested when dormant and when irritant is not present or is at the minimum (www.researchgate.net/publicatio). Species of strongly seasonal climates are included in this group and usually have elongated corms and are accustomed to heavy soil enriched with loam. This group includes all species with elongate corms, and those from Africa, China, Japan, India, Thailand, Indochina, the Philippines, Java (but not all), eastern Indonesia, New Guinea and Australia. Also, all three Asian species (*A. muelleri, A. paeoniifolius* and *A. prainii*) with a large geographical distribution are used to such soils (Hetterscheid and Ittenbach, 1996).
- **Group or Class-2** includes species from deep equatorial region that does not have a true dry period. Generally their seasonal periods

of growth and dormancy can be difficult to determine. They can be found in clay substrate overlaid with leaf litter, sometimes peat and it is almost always damp. These plants have no need to adapt to any dry period and a regularly defined period of dormancy would not be necessary (www.researchgate.net/publicatio). The species in this group have relatively smaller roots so prefer well-drained and aerated soil not containing loam. When soil becomes very compact due to excessive loam or waterlogging conditions, the roots will rot. It comprises nearly all the species from Sumatra, Borneo (Kalimantan, Sarawak, Sabah, and Brunei) and West Malaysia with the exception of only a few species (Hetterscheid and Ittenbach, 1996).

Land Preparation and Planting Method

Land should be prepared for planting by ploughing, harrowing and adding organic fertilizers to make it fertile. After one or two ploughings to a depth of 20-22cm, pits of 60×60×45cm are prepared at a spacing of 90×90cm during February. For small to medium sized tubers, distance between pits can be reduced to 60×60 cm. Pits are half-filled with top soil and well-dried farmyard manure @ 2-2.5kg/pit along with wood ash. Planting material is placed vertically in the pit. Since the roots emerge from the top, the top of an *Amorphophallus* corm must always be well below the soil surface to secure the corm to carry the long leaf or inflorescence. The growth of initial roots is very fast and horizontal, and starts contracting after reaching full length to secure and stabilize the corm and the developing leaf. This is necessary because the developing leaf depletes most or the entire old corm so its base tends to 'hang loose' in an empty hole where the corm initially was. This entire structure is very unstable and has a great need for the contractile roots (Hetterscheid and Ittenbach, 1996).

If the plants are being grown in pots, those species producing elongate corms need very deep pots to accommodate the strong vertical growth. When the pot is not deep enough, the base of the corm will get cramped against the bottom and become deformed. This however does not usually impair the health of the corm. The only problem may be that excess water that is often found in the bottom of the pot may cause local rotting of parts of the corm base. The diameter of the pot must be at least twice that of the corm that is planted, for obvious reasons (Hetterscheid and Ittenbach, 1996). Since most roots of *Amorphophallus* plants accumulate in the bottom of the pot, the soil there must be especially well prepared to transport water. Do not use a thick layer of coarse grit because then the roots will be unable to extract water and nutrients and will subsequently die. Adding bark to the bottom layer of the soil is a useful alternative (Hetterscheid and Ittenbach, 1996). Planting of *A. paeoniifolius* in South India is February-March, whereas March-April in Eastern and other parts of the country provided irrigation facilities are ensured as the crop needs irrigation during the summer months.

Manures and Fertilizers

Both the groups require a rich soil, especially in trace elements, which may be added with a fertilizer, and organic matter. Excess fertilizer may well kill off all roots. The ideal fertilizer is that with high phosphate content (*e.g.* 15-30-15 NPK) which stimulates corm growth. Group-1 *Amorphophallus* does not need frequent fertilizer or nutrients. However, Group-2 *Amorphophallus* needs additional soil nutrient inputs when soil is loose and contains only organic nutrients and fewer trace elements. Since lime (*e.g.* dolomite) has a high content of trace elements, adding limestone to the soil is important (Hetterscheid and Ittenbach, 1996). Apply fertilizer @ 40kg N, 60kg P_2O_5 and 50kg K_2O/ha after 45 days of planting along with mulching and application of well-rotten cow dung manure or compost after receiving of rains. This is followed by digging interspaces and light earthing up. Top dressing is done with 40kg N, 50kg K_2O again one month after, along with shallow intercultural operations such as weeding, light digging and earthing up.

Irrigation

Amorphophallus is mainly grown as a rainfed crop. Ample water is required during the growing season but the amount can be reduced dramatically during dormancy. During periods of late receipt of monsoon, a light irrigation is given during early stages of crop growth. Species of Group-1 are somewhat resistant to excess water with the exception of those with elongate corms, which usually show slight superficial rot even when they have not been overwatered (Hetterscheid and Ittenbach, 1996). These small spots usually dry well when the corm is taken out during dormancy. The large spots can be scraped off superficially and inspected for their deepness. Group-1 species are also quite resistant to dry conditions and they will stand neglect fairly well for up to 4 weeks. Group-2 species are sensitive to excess water, hence should be grown in quite well-drained soils (Hetterscheid and Ittenbach, 1996).

Interculture

Mulching immediately after planting is the most important operation in *Amorphophallus*. The pits are usually covered with organic mulches such as green leaves or paddy straw. It not only conserves soil moisture and regulates soil temperature but also suppresses weed growth.

Earthing up of the soil also brings about weed control besides improving the physical condition of the soil. Regular weedings and earthing up should be performed to keep the soil loose and to destroy the weeds.

Amorphophallus plants usually produce a single stem. In case of more numbers, it is advisable to remove these except the healthy one.

During initial period of 2-3 months after planting, leafy vegetables, green gram, black gram, cowpea, cucumber, *etc.* can be grown as an inter-crop, especially with 'elephant foot yam'. Intercropping of elephant foot yam in banana, coconut and other newly planted orchards gives additional income to farmers.

Dormancy

Amorphophallus species have a natural dormancy period with exception of a few evergreen species such as *A. arnautovii* and *A. pingbianensis*. In cultivation, all group-1 species show strong tendency of dormancy, whereas group-2 species may on occasion skip dormancy altogether though not as a rule. Dormancy may take place in two parts of the growing cycle. After the growing season, the leaf dies down and the tuber invariably goes to resting period which may be 3-7 months depending on the species after which a leaf may emerge. The resting period is usually shorter, *i.e.* a month or even less when the plant is mature enough to flower. During 'after-leaf' dormancy, corms of group-I species can be stored dry but not the seedling corms or thin offsets (Hetterscheid and Ittenbach, 1996).

Group-2 corms are usually prone to desiccation and must be left in the soil. A second type of dormancy may set in after flowering. This only holds true for non-evergreen Asian species which in nature invariably refuse to develop a leaf after flowering and effective pollination. Therefore, of these species fruiting plants are never found with leaves. In cultivation a similar behaviour is observed with most species but sometimes when the inflorescence is taken off, a plant may develop a leaf. Since this behaviour cannot be predicted, the corm is best left in the soil after the withered inflorescence has been removed. Watering must be lessened and the spot must be well observed for renewed growth. If this happens, the plant must be watered as usual; if not, it must be left to dry out until the next season (Hetterscheid and Ittenbach, 1996).

All African species develop a leaf in the same season as flowering, with or without pollination and fruit set. Most develop a leaf shoot directly alongside the peduncle; others take a short resting period. This leads to an important warning: when the old inflorescence of an African species is cut away, be sure not to cut the new leaf shoot which may be very close to the peduncle; remove the cataphylls first to locate the new shoot before cutting the peduncle. The African Savannah species such as *A. aphyllus* and *A. dracontioides* pose a special problem as these need a special treatment during dormancy. So far, none of these species have been brought to flower, even though plants have reached full size. Probably they need extra heat during resting, imitating the harsh conditions on the savannah during dry period (Hetterscheid and Ittenbach, 1996).

Harvest and Storage

The natural dormancy of *Amorphophallus* corms means that they can remain in the field for some months without quality loss. Planting material separated from the harvested product can best be stored under dry, dark and cool (10°C) conditions. Temperatures above 40°C as well as frosts may kill the corms planted in the open though under shade there may not be any problem.

Corms are harvested of 8-9 months of cropping by digging when the top is completely withered and fallen. However, on better market price the corms can be harvested 6 months onwards. Elephant foot yam is a long duration crop and generally matures in 6-7 months. Crop can be harvested at different stages of development starting from 6-7 months of plantation up to 4 years as per requirement. Harvested corms are cleaned and spread out at an airy place under shade cover for two days prior to storage, for healing of injuries and bruises on the corms. The corms can then be safely stored in a dry and well-ventilated storehouse facilitating diffused light. If corms are to be stored for seed purpose, the cut tubers should be immediately treated with Mancozeb (0.2 per cent). If storage is for planting purpose, it should be treated with Fenitrothion (0.05 per cent) + Mancozeb (0.2 per cent). The production potential of elephant foot yam is 50-60t/ha.

The species that have roots spread all over the healthy corm when the leaf has already died down, need continuous soil and/or watering. Usually the species from WC Africa, W. Malaysia, Borneo, Java, Sumatra, S and SE China and Vietnam need to be kept in soil all the time. They tend to have rather short dormancies or nearly none at all. They are from everwet tropical areas. Unfortunately, this behaviour makes them the most difficult species to grow (except the ones from N. Vietnam, S and SE China and *A. hottae*). The smallest of species also appreciate not to be borne dry, but they must also be watered carefully as these do not like wet feet. They too tend to have hardly any appreciable dormancy and flower immediately after leaf shed or alongside mature leaves (Hetterscheid and Ittenbach, 1996). The species that are usually without roots or have dead or dying roots when the leaf has died down, need dry conditions for storage. Such species with prolonged dormancy are usually much stronger in cultivation, requiring less humid conditions, therefore must be stored in dry *cum* ventilated storehouse with diffused light.

Diseases

Amorphophallus is free from major pests and diseases except **collar rot** or **foot-rot** caused by *Sclerotium rolfsii* or *Rhizoctonia solani*. Its symptoms appear as water-soaked lesions on the collar when the plant is about 2-3 months old, followed by whole plant turning yellow soon after, stem shrinking and collapsing due to rotting of the collar region. This pathogen is favoured by heavy soils, high organic matter in the soil, heavy rains, waterlogging, high relative humidity, poor drainage and mechanical injury at the collar region. It can be controlled by crop rotation, removal of plant debris and improving the drainage. Application of biocontrol agents like *Trichoderma harzianum* @ 2.5kg/ha mixed with 50kg of FYM is quite effective. Drenching soil with Captan (0.2 per cent) or Brasicol (0.1 per cent) is also quite effective in controlling this malady. Control measures recommended by CTCRI for collar rot include cultural practices such as crop rotation, removal of plant debris and improving drainage together with soil drenching around the plant with benzimidazol 0.05 per cent. **Dry rot** is caused by *Botryodiplodia theobromae* often under hot and humid conditions. In this case it is always better to destroy the infected corm so that other corms may be saved from infection, though infected corm may be saved if the top is still intact. A heavily infected corm must be cleaned and the largest scars sealed with a fungicide such as Topsin in order to prevent infection, *vis-à-vis* desiccation. Diseases can be managed by use of disease-free planting material, removal of infected plants, improving drainage, application of neem cake in soil, use of biocontrol agents such as *Trichoderma* and drenching soil with Captan 0.2 per cent. However, it would be always better to prevent diseases and pests rather than their curing. It is therefore recommended that corms should always be inspected for diseases and pests after the growing season. The group-2 corm species must be put back in the soil after examination. Some other maladies include **white leaf disease** and **bacterial leaf blight**. **Leaf blight (rot)** is caused by *Phytophthora colocasiaei* and is more common in warm *cum* highly humid areas. In its infection, in lower leaves small water-soaked lesions develop on the leaflets. After coalescing, these spots enlarge and give blighted appearance. Bacterial infections are recognized by the flesh of the corm which turns into a slimy substance, therefore, in such cases the infected parts should be cut away and treated with a bactericide. Remarkably few diseases and pests occur in the semi-natural 'jinenjo' cropping system in Japan.

'**Dasheen mosaic virus**' and '**konjac mosaic virus**' have been found infecting this crop so such plants should be burnt. In their attack the growth retards with poor corm yield. '**Yam mosaic potyvirus**'usually spreads through aphids (*Aphis gossypii, Myzus persicae*). The symptoms include mosaic mottling of leaves, distortion of leaf lamina, more proliferation of lateral buds, separation of buds from the mother corms, poor growth of roots and production of small corms. The infected leaves show yellow and green pattern mosaic between the veins or may show narrow green strips bordering the veins (called vein banding). In severe cases, the leaves become long, thin and strap-shaped (called shoe-string symptom) and whole plant becomes deformed and stunted. Such plants may produce a few small cormels with less starch content. Some plants may recover from the virus infection soon after first symptom but virus may survive in plant and causes reduction of the vigour. To get rid of this problem, corms from healthy plants should be used for seeding; fields must be kept free from weeds; crop debris should be destroyed; and systemic insecticides should be sprayed to prevent secondary spread. CTCRI has recommended integrated disease management as following:

- Use of healthy planting material with no symptom of corm rot or mosaic.
- Mulching of field with dry leaves, dry paddy straw or black polythene sheets.
- Before planting the corms should be treated with cow dung slurry mixed with *Trichoderma viridae* (100g in 20kg FYM or cowdung slurry).
- Application of *Trichoderma* enriched compost in pits before planting.
- Two preventive sprayings with Mancozeb (0.2 per cent)+Monocrotophos (0.05 per cent) at 60 and 90 days after planting.

Insect-Pests

Reported pests on elephant foot yam are *Galerucida bicolor* attacking the leaves, *Araecerus fasciculatus* attacking the corms, and various scale insects, caterpillars and nematodes. Two major pests of *Amorphophallus* in cultivation are nematodes, and root mealy bugs (*Rhizoecus amorphophalli* Betren) which are similar in appearance to mealy bugs and very resistant to most of chemical pesticides. As no biological control mechanism yet exists, therefore, best method is using a systemic pesticide with a broad spectrum activity. In case of mealy bug infestation during storage of seed-corms, dipping of corms before planting in 0.05 per cent Monocrotophos should be done to check the pest in field. Unfortunately use of pesticides is usually rather unhealthy for humans. However, cow urine, common salt (1,000ppm), cow dung (2:1) and clay slurry (1:1) treatments are very effective against mealy bugs during storage (Maniyam, 2019). Nematode infestations are easily recognized in their early stages as small to medium-sized hemispheric warts/bumps on the tuber surface are observed not associated with root scars or accessory buds. In such cases, Furadan at 2g/m^2 should be mixed in the soil before planting followed by watering. Other insect-pests include aphids, thrips, caterpillars and spider-mites. Certain lepidopterous larvae such as *Palpifer sexnotatus* and *P. sordida* feed on this crop. Control measures include use of pest-free seed-corms.

Corms that have been in wet soil too long may start rotting at the base of the roots. This type of rot spreads through the corm with tremendous speed and is usually detected too late. All rotting parts should be cut away and properly treated.

References

1. Bailey, A. 1950. The Standard Cyclopedia of Horticulture. The Macmillan Co., New York, 1: 276-7.
2. Chattopadhyay A, B. Saha, S. Pal, A. Bhattacharya and H. Sen, 2009. Quantitative and qualitative aspects of elephant foot yam. *J. Veg. Sci.*, **16**(1): 73-84.
3. Claudel, Cyrille; Sven Buerki; Lars W. Chatrou; Alexandre Antonelli; Nadir Alvarez; and Wilbert Hetterscheid, 2017. Large-scale phylogenetic analysis of *Amorphophallus* (Araceae) derived from nuclear and plastid sequences reveals new subgeneric delineation. *Bot. J. Linn. Soc.*, Oxford University Press (OUP), **184**(1): 32–45.
4. De, S., Y.N. Dey and A.K. Ghosh, 2010. Phytochemical investigation and chromatographic evaluation of the different extracts of tuber of *Amorphophallus paeoniifolius. Int. J. Pharm. Biol. Res.*, **1**: 150-157.
5. Eggli, Urs and Leonard E. Newton, 2004. Etymological Dictionary of Succulent Plant Names. Berlin, Heidelberg: Springer, p. 1 (ISBN 978-3-540-00489-9). Retrieved 14 November 2018.
6. Firdouse, S. and P. Alam, 2011. Phytochemical investigation of extract of *Amorphophallus campanulatus* tubers. *Intern. J. Phytomedicine*, **3**: 32-35.
7. Gu, Z. j., L. Wang and H. Li, 1992. Karyomorphological studies of some monocots in Dulongjiang area. *Acta Bot. Yunnan* (Suppl.), **5**: 77–90.
8. Hetterscheid, W.L.A. and S. Ittenbach, 1996. Everything you always wanted to know about *Amorphophallus*, but were afraid to stick your nose into!!!!!. *Aroideana*, **19**: 7-131.
9. Hetterscheid, W.L.A., S. Ittenbach and J. Bogner, 1999. Notes on the genus *Amorphophallus* (Araceae). 10. Revision of the endemic *Amorphophallus* species of Madagascar. *Bot. Jahrb. Syst.*, **121**(1): 1–17.
10. http: //eagri.org/eagri50/HORT281/pdf/lec26.pdf
11. *https: //www.researchgate.net/publication/318405135_The_genus_Amorphophallus_Cyto-histo-molecular_genesis_and_commercial_prospects.*
12. Ishida, G. 2004. Karyomorphological observations on some aroids cultivated in the Hiroshima Botanical Garden II. *Bull. Hiroshima Bot. Gard.*, **21**: 7–30.
13. Kuruvilla, K. M., B. Dutt and R. P. Roy, 1989. Karyomorphological investigations on aroids of North-Eastern Hills. *J. Cytol. Genet.*, **24**: 13–22.

14. Lekhak, M.M. and S.R. Yadav, 2011. Cytotaxonomy of some species of*Amorphophallus* sect. *Rhaphiophallus* (Schott) Engl. (*Araceae*) of the Indian Subcontinent. *Nucleus*, **54**(3): 169–76.

15. Li, B., J. Xia, Y. Wang and B.J. Xie, 2005. Grain-size effect on the structure and anti-obesity activity of konjac flour. *J agric. and Food Chem.*, **53**: 7404-07.

16. Liu, P.Y. 2004. Konjac. China Agriculture Press, Beijing (in Chinese).

17. Liu, P.Y., D.P. Zhang and L. Zhao, 1985. The karyotype analysis and protein study of two species of *Amorphophallus. J. South W. Agric. Univ.*, No. 4, pp. 39–43.

18. Maniyam, N. 2019. Elephant Foot Yam. Handbopok of Horticulture, vol. 1, pp. 663-6. ICAR, New Delhi.

19. Misra, R.S., M. Nedunchezhiyan and T.M.S. Swamy, 2002. *Amorphophallus paeoniifolius*; a high value cash crop for coastal areas. In: *National Conference on Coastal Agricultural Research*, ICAR Research Complex, Goa, April 6-7.

20. Misra, R.S. and S. Sriram, 2001. Medicinal value and export potential of tropical tuber crops. In: *Recent Progress in Medicinal Plants, Crop Improvement and Commercialisation*, **5**: 317-25.

21. Nataraj, H.N., R.L.N. Murthy and S. Ramachandra Setty. 2009a. *In vitro* quantification of flavonoids and phenolic content of Suran. *Int. J. Chem. Tech. Res.*,**1**(4): 36-39.

22. Nataraj, H.N., R.L.N. Murthy and S.R. Setty, 2009b. *In vitro* quantification of flavonoids and phenolic content of Suran. *Int. J. Chem.Tech. Res.*, **1**: 1063-7.

23. Nishinari, K. and M. Yoshimura, 1999. Dynamic viscoelastic study on the gelation of Konjac glucomannan with different molecular weights. *Food Hydrocolloids*, **13**: 227-33.

24. Niwa, T., H. Etoh, A. Shimizu and Y. Shimizu, 2000. Cis-N-(p-Coumaroyl) serotonin from Konnyaku, *Amorphophallus konjac. Biosci. Biotechnol. Biochem.*, **64**: 2269-71.

25. Petersen, G. 1992. Notes on the genus *Amorphophallus* (Araceae) I. Three new species from tropical Asia (Bogner, J. and W. L. A. Hetterscheid). *Blumea*, **36**: 467–75.

26. Sarker, De and W. L. A. Hetterscheid, 1997. Notes on the genus *Amorphophallus* (Araceae), 9: cytological investigation of *Amorphophallus* (*Plesmonium*) *margaritifer* (Roxb.) Kunth. *Aroideana*, **20**: 11-12.

27. Sun, Y.G. 1999. Study on the chromosome and karyotype of *Amorphophallus sinensis* Belval. *J. Anhui Norm. Univ., Nat. Sci. Ed.*, **22**(4): 329–31, 337.

28. Zemede, A. 1995. Conservation and use of traditional vegetables in Ethiopia Archived at the wayback machine. *Proc. IPGRI intern. Workshop on Gen. Res. of Trad. Veg. in Africa* (Nairobi, 29–31 August 1995).

29. Zheng, S. Q. and K. Y. Liu, 1989. Preliminary studies on chromosome band patterns and karyotypes of *Amorphophallus. J. Hunan Agric. Coll.*, **15**: 71–6.

14

Giant Taro (*Alocasia macrorrhizos* (L.) G. Don 1839

Family: Araceae.

Synonyms: *Alocasia cordifolia* (Bory) Cordem.; *A. grandis* N.E. Br.; *A. indica* (Lour.) Spach; *A. macrorrhiza* (L.) Schott; *A. macrorrhizos* var. *rubra* (Hassk.) Furtado; *A. macrorrhizos* var. *variegata* (K. Koch and C.D.Bouche) Furtado; *A. marginata* N.E.Br.; *A. metallica* Schott; *A. montana* (Roxb.) Schott; *A. pallida* K.Koch and C.D.Bouche; *A. plumbea* Van Houtte; *A. rapiformis* (Roxb.) Schott; *A. uhinkii* Engl. and K. Krause; *A. variegata* K. Koch and C.D.Bouche; ***Arum** cordifolium* Bory; *A. indicum* Lour; *A. macrorrhizon* L.; *A. montanum* Roxb; *A. mucronatum* Lam; *A. peregrinum* L.; *A. rapiforme* Roxb; ***Caladium** macrorrhizon* (L.) R.Br.; *C. metallicum* Engl.; *C. odoratum* Lodd.; ***Calla** badian* Blanco; *C. maxima* Blanco; ***Colocasia** boryi* Kunth; *C. indica* (Lour.) Kunth; *C. macrorrhizos* (L.) Schott; *C. montana* (Roxb.) Kunth; *C. mucronata* (Lam.) Kunth; *C. peregrina* (L.) Raf.; *C. rapiformis* (Roxb.) Kunth; ***Philodendron** peregrinum* (L.) Kunth; and *P. punctatum* Kunth.

Common names: Biga, Elephant's ear, Giant alocasia, Giant elephant's ears and Giant taro or Upright elephant ears. **Indian names**: Ghiya banda/Mankanda (Hindi), Maanaka, Maana, Maankandan, Kachalu/Kassalu, Hastikami and Kerukan kizhangu.

Introduction and Origin

The genus *Alocasia* got its name from *Colocasia*, another genus of the same family *Colocasia* derived its name from the Greek word *kolokasia* (lotus root). The species name is the combination of the Greek words *makros* (big) and *rhiza* (root).

Giant taro plant and rhizomes

Alocasia macrorrhizos is native to tropical rainforests of Maritime SE Asia, New Guinea, and Queensland. It may have been originated in India or Sri Lanka (Garcia *et al.*, 2008; Lebot, 2008) where it occurs in the wild, particularly in the humid Western and Eastern Ghats and in the north-east. It has become widely distributed in the SE Asia subcontinent, Malaysia, Indonesia and Polynesia, and has spread to parts of tropical America. Its seeds were probably naturally spread eastwards to Melanesia where wild forms were also found in natural habitats. The main centre of diversity of the genus is the island of Borneo where there are 23 species. It is a fast growing herb widely cultivated in tropical and subtropical regions of the world and it has the potential to become invasive (Randall, 2012; Pier, 2012) near to cultivated areas. It has escaped from gardens and cultivated lands and has been reported as naturalized in natural forests.

Giant taro is currently widely distributed and naturalized in many tropical and subtropical regions in NC and S America, the West Indies, tropical Africa and the Indo-Pacific Islands (Wagner *et al.*, 1999). It is cultivated in India, Sri Lanka, Bangladesh, Myanmar, Thailand and Peninsular Malaysia (Flach and Rumawas, 1996; Singh *et al.*, 2017) as well as in tropical America and in some parts of Africa as a minor crop (Lebot, 2008). It is grown extensively in Samoa, Tonga, the Wallis and Futuna Lau group of Fiji and parts of Vanuatu (Lebot, 2008).

The species has been introduced in many tropical and subtropical regions to be used as an ornamental, food crop and animal feed. It has been cultivated for many years now for its starchy rhizomes in the tropics as a potato-like substitute in areas where potatoes will not grow. Rhizomes and stems must first be thoroughly cooked (usually boiled) prior to eating in order to remove calcium oxalates which render uncooked rhizomes and stems inedible. Calcium oxalates can cause serious reactions if ingested raw; even the contact with the sap can cause allergic reactions. Stem pith is typically boiled, and then dried and ground into flour. Some similar plants include ~ taro (*Colocasia esculenta*), swamp taro (*Cyrtosperma merkusii*), and arrowleaf elephant's ear (*Xanthosoma sagittifolium*).

Botany and Genetics

Roots and leaf of Giant taro

Members of Araceae family are monocotyledonous flowering plants where flower merges out from the inflorescence spadix. *Alocasia macrorrhizos* is a monoecious, rhizomatous, fast-growing, evergreen herbaceous plant growing upto 4-5m tall with a spread of 2-3m under optimum growing conditions. This massive aroid has a spectacular cluster of upwardly pointing arrow-shaped leaf blades. The plant has green, thick, cordate, prominently-veined and wavy-edged leaves up to 0.9-1.8m long and 0.9-1.2m wide borne on rigid stalks (long-sheathed petioles) which emerge from a stout upright trunk that become woody with age and can eventually grow to as much as 1.8m tall. The spathe (floral bract) is about 13-35cm long, with the lower part folded into a green tube. The spadix is slightly shorter than the spathe. Flowers bloom intermittently throughout the year. The flowering parts are partly concealed by the floral bract. This bends back and then falls off at maturity though to reveal a spike of berries (fruits) that are globose or ovoid, fleshy and red when mature. Each berry contains several pale brown seeds. The berries are infrequently formed in cultivation. It contains 2n=28 chromosome numbers.

Lower 1-2cm of the spadix forms the female zone covered with about 30 pistils; each consisting of a pale green ovary with a yellow 3-5-lobed stigma sitting directly on it. Above the female zone is a zone of about four whorls of white sterile organs corresponding in position to the spathe constriction. The next floral zone is male covered with tightly packed white rhomboid to hexagonal male flowers. At the tip of the spadix is a yellowish appendix, which is at least half the spadix length (www.kew.org).

The leaves of *Alocasia* and *Xanthosoma* are usually not peltate, unlike *Colocasia*, and are held more upright. These plants generally grow from rhizomes, but certain types also produce long and slender stolons (aboveground runners). The lumpy rhizomes with rough ridges have a brown skin and a white or pink interior. Some types of elephant ears also produce smaller cormel-like lateral rhizomes (also called 'eddos') which grow off the sides of the main rhizome.

Constituents

Plant yields flavonoids, cynogenetic glycosides, ascorbic acid, gallic acid, mallic acid, oxalic acid, alocasin, amino acids, succinic acid, and β-lectins. Rhizomes contain phytosterols, alkaloids, glucose and fructose. Root tuber contains neurotoxin and sapotoxin. This plant contains alocasin ~ an antifungal and trypsin inhibitor. A new ceramide, alomacrorrhiza A was isolated from an ethanolic extract which was elucidated as (2S, 3S, 3R)-2A/-[(2'R)-2'-hydroxy-hexacosanoyl]-tetradecane-1,3,4-triol (Nguyen Quyet Tien and PhamHoang Ngoc, 2004). Investigation for chemical constituents yielded 9 compounds identified as linoleic acid, glyceryl-palmitate-sitosterol, daucosterol, gigantine A, Apigenin-8-C-β-D-glucoside, Quercetin-3-O-α-L-rhamnoside, 6-O-palmitoly glucose ester and gigantine B (ZhaoYin, 2012). Zhu *et al.* (2012-13) reported that rhizomes yield nine compounds *viz.*, glycosmisic acid, N-trans-feruloyltyramine, grossamide, protocatechuic acid, borneol acetate, vanillic acid, methyl 4-hydroxybenzoate, β-daucosterol, and β-sitosterol. The leaves when broken emit a smell of bitter almonds as also in case of stems due to the presence of prussic acid (also known as hydrocyanic acid or cyanide) which in leaf is about 30 mg of HCN/kg though in stems quite less. Prussic acid is a rapidly acting lethal toxin.

Toxicity in edible aroids including taro (*Colocasia esculenta*), tannia (*Xanthosoma violaceum*), giant taro (*Alocasia macrorrhizos*), giant swamp taro (*Cyrtosperma chamissonis*) and elephant foot yam (*Amorphophallus paeonifolius*) has been shown to contain varying amounts of cyanide in their leaves and stems using a screening assay and a more accurate acid hydrolysis method. The cyanogenic glucoside triglochinin (previously identified in giant taro) was found to be less stable to acid hydrolysis (requiring only 15 minutes) compared with 50 minute hydrolysis for linamarin and lotaustralian (from *Cassava*) under the same experimental conditions. The amounts of cyanide present in the leaves (0-3mg HCN/100g fresh weight) and in the stems (0—0.3mg HCN/100g fresh weight) of *Colocasia* and *Xanthosoma* are only about 1-5 per cent of that found in cassava leaves and tubers and are not a cause for concern for human nutrition. *Cyrtosperma* stems and *Amorphophallus* leaves contain only small amounts of cyanide. In contrast, all *Alocasia* varieties contain much more cyanide in leaves (2-30mg HCN/100g fresh weight) and in stems (0.5-4mg HCN/100g fresh weight) (Bradbury *et al.*, 1995).

Like most other members of the family, the plant contains an irritant which causes intense discomfort to the lips, mouth and throat. This acridity is caused in part by microscopic needle-like raphides of calcium oxalate monohydrate. Calcium oxalates can cause serious reactions if ingested raw; even the contact with the sap can cause allergic reactions. Calcium oxalate is broken down by thoroughly cooking the plant or by fully drying it, rendering it safe to eat. People with a tendency to rheumatism, arthritis, gout, kidney stones and hyperacidity should take special precaution if including this plant in their diet. Sapotoxin is also present, and the toxic effects include gastroenteritis and paralysis of the nerve centres.

Nutritional Value

Nutritional value of raw giant taro rhizomes (*Source*: https://www.healthbenefitstimes.com/giant-taro):

Serving Size: 100 g; **Calories** 100 Kcal. **Calories from Fat** 0.9 Kcal.

Proximity	*Amount*	*Per cent DV*
Water	70 g	N/D
Energy	100 Kcal	N/D
Energy	412 kJ	N/D
Protein	2.2 g	4.40 per cent
Total Fat (lipid)	0.1 g	0.29 per cent
Carbohydrate	23 g	17.69 per cent
Total dietary Fiber	1.9 g	5.00 per cent
Minerals	***Amount***	***Per cent DV***
Calcium, Ca	38 mg	3.80 per cent
Iron, Fe	0.8 mg	10.00 per cent
Magnesium, Mg	52 mg	12.38 per cent
Potassium, K	267 mg	5.68 per cent
Sodium, Na	30 mg	2.00 per cent
Zinc, Zn	1.6 mg	14.55 per cent

Vitamins	*Amount*	*Per cent DV*
Water soluble Vitamins		
Vitamin B1 (Thiamin)	0.02 mg	1.67 per cent
Vitamin B2 (Riboflavin)	0.02 mg	1.54 per cent
Vitamin C (Ascorbic acid)	17 mg	18.89 per cent
Fat soluble Vitamins		
Vitamin E (alpha-tocopherol)	2 mg	13.33 per cent

Source: https://www.healthbenefitstimes.com/giant-taro.

*Above mentioned 'per cent daily values' (per cent DVs) are based on 2,000 calorie diet intake. Daily values (DVs) as per USDA (and not as per the healthbenefitstimes.com recommendations) may be different depending upon daily calorie needs of the body. Calculations are based on average age of 19 to 50 years and 71.78kg weight.

Uses

The rhizomes and stems are edible but must be first thoroughly cooked (usually boiled, baked or roasted) prior to eating in order to remove calcium oxalates which render uncooked rhizomes and stems inedible. In SE Asia, the rhizomes are chopped into pieces and consumed as a vegetable after being cooked in curries or stews. The rhizomes and leaves are also cooked as food. Giant taro is used as a source of starch or flour.

Leaf of the plant is considered astringent, styptic and antitumor. The rootstock is considered laxative and diuretic. It is experimentally considered antimicrobial, antifungal, antioxidant, hepatoprotective, antidiarrheal, antiprotozoal and anticancer. The plant has been reported to show antioxidant, antifungal, antiviral, neurotoxic, anticancer, antidiabetic, diuretic, hepatoprotective, larvicidal, antimicrobial and anthelmintic properties.

Boiled stems of *A. macrorrhizos* are used as a laxative, chopped-up roots and leaves as a rubefacient, and juice from the petiole against a cough. The plants are applied for stimulating the skin, *e.g.* in cases of fever and to remove blotches. The rhizome is sometimes used as a poultice to treat furuncles. The pounded stems are applied as a paste to snake-bites and scorpion stings. In India, many people have been using rhizome paste to treat wounds, cure heel cracks and kill worms in domestic animals. In India, the Zeliang tribe of Nagaland applies exuded liquid from rhizomes and leaves mixed with cow or buffalo's milk on snake bite to remove poison (Singh *et al.*, 2015).

The poisonous substances can be removed by repeated cooking, but the rhizomes and bases of petioles of *A. macrorrhizos* which are sometimes used for food usually contain few poisonous substances. A lectin has been isolated from the rhizome of *A. macrorrhizos* which showed potent mitogenic activity on human peripheral blood lymphocytes in the [3H]-thymidine uptake assay. This species also contains a protein which inhibits both the enzymes trypsin and chymotrypsin. The seed extract showed antifungal activity. There is also a report on HIV-1 protease inhibitory activity (Hitchon and El-Gabalawy, 2004).

Alocasia macrorrhizos is a popular ornamental plant grown for its large foliage and striking aroid inflorescences. It can also be used in sewerage treatment as it grows rapidly in wetland conditions and has a propensity to accumulate metal contaminants such as zinc.

Propagation

Giant taro can reproduce sexually by seeds as well as vegetatively by basal offshoots (*aka* suckers or rhizomes), and can grow in a great variety of substrates and habitats ranging from full sun to deep shaded areas (Flach and Rumawas, 1996). Propagation of alocasia should be done in spring or early summer when the plant is just coming out of dormancy from the winter. Seeds are best sown as soon as they are ripe and germinate best at 24°C.

Vegetatively the alocasia plants can be propagated through rhizome or tuber cuttings (cut setts). Other methods of propagation include: stem cuttings of mature plants having two or three growth points or nodes, by division of the rootstock as the plant is coming into growth, or by new off-sets that appear at the base of the mature plant.

The planting should be done in partial shade and damp and fertile loamy soil. Planting holes must be 15-25cm deep for suckers or 8-15cm deep for offsets (lateral rhizomes). The interplant distance should be at least 91cm. *Alocasia* is commonly an intercrop with yams, and the spacing is usually 3.5×3.5m. In pure stand 60×60cm to 1.5×1.5m distance is maintained. The new plant prefers plenty of water. If alocasia plants are being propagated in containers, growers must make sure to keep them somewhere in warm and humid areas and by giving them plenty of water.

To propagate *Alocasia* through rhizome cutting, the plant should be mature with a long rhizome having multiple growth points. A mature plant may have multiple rhizomes shooting multiple leaves. The rhizomes should be cut horizontally; each cut sett should include nodes. Each sett should be treated with baking soda, ash powder or an anti-fungal powder. Although these setts can be directly planted in the fields, it would be better to plant them in the nursery bed or in a pot. New leaves will appear in 3 to 4 weeks, and once they have enough roots they can be planted in the main field or in separate pots.

Growth and Development

The giant taro is not very difficult to grow provided minimum favourable conditions are met. Although environmental factors such as temperature, moisture, light and soil type influence the growth and quality, other factors such as varieties, quality of planting material, planter operation, irrigation, nutrition, pest management, planting date and harvest date, *etc.* are equally responsible and can be controlled by the grower. Also a good ground preparation, harrowing and ploughing, and planting method are always needed. The most profitable yields can be attained only when all factors are at optimum levels.

Alocasia macrorrhizos prefers humid tropical and subtropical climates with temperatures ranging from 25-35°C and more than 1,700mm of evenly-distributed rainfall at 600–800m elevations (Smith, 1979; Wagner *et al.,* 1999). It is the most drought-tolerant of the edible aroids, though not tolerating a long period of drought. It can tolerate temperatures down to 10°C though even lower temperature is harmful. Freezing temperatures damage the leaves, but new ones sprout readily (Kay, 1987). The species can be found growing in a wide variety of soil types ranging from freely drained sandy soils to deep, well drained clayey soils. Frequently it is found growing naturally along river banks and can tolerate shallow flooding but cannot stand waterlogging. It prefers a slightly acidic pH, *i.e.* 5.5-6.5. *A. macrorrhizos* has the capability to grow in habitats ranging from full sunlight to deep shade and also tolerates up to 4 months of drought (Flach and Rumawas, 1996). It can be found growing in limestone rocky soils with low water-holding capacity and holes in the exposed limestone substrate (Kay, 1987). It also grows in gaps in the forest understorey, margins of wet fields, alongside streams or in roadside ditches. It is moderately salt tolerant and occurs on many South Pacific atolls (Lim, 2015).

Indoor plants which may not get enough light can be augmented by fluorescent lighting from overhead. Dappled sunlight or light shade will often benefit outdoor plants in more arid or sunny climates. Being a native to tropical rainforests, high humidity is important. When growing indoors, the plant should be augmented with a pebble tray or humidifier. Grouping with other plants will also help keeping the ambient humidity up.

Land Preparation, Planting Method and Maintenance

Land should be prepared for planting by ploughing, harrowing and adding organic fertilizers to make it fertile. After 1-2 ploughings to a depth of 20-30cm, pits some 15-25cm deep for suckers or 8-15cm deep for offset rhizomes are made at a spacing of 90×90 cm during spring or late spring. *Alocasia* rhizomes should be planted with the narrow-end pointing up and the root zone facing down. The new plant prefers plenty of water.

Applying a half-strength 20-20-20 (NPK) fertilizer every three weeks during their active growth period promotes the exceptional growth each season.

For consistent moisture during growing season, it requires regular watering especially when there is less or no rain. During winter its water requirement is almost nil. It does not tolerate soggy soil. During warmer months, alocasias are watered thoroughly, but during winter months, however, when the temperature is cooler and there is less light, these plants will require significantly quite less water. The top soil should be allowed to get dry before next watering.

Plants are **mulched** to a depth of 5cm in the beds and 2.5cm in pots. **Earthing up** of the soil also brings about weed control besides improving the physical condition of the soil. Regular weedings and earthing up should be performed to keep the soil loose and to destroy the weeds. The foliage should be removed when the leaves turn yellow and die back as the plant slips into dormancy. Not all alocasias go dormant; some varieties are evergreen depending on the location. Alocasia is commonly an **intercrop** with yams, and then spacing is usually 3.5×3.5m. In pure stand, 60×60 cm to 1.5×1.5m is used.

Harvest and Storage

Mechanical removal of *A. macrorrhizos* is effective but is labour intensive. All rhizomes and lateral offsets should be removed to prevent spread into new areas. Dense stands should be removed using specialized machinery. For cultivated *A. macrorrhizos*, the crop life is usually 12–18 months, but harvesting can be delayed for up to 4 years. Plants can live for several years and flowering occurs sometime during the second year of growth (Manner, 2011). If the crop is being grown in a region colder than international zone 8, it is harvested after the first frost. The dirt is removed and rhizomes are allowed to dry in the shade for a day or two and then stored in a cool and dry place in barely moist peat moss to replant in the spring.

Pests and Diseases

Alocasia macrorrhizos is resistant to most pests and diseases affecting the aroid family (Kay, 1987). For islands in the Pacific where this species is commonly cultivated, pests include the black cutworm (*Agrotis ipsilon*), taro sphinx moth (*Hippotion celerio*), and the cluster caterpillar (*Spodoptera litura*). Diseases caused by *Cladosporium colocasiae* and *Mycosphaerella colocasiae* are reported as minor pathogens (Manner, 2011). This plant is also susceptible to *Botrytis*, multiple forms of rot, leaf spot, powdery mildew, rust, southern blight, and *Xanthomonas*

Agrotis ipsilon and its larva

but not of much economical importance. A few insect pests of the *Alocasia* include spider mites, mealybugs, aphids, and scale. These buggers can be found on both sides of the leaves, also on the stems or in between leaf sheaths near the crown. They can be wiped away with a damp cloth and/or sprayed with mild pest spray. The cracks and creases must also be checked carefully.

Hippotion celerio Caterpillars of *Spodoptera litura*

Phyllosticta leaf spot or giant taro leaf spot (*Mycosphaerella alocasiae*; the asexual name is *Passalora colocasiae*) is a fungal infection that takes place when the leaves are young but the spots appear only when the leaves become mature. They start as small round flecks, sunken and grey with a raised brown border, increasing up to 8mm in diameter, but much smaller when the leaf is heavily infected. Sometimes they have a yellow halo. Small black dots are present in the middle of the grey centres, especially on the upper surface of the leaves. These are the fruiting bodies containing large numbers of spores. Though the leaves with severe infections die prematurely, it does not affect the crop yield to any great extent. It can be identified as small round spots with grey centres but inside these in the centre with black dots, and brown borders on the oldest leaves.

Giant taro leaf spot

The most common elephant ear plant disease is fungal leaf blight (*Phytophthora colocasiae*). It produces tiny round lesions on the ornamental leaves that may ooze fluid and turn purple or yellow when dry. When the fungus is in full bloom, there is also fuzzy growth. Over time the entire leaf collapses on itself and the disease travels down the rhizome. Pythium rot (*Pythium myriotylum*) can cause plants to die. It is most common in areas with too much water and humidity.

Crown rot spreads by a soil-borne fungus. It is another disease of concern and is supported by a heavy and water-logged soil mix. Yellowish leaf margins or dark spots on the leaves are major symptoms of disease. As the rot spreads, the parts near the soil surface start rotting. The tissue in the infected areas will die causing them to turn dark brown. Younger plants have difficulty dealing with crown rot but even the well-established mature plants will start wilting as the rot progresses. It can be prevented to some extent by avoiding overwatering and providing air circulation around the plant. The leaves should not remain wet for too long. Fungicide can be used for early infections.

References

1. Bradbury J. H., S.V. Egan and P.J. Matthews, 1995. *Phytochemical Analysis*, **6**(5): 268–71.
2. Flach M. and F. Rumawas, 1996. Plant Resources of South-East Asia. Bull. No. 9. Plants yielding non-seed carbohydrates. Backhuys Publishers, Leiden, Netherlands:; 237 pp
3. Garcia, J. Q., A. Ivancic and V. Lebot, 2008. Morphological variation and reproductive characteristics of wild giant taro (*Alocasia macrorrhizos*, Araceae) populations in Vanuatu. *N. Z. J. Bot.*, **46**(2): 189-203. http: //www.rsnz.org/publish/abstracts.php.
4. Hitchon, C.A. and H.S. El-Gabalawy, 2004. Oxidation in rheumatoid arthritis. *Arthritis Res Ther.*, **6**(6): 265–78.
5. https: //www.healthbenefitstimes.com/giant-taro
6. http: //www.kew.org/science-conservation/plants-fungi/alocasia-macrorrhizos-elephant-ear-taro.
7. Kay, D.E., 1987. Root Crops. Tropical Development and Research Institute, London, UK:
8. Lebot, V., 2008. *Tropical Root and Tuber Crops: Cassava, Sweet potato, Yams and Aroids* [eds Atherton, J., and A. Rees]. Wallingford, UK: CABI.xv-xix + 404 pp. http: //www.cabi.org/cabebooks/ebook/20093020206.

9. Lim, T. K., 2015. Edible Medicinal and Non-medicinal Plants: volume 9, Modified Stems, Roots, Bulbs, 1036 p, Springer, Dordrecht, Netherlands.

10. Manner, H.I., 2011. Farm and Forestry Production and Marketing Profile for Giant Taro (*Alocasia macrorrhiza*). Specialty Crops for Pacific Island Agroforestry [ed.Elevitch, C. R.]. Holualoa, Hawaii, USA: Permanent Agriculture Resources (PAR). http: //agroforestry.net/scps.

11. Nguyen Quyet Tien and PhamHoang Ngoc, 2004. New Ceramide from *Alocasia macrorrhiza. Arch Pharm Res.*, **27** (10L): 1020-22.

12. Pier, 2012. Pacific Islands Ecosystems at Risk. Honolulu, USA (HEAR). University of Hawaii. http: //www.hear.org/pier/index.html.

13. Randall, R.P. 2012. A Global Compendium of Weeds. Perth, Australia, Department of Agriculture and Food, Western Australia, 1124 pp. http: // www.cabi.org/isc/FullTextPDF/2013/20133109119.pdf.

14. Singh, Premkumar, P.R. Gajurel and P. Rethy, 2015. Ethnomedicinal value of traditional food plants used by the Zeliang tribe of Nagaland. *Indian J. tradition. Knowl.*, **14**(2): 298-305.

15. Singh, S. K., J.R. Patel, A. Dangi, D. Bachle and R.K. Katariya, 2017. A review paper on *Alocasia macrorrhiza*: traditional Indian medicinal plant. *European J. Pharmacy and medic. Res.*, **4**(2), 366-75.

16. Smith, A.C. 1979. Flora Vitiensis *nova*: A new flora of Fiji (Vol. I). National Tropical Botanical Garden, Lawai, Kauai, Hawaii, USA, 494 pp.

17. Wagner, W.L., D.R. Herbst and S.H. Sohmer, 1999. Manual of the Flowering Plants of Hawaii (rev. ed.). Honolulu, Hawaii, USA: University of Hawaii Press/Bishop Museum Press; 1919 pp.

18. ZhaoYin, 2012. Studies on the Quality Analysis and Chemical Constituents of Crude and Processed Products of *Alocasia macrorrhiza* (L.) Schott. Master Thesis, 2012, China.

19. Zhu Ling-hua, Xiao-sheng Huan, Y.E. Wen-cai and Guang-xiong Zhou, 2012-13. Study on Chemical Constituents of *Alocasia macrorrhiza* (L.) Schott. *Chinese pharmaceut J.*, 2012-13.

15

Guinea Arrowroot (*Goeppertia allouia* Lindl.)

Family: Marantaceae.

Synonyms: *Allouya americana* (Lam.) A.Chev., ***Calathea allouia*** (Aubl.) Lindl., *Curcuma americana* Lam., *Maranta allouia* Aubl., *M. niveiflora* A.Dietr., *M. semperflorens* Horan., *Phrynium allouia* (Aubl.) Roscoe, and *Phyllodes allouia* (Aubl.) Kuntze.

Common names: Alleluia, Allouya, Aria and Guinea arrowroot. **Other names**: Leren or Lairen (South America), Sweet corn root (Caribbean), Topi-tambu or topinambour (Spanish) and Topi-tambo or tambu (West Indies).

Introduction and Origin

Genus *Goeppertia* was named in honour of Heinrich Goeppert (1800–1884), a German botanist and paleontologist (Burkhardt, 2018). The name *allouia* is derived from the Carib name for the plant (Topi Tambo, 2016). *Goeppertia allouia*, native to northern South America and the Caribbean, was one of the earliest plants domesticated by pre-historic Amerindians in South America. Although it is traditionally cultivated only on a small scale and that too only by some traditional growers and indigenous populations, guinea arrowroot can be found practically throughout the Amazon region. It is distributed through Puerto Rico, the Antilles and countries situated in northern South America (the Guyanas, Venezuela, Colombia, Ecuador, Peru and Brazil) and these are the countries where it is assumed to have originated (Hernando and León, 1994). There are records of Guinea arrowroot's introduction into India, Sri Lanka, Malaysia, Indonesia and the Philippines.

By Kalamazadkhan - Own work, CC BY-SA 3.0, https://commons.wikimedia.org/w/index.php?curid=20671557

By Johabb - Own work, CC BY-SA 3.0, https://commons.wikimedia.org/w/index.php?curid=16845234

Its cultivation has been limited due to this being intolerant to both the conditions ~ drought and waterlogging, *vis-à-vis* introduction of certain other more economical crops. Moreover, its starch presents some physicochemical properties that are not interesting for the food industry such as low solubility and high retrogradation when compared to corn and potato starches. This has been introduced as a minor root crop in tropical regions around the world (Martin and Cabanillas, 1976).

Botany

It is a tropical perennial rhizomatus herb growing to a height of about 1.5m with a fleshy rootstock bearing erect to almost erect leaves with long and grooved petioles *cum* elongated and oval blades. The leaves are large up to 60cm long and 20cm wide. Old plants have a pseudostem consisting of a 10-30cm long stalk bearing a few greenish to yellow or white flowers which rarely set seeds. At the end of fibrous roots, it produces 2-8cm long with 2-3.5cm diameter egg-shaped rhizomes which resemble small potatoes. These are covered with a thin yellowish-grey parchment-like skin. Usually this reproduces itself through rhizomes which produce shoots and new plants.

Nutritional Values

An analysis of the edible portion of the mucilaginous tuberous roots reveals this containing energy 395 kJ/100g; water 75.7 per cent; protein 1.5 per cent; fat 0.3 per cent and carbohydrate 21.4 per cent with about 70 per cent being starch. A carbohydrate similar to laevulose is present.

Uses

This has high starch content and is consumed only after cooking. Young tender flower spikes are also edible but only after cooking. Initial raw flavour of the tuber gets neutralized after about 15 minutes of boiling. It retains its crispness even after being cooked and its flavour resembles somewhat similar to sweet corn though with slightly bitter but pleasant aftertaste. Longer cooking, *i.e.* up to about 1 hour makes the texture more floury similar to that of potato. Cooked rhizome is covered with a thin edible skin which is most easily peeled after cooking. Mostly it is eaten as 'hors d'oeuvre' or appetizer.

Indigenous people of the Americas have used the durable leaves to make traditional medicines and as baby clothing. A tincture of the leaves is reported to be used in traditional South American medicine for the treatment of cystitis and as a diuretic.

Propagation

The species is propagated by short sections of the rhizome each having an upright terminal bud. These are obtained by breaking up the clump of rhizomes that forms at the basal portion of an old plant. After harvest the undivided rhizomes are normally stored in a cool dry place until required for planting. Sprouting is often unpredictable but immersing the sections in water for 10 minutes at 48°C gives satisfactory sprouting percentage.

The rhizomes may also be left in the ground until the approach of the planting season but intense shoot competition arising under such conditions leads to poor yields unless they are separated. The yield from a single replanted offshoot having large tuber is often greater than that of a whole clump.

Growth and Development

Depending on the climatic conditions, the growth cycle of this species is between 9 and 14 months. Rhizome yields of 2-15t/ha have been reported. Water shortages can reduce the plant's cycle at the cost of rhizome production.

It is a plant mainly of the moist lowland tropics where it is found at elevations up to 1,400m. It is adapted to a tropical climate of alternating wet and dry seasons with the dry season occurring during the short day season of the tropical winter. A moderate annual rainfall of 1,500-2,000mm is required for its successful growing though can tolerate 700-2,800mm. It grows best in areas where annual daytime temperatures occur 22–31°C though can tolerate 16-35°C.

It can grow in the open or in semi-shaded conditions (light woodland). Shade may facilitate the growth of the plants but best growth is achieved only under full sunlight exposure when the humidity, nutrients and soil drainage are not the limiting factors. It is planted before start of rainy season.

Soil

It prefers a moist but well-drained humus-rich fertile soil having medium texture with a pH range of 5.5-7.5, optimum being 6-7. Quite clayey soils impair the development of rhizomes while in more sandy soils its growth remains deficient (Hernando and Leon, 1994). The right soil for its growing is clayey-loam which retains nutrients and allows good drainage though organic matter in the form of farmyard manure or compost also needs to be added (Hernando and Leon, 1994) at the rate of 200-250 quintals per hectare at the time of field preparation.

The field should be ploughed deeply or dug thoroughly and mixed with compost. All the weed rootstocks, and pieces of bricks, stones, wood chips, shredded polythene pieces, *etc.* should be taken out thoroughly with forks. Then it is again ploughed with repeating all the above operations, followed by planking. At this time soil should have sufficient moisture otherwise the field is watered shallowly. When soil matures in case it has been watered, once more soil should be dug up, then planked and lay out is made, keeping the bed size as per convenience and bunds and channels in the manner that after planting there may not be any hinderance in walking, field operations and irrigation. In regions with regular and heavy rainfall planting should be on ridges but the regions experiencing moderate rainfall the planting is carried out on the flat beds. If rainfall is likely to be limited and the soil has poor water-holding capacity planting should be in small pits to which organic matter should be added before planting. Weeding in the early stages of growth is important.

Recommended planting distances are about 40cm between the plants on ridges spaced 80-100cm apart which will accommodate 25,000-31,000 plants/ha or on the flat beds at 40-80cm apart (16,000-62,000 plants/ha). Usually in ordinary cases the planting distance of 1m between rows and 50cm between the plants are kept. However, more people prefer denser plantings (Hernando and Leon, 1994). These are usually interplanted with cassava, plantain or certain fruit trees. In Puerto Rico, it is planted with coffee plantation as total or partial shade is important for goot vegetative growth.

Irrigation

Drought affects the rhizome yield adversely. Low yield occurs mainly due to drought at the end of the rainy season. At a time when there had been no rain but crop requires watering, it should immediately be performed for optimum rhizome yield. Hernando and Leon (1994) stated that in Manaus, rhizome yield to a tune of 15t/ha with more uniform product can easily be obtained by irrigating the plantation during critical periods.

Harvest

If all the requirements are met with properly a good crop of rhizomes, *i.e.* up to 12t/ha is obtained in 9-14 months after planting. Small rhizomes are usually dug up by hand with a fork. Low yields are associated with drought conditions occurring usually towards the end of the wet season so under such cases irrigation is necessary for maximizing the yield. These harvested rhizomes are first dried for about one week in a well-aerated room and then these can be stored for about three months under room temperature as these can not tolerate cold storage well.

Pests and Diseases

After planting, guinea arrowroot is not so demanding. In areas infested with phytoparasitic nematodes, this plant shows no symptom of pest attack. It is antagonistic to the gall nematode, *Meloidogyne incognita* because of its root secretions which impair the larvae's hatching, penetration and reproduction (Hernando and Leon, 1994).

The underground part may be subject to sporadic insect attack. The insect *Calopodes ethlius* is reported to attack the rhizomes. The larvae of *Coleoptera* and *Lepidoptera* cause lesions on the rhizomes. Lindane or primiphos methyl may be used against these insects in mid-summer. Fungal or bacterial rotting of rhizomes occurs when plants suffer excessive flooding, therefore care should be taken that there is no water-logging in any case in the field. Before planting, the rhizomes should be dipped in Bordeaux mixture or copper oxychloride as a precautionary measure against any serious rhizome problem. Mites infest the leaves of this plant and its serious infestation causes the plants to die (Hernando and Leon, 1994). To control the mites, the plants are sprayed thrice at weekly intervals thoroughly with bifenthrin, primiphos methyl or fenitrothion.

References

1. Burkhardt, Lotte, 2018. Verzeichnis eponymischer Pflanzennamen – Erweiterte Edition [Index of Eponymic Plant Names – Extended Edition] (pdf) (in German). Berlin: Botanic Garden and Botanical Museum, Freie Universität Berlin.

2. Hernando Bormejo, J.E. and J. León (eds), 1994. Plant Production and Protection (Series No. 26), pp. 239-44. FAO, Rome, Italy.
3. Martin, F.W. and E. Cabanillas, 1976. Leren (*Calathea allouia*), a little known tuberous root crop of the Caribbean. *Econ. Bot.*, **30**(3): 249-56.
4. Topi Tambo, 2016. Leren, Guinea Arrowroot, Eat the Weeds, [1], accessed 17 Feb 2016.

16

Indian Arrowroot [*Tacca leontopetaloides* (L.) Kuntze, 1891]

Family: Dioscoreaceae.

Synonyms: *Chaitaea tacca* Sol. ex Seem., *Chaitea tacca* Solander ex Parkinson, *Leontice leontopetaloides* L., *Tacca abyssinica* Hochst. ex Baker, *T. artocarpifolia* Seem., *T. brownii* Seem., *T. dubia* Schult. and Schult.f., *T. gaogao* Blanco, *T. guineensis* G. Don ex Loudon, *T. hawaiiensis* H.Limpr., *T. involucrata* Schumach. and Thonn., *T. maculata* Zipp. ex Span. nom. inval., *T. madagascariensis* (H.Limpr.) H.Limpr., *T. madagascariensis* Bojer, *T. oceanica* Seem., *T. phallifera* Schult. and Schult.f., *T. pinnatifida* J.R.Forst. and G.Forst., *T. pinnatifolia* Gaertn., *T. quanzensis* Welw., *T. umbrarum* Jum. and H.Perrier, *T. viridis* Hemsl.

Other common names: East Indies arrowroot, Fiji arrowroot, Indian arrowroot, Pia, Polynesian arrowroot, Tacca, Tahiti arrowroot and William's arrowroot. **Indian names**: **Hindi**: Bagh-moochh and Devkanda; **Sanskrit**: Devakanda and Surna; **Asomese**: Keturi; **Kannada**: Daeva kaanda, Handigedde, Suranagadde; **Konkani**: Devakamdo, Kolyacho mad, Sardecho maad, Surnu; **Malayalam**: Kattuchena; **Marathi**: Devakanda, Jatashankar, Penghagara, Towkiracha jhaad; **Odia**: Dhoi; **Tamil**: Cenai, Kakanam, Kara-k-karunai, Kattu-k-karuna; **Telugu**: Adavi dumpa, Kanda, Pedda kanda gadda, Ritta kanda; and in **Tulu**: Kattukene.

By Forest & Kim Starr, CC BY 3.0, https://commons.wikimedia.org/w/index.php?curid=6170235

Introduction and Origin

Generic name is from New Latin *tacca*, from Malay *takah* (meaning 'notch'). The specific epithet *leontopetaloides* is a Latin word that probably refers to 'petals similar to that of a lion'. *Tacca leontopetaloides*, native to SE Asia, was introduced throughout the Indo-Pacific tropics during prehistoric times and is naturally distributed from western Africa through SE Asia to northern Australia and Oceania (Anon., 2009). Generally it is found growing in open semi-evergreen forests, and is widely naturalized throughout the tropical Pacific as an escape from cultivation. Sometimes it can also be found in monsoon forest as well as in or on the margins of lowland rain forests.

The East Indian arrowroot is a wild perennial herbaceous and underutilized tuber crop. It was at one time widely cultivated, especially in the Pacific Islands for its very starchy edible tubers, the starch was extracted and sold in quantity to Europe where it was used in bread-making. In the past, the tubers were a staple foodstuff in Polynesia, and in the 19th century this was used as a source of starch similar to that of more familiar arrowroot (*Maranta arundinacea*), often given in the treatment of dysentery and for feeding infants. Though its importance has declined, this widespread species is still cultivated for local use in several areas. The plant has various local medicinal uses and is also grown as an ornamental for its handsome foliage and extraordinarily beautiful blooms.

Botany

Tacca leontopetaloides is a perennial herbaceous plant with tuberous rootstock. The true stem consists of an underground tuber producing 60-90cm long petioles bearing deeply lobed leaf blades, consisting of three main segments, each further divided in a pinnate manner; and the blades are about 30cm across (Humanity development library 2.0). The leaves are palmately incised and/or divided into 3-13 lobes, each lobe pinnately divided into numerous smaller ones (Drenth, 1972). The leaf's upper surface has depressed veins and the under-surface is shiny with bold yellow veins. Greenish purple flowers with long trailing whisker-like bracts are borne in clusters on top of tall stalks that also arise from the basal tuber. The fruits are mostly round berries, 3.5×1.5–2.5cm, pale orange when mature, and containing many-ribbed seeds that are yellowish-brown, flat, round and 3–5mm wide. The seeds are also covered by a white and spongy covering.

Inflorescence (umbel)

Flower with 6 tepals, 6 stamens and 3-lobed stigma

By - Ton Rulkens from Mozambique

The plant usually dies down to the ground and remains dormant for part of the year, *i.e.* 8-10 weeks in warm, almost dry conditions but on completion of a new tuber. When favourable conditions arise new leaves start emerging from round underground tubers. There are two distinct types from the Pacific Islands ~ one producing a single large tuber, the other with a number of smaller (potato-sized) tubers. The tubers are usually 10-15cm in diameter but they can reach 30cm on rich soils. These normally weigh 70 to 340g but can reach up to 1kg (Kay, 1987). These potato-like tubers have eyes, a pale-yellow skin, dull-whitish flesh and are usually bitter and almost inedible when raw. It has chromosome no. 2n=30.

Nutritional Value

Tacca leontopetaloides contains various nutritional components such as vitamins B and C; phytate, oxalate, saponins, calcium, copper, iron, magnesium, manganese, phosphorus, potassium, sodium, sulphur and zinc. Tubers contain 20-30 per cent of starch which can be easily extracted in a pure state. An analysis of the Tahitian tuber has shown 2.5 per cent skin, 60.59 per cent water, 30.6 per cent starch and 6.3 per cent fibre. African tuber analysis on a dry weight basis showed protein 5.1 per cent, ether extract 0.2 per cent, carbohydrate 89.4 per cent, cellulose 2.1 per cent, fibre 8.8 per cent, ash 3.2 per cent, calcium 0.27 per cent and phosphorus 0.2 per cent (Anon., 2016). Principal amino acids present in the protein are arginine, glutamic and aspartic acids, leucine, lysine and valine (Humanity development library 2.0). Starch obtained from the tubers is very white and similar in many respects to that of cassava or arrowroot ~ the grains being simple polyhedrons or semi-hemispheres with diameters ranging from 8 to 40microns, average being 20microns. In addition, the tubers contain about 2.2 per cent of a bitter extract, and a bitter principle taccalin has been isolated from the dried tubers (Humanity development library 2.0). An analysis of tubers collected in the wild from Angwan Zam and Ntuer in Shendam Local Government Area (L.G.A) of Plateau State (Nigeria) revealed that 100g of *Tacca leontopetaloides* tuber contains tannins 2.5mg, phylate 49.77mg, oxalate 15.51mg, cyanamide 0.18mg, alkaloids 42.9mg, saponins 14.67mg, flavonoids 1.46mg, Na 34.71mg, K 40.18mg, Ca 0.25mg, Mg 1.4mg, Fe 1.37mg, Zn 1.64mg, Mn 0.72mg, Cu 0.68mg, P 0.06mg, moisture in tuber starch 8.66 per cent, crude protein 6.69 per cent, crude fibre 5.44 per cent, crude fat 0.51 per cent, ash 0.41 per cent and total carbohydrate 78.19 per cent (https://www.researchgate.net; Ogbonna *et al.*, 2017).

Good washing is essential because of the presence of the bitter substance (taccalin) which may be poisonous in large quantity. The anti-nutrient elements present are tannins, phytate, oxalate, cyanide, alkaloids, saponins and flavonoids. It was observed that steeping in water for sometimes had effects on the concentration of the anti-nutritional components (Ogbonna *et al.*, 2017).

A bitter extract (about 2.2 per cent) was also isolated from the tuber, containing β-sitosterol, cerylic alcohol, taccalin (a rather unusual, bitter principle in plants), alkaloids and steroidal sapogenins. Small young tubers are said to be bitterer than large older ones. The sapogenins have shown a strong activity against snails (*Bulinus truncatus* and *Biomphalaria pfeifferi*) (Anon., 2016).

Culinary Uses

Although fresh tubers are inedible because of the presence of bitter toxic substances, they can be eaten after thorough preparation or more important an edible starch can be extracted from them. Starch is similar to that of more familiar arrowroot (*Maranta arundinacea*) and can be used for baking bread, making pastes and puddings, often mixed with other ingredients such as sugar, coconut juice, fruit pulp, wheat flour and flavourings. Cooked starch is easily digestible and suitable for patients with digestive problems (Anon., 2016). In Tahiti, it is used to make 'poi' ('poke' in the Cook Islands), a traditional food which consists of a mixture of fruit pulp and starch, flavoured with vanilla and lemon and cooked in an oven. Wild plants are regarded as a famine food in parts of West Africa (Humanity development library 2.0).

Medicinal

In traditional Hawaiian medicine the raw tubers were eaten to treat stomach ailments. Mixed with water and red clay, the plant was consumed to treat diarrhoea and dysentery. This combination was also used to stop internal haemorrhaging in the stomach and colon and applied to wounds to stop bleeding.

Starch

The tubers are peeled and grated and then resultant pulp is washed in water several times with final washing through a sieve or cloth. The aqueous starch solution is collected and the starch grains are allowed to settle out, and then collected and dried under sun. Starch extracted from the tubers is utter white which resembles as cassava starch. Its grains are simple polyhedrons or hemispheres and (8-)20(-40) µm in diameter [*Tacca leontopetaloides* (PROSEA), 2016].

Other Uses

The plant is cultivated as an ornamental herb for its leaves and attractive flowering shoot. The stems are used to manufacture hats as they make good braiding material. In 19th century the leaves were marketed in Europe for manufacture of hats. The leaf stalks and floral scapes make an excellent straw which can be used as a plaiting material. On some islands, the stem's bast fibres are woven into mats. The fresh starch extracted from the roots is used as a starch for clothes and as glue.

In Nigeria, water in which tuber gratings have been washed is used as a detergent. Tubers are potential starting point for making alcohol. Also, the plant is used in traditional worship and sacrifices. In India, it is used against dysentery and the tubers against piles.

Propagation

The plant can be propagated by seed or tubers. Seeds are generally formed about 3 years after planting which are usually surface-sown at 22-27°C temperatures. In general, it is usually propagated through division of small tubers formed at the base of the plant which often remains in the soil when the larger ones are harvested.

Growth and Development

In Hawaii, the leaves appear in early spring and the crop is mature by the end of summer, approximately 8 months from planting, though under some conditions the crop can take up to 10-12 months to reach maturity (Kay, 1987). Plants usually set seeds three years from being a seedling. The tuber gets replaced during the year by a main new tuber formed underground at a lower level which remains dormant after the yearly death of the aerial parts of the original plant.

Factors Influencing Growth and Quality

East Indian arrowroot is a plant of low elevations in the moist tropics where it is most commonly found near the sea and below elevations of 200m (Kay, 1987). It grows best under tree-shades in a humus-rich fertile soil (Huxley, 1992). For its planting, the land should be prepared well by mixing well decayed organic matter in the field similar to that of potato. The tubers are often planted about 15cm deep at a distance of 45cm in rows 75-90 cm apart. The field should be weed-free thoughout the growth period.

Harvest

Harvesting of tubers usually takes about 8 months from planting though sometimes it can take as much as 10-12 months (Kay, 1987). Tubers are harvested when aerial plant parts die. Tubers are dug up and usually stored in pits though there these are liable to sprout.

Diseases and Pests

No literature is available on pests and diseases, though red spider mite, tarsonemid mite and *Botrytis* have been observed attacking this crop in the wild.

References

1. Anon. 2009. *Tacca leontopetaloides.* Germplasm Resources Information Network. Agricultural Research Service (ARS), USDA. Retrieved 2009-11-17.
2. Anon. 2016. *Tacca leontopetaloides.* PROSEA. *PlantUse English,* Retrieved 12: 31, October 23, 2022. https: //uses.plantnet-project.org/e/index.php?title=Tacca_leontopetaloides_(PROSEA) and oldid=219042.
3. Drenth, E. 1972. A Revision of the Family Taccaceae. *Blumea*, Vol. 20. p. 401.
4. https: //en.wikipedia.org/wiki/Tacca_leontopetaloides
5. https: //flickr.com/photos/47108884@N07/6659237845. reviewed on 3 November 2018. Ton Rulkens from Mozambique.

6. http: //www.archive.org. Flora Malesiana Series 1. Publisher: Nationaal Herbarium Nederiand, Universiteit Leiden branch.
7. https: //www.researchgate.net/publication/326377694_Root_tuber_of_Tacca_leontopetaloides_L_kunze_for_food_and_nutritional_security
8. Humanity development library 2.0. Root crops: East Indian arrowroot (*Tacca leontopetaloides*). nzdl.org.
9. Huxley A. 1992. The New RHS Dictionary of Gardening. MacMillan Press, England.
10. Kay, D.E. (rev. by B. Gooding), 1987. Root Crops. Tropical Development and Crop Research, London.
11. Ogbonna A., S.O. Adepoju, C.I.C. Ogbonna, T. Yakubu, Janet Itelima and V.Y. Dajin, 2017. Root tuber of *Tacca leontopetaloides* L. (kunze) for food and nutritional security. *Microbiol.Cur. Res.*, 01: 10.4066/2591-8036.16-2241.

17

Indian Shot (*Canna indica* L.)

Family: Cannaceae.

Common names: African arrowroot, Australian arrowroot, Canna lily, Edible canna, Indian shot (Anon., 2020), Purple arrowroot, Sierra Leone arrowroot (Anon. 2016a) and Queensland arrowroot; In Hispanic America: Achira (Anon., 2016a); in Brazil: Cana-da-India (en.wikipedia.org). **Indian names**: Kardal (Marathi), Kele phool (Konkani), Keli and Vijayanti (Hindi), Laphoorit and Sarvajjaya (Manipuri) and Sarbajaya (Bengali) (en.wikipedia.org).

Synonyms: *Canna annaei* Andre., *C. aurantiaca* Roscoe., *C. barbadica* Bouche, *C. bidentata* Bertol., *C. carnea* Roscoe, *C. cearensis* Huber., *C. chinensis* Willd., *C. cinnabarina* Bouche, *C. coccinea* var. *bicolor* Kraenzl., *C. commutata* Bouche, *C. compacta* Bouche, *C. concinna* Bouche., *C. crocea* Lag. ex Rchb., *C. crocea* Roem. and Schult., *C. densifolia* Bouche, *C. edulis* Ker Gawl., *C. ehrenbergii* Bouche, *C. ellipticifolia* Stokes, *C. esculenta* Loudon, *C. exigua* Bouche, *C. flavescens* Link, *C. floribunda* Bouche, *C. formosa* Bouche, *C. fulgida* Bouche, *C. glauca* var. *annaei* Petersen., *C. heliconiifolia* Bouche, *C. heliconiifolia* var. *xalapensis* (Horan.') Kraenzl., *C. indica* var. *edwardsii* Regel., *C. indica* var. *lutea* (Mill.) Aiton, *C. indica* var. *napalensis* (Wall. ex Bouche) Baker, *C. indica* var. *orientalis* (Roscoe) Baker, *C. indica* var. *rubra* Aiton, *C. juncea* Retz., *C. laeta* Bouche, *C. lagunensis* Lindl., *C. lambertii* Lindl., *C. leptochila* Bouche, *C. lutea* Mill., *C. lutea* Larrañaga, *C. lutea* var. *aurantiaca* Regel, *C. lutea* var. *genuina* Kraenzl., *C. lutea* var. *maculata* (Hook.) Petersen, *C. lutea* var. *pallida* Kraenzl., *C. maculata* (Hook) Link., *C. maxima* Lodd. ex Roscoe, *C. moritziana* Bouche, *C. nepalensis* Wall. ex Bouche, *C. nepalensis* D.Dietr., *C. occidentalis* Roscoe, *C.*

orientalis Bouche, *C. orientalis* Roscoe, *C. orientalis* var. *flava* Roscoe, *C. pallida* Roscoe., *C. pentaphylla* D.Dietr., *C. plurituberosa* T.Koyama and Nob.Tanaka, *C. poeppigii* Bouche, *C. polyclada* Wawra, *C. polymorpha* Lodd. ex Loudon, *C. portoricensis* Bouche, *C. pulchra* Hassk., *C. pulchra* Bouche ex Horan., *C. rubra* (Aiton) Willd., *C. rubricaulis* Link, *C. sanctae-rosae* Kraenzl., *C. sanguinea* Warsz. ex Otto and A.Dietr., *C. sanguinea* Bouche, *C. saturate-rubra* Bouche ex K.Koch, *C. seleriana* Kraenzl., *C. sellowii* Bouche, *C. speciosa* Hegetschw, *C. speciosa* Roscoe ex Sims, *C. spectabilis* Bouche, *C. sulphurea* Bouche, *C. surinamensis* Bouche, *C. tenuiflora* Bouche, *C. textoria* Noronha, *C. thyrsiflora* Hegetschw, *C. tinei* Tod., *C. variabilis* Willd., *C. variegata* Besser, *C. variegata* Bouche, *C. variegatifolia* Ciciar., *C. ventricosa* Bouche, *C. warszewiczii* var. *flameus* Ram. Goyena, *C. xalapensis* Horan (en.wikipedia.org).

Introduction and Origin

The definite derivation of the genus name is not known as it is of oriental origin but it is presumed that the word *Canna* derives from the Greek or Latin word *kanna* meaning a 'reed' as some species are tall with reed-like stems. (Misra and Misra, 2013). The specific epithet *indica* refers its origin as India. Despite its botanical name, *Canna indica* is not native to India but from tropical America and the Caribbean. It originated in the Andean region or Peruvian coast extending from Venezuela to northern Chile in South America. It is native to much of South America, Central America, the West Indies and Mexico, and is now cultivated pantropically and in other warmer regions of the world [*Canna indica* (PROSEA), 2016]. It has been naturalized in the SE United States (Florida, Texas, Louisiana and South Carolina), and much of Europe, sub-Saharan Africa, SE Asia, and Oceania (GRIN, 2016; en.wikipedia.org). It is rhizomatous plant widely distributed in tropics and sub-tropics. Commercially it is cultivated in Australia for production of starch. *C. indica* can escape from cultivation and become a weed. It has been included in the Global Invasive Species Database (CABI) and has been declared as invasive in South Africa, Australia (New South Wales and South Eastern Queensland), Tanzania and Ghana.

Canna has been a minor food crop cultivated by indigenous peoples of the Americas for thousands of years (Anon.,2000; Hokche *et al.*, 2008). The fleshy rhizomes of *C. indica* and certain other species such as *C. tuerckheimmii* (syn. *C. gigantea*) yield edible starch. In West Indies, *C. indica* starch is called *tous-les-mois* meaning 'every month' (Misra and Misra, 2013). It is most important in South America. In Australia, canna starch is traded internationally as 'Queensland arrowroot', though in SE Asia the edible cannas are common home garden plants.

Botany

Almost all the available cannas are herbaceous, perennial and clump-forming hybrids with fleshy rhizomes. *Canna indica* is an erect, robust and rhizomatous perennial herb, and depending on the variety, it grows to a height of 0.5-3m forming about 60cm long branched rhizomes that are around 10cm in diameter with fleshy bulbous segments resembling corms, covered in two lines by pale green or purple scale leaves and thick fibrous roots (en.wikipedia.org). The surface of the rhizome is carved by transverse grooves which mark the base of scales that cover it (Wu *et al.*, 2000). Very large grains of starch stored there can sometimes be seen with the naked eye.

Source: (en.wikipedia.org)

The plants form an unbranched and upright fleshy stem or the overlapping leaf sheaths form a pseudo trunk (Funk *et al.*, 2007). The plants appear as small banana plant. Very large and simple leaves that are divided into leaf sheaths, short petioles and narrowly ovate to narrowly elliptical leaf blades are alternate and spiral or arranged in two lines. The blades are up to 30-60cm long and 10-25cm wide, entire, base rounded to cuneate, gradually attenuate to the sheath, apex acuminate, midrib prominent and the underside often slightly purplish. The parallel leaf veins arise from the midrib (not typical of monocots).

Inflorescence appearing from August to October is terminal, racemose, usually simple but sometimes branched, bearing large, red or yellow-orange, single or paired, zygomorphic, hermaphrodite and threefold flowers (Anon., 2016b; en.wikipedia.org). These may be 4.5–7.5cm long with three closely triangular and usually green sepals that are 1–1.7cm long, and three 4–6.5cm long erect petals that are green or shades of yellow to orange and red to pink depending on the variety (en.wikipedia.org). The base of the petals is fused with the staminodien to a stamen column. There are two circles each with originally three stamens. The petals and staminodes are usually yellow to red. The three carpels are adherent under the constant syncarpous ovary which has a soft-spiky surface and with many centrally angled ovules. The pollen is deposited on the abaxial (off-axis) surface of the stylus. The pollination usually occurs through insects that pick up the pollen from the flattened style.

The fruit-capsules are ellipsoid to globose, warty, 1.5-3cm long, chestnut coloured (Nobuyuki Tanaka, 2001) with numerous seeds that are globose, 5mm in diameter, hard, smooth and blackish to very dark brown (Anon., 2016b). The seeds resemble shotgun pellets giving rise to the plant's common name of Indian shot (Anon., 2020, e2000). The seeds are hard enough to shoot through wood and still survive and later germinate (en.wikipedia.org). Its diploid chromosome number is 2n=18, along with even triploids with 2n=27.

Fruit

Flower of *C. indica* var. *flava*

Seeds

Cultivars

C. indica is a problematic species-complex comprising numerous, mainly unnamed cultivars varying in flower colour and length, where number and shape of staminodes are also extremely variable. In the Andes of South America, two cultivars are well known 'Verdes' with dirty white rhizomes and bright green foliage, and 'Morados' where rhizomes are covered with violet-coloured scales.

Of the more than 1,000 cultivars, most fall into two main groups of complex hybrids:

- *Canna* × *generalis* L.H. Bailey: Principal progenitors are *C. indica*, *C. glauca* L., *C. iridiflora* Ruiz and Pavon and *C. warszewiczii* A. Dietr. Their flowers are usually up to 10cm in diameter, basally non-tubular, petals unreflexed, and staminodes *cum* labellum erect or spreading.
- *Canna* × *orchiodes* L.H. Bailey: Principal progenitors are *C. flaccida* Salisb., *C.* × *generalis* cv. Crozy. Their flowers are up to 20cm in diameter and basally tubular, petals are reflexed and the staminodes wavy and exceeded by the labellum.

Many of these cultivars have handsome yellow, pink, orange, red or variegated flowers and green, crimson, purple or variegated foliage.

Nobuyuki Tanaka from Japan recognises several varieties of *C. indica* (Source: en.wikipedia.org):

- *Canna indica* var. *indica* L.: Its purple rhizomes are thick with up to 3cm in diameter. It is a medium-sized species with prolific tillering, spreading habit, triangular floral stems, green and oblong foliage, erect floal spikes with self-coloured red florets, staminodes long and narrow, edges regular, petals red, self-pollinating and also true to type, and capsules are globose. It was introduced by Linnaeus (Tanaka, 2001).
- *Canna indica* var. *flava* (Roscoe ex Baker) Nb. Tanaka: Bloom is yellow. Many plants previously offered as *C. lutea* fall into this subspecies (Tanaka, 2001).
- *Canna indica* var. *maculata* (Hook) Nb. Tanaka: It bears white to pink rhizomes up to 3cm thick with average tillering. A medium-sized shy-blooming branching species having ovoid-shaped green foliage, erect spikes with spotted-red yellow self-pollinating florets, long and narrow staminodes, regular edges and green petals, and the capsules are globose. It was introduced by Hook. Many plants previously offered as *C. lutea* fall into this subspecies (Tanaka, 2001).
- *Canna indica* var. *sanctae rosea* (Kraenzl) Nb. Tanaka: A small branching species bearing up to 3cm thick white to pink rhizomes; oval-shaped white-margined green foliage; erect flowering spikes; self-coloured pink flowers; long and narrow staminodes with regular edges; labellum, stamens and styles being pink; petals are red with farina *cum* fully self-cleaning; self-pollinating *cum* true to type; and the capsules are ellipsoid (Tanaka, 2001).
- *Canna indica* var. *warszewiczii* (A.Dietr.) Nb.Tanaka: This is distinguishable from *C. indica* var. *indica* by having purple-red-margined leaves, purple-red fruits and slightly corm-like thickened terrestrial stem at the base. In addition to this, there are normally two backward recurved staminodes, where stamen is often strongly reflexed at the apex. Sometimes, this variety is confused with *C. discolor* Lindl. from which it differs in having much smaller deep-red flowers and short *cum* slender rhizomes. *C. indica* var. *warszewiczii* is diploid with 2n=18 chromosome numbers though *C. indica* var. *indica* is triploid with 2n=27 (Tanaka, 2001).

Nutritional Values

As per various sources (Humanity development library 2.0; en.wikipedia.org; Anon.,2016b), per 100g edible portion of rhizomes of *C. indica* contains 67-75 per cent water, 1-1.7 per cent protein, 0.1 per cent fat, 0.6 per cent crude fibre, 22.6-30 per cent carbohydrates, 1.4 per cent Ca, 63-70mg P, 20mg Fe, 0.1mg vit.

B and 7-10mg vit. C. The carbohydrates consist of more than 90 per cent starch and about 10 per cent sugar (glucose and sucrose).

The percentage of starch varies with the age of the rhizomes and is usually at a maximum between 6 to 15 months when the sucrose content is also high. Highly soluble and easily digestible starch is a shiny yellowish powder with irregularly shaped large grains (125-145μm×60μm). After cooking, the starch becomes glossy and transparent (Anon., 2016b). The cysteine level in the protein is very low (Humanity development library 2.0).

Approximate composition of fresh leaves used as fodder is water 90 per cent, protein 1 per cent, fat 0.2 per cent, carbohydrates 7 per cent, ash 1.4 per cent and the digestibility is about 20 per cent (Anon., 2016b).

Uses

The sweet but fibrous young rhizomes can be eaten raw but are usually cooked in various ways (boiled, roasted or fried). After cooking, the rhizomes become white, translucent, fibrous and a somewhat mucilaginous mass with a sweetish taste. Peeled and dried rhizomes are milled to produce canna flour. Young shoots can be eaten as a green vegetable [en.wikipedia.org; Anon., 2016b].

Juice extracted from grated rhizomes is locally used against diarrhoea. In Java, pounded seeds are traditionally used in a poultice to relieve headache. In Cambodia, mush made from the rhizome is taken as a remedy for yaws. In Hong Kong, a decoction of fresh rhizomes is prescribed in acute hepatitis. Crushed fresh rhizomes are applied topically for traumatic injuries in traditional medicine in Indo-China. In the Philippines a decoction of the rhizomes is used as a diuretic and macerated rhizomes in water are applied to alleviate nose bleeding (Anon., 2016b).

Canna starch is obtained by grinding or pounding the roots, soaking them in water and then starch granules are separated from root fibers. The starch granules of *C. indica* are translucent and the largest known from any plant. The starch is occasionally marketed commercially as 'arrowroot', a name also applied to the starch of other similar root crops such as *Maranta arundinacea*. The starch is in very large grains, about three times the size of potato starch grains, and can be seen with the naked eye. This easily digestible starch is used in foods and is also well suited as a health and baby food. The starch is also suitable for baking. It is also used as laundry starch (en.wikipedia.org; Anon., 2016b).

The leaves are suitable for wrapping and as plates. Both the leaves and the rhizome can be used as cattle feed. Canna is also well known as a garden plant for its beautiful flowers and foliage of various colours. Fumigated stems and leaves are used as an insecticide. The black and hard-coated seeds are used as beads. They are also used in percussion instruments and rattles, especially in Africa [en.wikipedia.org; Anon., 2016]. *Canna indica* can be used for treatment of industrial waste waters through constructed wetlands. It is effective for the removal of high organic load, colour and chlorinated organic compounds from paper mill wastewater (Choudhary, 2011).

Propagation

Manual cross-pollination for production of new hybrid cultivars is possible. Seeds are presently used solely for the purpose of producing new varieties. Before sowing, they should be soaked for 24 hours in warm water or seed coat should be chipped and sown in late winter into pots filled with soilless medium or in sandy-peat mix at an environmental temperature of 18-22°C and of the soil above 16°C until these germinate or soaking these for 40 minutes in sulphuric acid causes up to 80 per cent germination (Misra and Misra, 2013). After germination these are repotted gently in rich soil mix (half good garden soil and half peat + sand) without breaking or damaging their brittle roots and the pots are kept at 16°C for further growing and hardening.

In cultivation, cannas are usually propagated through their rhizomes (cuttings or the whole rhizomes) to maintain genetic identity of the clones (Anon., 2016b). Whole rhizomes or young tips of rhizomes without old brown parts having at least two healthy buds are used as planting materials. As a precaution against rotting, the rhizomes are sometimes dipped in a 10 per cent solution of copper sulphate before planting.

Growth and Development

After planting of rhizome cuttings under optimal conditions, seedlings start emerging 20–30 days after, which develop into harvestable plants in 6-8 months. In tropical regions, flowering starts a few months after planting and flowers continue to appear throughout the plant life but not during winter months. In regions where frost can be expected, the rhizomes are lifted and overwintered at about 7°C. A rhizome is considered mature when the triangular slit in the outer scale leaf of the rhizome has turned purple.

Environmental Factors

Canna grows well under various climatic conditions. In regions where temperature is usually uniform it can be planted anytime except during drought, otherwise during rainy season to avoid frequent watering. It grows from sea level to 2,900m a.s.l. and thrives in temperate, tropical or subtropical mountain climates between 1,000 and 2,000m a.s.l. (in humid tropical climates for higher elevations) at a mean temperature of 14 to 27°C. Normal growth occurs at temperatures above 10°C as well as survives high temperatures of 30-32°C. The top growth tolerates even light frosts and the plants can be grown in areas with winter snow. Mature top growth can be killed by temperatures of -2°C though fresh young growth is severely damaged at 0°C. A well distributed annual rainfall of 1,000-1,200mm is satisfactory though can tolerate 500–5,000mm per year. It is affected by drought but tolerates excessive moisture (but not waterlogging). It is very tolerant of shade and grows and flowers under a broad range of photoperiodic conditions. It is a tall plant therefore should be cultivated at locations with relatively low wind speed to prevent bending over.

Soil

It thrives on various soil types including the marginal ones as in case of most other tuber crops *e.g.* weathered, acidic latosols, *etc.* The optimal substrate consists of a deep sandy loam and well-drained light, rich and humiferous soil at a sunny place with a pH of 5.5–6.5 though can tolerate the range of 4.5-8.

Land Preparation and Planting

It is planted in beds that have been ploughed or dug thoroughly, cleaned of weeds and other unwanted materials, and mixed with plenty of well-rotten farmyard manure or compost.

Since the rhizomes are quickly perishable these should be stored properly between the harvest and the next planting season (Gade, 1966). Large rhizomes can be divided in spring before the new shoots appear (Cook, 2001). The rhizomes should be planted 60-70cm apart and about 10-15cm deep after the last frost. The spacing between rows should be 70-80cm (en.wikipedia.org/wiki/Canna_indica) as too close planting will cause the plants becoming soon too crowded and therefore giving poor performance. When putting the rhizomes in the ground, the sprouts should point upwards. To avoid frequent watering, it would be better to plant the rhizomes during rainy season.

The quantity of rhizomes required for planting one hectare of land is usually 3,000–4,500 kg with a planting density of 22,500 plants/ha (en.wikipedia.org/wiki/Canna_indica).

Manure and Fertilizer

Fertilizer should be applied during first tillage operation and spread according to plant conditions. Up to 750 kg/ha of ternary compound fertilizer (N, P, K) can be applied. Direct contact of fertilizer with the base and leaves should be avoided. Before flowering, the second tillage can be combined with the second fertilizer application to promote the growth of plants and underground parts. Fertilizer in the soil should be applied evenly to both the sides of the growing plant for uniform growth. It requires ample water in the growing season.

Weed Control

Fast growing weeds have a negative impact, especially before the appearance of fourth leaf. Either herbicides can be applied to eradicate weeds or mechanical operations can be carried out, depending on the weed growth.

Harvest

The crop grown from rhizome tips can be harvested in about 4-6 months of planting when the rhizomes are tender though harvesting after 8 months gives higher yields as then the rhizomes have swollen to their maximum. Rhizomes should not be allowed to become much older than 10 months as they become tough and less suitable for starch production *vis-a-vis* consumption. Harvesting is carried out manually by pulling out the crop with digging tools, shaking off the soil and then cutting the stems to separate the rhizomes (Caicedo *et al.*, 2003; Gade, 1966; Anon., 2016b).

Yield

The rhizome yield varies depending on the climate, soil conditions, planting season and harvest time (Anon., 1989; Kay, 1973). Highest rhizome yield does not necessarily correspond to the highest starch yield (Caicedo *et al.*, 2003). Average rhizome yield is generally around 22–50t/ha, whereas the starch yield is about 2–5t/ha which can reach up to 10t/ha (Anon., 1989).

Postharvest

Cleaned and surface-dried rhizomes can be stored safely for several weeks under cool and dry conditions. To store rhizomes for longer periods, these should be stored at a frost-free place which is not too dry. Japanese store these over winter in 30cm deep field pits (Humanity development library 2.0). For commercial manufacturing of starch (flour), rhizomes are processed immediately after harvesting (Anon., 2016b).

Pests and Diseases

Canna is almost disease- and pest-free plant. A few attacking this are of little importance. *Fusarium*, *Puccinia* and *Rhizoctonia* spp. are possible fungal diseases. Canna rust (*Puccinia thaliae*) causes orange spots on the leaves. *Botrytis* mould sometimes may affect the flowers during severe winter (Cook, 2001) which can be controlled by applying 0.2 per cent maneb. 'Hippeastrum mosaic virus', 'Tomato aspermy virus', 'Canna yellow mottle virus' and 'Canna yellow streak virus' can cause mild to strong symptoms from streaked leaves, stunted growth to distorted blooms. In such cases such plants can be uprooted and burnt. Some canna varieties may be resistant to one or certain of these diseases so such varieties may be grown if possible. To prevent mould, the soil should be well-drained without too much soil moisture or stagnating water (Reddy, 2015). Dead and infected leaves should be removed to control the diseases from spreading.

Japanese beetles (*Popillia japonica*) and grasshoppers have occasionally been found feeding on the foliage, and cutworms (*Agrotis* spp.) the rhizomes in Hawaii. Japanese beetle, a leaf ragging pest of Japanese origin feeds on the part of the leaves between the veins. Although this beetle does not cause serious damage in Japan, it can be a serious problem in the US where it has no natural predator (Reddy, 2015). Stored rhizomes are sometimes attacked by 'bird cherry-oat aphid' (*Rhopalosiphum padi*). Although this pest has not caused any serious damage yet, it can particularly affect plants grown in greenhouses, and this can be combatted with parasitical wasps. It is a more common pest on cereals (Reddy, 2015).

The canna leaf roller moth (*Calpodes ethlius*) and *Quinta cannae* have been observed infesting in USA (Cook, 2001), primarily found in the southern United States. This pest causes damages by laying its eggs in the bud of developing stalks. To protect the eggs from predators the females use sticky webs to keep the leaves from unfurling. The caterpillars feed on the leaves which can lead to losses of yield due to reduced photosynthesis (Reddy, 2015).

References

1. Anon. 2001. *Canna indica* Factsheet. CABI Invasive Species Compendium. Archived from the original on 2016-03-04. Retrieved 6 March 2021.
2. Anon. 2016a. *Canna indica* L. Germplasm Resources Information Network (GRIN). Agricultural Research Service (ARS), USDA. Retrieved 12 Aug 2016.
3. Anon. 2016b. *Canna indica* (PROSEA). PlantUse English, Retrieved 02: 46, October 11, 2022 from https: //uses.plantnet-project.org/e/index.php?title=Canna_indica_(PROSEA) and oldid=218993.
4. Anon. 2020. *Canna indica*. USDA, NRCS (n.d.). The PLANTS Database (plants.usda.gov). Greensboro, North Carolina: National Plant Data Team. Retrieved 23 March 2020.
5. Anon., e2000. Magnoliophyta: Alismatidae, Arecidae, Commelinidae (in part), and Zingiberidae. Flora of North America Editorial Committee. Flora of North America, **22**: i-xxiii, 1-352.

6. Anon. 1989. Lost Crops of the Incas: Little-known Plants of the Andes with Promise for Worldwide Cultivation (National Research Council). National Academies Press.
7. Caicedo DIaz, G. E., L.S. Rozo Wilches and G. Rengifo BenItez, 2003. La achira: alternativa agroindustrial para áreas de economIa campesina. (in Spanish). Flora of North America Editorial Committee, 2000, USA.
8. Choudhary, H.B. 2011 Performance of constructed wetland for the treatment of pulp and paper mill wastewater. Proceedings of World Environmental and Water Resources Congress, May 22-26, 2011. Bearing Knowledge for Sustainability, Palm Springs, California, USA, pp.4856-65.
9. Cook, I. 2001. The Gardener's Guide to Growing Canna. RHS, Great Britain.
10. Funk, V.A., P.E. Berry, S. Alexander, T.H. Hollowell and C.L. Kelloff, 2007. Checklist of the Plants of the Guiana Shield (Venezuela: Amazonas, Bolivar, Delta Amacuro, Guyana, Surinam, French Guiana). *Contr. US Natl. Herb.* **55**: 1-584.
11. Gade, D. W. 1966. Achira, the edible canna, its cultivation and use in the Peruvian Andes. *Econ. Bot.*, **20**(4): 407-15.
12. Hokche, O., P.E. Berry and O. Huber (eds), 2008. *New Cat. Fl. Vasc. Venezuela*, pp. 1-860. Botanical Institute of Venezuela Foundation, Caracas.
13. https: //en.wikipedia.org/wiki/Canna_indica
14. Humanity development library 2.0. Root crops: Queensland arrowroot (*Canna indica*). nzdl.org.
15. Kay, D. E. 1973. *Root crops* (No. 2). Tropical Products Institute, London.
16. Misra Sanyat and R.L. Misra, 2013. Canna. Commercial Ornamental Bulb Science. Westville Publishing House, New Delhi.
17. Nobuyuki Tanaka, 2001. Taxonomic revision of the Cannaceae family in the New World and Asia. *Makinoa*, **1:** 1–74.
18. Reddy, P. P. 2015. Achira, *Canna edulis*. In: Plant Protection in Tropical Root and Tuber Crops. Springer, New Delhi.
19. Tanaka, N. 2001. Taxonomic revision of the family Cannaceae in the New World and Asia. Makinoa (ser. 2), **1**: 34–43.
20. Wu, C.Y., P.H. Raven and D.Y. Hong (editors of Flora of China Editorial Committee), 2000. Flora of China (Flagellariaceae through Marantaceae), **24**: 1-431. Science Press and Missouri Botanical Garden Press, Beijing and St. Louis.

18

Jerusalem Artichoke (*Helianthus tuberosus* L.)

Family; Asteraceae.

Synonyms: *Helianthus esculentus* Warsz., *H. serotinus* Tausch, *H. tomentosus* Michx. and *H. tuberosus* var. *subcanescens* A.Gray.

Common names: Earth apple, Girasole, Jerusalem artichoke, Root artichoke, Sunchoke, Sunroot, Topinambour and Wild sunflower. In **Japan**: Kiku-imo and in **India**: Hathipick or Hatichuk.

Introduction and Origin

The name *Helianthus* is derived from the Greek words *helios* meaning 'sun', and *anthos* meaning 'flower'. The specific epithet *tuberosus* is from Latin meaning 'tuberous' in reference to the edible underground tubers. The common name Jerusalem artichoke is probably from the Italian name for this plant, *girasole articiocco*, which means 'sunflower artichoke' in reference to flower appearance and tuber taste, with 'Jerusalem' probably being a mispronunciation of *girasole*.

Helianthus tuberosus, a native to temperate North America, is a species of sunflower cultivated widely across the temperate zone for its tuberous rhizomes used as a root vegetable. It was cultivated by the Indians in the north-eastern part of the continent in pre-Columbian times. The species was introduced into Europe in the early 17th century and is now widely grown in both the hemispheres, but in general it is regarded as of rather minor importance. Although it originated in the United States, it gets extensively cultivated in the United Kingdom.

By Gilabrand at English Wikipedia, CC BY-SA 3.0, https://commons.wikimedia.org/w/index.php?curid=8001862

The tuberous rhizomes of the plant are a carbohydrate-rich staple valued as food since before European contact. Unfortunately the plant can be quite weedy in cultivation and difficult to eradicate. Furthermore, indigestible inulins in the tuber may cause flatulence (en.wikipedia.org).

Botany

Jerusalem artichoke is an erect, hardy and rhizome-forming perennial herb that grows 1.5-3m (sometimes up to 4 m) tall in temperate areas though in the tropics it attains a height of 0.6-1.2m or even more. The fibrous root system develops cord-like rhizomes that can reach more than 1m in length with fleshy swollen tips forming tubers, which somewhat resemble potatoes, but have much larger eyes and are often knobbly. The elongated and uneven rhizomes similar to ginger in shape are usually 7.5–10cm long and 3–5cm thick and varying in colour from pale brown to white, red or purple (Huxley, 1992) with a crisp and crunchy texture when raw. Jerusalem artichoke is mainly grown for its edible rhizomes which vary in size and shape; some are potato-like, small, round and knobby while others long, slender and smooth. Usually cultivated varieties yield larger and thick rounded rhizomes that are clustered near the main stem, whereas, the invasive plants in Central Europe and also wild varieties in North

America usually produce narrow elongated tubers at the ends of longer rhizomes (Wagenitz, 1980).

By H2ase - Own work, CC BY-SA 3.0, https://commons.wikimedia.org/w/index.php?curid=24805030

The stems are generally hairy with branching in their lower part. Per plant stems and branch numbers are usually variable. Ovate to lanceolate, 3-30cm long, 5-8cm broad and toothed leaves with rough *cum* hairy texture are opposite on the lower part of the stem but alternate upwards (Dickinson *et al.*, 2004). Larger leaves on the lower stem are broad ovoid-acute and up to 30cm long though in the upper part these are smaller and narrower (Gibbons, 1962).

Capitate flowerheads some 5-10cm in diameter bear yellow flowers which are 5–10cm in diameter with 10-20 showy, petal-like and bright sterile yellow rays surrounding 60 or more yellow, small and tubular fertile disc florets. Mildly fragrant flowers give off a light vanilla-chocolate aroma (Gibbons, 1962). The fruit is a hairy achene containing a mottled black or brown seed some 5mm long × 2mm wide. Its chromosome number is 2n (6x)=102 (Kulshreshtha and Gupta, 1979; Swanton *et al.*, 1982).

CC BY-SA 3.0, https://commons.wikimedia.org/w/index.php?curid=81355

Cultivars

Well-known cultivars include 'Fuseau' (with purple, easy to peel and smooth tubers), 'New White' (tubers white), 'Boston' (tubers red), and 'Dwarf Sunray' (plants relatively small). Cultivated plants usually have thick rounded tubers though invasive ones in central Europe as well as the wild ones in North America have narrow elongated tubers (Wagenitz, 1980). There is a limited range of cultivars, one of which was developed in France and has long fusiform tubers of good quality and flavour which is often designated as *H. tuberosus* var. *fusiformis*.

Nutritional Value

The rhizomes/tubers of Jerusalem artichoke are almost oilless but with about 2 per cent protein and little starch. They store their carbohydrate as inulin which is an important source for use as a dietary fibre in food manufacturing (Flamm *et al.*, 2001). Tubers convert their inulin into its component fructose which is about 1.5 times sweeter than sucrose. Temperature variances have been shown to affect the amount of inulin the Jerusalem artichoke can produce. It makes less inulin in a colder region than when it is in a warmer region (Puangbut, 2012). When used for animal feeding, the nutritive value of the tubers has been quoted as digestible protein 1.2 per cent; total digestible nutrients 15.9 per cent and nutritive ratio 12:3 (Humanity development library 2.0).

Green tops have given analytical figures as dry matter 27.2 per cent, protein 1.4 per cent, fat 0.3 per cent, nitrogen-free extract 18.5 per cent, minerals 2.1 per cent, calcium 0.44 per cent, phosphorus 0.09 per cent, potassium 0.37 per cent, digestible protein 0.8 per cent, total digestible nutrients 18.1 per cent and nutritive ratio 21:6 (Humanity development library 2.0).

Approximate Nutritional Value per 100g of Raw Jerusalem Artichokes

Energy	304 kJ (73 kcal)
Carbohydrates	17.44 g
Sugars	9.6 g
Dietary fiber	1.6 g
Fat	0.01 g
Protein	2 g
Vitamins	***Quantity (per cent DV†)***
Thiamine (B1)	17 per cent (0.2 mg)
Riboflavin (B2)	5 per cent (0.06 mg)
Niacin (B3)	9 per cent (1.3 mg)
Pantothenic acid (B5)	8 per cent (0.397 mg)

Vitamin B6	6 per cent (0.077 mg)
Folate (B9)	3 per cent (13 μg)
Vitamin C	5 per cent (4 mg)
Minerals	***Quantity (per cent DV†)***
Calcium	1 per cent (14 mg)
Iron	26 per cent (3.4 mg)
Magnesium	5 per cent (17 mg)
Phosphorus	11 per cent (78 mg)
Potassium	9 per cent (429 mg)

Units: μg = micrograms, mg = milligrams, IU = International units

Percentages are roughly approximated using US recommendations for adults. *Source*: USDA Food Data Central

Source: https://en.wikipedia.org/wiki/Jerusalem_artichoke

Jerusalem artichoke is a sunflower species with a rich nutritional profile having a balanced combination of almost all essential nutrients, minerals and vitamins. Artichokes contain a good amount of potassium, iron and other vitamins. It increases metabolism thus increasing perspiration or sweating. In addition, the fibre content in Jerusalem artichokes helps breaking down the triglycerides. As inulin does not get assimilated in the intestine, it does not increase the body's glucose level. Since it aids in cutting down on unnecessary calories so helping in controlling obesity and therefore this can prove a best possible replacement for potatoes.

Being a good source of potassium, artichoke can mainly contribute to the daily potassium intake level. The presence of inulin and potassium helps improving heart health, and potassium also aids in proper bone development, maintaining cardiac health, *vis-a-vis* in smooth muscle contraction. Artichokes also contain a good amount of iron that is a vital part of the protein haemoglobin.

The amount of proteins in Jerusalem artichokes is more than other roots and tubers. These proteins comprise sulphur-containing amino acids such as cysteine, methionine, taurine and homocysteine. These amino acids aid the liver in detoxifying and maintaining the flexibility of the connective tissue in the skin and the organs.

Jerusalem artichokes can trigger the urinary bladder to pass on more urine, thus detoxifying the liver regularly by flushing out all the toxins along with the urine. Most artichokes are diuretic in nature.

Inulin is a fructose-based polysaccharide used either as dietary fibre or for production of bioethanol. The carbohydrate in Jerusalem artichoke consists almost entirely of inulin and inulides with small amounts of starch, fructose and glucose; occasionally the starch reaching up to 30 per cent of the carbohydrate but is usually considerably less. Its inulin protects the liver, bile, and gallbladder. The primary function of inulin is to detoxify the liver by brushing out the toxins while passing through the digestive tract. It also helps capturing the excess fats and cholesterol while passing through the alimentary canal. As inulin is not digestible by human beings, utilisable energy intake is much lower than suggested, and allows the tubers to be used as a low energy food which still satisfies volume intake and hunger. Therefore, its tubers have also been recommended for diabetics. Jerusalem artichokes contain a good amount of inulin fibre that softens the stools and regulates bowel movement. This fibre also acts as a prebiotic as it multiplies good bacteria which in turn reduces the number of harmful bacteria in the stomach and flushes out the toxins from the intestine. It also possesses anti-inflammatory properties. Jerusalem artichoke also improves the body's overall immune system by reducing the risk of gut-related *cum* gastrointestinal diseases. The inulin in Jerusalem artichokes is a short-chain polysaccharide that does not break down into sugars, so regulating glucose metabolism in the body without adding any sugar content, hence sugar level in blood remains the same. Reduced sugar and small inulin can help in controlling type 2 diabetes. It is also helpful in the prediabetes state. Inulin in Jerusalem artichokes is not broken down into simple sugars and does not accumulate in the gut lining, hence, it does not increase the calorie content of the body so cholesterol level remains the same, unlike potatoes.

Adverse Effects

Significant side effect of Jerusalem artichokes is the presence of non-digestible dietary inulin fibre as people may be intolerant to it, therefore, its overconsumption may lead to irregular bowel movement, diarrhoea and gas formation or flatulence and in some cases gastric pain. People intolerant to inulin can experience abdominal pain and irregular bowel movements. People allergic to birch pollen can be allergic to Jerusalem artichokes. These people may experience 'pollen food allergy syndrome'. People having 'hereditary fructose intolerance' (HFI) should avoid Jerusalem artichokes as they may experience extremely low blood pressure and accumulation of substances in the liver, even if they consume a small quantity of fructose.

Uses

The tubers are used for cooking and baking in the same ways as potatoes (Reynolds, 1921) but unlike the potato they have sweeter and nuttier flavour and can also be eaten raw. Its thin raw slices can be used in salads. Their inulin form of carbohydrates gives the tubers a tendency to become soft and mushy if boiled but retains its texture better when steamed. These are good as pickles as the fermentation process reduces the risk of gastrointestinal issues. These are pretty famous in the form of soups as cooking and mashing prevent gas formation in the body.

It is used in the food industry due to its interesting nutritional and functional properties. It is being used as a supplement for sugar as high-fructose syrup. Since inulin is not assimilated in the intestine, it doesn't cause a glycemic spike as potatoes do. It has also been reported as a folk remedy for diabetes (Levetin and Karen McMahon, 2012). It has been used as a feedstock for ethanol production using inulin-adapted strains of yeast for fermentation.

It can be used as animal feed but only after proper washing. Pigs can forage and safely eat this directly from the ground. The stalks and leaves can be harvested and used for silage though cutting the tops greatly reduces the harvest of the roots.

Propagation

Reproduction by seed is quite uncommon as it requires considerable time for rhizome formation which makes it uneconomical, but is a means of dispersal for wild populations. It is vital when evolving new improved varieties through crossing. For quick multiplication, usually the propagation is through rhizomes, slips and cuttings. Jerusalem artichoke is amenable to propagation by tissue culture, being a model species used in pioneering micropropagation studies (Kays and Nottingham, 2008). Vegetative propagation is usually done by using disease-free setts (small tubers or pieces of larger tubers) weighing approximately 45-50g and having at least two or three eye buds.

Growth and Development

Tubers start forming with a drastic drop (dormancy) of the elongation rate near the stolon apex. If elongation rate drops to zero the tuber will be pear-shaped, tapering backwards. The top end of the tuber will be oval if dormancy is shallow. Irregular tuber shapes arise if the lateral eyebuds on the developing tuber retain or regain some elongation activity. Although the dormancy behaviour of the different buds and their interaction depend on the cultivar, it is also influenced by soil temperature and texture. Short stolons and early tuber initiation lead to an unwieldy tightly-packed bunch of tubers. Long stolons and late initiation of dormancy cause a very loose pattern of tubers wide apart that easily become lost at lifting. Swanton (1982) mentions that as many as 75 tubers per plant can be produced under certain conditions though at all there may not be any tuber formation (Wyse and Wilfhart, 1982). Tubers may be of various shapes and in the colours of white, pink or red. Usually, a crop reaches maturity in 3-6 months. Under reduced light intensity or cold conditions, no flowerheads may appear so fruits (seeds) may rarely form.

Factors Influencing Growth and Quality

Environmental Factors

Although *Helianthus tuberosus* is of temperate origin but it can be grown even in the tropics preferably at 300-750m altitude, but in India it is cultivated up to 1,200m. Most cultivars require a growing season of at least 125 frost-free days, with preferable average temperatures of 18-27°C and an evenly distributed rainfall of up to 1,250mm. In dry areas, irrigation may be necessary for starting sprouting. Tubers apparently exhibit a post-harvest dormancy that can be broken with gibberellic acid or a cold treatment for 2-3 months (Russell, 1979; Swanton *et al.*, 1992). *H. tuberosus* is drought-tolerant and can recover even after wilting, and even surviving short periods of flooding. Tubers are frost-resistant in the soil as well as in storage and may be improved rather than harmed by frost. Tubers remaining in the ground lie dormant over winter and can handle temperatures as low as –30°C. The first frost in autumn kills the aboveground parts of the plant, and only tubers can withstand freezing temperatures. Temperature variances have been shown to affect the amount of inulin the Jerusalem artichoke can produce. It makes less inulin in a colder region than when it is in a warmer region (Puangbut *et al.*, 2012).

Short day-lengths (photoperiods of around 12h) are reported to favour tuber formation but increased time for photosynthesis under the longer days of more temperate climates gives somewhat greater yields though as noted earlier this is partially compensated for by the more rapid growth under the high temperatures of the tropics.

Soil

The plant is adapted to grow in wide range of soil types and pH levels but prefers well-cultivated light to medium little alkaline loamy soils. Yields are poor on heavy clays particularly if there is waterlogging. Tuber and top yields are limited if soil moisture is less than 30 per cent of field capacity during the tuber formation period (early September to November). The soil should preferably not be too heavy and should be well-drained and friable in order to facilitate harvesting of the tubers.

Land Preparation and Planting Methods

Before planting, the field should be ploughed thoroughly to a depth of 15-20 cm, cleaned of weeds, and mixed with plenty of manure and compost. The setts are usually planted in rows in flat beds, on small hills or on ridges about 7.5-10cm deep and are earthed up in the same manner as potatoes when the plants are about 0.3m tall. Once planted out, the crop usually requires weeding only at early stages and as plants grow and spread rapidly, later weeding should be avoided as the developing tubers are likely to be damaged easily. Once established the crop is difficult to eliminate from the soil as tubers or parts of tubers are frequently left in the ground and often replanting for subsequent crops is unnecessary (Humanity development library 2.0). For maximum yield in USA, a planting distance of 1-1.3m between the rows and 60cm between the plants has been recommended. In the tropics, *e.g.* in India, a closer spacing of 0.3-0.5m between the plants and 0.6-1m between the rows is often used, while various intermediate spacings are noted in the literature. In India, nearly 550kg of setts are used to plant one hectare of land though in many other localities the seed rate is approximately 320kg/ha.

Manure and Fertilizers

University of Minnesota trials on irrigated, leached sandy soil have found increases in tuber yield in response to nitrogen rates of 67 and 135kg/ha. Yield increases in response to potassium were observed at rates of 170kg/ha but only at high nitrogen rates. Fertility programmes similar to that of potatoes are suggested as a starting point for artichokes. However, application of 500 kg/ha of superphosphate before planting is quite effective followed by a side dressing of 250 kg/ha of ammonium sulphate when the top growth reaches 0.3-0.6m.

Harvest

The crop usually matures in 120-180 days though there are a few cultivars maturing even in 90 days. The tubers are ready for harvesting when the leaves begin to wither and die. They are usually lifted manually with a fork only as required since they can be 'field-stored' without any deterioration in their quality or flavour. When grown for pig feed the animals are often turned loose on the plot and root out the tubers. When cultivated on a large-scale the tops are frequently cut off before harvesting and sometimes the tubers are ploughed out but this is not as efficient as manual harvesting owing to irregular shape, small size and distribution of the tubers (Humanity development library 2.0).

Postharvest

Once harvested the tubers quickly shrink and deteriorate if kept at ambient temperatures. Tubers can be harvested in the fall or left in the ground for winter storage and spring harvest. Tubers should be stored at 0°C and at very high humidity (90-95 per cent RH). Spoilage is more prevalent with diseased, bruised or skinned tubers. Tubers that are kept for seed should not be frozen in storage.

Yield

Crop yields are high, typically 16–30t/ha for tubers and 18–28t/ha green weight for foliage. On sandy soils in Europe, average tuber yield is about 30t/ha; in India the yields are usually 12-25t/ha, although under favourable conditions 37t/ha can be obtained and yields as high as 150-160t/ha have been reported from the northern Caucasus (Humanity development library 2.0).

Pests and Diseases

Jerusalem artichoke is relatively free from serious attack of pests and diseases in the field but under poor drainage root rot occurs due to *Corticium rolfsii* which is quite troublesome so care should be taken that field is well-drained. Powdery mildew infection due to *Erysiphe chicorianum* f. sp. *helianthi* has been reported from Bulgaria which can be controlled through sulphur dusting of the crop. Common diseases of sunflowers (*Helianthus annuus*) such as downy mildew, *Septoria* leaf spot, rust and *Sclerotinia* wilts are regarded as potential problems which can be controlled by use of zineb, maneb, Captafol and Bavistin. The tubers are susceptible to rots during storage so care should be taken that there is no build up of humidity in the storage. *Botrytis cinerea*, *Rhizopus stolonifer* and *Sclerotinia sclerotiorum* are reported to be fairly common where the last two can cause severe losses even when the tubers are stored at low temperatures. In addition, sometimes the tubers also rot probably due to infection of *Fusarium acuminatum* (Humanity development library 2.0) which can be controlled by the use of Captafol and/or zineb/maneb as the case may be.

References

1. Dickinson, T., D. Metsger, J. Bull and R. Dickinson, 2004. ROM Field Guide to Wildflowers of Ontario, 170p. Royal Ontario Museum, Toronto, Canada.
2. Flamm, G., W. Glinsmann, D. Kritchevsky, L. Prosky and M. Roberfroid, 2001. Inulin and oligofructose as dietary fiber: a review of the evidence. *Crit. Rev. Food Sci. Nutr.*, **41** (5): 353–62.
3. Gibbons, Euell. 1962. Stalking the wild asparagus. David McKay, New York
4. https: //en.wikipedia.org/wiki/Jerusalem_artichoke

5. Humanity development library 2.0. Root crops: Jerusalem artichoke (*Helianthus tuberosus*). nzdl.org.
6. Huxley, A.J., Mark Griffiths and Margot Levy, 1992. *The New Royal Horticultural Society Dictionary of Gardening*. Macmillan Publishers, London.
7. Kays S.J. and S.F. Nottingham, 2008. Introduction: An underutilized resource. In: *Biology and Chemistry of Jerusalem Artichoke - Helianthus tuberosus* L, pp. 1-6. Taylor and Francis Group, CRC Press, USA.
8. Kulshreshtha, V.B. and P.K. Gupta, 1979. Cytogenetic studies in the genus *Helianthus* L. *Cytologia*, **44**, 325-34.
9. Levetin, E. and Karen McMahon, 2012. *Plants and Society*, 231.
10. Puangbut, D. 2012. Influence of planting date and temperature on inulin content in Jerusalem artichoke (PDF). *Aust. J. Crop Sci.*, **36**: 1159–1165.
11. Reynolds, F.J. (ed.), 1921. Artichoke. *Collier's New Encyclopedia*. P. F. Collier and Son Company, New York.
12. Russel, W.E. 1979. The Growth and Reproductive Characteristics and Herbicidal Control of Jerusalem Artichoke (*Helianthus tuberosus*). Ph.D. Thesis, Ohio State University, 86p. USA.
13. Swanton, C.J, 1982. Jerusalem artichoke - crop or weed?. *Highlights of Agricultural Research in Ontario*, **5**(1): 17-19.
14. Swanton, C.J., P.B. Cavers, D.R. Clements and M.J. Moore, 1992. The biology of Canadian weeds. 101. *Helianthus tuberosus* L. *Canadian J. Pl. Sc.*, **72**(4): 1367-82.
15. Wagenitz G, 1980. Compositae I (Illustrierte Flora von Mitteleuropa, ed. Gustav Hegi), Band VI, Teil 6. Parey, Berlin, Germany.
16. Wyse, D.L. and L. Wilfahrt, 1982. Today's weed: Jerusalem artichoke. *Weeds Today* (Spring issue), pp. 14-16.

19

Lady's Sorrel (*Oxalis tuberosa* Molina)

Family: Oxalidaceae

Common names: Lucky clover and Oca/Oka.

Introduction and Origin

The name *Oxalis* originates from the Greek word *oxys* meaing 'sharp' and *als* referring to the usual 'acid taste', which is in reference to its oxalic acid content that gives it a sour flavour *cum* taste, and which also gives the common name 'sorrel'. The specific epithet *tuberosa* is from Latin meaning 'tuberous' in reference to the edible underground tubers which are known as Oca/Oka in Spanish, Uqa in Quechua (Anon., 2007), Yam in New Zealand and a number of other vernacular names elsewhere.

Oxalis tuberosa (formerly *O. crenata* Jacquin), is not known in the wild, but populations of wild *Oxalis* species bearing smaller tubers are known from four areas of the central Andean region (Bradbury and Emshwiller, 2011). The plant was brought into cultivation in the C and S Andes for its tubers which are used as a root vegetable. Oca is grown in the high Andean region at an altitude of 2,800-4,100m (Emshwiller, 2006) but is limited to southern Colombia, Bolivia, Ecuador, Peru and northern Chile, between 4°N and 17°S. Its greatest diversity is in central Peru and northern Bolivia, the probable area of its domestication. Oca was introduced to Europe in 1830 and to New Zealand as early as 1860 where it has become a popular table vegetable and is simply called yam or New Zealand yam.

By Izawa Ryu - Own work, CC BY-SA 4.0, https://commons.wikimedia.org/w/index.php?curid=106795555

Oca is one of the important staple crops of the Andean highlands. Among all Andean root and tuber crops, oca is second only to potato in area planted within the central Andean region (Anon., 1989) and has been a staple of rural Andean diets for centuries.

Botany

Oxalis tuberosa is a small herbaceous *cum* succulent perennial growing to a height of 20-30cm with cylindrical stems varying in colour from various shades of green to a purplish-red, which arises usually from the base of the plant in the form of a cone or hemisphere. Oca is a perennial due to regrowth from tubers though it has an annual life cycle, senescing in winter. It is erect in the first stages of its growth but decumbent to prostrate towards maturity. Spirally arranged green or purple leaves are phyllotaxis and with or without hairs, *vis-a-vis* showing great variation in form (Humanity development library 2.0). Leaves are trifoliate with petioles of varying lengths (2-9cm) and the inflorescences comprise of 4-5 trimorphic flowers having long, short or mid-length styles. In the flowers, calyx comprises of five pointed green sepals, corolla of five purple-striped yellow petals, 10 stamens in two groups of five and a pistil which may be shorter or longer than stamens. The flower structure has an efficient mechanism which facilitates cross-pollination through insects. Fruits are 5-celled capsules each containing 1-3 tiny seeds though these are rarely formed (Carrion *et al.*, 1995).

The plants overwinter as underground starchy stem tubers that are generally ovoid or elongated but are highly variable in shape, size and colour. These are characterised by long transverse shallow depressions in which the eyes are situated on the whole surface. Ranging from 2.5-15cm in length and about 2.5cm in width (Martin *et al.*, 2005) the tubers are available in a range of skin and flesh colours including white, cream, yellow, orange, pink, apricot, red and/or purple and with distributed range of patterns. These vividly coloured tubers often have contrasting eyes. Cooked tubers are somewhat softer than potatoes and have a variety of flavours ~ from mild to acidic and a bit sweet to starchy.

Genetics

Oxalis tuberosa alliance was first described on cytological grounds by de Azkue and Martinez (1990) as a group of morphologically similar *Oxalis* species related to the Andean tuber crop oca (*O. tuberosa* Molina) that shares the basic chromosome number x=8 which is rare in *Oxalis*.

Chromosome Numbers Reported for *Oxalis tuberosa* Molina

2n	*Reference*
14	Heitz (1927) (as *O. crenata* Jacquin); Taffedo and Escobar (1995)
63–64, 68–70	Kostoff *et al.* (1935)
66	Cardenas and Hawkes (1948)
58–66	Gibbs *et al.* (1978)
64	de Azkue and Martinez (1990); Valladolid *et al.* (1994); Medina Hinostroza (1994); Valladolid (1996); Vinueza Vela (1997)
24, 32, 40, 48, 64	Guaman Calderon (1997)
49	Hayano Kanashiro (1998)

Source: https://www.ncbi.nlm.nih.gov/pmc/articles/PMC4233841/

Basic number of chromosomes has been established as x=11 though there are also reports that this has diploid number even of 14. There are reports of ocas being aneuploids (2n=24, 32, 40, 48, 49, 58, 63, 64 and 68-70) and hexaploid (2n=6x=66). Frequency of diploids, triploids, tetraploids, pentaploids and hexaploids as well as those that are not exactly euploid need to be explained clearly.

Nutritional Value

Oca is a source of carbohydrates, dietary minerals and protein (Hermann and Erazo, 2000; King and Gershoff, 1987). Its cultivars vary substantially in their nutritional contents. The bitter forms of oca contain appreciable amounts of calcium oxalate. This is an important source of calcium and iron in the diet though a few available figures do not show exceptionally high proportions of these elements.

As per en.wikipedia.org/wiki, edible part of oca tuber is composed of water (83.8-84.1 per cent), energy (255-264kJ/100g), protein (1 per cent), fat (0.6 per cent), carbohydrate (13.3-13.8 per cent), fibre (0.8 per cent), ash (0.8-1 per cent), calcium (2-4mg/100g), iron (0.8-1.6mg/100g), phosphorus (34-36mg/100g), vitamin A (trace), thiamine (0.05mg/100g), riboflavin (0.07-0.13mg/100g), niacin (0.4mg/100g) and ascorbic acid (37-38.4mg/100g) (en.wikipedia.org/wiki).

Nutritional Value

Nutrient per 100g	*Fresh*	*Dried*
Energy	255kJ (61kcal)	1,360kJ (330kcal)
Water (g)	84.1	15.3
Protein (g)	1.0	4.3
Carbohydrates (g)	13.3	75.4
Ash (g)	1.0	3.9
Calcium (mg)	2	52
Phosphorus (mg)	36	171
Iron (mg)	1.6	9.9
Retinol (μg)	1	0
Riboflavin (mg)	0.13	0.08
Niacin (mg)	0.43	0.85
Vitamin C (mg)	38.4	2.4

Source: https://en.wikipedia.org/wiki/Oxalis_tuberosa

Using proximal chemical analysis, Chirinos *et al.* (2009) compared and described the chemical composition of *O. tuberosa* when fresh and when left in the sun.

Parameters	*Fresh Oca*	*Dehydrated (80°C)*	*Sunlight Oca*	*Dehydrated (80°C)*
Moisture (per cent)	80.1	15.1	51.0	13.0
Ash (per cent)	3.9	7.3	4.3	8.1
Total sugar (per cent)	14.1	59.1	36.4	61.8
Reducing sugars (per cent)	4.7	21.9	21.1	34.4
Nonreducing sugars (per cent)	9.4	37.2	15.6	27.2
Fiber (per cent)	0.8	6.8	3.0	7.3
Protein (per cent)	1.1	8.6	5.3	9.8
pH	4.54	6.30	5.70	6.0
Starch (per cent)	10.6	23.7	7.2	17.4
Oxalic acid (mg/100g)	135	31.4	72	54.4
Ascorbic acid (mg/100 g)	187.19	71.93	65.34	41.96

Chemical composition of *O. tuberosa* when fresh, subjected to sunlight and dehydrated. (Chirinos *et al.*, 2009).

Oxalates

Oca and other members of the genus contain an antinutritive factor called oxalate. Twelve South American and two New Zealand cultivars of oca were analyzed for total and soluble oxalate contents of the tubers. The range of total oxalate levels was 92-221mg/100g of fresh weight. Levels of soluble and total oxalate extracted from the tubers were not significantly different, suggesting that no calcium oxalate is formed in the tubers. The oxalate concentrations obtained in this study for oca suggest that previously reported values are too low and that oca is a moderately high oxalate-containing food. This is the first report of a tuber crop containing moderate to high levels of soluble oxalates in the tubers and no insoluble oxalates (Ross *et al.*, 1999).

Cultivars

Andean farmers cultivate numerous varieties of oca (Pete Wells, 2022) and often name them based primarily on tuber morphology (Terrazas and Valdivia, 1998) and secondarily on flavour (Bradbury and Emshwiller, 2011). The common names may include 'Ushpa Negra' (black ash) or 'Puka Panti' (red *Cosmos peucedanifolius*). Oca morphotypes are distinguished by foliar, floral, fruit, stem, and tuber characteristics, as described in the International Plant Genetic Resources Institute's document on oca descriptors (Anon., 2001).

Sweet and Sour Cultivars and their Uses

There is significant variation in oxalate concentration among varieties. The cultivars with high levels of oxalic acid are called 'Sour oca' whereas the cultivars with lower levels of oxalic acid are called 'Sweet oca' (Bradbury and Emshwiller, 2011).

- ☆ 'Sour oca': The tubers are usually processed to form a usable storage product locally called as 'khaya' in Quechua which is prepared first by soaking the tubers in water for about one month and then they are removed from water and left outside during hot sunny days and cold freezing nights until they become completely dehydrated. This process is similar to the preparation of *chunu* from bitter potatoes. These cultivars are called as *khaya* (name of the dried and processed one) or *p'usqu* (sour/fermented) in Quechua (Emshwiller, 2006) and *luk'i* in Aymara (Bradbury and Emshwiller, 2011). For use, the dried tuber is soaked in water and then eaten in soup or stews.
- ☆ 'Sweet oca': The oxalate levels of these tubers are also reduced by exposure to sunlight, without dehydration. Exposure to sunlight decreases the organic acid content and thereby increases the sweet taste

of oca tubers. Once exposed to sunlight, oca can be boiled, baked or fried. In the Andes it is used in stews and soups, served like potatoes, or can be served as a sweet. These cultivars are called as *wayk'u* (boiling) or *misk'i* (sweet/delicious) in Quechua and *q'ini* in Aymara (Bradbury and Emshwiller, 2011). Sweet types of oca are sometimes dried to form a product known as 'cavi' which is cooked with honey or cane sugar syrup.

Based on AFLP data, sour oca and sweet oca forms have distinct genetic clusters (Emshwiller *et al.*, 2009), suggesting the possibility of distinct evolutionary histories for each type of cultivars. AFLP-PCR (or AFLP) is a PCR-based tool used in genetic research, DNA fingerprinting, and in the practice of genetic engineering.

Oca diversity on one farm in Peru

By Laurenjm - Own work, CC BY-SA 3.0, https://commons.wikimedia.org/w/index.php?curid=18506888

'Apricot' *O. tuberosa* tubers

By Nzfauna - Own work, CC BY-SA 3.0, https://commons.wikimedia.org/w/index.php?curid=19354058

'Pink' *O. tuberosa* tubers

By fr:Utilisateur:Jonathaneo - French Wikipedia fr:Image:Oca.jpg, CC BY-SA 3.0, https://commons.wikimedia.org/w/index.php?curid=166437

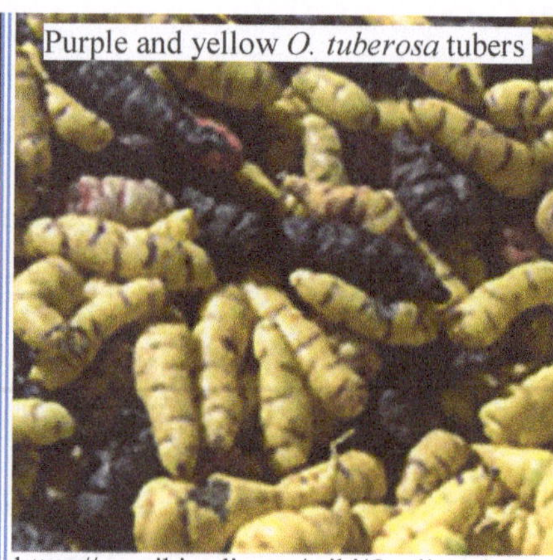

Purple and yellow *O. tuberosa* tubers

https://en.wikipedia.org/wiki/Oxalis_tuberosa

In the absence of sweetness and tartness, oca tastes very much like a potato. Both sweet and tart taste something like combination of potato and berry. When sweet but not tart, it is more like a combination of squash and potato. Most varieties increase in sweetness with long storage or exposure to light.

The tubers are one of the principal carbohydrate foodstuffs amongst the native communities of the high Andes and are boiled, roasted or candied. Mature stems can be used similarly to rhubarb (Oca: Pie, 2014:07:26). Andean communities have various methods to process and prepare tubers. In Mexico it is eaten raw with salt, lemon and hot pepper (Anon., 1989). The flavour is often slightly tangy. There is a significant difference in flavours between varieties ~ some being very acidic though certain others not at all acidic. The texture of tubers ranges from crunchy when raw or undercooked, to starchy or mealy when fully cooked (en.wikipedia.org/wiki).

Although oca is cultivated primarily for its edible stem tubers, the leaves and young shoots can also be eaten as a green vegetable. Leaves can be eaten raw and provide a sour flavour similar to other sorrels. Stems are edible, although more tarter than the tubers. Flowers are small and mild tasting.

Cultivation of Andean Tubers

Three Andean tubers: oca, mashwa and ullucu are grown in the same agro-ecological zone and their soil requirements and cultivation practices are almost similar to those of the potato, therefore, they are dealt with together. In the northern area of the Peruvian sierra the traditional form of cultivation is on Melgas. After taking the potato crop, the land is divided into 3-5 plots, each planted with one of the Andean tubers. On the high plateau of Puno and in the agro-ecological zone of the semi-humid puna, the mixture of tubers is planted. By contrast, in the Quechua agro-ecological zone, the oca and ullucu are planted together with maize. These crops show a high response to agricultural work such as hoeing, earthing up, fertilization and above all the control of pests and diseases and thus their production increases up to 40-50 tonnes per hectare which are well comparable to the highest potato yield (https://hort.purdue.edu/newcrop/1942/tubers.html).

Propagation

Though propagation by seed is possible, it is rarely used in practice because like several other species in its genus, oca flowers too exhibit tristylous heterostyly and are therefore subject to auto-incompatibility (Carrion *et al.*, 1995). Furthermore, oca plants not only produce fruits rarely but their loculicidal capsules dehisce spontaneously making quite difficult the harvesting of seeds

(Carrion *et al.*, 1995). True seeds are produced only under favourable conditions only for breeding purpose. Oca is usually propagated vegetatively by planting whole tubers or cut pieces of tuber bearing 1-3 eyes.

Growth and Development

The seed tubers are planted in the Andes in August-September and harvested from April to June (Gonzales *et al.*, 2003). Flowering and tuber formation begin around 3-4 months after planting (Carrion *et al.*, 1995). It needs a long growing season and is day length dependent.

Factors Influencing Growth and Quality

Oca is easy to propagate and has tolerance for poor soil, high altitude and harsh climates. The cultural practice for oca crop is similar to potatoes.

Being a high altitude crop, oca's growth is favoured by moderately cool day conditions and cold (possibly frosty) nights, frequently misty or cloudy weather. Oca requires short days in order to form tubers, when the day length shortens in autumn (around March in the Andes). Outside the tropics, it will not begin to form tubers until approximately the autumn equinox. If frosts occur too soon after the autumn equinox, the above ground parts of the plant will die before tubers are produced. Oca requires climates with average temperatures of approximately 10-12°C (ranging between 4 and 17°C) and average precipitation of 700 to 885mm per year (Gonzales *et al.*, 2003).

Usually oca grows on plots of marginal **soil** quality, tolerating soil pH between 5.3 and 7.8. In traditional Andean cropping systems, it is often planted after potato and therefore benefits from persisting nutrients applied to or left over from the potato crop (Gonzales *et al.*, 2003). **Land preparation and planting** are quite similar to potatoes. Before planting, the field should be ploughed or dug thoroughly 2-3 times, followed by plankings, every time the weed-rootstocks and other foreign materials should be taken out with the help of forks and mixed with FYM or compost @ 200-250 quintals per hectare of land. Now the soil is ready for bed preparation, keeping provision for irrigation channels and bunds. Planting is usually done by hand in rows or hills 80–100cm apart with plants spaced 40–60cm apart in the rows (Frere *et al.*, 1978). Emerging weeds are immediately removed to save the crop from being robbed of its nutrients. Afterwards it requires only a little attention of weedings, hillings (Gonzales *et al.*, 2003) and timely irrigations.

Oca is usually planted in a field directly after the potato harvest. A common sequence in this **rotation** system may be one year of potato, one year of oca, one year of oats or faba beans, and 2-4 years fallow (Gonzales *et al.*, 2003). Though monoculture predominates, **interplanting** with several other tuber species such as mashua and ullucu in the same field is together common in Andean production. This intercoppng often consists of different varieties of each species. Such mixed fields may later be sorted into tuber types during harvest or before cooking.

Harvest and Yield

Oca tubers mature in approximately 8-9 months and are dug by hand. Harvested tubers, especially those of the bitter varieties are generally left to cure for several days in the sun to reduce the bitterness due to calcium oxalate. Storage life is markedly affected by temperature. Storage life at 4°C is stated to be 20 weeks though at 21°C sprouting occurs and the flavour deteriorates and so this limits the storage life only to about 12 weeks. Yields vary as per cultivation practices applied. The tuber yield per hectare is about 7-10 tonnes for *O. tuberosea* though with adequate inputs and planting of virus-free propagation materials the oca production may range from 35 to 55 tonnes per hectare (Sperling and King, 1990). In Peru, the average tuber yield per hectare is 4-5t, however, with improved cultivation practices and by use of disease-free planting material this could be raised to 20t/ha.

Insect-Pests and Diseases

Though only a few insect-pests, nematodes and fungi attack this crop but in most cases these are not of much significance. Cultivation is mostly constrained by attack of Andean potato weevil (*Premnotrypes* spp.), ulluco weevil (*Cylydrorhinus* spp.) and oca weevil (possibly *Adioristidius, Mycrotrypes* or *Premnotrypes*). These **weevils** often destroy entire crop. In Peru, **tuber borer** (*Crisomelidas* spp.) also infests the tubers. For controlling all these pests, soil application of 5 per cent Heptachlor dust @ 45kg/ha at the time of planting or earthing up will not allow any underground insect to develop and damage the crop. However, **viruses** may pose a problem. A strain of 'arracacha virus B' has been found to infect the plant though no vector has been identified, and mycoplasma-like organisms (MLOs) have been found in seriously diseased material. Virus or MLOs when first observed on any plant, this should immediately be uprooted with adhering soil and buried or burnt. Further notable pests are **root-knot nematodes** (Sperling and King, 1990) which can be kept at bay by application of Furadan 3G granules @ 20kg/ha at planting and the same quantity at the time of earthing up. ***Cercospora oxalidiphila*** usually causes circular and grey to brown leaf spots which after coalescing develop large dead areas, and sometimes this even kills the leaves. In its infection the field should be kept clean of plant debris, and affected leaves should be removed. Carbendazim or mancozeb at 0.2 per cent can be sprayed in its infection. ***Colletotrichum*** spp. causes sunken brown longitudinal marks on stems with discolouration on leaves. This is mostly a seed-borne fungus which may develop slimy fungal growth during wet weather. Its spores spread through rains or irrigation splash from young plants to adjacent plants. Such plants are removed promptly. Resistant cultivars against this problem are required to be developed. ***Phoma oxalidicola*** infection causes tuber gangrene in oxalis as symptomized by little sunken areas with little wrinkling of the skin. Infected tubers rot, become wet and finally darken. While lifting, care should be taken not to damage the tubers. Any tuber at lifting if showing any such symptom should be disposed of. ***Phyllosticta*** spp. causes brown to grey near-circular leaf spot, mostly surrounded by a narrow band of reddish-purple discolouration. Though this disease is not so serious even then infected leaves should be removed to limit its spread. ***Septoria*** sp. is a weak parasite which in its infection causes leaf spots. Proper sanitation in the crop is required along with 0.2 per cent spraying of Indofil M-45. ***Puccinia oxalidis*** is mostly a seed-borne disease which overwinters on the plant remnant. Its infection to oxalis causes concentric rings of pustules full of brown spores. It develops chlorotic yellow halo around each patch of spores. If at leaf stage it is not checked, its infection may spread even to stems and floral organs and such plants are likely to die. For its control, severely infected plants should immediately be removed. Such crops should be sprayed with a fungicide containing penconazole or mancozeb at 0.2 per cent. At least for next two years, same site should not be used for oxalis cultivation. *Urocystis oxalidis* spreading through water splashing where stems and leaves develop dark raised blisters. In its infection affected parts should promptly be removed and severely affected plants should be destroyed. Following year, oxalis should be planted at a new site.

Production and Trade

Oca is consumed mainly by the rural population of the Sierra regions of the Andes and there is little information relating to production. It has been estimated that in Peru production is about 81,000t/ha, of which approximately 55,000t are used for human consumption and 12,000t for seed purposes, apart from its spoilage amounting to about 14,000t (Humanity development library 2.0).

References

1. Anon. 2001. Descriptores de oca. International Plant Genetic Resources Institute (CIP, Lima, Peru). Rome, Italy.
2. Anon. 1989. Lost Crops of the Incas: Little-known Plants of the Andes with Promise for Worldwide Cultivation. *National Academies*, pp. 83–92. National Research Council Advisory Committee on Technology Innovation, USA.

3. Anon. 2007. Teofilo Laime Ajacopa, Diccionario Bilingue Iskay simipi yuyayk'ancha, La Paz, 2007 (Quechua-Spanish dictionary).
4. Arbizu, C. and M. Tapia, 1992. Tuberculos andinos. In: Cultivos marginados: otra perspectiva de 1992 (eds Hernandez Bermajo, J.E. and S. Leon), pp. 147-61.
5. Bradbury, E.J. and E. Emshwiller, 2011. The role of organic acids in the domestication of *Oxalis tuberosa*: a new model for studying domestication resulting in opposing crop phenotypes. *Econ. Bot.*, **65**(1): 76–84.
6. Cardenas, M. and J.G. Hawkes, 1948. Numero de cromosomas de algunas plantas nativas cultivadas por los indios en los Andes. *Revista de Agricultura Bolivia*, **5**: 30–32.
7. Carrion, S., M. Hermann and B. Trognitz, 1995. La biologia reproductiva de la oca. *Boletin de Lima*, Lima, pp. 48–68.
8. Chirinos, R., L. Betalleluz-Pallardel, A. Huaman, C. Arbizu, R. Pedreschi and D. Campos, 2009. HPLC-DAD characterization of phenolic compounds from Andean oca (*Oxalis tuberosa* Mol.) tubers and their contribution to the antioxidant capacity. *Food Chem.*, **113**: 1243–51.
9. de Azkue, D. and A. Martinez, 1990. Chromosome number of the *Oxalis tuberosa* alliance (Oxalidaceae). *Pl. Syst. and Evol.*, **169**: 25–9.
10. Emshwiller, E. 2006. Evolution and conservation of clonally propagated crops: Insights from AFLP data and folk taxonomy of the Andean tuber oca (*Oxalis tuberosa*). Darwin's Harvest: New Approaches to the Origins, Evolution, and Conservation of Crops (eds Motley, T.J., N. Zerega and H. Cross), pp. 308-346. Columbia University Press.
11. Emshwiller, E., Terra Theim, Alfredo Grau, Victor Nina and Franz Terrazas, 2009. Origins of domestication and polyploidy in oca (*Oxalis tuberosa*: Oxalidaceae) 3: AFLP data of oca and four wild, tuber-bearing taxa. *Am. J. Bot.*, **96**(10): 1839–48.
12. Frere, M., J. Rea and J.Q. Rijks, 1978. *Ullucus tuberosus. Estudio agroclimatologico de la zona andina. Nota Tecnica.* Vol. 161. Organizacion Metereologica Mundial. ISBN 9263305064. OCLC 991673597.
13. Gibbs, P.E., D. Marshall and D. Brunton, 1978. Studies on the cytology of *Oxalis tuberosa* and *Tropaeolum tuberosum*. *Edinburgh Royal Botanical Garden - Notes*, **37**: 215–20.
14. Gonzales, S., F. Terrazas, J. Almanza and P. Condori, 2003. Produccion de oca (*Oxalis tuberosa*), papalisa (*Ullucus tuberosus*), e isano (*Tropaeolum tuberosum*). In: *Importancia, zonas productoras, manejo y limitantes* (eds

Cadima, X., W. Garcia and J. Ramos). Vol. 20., pp.1-46. Cochabamba, Bolivia: PROINPA.

15. Guaman Calderon, S. 1997. Conservacion *in situ* caracterizacion y evaluacion de la biodiversidad de oca (*Oxalis tuberosa*) y papalisa (*Ullucus tuberosus*) en Candelaria (Chapare) y Pocanche (Ayopaya). Ing. Agr. Thesis, Universidad Mayor de San Simon, Cochabamba, Bolivia.
16. Hayano Kanashiro, A.C. 1998. Citogenetica de *Oxalis tuberosa* Mol. Oca: Numero Cromosomico y Analisis Cariotipico. Lic. Biol. Thesis, Universidad Ricardo Palma, Lima, Peru.
17. Heitz, E. 1927. Uber multiple und aberrante Chromosomenzahlen. *Abhandlungen aus dem Gebiete der Naturwissenschaften, Hamburg*, **21**: 47–57.
18. Hermann, M. and C. Erazo, 2000. Compositional changes of oca tubers following post-harvest exposure to sunlight. *CIP Program Report*, pp. 391–6.
19. https: //en.wikipedia.org/wiki/Oxalis_tuberosa.
20. https: //www.ncbi.nlm.nih.gov/pmc/articles/PMC4233841/
21. Humanity development library 2.0. Root crops: Oca (*Oxalis tuberosa*).nzdl. org.
22. King, S.R. and S.N. Gershoff, 1987. Nutritional evaluation of three underexploited Andean tubers: *Oxalis tuberosa* (Oxalidaceae), *Ullucus tuberosus* (Basellaceae) and *Tropaeolum tuberosum* (Tropaeolaceae). *Econ. Bot.*, **41**(4): 503–11.
23. Kostoff, D., H. Dogadkina and A. Tichonowa, 1935. Chromosome number of certain Angiosperm plants (*Nicotiana*, *Petunia*, *Oxalis*, *Secale*, and *Punica*). *Doklady Academiia Nauk URSS,* **3**: 401.
24. Medina Hinostroza, T.C. 1994. Contaje cromosomico de la oca (*Oxalis tuberosa* Molina) conservada *in vitro*. Ingeniero Agronomo Thesis, Universidad Nacional del Centro del Peru, Huancayo, Peru.
25. Oca: Pie. 2014-07-26.
26. Oca: What does it taste like?. 2015-03-26.
27. Pete Wells, 2022 (Feb. 6). What's for Dinner? Probably Not London Porker. *New York Times*. Review of Dan Saladino, *Eating to Extinction: The World's Rarest Foods and Why We Need to Save Them.*
28. Ross, A.B., G.P. Savage, R.J. Martin and I. Vanhanen, 1999. Oxalates in oca (New Zealand yam) (*Oxalis tuberosa* Mol.). *J. agric. Food Chem.*, **47**(12): 5019-22. https: //doi.org/10.1021/jf990332r

29. Sperling, C.R. and S.R. King, 1990. Andean Tuber Crops. Worldwide Potential Advances in New Crops: Proc. of the First Nat. Symp. New Crops - Research, Development, Economics (eds Janick, J. and J.E. Simon). held on Oct. 23-26, 1988 at Indianapolis, Indiana. Timber Press, USA.
30. Talledo, D, and C. Escobar, 1995. Citogenetica de *Oxalis tuberosa*: ciclo celular y numero cromosomico. Biotempo. *Universidad Ricardo Palma, Lima* (*Peru*), 2: 33–46.
31. Terrazas, F. and G. Valdivia, 1998. Spatial dynamics of *in situ* conservation: handling the genetic diversity of Andean tubers in mosaic systems. *Pl. Genet. Res. Newsl.*, **114**: 9–15.
32. Valladolid, A, C. Arbizu and D. Talledo, 1994. Niveles de ploidia de la oca (*Oxalis tuberosa* Mol.) y sus parientes silvestres. *Agro Sur* (*Universidad Austral de Chile, Facultad de Ciencias Agrarias*) 22 (numero especial), pp. 11–12.
33. Valladolid, A. 1996. Niveles de ploidia de la oca (*Oxalis tuberosa* Mol.) y sus parientes silvestres. M.Sc. Thesis, Universidad Nacional Agraria La Molina, Lima, Peru.
34. Vinueza Vela, J.F. 1997. Evaluacion y Caracterizacion Citogenetica de 20 Entradas de Oca (*Oxalis tuberosa* Molina) Recolectadas en Ecuador. Ing. Agr. Thesis, Universidad Central del Ecuador, Cutuglahua, Pichincha, Ecuador.

20

Lotus [*Nelumbo* (*Nelumbium*) Adans.]

Family: Nelumbonaceae. **Species**: *Nelumbo lutea* Willd. and *N. nucifera* Gaertn.

Common name: Lotus.

Common names of *Nelumbo lutea* (syn. *Nelumbo pentapetala*): American lotus, Duck acorn, Lotus lily, Pondnut, Volee, Water-chinquapin, Winkapin, Yanquapin and Yellow lotus.

Common names of *Nelumbo nucifera* (syn. *N. indica, N. komarovii* Grossh., *N. speciosa, Nelumbium speciosum* Willd., *N. nelumbo*): Chinese waterlily, East Indian lotus, Indian lotus, Laxmi lotus, Magnolia lotus (*Nelumbo nucifera* var. *alba*), Nelumbo, Nelumbium, Sacred lotus and Sacred lotus of the Nile. **Hindi and Sanskrit names**: Jalaj, Kamal, Neeraj, Padma, Pankaj, Pundareek, Purain, Pushkar, Raajeev, Saroj, Toyaj. Vaarij, *etc.*; **Asomese**: Podum; **Bengali**: Padma; **Gujarati**: Suriya kamal; **Kashmiri**: Pamposh; **Khasi**: Soh-lapuding; **Kanarese**: Kamola and Tarareddade; **Malayalam**: Senthamara and Thamara; **Marathi**: Kamal; **Mundan**: Kombolba, Salukidha and Upalba; **Oriya**: Padam; **Punjabi**: Kanwal and Pamposh; **Tamil**: Ambal and Thamarai; and **Telugu**: Kalang and Erra-tamra.

Introduction and Origin

The genus name is derived from the Tamil word *nelum* (means blue), the name for *Nelumbo nucifera*. Some literature sources also mention it native to

Fresh lotus seeds ready to eat (*N. nucifera*)

By Basile Morin - Own work, CC BY 4.0, https://commons.wikimedia.org/w/index.php?curid=64302105

Sri Lanka. The name 'lotus' is also applied to various other plants and plant groups, including the unrelated genus *Lotus*. *Nelumbo*, the only living genus in Nelumbonaceae, comprises only two known living species of aquatic plants: *Nelumbo lutea* (American lotus) a native to the New World, *i.e.* North America (from New York and South Ontario to Minnesota, Nebraska, Florida, Texas and Louisiana) and South America (West Indies and Colombia) whereas *Nelumbo nucifera* (syn. *N. speciosa*; commonly known as sacred or Indian lotus) is native to the Old World, *i.e.* Asiatic, distribution ranging through southern Asia, the Philippines, Indonesia, Combodia, Myanmar, Korea, Japan, China, Taiwan, India, Thailand, Sri Lanka and probably northern Australia (Misra and Misra, 2013). It is also found in Hawaii. It is naturalized in the Volga River Delta where the river flows to the Caspian Sea. The first fossil record of lotus is available with Pleistocene epoch of Kashmir, and subsequently from the early tertiary period rocks of Asom as had been evident through the impressions of leaves and rhizomes of lotus found in a collection from the Eocene bed near Damalgiri, which reinforces the belief that lotus is indigenous to Indian subcontinent. Horticultural hybrids have been produced between these two allopatric species.

Botany

Nelumbo is a perennial aquatic herb that roots in mud with several metres (up to 12m) long chains of creeping white globulous rhizomes that produce ~ at intervals ~ a single leaf and a single flower. The first young leaves when emerging out of the water are tightly rolled and come out forcefully even piercing its own parts and on opening these are round, flat and surface-floating, followed by waxy and bluish-green adult leaves attached with quite long (0.6-2.4m) cylindrical leaf stalks in its centre, on which leaves of inverted parasol-shape remain intact usually well above the water level. The diameter of adult leaves is from 45cm to 90cm with frilled edges, being larger in *Nelumbo nucifera* than *N. lutea.* Sprinkling of water on the leaves appears as mercury beads, and during rains the rain drops merge and collect in the centre but when it becomes heavy, leaves tilt to one side to expel the collected water. Single flowers are large *cum* magnificent, quite showy, fragrant, velvety white, cream, greenish-white, yellow, sulphur-yellow, pink, rose, rosy-carmine, red, mauve, *etc.* but the true species bears sulphur-yellow flowers with dark yellow stamens in *Nelumbo lutea* (the American species) and deep rose (on opening) to pale-rose and then creamy-white (with age) with yellow stamens in *N. nucifera* (the Asiatic species), flowers being larger in the Asiatic species and are borne on a strong peduncle which is higher than the leaves. Sepals are 4-5, petals many (usually 16-36; rarely up to 160 as in *N. nucifera* var. 'Krishna', a pink double race collected by NBRI, Lucknow from a pond in Midnapore, W.B.), erect-spreading, stamens many with short filaments and fruit a large, very attractive, funnel-shaped but flat-topped with perforated receptacle containing many carpels. The funnel-shaped seed pod which is up to 10cm in diameter looks like the rose of a watering can but with flat and cushioned tops, initially yellow but maturing to a brown woody texture and this is used in dry flower arrangements. The surface of these pods contains many (7-15) edible nut-like seeds individually in its sunken cavities. Viability of its seeds is enormous as a paleobotanist of Japan collected its seeds from Manchuria estimated to be 1,000 years old and these successfully germinated. Its centuries old seeds collected from various tombs also germinated. The basic chromosome number in lotus is n=8 and the diploid number in *Nelumbo nucifera* being 2n=16.

Species and Varieties

What details have been given above about *Nelumbo* holds true for the species, *viz.*, *N. lutea* and *N. nucifera* including the colour of the flowers except that the leaves, the flowers, the petioles, the peduncles and the rhizomes in latter species are significantly larger as compared to the former. Many literature sources mention certain other species but those are in fact the various forms of these elemental species. *N. lutea* is though bold and quite beautiful but could not gain that popularity which it deserved. *N. flavescens* introduced by Sir Marliac bearing a strongly scented soft lemon-yellow flower is a smaller form of *N. lutea* and has a red spot on the base of each petal and centre of each leaf, and var. 'Grossherzog Ernst Ludwig' (*N. flavescens* × *N.* 'Osiris') bears salmon-rose flowers. *N. alba* is in fact a form of *N. nucifera* having white flowers, *N. n.* var. *alba grandiflora* (syn. *N. grandiflora*) has white flowers with golden stamens but larger with flowers

Flower of *Nelumbo nucifera*

By T.Voekler - Own work, CC BY-SA 3.0, https://commons.wikimedia.org/w/index.php?curid=8142220

Nelumbo lutea

By liz west from Boxborough, MA - American lotusUploaded by Epibase, CC BY 2.0, https://commons.wikimedia.org/w/index.php?curid=8400572

(more than 30cm in diameter), *N. n.* var. *alba striata* (syn. 'Empress') produces lovely fragrant and globular flowers where petals are tipped red with red stripes on edges, *N. n.* var. *alba plena* (syn. 'Shiroman') is a Japanese variety with large, globular, fully double and creamy flowers tinged with faint green but become snow-white with age, the flower size under optimum conditions is almost 30cm in diameter and most suitable as cut flower when the cut end is sealed, *N. n.* var. *pygmaea alba* is a dwarf variety with about 15cm diameter of leaves and 10cm across flowers of white colour, *N. n.* var. *pygmaea alba plena* is a dwarf and double

Lotus rhizomes

By FotoosRobin - originally posted to Flickr as Lotus root, CC BY-SA 2.0, https://commons.wikimedia.org/w/index.php?curid=4826529

variety of white colour, *N. n.* var. *pygmaea rosea* is a dwarf variety with flowers of bright rose colour, *N. n.* var. *rosea* with deep rosy-pink flowers, *N. n.* var. *rosea plena* with double and deep rosy-pink flowers, *N. n.* var. 'Osiris' with deep rose flowers, *N. n.* var. *pulchra* with dark rosy-red flowers, *N. n.* var. *pekinensis rubra* with rosy-carmine flowers, *N. n.* var. *pekinensis rubra plena* with very large, double and rosy-carmine flowers, *N. n.* var. *gigantea* bears large and rose-purple flowers, *N. n.* var. 'Kinshiren' with white flowers but shaded pink, *N. n.* var. *kermesina* with light rose flowers, vars. 'Angel Wings' with tulip-shaped, fragrant, white and large (20-25cm flower diameter) flowers where inner petals are beautifully rolled, 'Chawan Basu' is a semi-dwarf and free-flowering variety suitable for tub culture where white flowers are edged with pink, 'Charles Thomas', dwarf pink variety suitable for small tubs and pools, 'Maggie Belle Slocum', large (25-30cm across) mauve flowers with inner petals beautifully rolled and suitable for tub culture, 'Momo Botan', small plants with 30-45cm across leaves, flowers fully double, paeony-shaped and carmine-red, suitable for small pools and tubs, 'Mrs Perry D. Slocum', most beautiful and vigorous among all the existing varieties with double flowers about 30cm across, opening rose-pink but changing to creamy-yellow with pink edges and then to complete creamy-yellow, all the stages appealing quite good but the second stage is most attractive, *etc.*

Across the country from tropical to subtropical Gujarat to sub-temperate and temperate Sikkim, and absolute tropical Andaman and Nicobar Islands, and from temperate, subtemperate and subtropical Jammu and Kashmir to absolute tropical Cape Comorin, the lotus plants can be found growing naturally in plenty in the ponds, lakes, water reservoirs and even in quite slow moving waters predominantly with rose colour though white to purple forms are also available frequently. These are mostly found growing along with waterlilies. In West Bengal and Odisha, almost every pond has this plant and that too with great variation in flower colours. Most of these races (some 35) of *Nelumbo nucifera* collected from 11 states of India and 25 races collected from the botanical gardens of Australia, Brazil, Japan, Thailand, United Kingdom and United States of America are being maintained at NBRI, Lucknow. There is much scope of improvement in this crop using both the species *vis-à-vis* their natural existing varieties and the result will be overwhelming when both the species are combined together.

As mentioned in en.wikipedia.org/wiki/Nelumbo, the recorded fossil species include:

- *Nelumbo aureavallis* Hickey – Eocene (North Dakota), described from leaves found in the Golden Valley Formation in North Dakota, USA (Hickey, 1977).
- *Nelumbo changchangensis* – Eocene (Hainan Island, China), described from several fossils of leaves, seedpods, and rhizomes from the Eocene-

aged strata in the Changchang Basin of Hainan Island.

- ☆ *Nelumbo minima* – Pliocene (Netherlands), described from leaves and seedpods that suggest a very small plant. Originally described as a member of the genus *Nelumbites*, as *Nelumbites minimus*.
- ☆ *Nelumbo nipponica* – Eocene-Miocene, fossil leaves are known from Eocene-aged strata in Japan, and Miocene-aged strata in Russia.
- ☆ *Nelumbo orientalis* – Cretaceous (Japan), one of the oldest known species, fossils are found in Cretaceous-aged strata of Japan.
- ☆ *Nelumbo protolutea* – Eocene (Mississippi), fossils of leaves strongly suggest a plant similar in form to the American lotus.

Chemical Composition

The leaves of *N. nucifera* comprises the flavonol miquelianin as well as the alkaloids (+)-(1R)-coclaurine and (–)-(1S)-norcoclaurine (Kashiwada *et al.*, 2005; en.wikipedia.org/wiki/ullucus). Plant also contains nuciferine, neferine and many other benzylisoquinoline alkaloids with medicinal properties (Menendez-Perdomo and Facchini, 2018, 2020). Lotus rhizome is used to add seasoning to food. It is a moderate calorie root vegetable (100g of rhizome provides about 74 calories), composed of several vitamins, minerals and nutrients. It contains about 83.8 per cent water, 0.11 per cent fat, 1.56 per cent reducing sugar, 0.41 per cent sucrose, 2.7 per cent crude protein, 9.25 per cent starch, 0.8 per cent fibre, 0.1 per cent ash and 0.06 per cent calcium (Mukherjee *et al.*, 1996c). Its 100g of root provides 44mg of vitamin C or 73 per cent of daily recommended values (RDA). Lotus rhizome and its extracts have shown diuretic, psychopharmacological, anti-diabetic, anti-obesity, hypoglycemic, antipyretic and antioxidant activities (Huralikuppi *et al.*, 1991; Mukherjee *et al.*, 1996a,b,c; 1997). Lotus seeds contain high quality proteins and are rich in variety of essential amino acids including high contents of albumin (42 per cent) and globulin (27 per cent). These contain unsaturated fatty acids, carbohydrates, vitamins [VB1, VB2, VB6 and E (Zheng *et al.*, 2003; Wu *et al.*, 2007)], calcium, iron, zinc, phosphorus and other trace elements (Tang *et al.*, 1998; Shad *et al.*, 2013). These also provide water-soluble polysaccharides, alkaloids, flavonoids, superoxide dismutase and other bioactive components. (Zheng *et al.*, 2004). The functional components (polyphenols, protein, polysaccharides) in *N. nucifera* seeds can help combatting high blood pressure, diabetes and gallstones (Wu *et al.*, 2013). Water-soluble polysaccharides of lotus-seeds have also shown to promote lymphocyte transformation and enhancing the immune function (Sung *et al.*, 2011). After lotus seed germination, crude protein and fat levels in the endosperm significantly increase. It is therefore an important method to enhance its nutritional quality (Xia, 2012; Zhang *et al.*, 2015). However, the nelumbine alkaloid obtained from its petiole, pedicel and seed embryo is a cardiac poison.

Lotus Root, Cooked, No Salt

Nutritional Value per 100g

Energy	278 kJ (66 kcal)
Carbohydrates	16.02g
Sugars	0.50g
Dietary fiber	3.1g
Fat	0.07g
Protein	1.58g
Vitamins	***Quantity (Per cent DV†)***
Thiamine (B1)	11 per cent (0.127mg)
Riboflavin (B2)	1 per cent (0.01mg)
Niacin (B3)	2 per cent (0.3mg)
Pantothenic acid (B5)	6 per cent (0.302mg)
Vitamin B6	17 per cent (0.218mg)
Folate (B9)	2 per cent (8μg)
Choline	5 per cent (25.4mg)
Vitamin C	33 per cent (27.4mg)
Minerals	***Quantity (Per cent DV†)***
Calcium	3 per cent (26mg)
Iron	7 per cent (0.9mg)
Magnesium	6 per cent (22mg)
Manganese	10 per cent (0.22mg)
Phosphorus	11 per cent (78mg)
Potassium	8 per cent (363mg)
Sodium	3 per cent (45mg)
Zinc	3 per cent (0.33mg)
Other Constituents	***Quantity***
Water	81.42g

Units: μg = micrograms; mg = milligrams; IU = International units.

†Percentages are roughly approximated using US recommendations for adults. *Source*: USDA Food Data Central

Source: en.wikipedia.org/wiki/Nelumbo_nucifera

Uses

All parts of *Nelumbo nucifera* are edible with the rhizome and seeds being the main consumption parts. Lotus rhizomes and seeds and their processing by-products are widely consumed in Asia, Americas, and Oceania for high content of physiologically active substances (Mukherjee *et al.*, 2009).

Its rhizomes are white to cream or orange and are known as *kamal-kakdi* (in U.P., M.P., Bihar, Rajasthan), *pein* or *bhein* (in H.P. and Punjab), *etc.* which are rich in vitamins, fibres, starch, sugar, proteins, fat and calcium and are used for pickling and for cooking vegetables. Rhizomes also yield a type of edible flour or arrowroot which is aromatic and sweet in taste. The flour is rich in ascorbic acid, iron, calcium, phosphorus, fibre, starch, sucrose, fat and protein. Its tender leaves, pedicels (petioles), floral stalks (peduncles) and even flowers are also cooked as vegetables. The torus (fruiting body) embedded with edible carpels is also sold in the market as *kamalgatta* because of the edible carpels which are sweet for taking raw or by roasting or boiling. The seeds are also dried and ground for making a typical flour to be eaten during fast with milk and water or for loaf making.

Traditionally rhizomes, leaves and seeds have been used in folk medicines, Ayurveda, Chinese traditional medicine and Oriental medicine (Khare, 2004; Sridhar and Bhat, 2007). Leaves are used for hematemesis, epistaxis and hematuria; the flowers for lowering blood sugar levels (Lotus, 2020), *vis-à-vis* curing diarrhoea, cholera, fever and hyperdipsia (Khare, 2004; Sridhar and Bhat, 2007); and the rhizomes are reported to contain purported diuretic, antidiabetic and anti-inflammatory properties (Mukherjee, 1996a; 2009; Khare, 2004; Sridhar and Bhat, 2007). Rhizome flours offer a remedy against diarrhoea and dysentery in children and its paste being anti-fungal is used against ringworm and cutaneous infections. Milky and viscid juice obtained from leaves and floral stalk is effective against diarrhoea. Its plants, to a limited extent, have the ability to purify water in ponds, lakes and other water bodies by absorbing heavy metals from the water polluted due to industrial effluents. Rural folks use its leaves as dining plates on various social gatherings.

Propagation

Lotus is propagated vegetatively through rhizomes and sexually through seeds. The multiplication through rhizome is very easy as the rhizomes having three nodes with at least one active bud are planted horizontally 7-10cm deep in the soil of natural ponds or lakes as being normally most suitable for their growth and multiplication. In case of artificially created pools, the bottom should be filled to a depth of 45-50cm with 3 parts clayey soil being rich in organic matter and 1 part of FYM, plus 100g each of *neem* cake and diammonium phosphate per square metre area. The water level at any time should not be less than 30cm.

Through seeds these are grown in February-March in tropical and subtropical conditions but in April-May in subtemperate and temperate conditions. The viability of lotus seeds is extremely high and manifold than any other flowering plant. Scarified (sand paper rubbing at both the ends) and stratified (in H_2SO_4), seeds are sown for better results. Properly scarified seeds germinate within a week at room temperature and after next 10-20 days the seedlings become ready for transplanting. At 10 days of germination these attain 2-leaf stage. Transplanting is done in the pools having very shallow water but with the growth the level of the water will have to be increased.

Approximately 10-12kg of seeds or 45kg of rhizome pieces are required for planting one hectare of pond. Planting density is low because of the very rapid growth of the rhizomes, reported as up to 15m^2 per year (Humanity development library 2.0).

Growth and Development

Timely planted pools may start producing flowers from April-end and may continue blooming up to September. Seed setting also takes place by September-end to October. The flowers last for about four days and then one by one petals start shedding causing protrusion of the yellow torus with flat but spongy top which turn to green after sometime and then woody-brown at maturity bearing 7-15 seeds per fruit. Rhizomes can be harvested by October end.

Season

The best time for planting in the subtropical, subtemperate and temperate areas is April when water temperature has risen substantially. However, in the tropical climates it can be planted throughout the year, preferably when most of the plants from a lotus pool start the new growth. This period normally occurs under tropical conditions after about one month of cessation of flowering of most of the plants. It should not be planted at a situation where water may freeze up to the bottom as their rhizomes rot under such conditions. Even in 'Dal Lake' of Srinagar there is one lotus garden. Wooler and many other lakes in Kashmir harbour lotus gardens in them.

Cultural Practices

Cultivation of lotus is most simple. Wherever these are to be cultivated, it should be at an open and sunny situation with the pH reaction of the water as well as of soil not more than 8.5, preferably 5.5-7. Rhizomatous rootstocks are planted when growth on them is visible. However, if rhizomes are dug up before initiation of growth, they should be kept at a warm place for starting the growth or should immediately be planted in an open pond having warmer water.

The rhizomes having at least three nodes (at least one node should have started) are set horizontally 7-10cm below the soil at a shallow bank of the lake or pond in a manner that these may not move from its place of planting so after planting a brick or stone should be laid over the rhizome to keep the rhizome securely in position. If by any reason the planting is delayed, these may be planted timely in pots or other containers from which these plants should be shifted in the pond when congenial time approches. The natural soil and deposits in the lake is normally the same which these plants require. In created ponds, the bottom of the pool should be filled with about 45cm thick layer of 2 parts fibrous loam mixed with 1 part of well-rotten cowdung or farmyard manure or 3 parts clay soil and 1 part farmyard manure and per square metre of area some 100g each of *neem* cake and diammonium phosphate should be mixed. In case if properly decomposed manure is not available and there is sufficient time for planting, the fresh dung and the litters should be stacked in layers with clayey soil in the prepared pit which may be turned over twice or thrice during spring for proper decomposition. For tub-planting also 2:1 fibrous loam and well-rotten cow dung manure should be used but during active growth the plants should frequently be fed with liquid fertilizers.

Harvest and Yield

When flowers start emerging these are pulled up at bud stage to divert their energy towards rhizome formation. By the end of October the rhizomes may be lifted and kept at a well-aerated *cum* warmer place for starting the buds. These are now ready for marketing. The yield of rhizome is approximately 30-50 quintals per hectare depending upon the variety, climate and soil conditions.

Pests and Diseases

Weed is not a problem when lotus has been grown in tubs as the growing weeds are pulled up gently but in ponds since several livestocks such as fish, frogs, snails and certain other animals are there so spray of chemicals may prove disastrous to such livestocks. However, Simazin 3-6ppm or Monuron 4-12ppm is safe for lotus *vis-a-vis* livestocks and will also kill the weeds. In ponds, water hyacinths (*Eichhornia crassipes*), water lettuce (*Pistia stratiotes*), azolla, *Typha*, and *Cyperus papyrus* may sometimes create the problem. Out of these, *Typha* and *Cyperus* can easily be removed by pulling them whole, and to a greater extent even water hyacinths may also be collected and taken out but azolla algae and the blanket weeds can be controlled only by the use of $KMnO_4$ and if there is no fear of livestocks even $CuSO_4$ can also be used.

Lotus leaf spot is caused by *Cercospora* and *Ovularia* species, starting as dark patches on the leaves or drying of the leaf margins. Such leaves should immediately be removed and burnt. At initial attack a mild solution of Bordeaux mixture will also prove effective if sprayed at every 2-3 days. Sometimes even root rot appears due to *Phytophthora* infection which is very difficult to control, however, such plants should be uprooted alongwith adhering soil and destroyed.

Small dark brown beetles (*Galerucella nymphaeae*) which are almost double the size of ladybird, attack the lotus during summer. Though its attack is not so common but when attacked these should immediately be controlled. If these are not checked in time their population build up being enormous (several generations occuring in a season), these devastate the crop as larvae feed on the leaves voraciously. These can be dislodged in the water by jetting the foliage with fresh water where fishes and frogs will swallow them. A solution containing powdered *Bacillus thuringiensis* bacteria when sprayed on such plants parasitize the larvae and kill them. Various aphids such as *Rhopalosiphum* species or greenflies feed on the leaves sucking the sap which hampers the vigour of the plants. Jetting the foliage with fresh water will dislodge them in water to be swallowed by the water inhabitants. Slugs and snails may also feed on the leaves and flowers which can be trapped by floating lettuce or cabbage leaves for 24 hours in the vicinity where all these pests will start feeding on these leaves and then these are collected and destroyed. Leaf stalk borers are quite damaging for the crop which can be controlled by cutting back the affected parts and destroying them. In this case the new growth will appear soon.

References

1. Hickey, L 1977. Stratigraphy and Paleobotany of the Golden Valley Formation (Early Tertiary) of Western North Dakota. Geological Society of America, Boulder, Colorado; pp.110 + 5 Plates.
2. https: //en.wikipedia.org/wiki/Nelumbo_nucifera.
3. Humanity development library 2.0. Root crops: Lotus root (*Nelumbo nucifera*). nzdl.org.
4. Huralikuppi, J.C., A.B. Christopher and P. Stephen, 1991. Antidiabetic effect of *Nelumbo nucifera* (Gaertn): Part I: Preliminary studies in rabbits. *Phytother. Res.*, **5**(2): 54–8.
5. Kashiwada, Y., A. Aoshima, Y. Ikeshiro, Y.P. Chen, H. Furukawa, M. Itoigawa, T. Fujioka, K. Mihashi, L.M. Cosentino, S.L. Morris-Natschke and K.H.

Lee, 2005. Anti-HIV benzylisoquinoline alkaloids and flavonoids from the leaves of *Nelumbo nucifera*, and structure–activity correlations with related alkaloids. *Bioorganic and medic. Chem.*, **13**(2): 443–8.

6. Khare, C.P. 2004. Indian Herbal Remedies: Rational Western Therapy, Ayurvedic, and Other Traditional Usage. Botany (1st edn), Springer, USA, pp. 326–7.

7. Menendez-Perdomo, I.M. and P. J. Facchini, 2018. Benzylisoquinoline Alkaloids Biosynthesis in Sacred Lotus. *Molecules*, **23**(11): 13-7.

8. Menendez-Perdomo, I.M. and P.J. Facchini, 2020. Isolation and characterization of two O-methyltransferases involved in benzylisoquinoline alkaloid biosynthesis in sacred lotus (*Nelumbo nucifera*). *J. Biol. Chem.*, **295**(6): 1598-1612.

9. Misra, S. and R.L. Misra, 2013. *Nelumbo* (*Nelumbium*). Commercial Ornamental Bulb Science, pp. 306-12. Westville Publishing House, New Delhi.

10. Mukherjee, P.K., J. Das, K. Saha, M. Pal and B.P. Saha, 1996a. Diuretic activity of the rhizomes of *Nelumbo nucifera* Gaertn (Fam. Nymphaeaceae). *Phytother. Res.*, **10**(5): 424–5.

11. Mukherjee, P.K., K. Saha, R. Balasubramanian, M. Pal and B.P. Saha, 1996b. Studies on psychopharmacological effects of *Nelumbo nucifera* Gaertn. rhizome extract. *J. Ethnopharmacol.*, **54**(2): 63–7.

12. Mukherjee, P.K., J. Das, K. Saha, S.N. Giri, M. Pal and B.P. Saha, 1996c. Antipyretic activity of *Nelumbo nucifera* rhizome extract. *Indian J. exper. Biol.*, **34**(3): 275–6.

13. Mukherjee, K., K. Saha, M. Pal and B. Saha, 1997. Effect of *Nelumbo nucifera* rhizome extract on blood sugar level in rats. *J. Ethnopharmacol.*, **58**: 207–13.

14. Mukherjee, P.K, D. Mukherjee, Amal K. Maji, S. Rai and Michael Heinrich, 2009. The sacred lotus (*Nelumbo nucifera*) - phytochemical and therapeutic profile. *J. Pharm. and Pharmacol.*, **61**(4): 407–22.

15. Shad, M.A., H. Nawaz, F. Siddique, J. Zahra and A. Mushtaq, 2013. Nutritional and functional characterization of seed kernel of lotus (*Nelumbo nucifera*): application of response surface methodology. *Food Sci. Technol. Res.*, **19**(2): 163–72.

16. Sridhar, K.R. and R. Bhat, 2007. Lotus: a potential nutraceutical source. *J. Agri. Technol.*, **3**: 143–55.

17. Sung, J., J. Sung and H. Shin, 2011. Cytoprotective effects of lotus (*Nelumbo nucifera* Gaertner) seed extracts on oxidative damaged mouse embryonic fibroblast cell. *Food Sci. Biotechnol.*, **20**: 1533–7.

18. Tang, P., Z. Jiang, C. Mei and H. Jiang, 1998. The composition, solubility and quality of lotus seed. *J. Beijing Norm. Univ.* (*Nat. Sci.*), **34**: 532–7.

19. Wu, C., Y. Xia and X. Tang, 2013. Optimization of fermentation process for lotus rice-wine production by response surface methodology. *Mod. Food Sci. Technol.*, **29**: 1675–9.

20. Wu, J., Y. Zheng, T. Chen, J. Yi, L. Qin, K. Rahman and W. Lin, 2007. Evaluation of the quality of lotus seed of *Nelumbo nucifera* Gaertn. from outer space mutation. *Food Chem.*, **105**: 540–47.

21. www.webmd.com. Retrieved 2020-02-18.

22. Xia, Y. 2012. Effect of germination on chemical and functional properties of lotus seeds. *Food Sci.*, **33**: 91–8.

23. Zheng, B., J. Zheng and S. Zeng, 2003. Analysis of the nutritional composition in Chinese main lotus seed varieties. *Acta Nutr. Sin.*, **25**: 153–6.

24. Zheng, B., J. Zheng and C. Zeng, 2004. Study on three functional components of Chinese main lotus seed varieties. *Acta Nutr. Sin.*, **26**: 158–60.

25. Zhang, Yi, Xu Lu, Shaoxiao Zeng, Xuhui Huang, Zebin Guo, Yafeng Zheng, Yuting Tian and Baodong Zheng, 2015. Nutritional composition, physiological functions and processing of lotus (*Nelumbo nucifera* Gaertn) seeds: A review. *Phytochem. Rev.*, **14**(3): 321.

21

Nasturtium (*Tropaeolum tuberosum* Ruiz and Pav.)

Family: Tropaeolaceae.

Common names: Mashua or Tuberous nasturtium; **Ecuador and Peru**: Mashua/Mashwa (Anon., 2008); **Peru**: Maswallo, Mazuko, Mascho; **Peru** and **Bolivia**: Anu; **Colombia**: Isano and Cubio; **Boyaca and Colombia**: Nabu.

Introduction and Origin

The genus *Tropaeolum* has a wide geographical distribution and seems to be very variable. There are an estimated 50 species in Mexico and C and S America. Wild species in Peru can be found on the low ridges of the Peruvian coast, on the edges of forests or growing sympatrically with cultivated mashwa in the Andes (Hernando Bermejo and Leon, 1994). *Tropaeolum tuberosum* is an edible tuber crop native to the high Andes of South America up to an elevation of 2,800-4,000m a.s.l. but with a greater concentration being at 3500-3800m elevations. It is little known outside of its native production areas. It is cultivated for its edible tubers mainly in the Andean valleys of Peru and Bolivia and to a lesser extent in Ecuador as well as in some areas of Colombia. It is of economic value as a food and medicinal crop, ranking fourth in importance in the Andean region after potatoes, oca and olluco (Anon., 1989). Though mashua is fully domesticated but still it persists in wild vegetation due to its aggressive growth and robustness.

In spite of the poor-quality soils, extreme temperatures *cum* radiation, variation in precipitation and the winds of the Andes, the plant grows quickly, managing to repel insects and nematodes, suppressing the weeds and maximizing photosynthesis. The proportion of dry matter transferred to the tubers can be as high as 75 per cent (Hernando Bermejo and Leon, 1994). Despite its hardiness, no reference is available to its introduction into other countries as flavour of the tuber is not so pleasant when eaten for the first time (Hernando Bermejo and Leon, 1994).

Botany

Mashua resembles its close relative ~ the ornamental nasturtium but is more compact and has smaller flowers. It is a perennial herbaceous climber growing 2-4×2m with green or reddish-green stem having leaves that show a considerable variation in form but are usually 5-20cm long and peltate with 3-5 lobes. The plant has erect growth when young and it has prostrate stems with compact foliage when mature (Hernando Bermejo and Leon, 1994). The flowers have 10-19cm long peduncles, the red calyx forms 1 (rarely 2 spurs) and there is considerable variation in the size and shape of the orange petals (Humanity development library 2.0).

Various mashua tubers of different shapes and pigmentations

Mashua tubers have shapes ranging from conical or ellipsoid to carrot-like, generally deeply furrowed, each furrow containing a bud. They vary in colour from white to yellow but occasionally with variants that are purple or red. Often the peel is mottled or striped with red or purple; especially below the eyes (Anon., 1989). At first sight, the tubers may be confused with oca tubers but can be distinguished by their conical shape, dark markings and a greater concentration of buds on the distal part as well as by their sour taste (Humanity development library 2.0).

About 100 clones have been recognized and on the basis of chemical examination and cytological evidence some authorities have suggested that there are two distinct types so subspecies status has been suggested as *T. tuberosum* ssp. *tuberosum* and *T. tuberosum* ssp. *silvestre* (Humanity development library 2.0). Colombian species are characterised by long deeply furrowed white tubers or white with pink extremes. Peruvian/Bolivian species have yellow tubers often with dots and lines on them.

Its basic chromosome number is x=13. Cultivated forms are clearly tetraploids (2n=4x=52); however, the frequency of diploids, triploids and tetraploids is not known (Hernando Bermejo and Leon, 1994).

Nutritional Value

Published figures give composition of the edible portion of tubers as water 86 per cent, energy 218 kJ/100g, protein 1.6 per cent, fat 0.6 per cent, carbohydrate 11 per cent, fibre 0.8 per cent, ash 7 per cent, calcium 7mg/100g, iron 1.2mg/100g, phosphorus 42mg/100g, vitamin A 0.015mg/100g, thiamine 0.06mg/100g, riboflavin 0.08mg/100g, niacin 0.6mg/100g and ascorbic acid 67mg/100g. Exceptionally high ascorbic acid content is noteworthy. The plant contains isothiocyanates and thiourea. Differences in composition correspond with the cultivated and wild forms and a division into two proposed subspecies, *e.g. Tropaeolum tuberosum* ssp. *tuberosum* and *T. tuberosum* ssp. *silvestre* (Humanity development library 2.0).

Uses

Apart from its edible tubers but being related to garden nasturtium it is occasionally cultivated as an ornamental for its brightly coloured tubular flowers (Grau and Hermann, 2003). Planting among other crops is reported to bring protection against nematodes (Humanity development library 2.0).

The tubers being minor food source especially to native Amerindian population, these are eaten cooked or roasted as a vegetable. Raw mashwa tuber is bitter due to glucosinolates but bitterness diminishes after cooking, freezing, or pounding (Sanderson, 2005). Mashwa is important for fulfilling the food requirements of resource-poor people in marginal rural areas of the high Andes (Hernando Bermejo and Leon, 1994). It is prepared in the form of a stew, as a roast or in the form of *thayacha*. For the latter preparation, the tubers are exposed overnight to frost and are eaten the following day accompanied by sugar-cane syrup (Hernando Bermejo and Leon, 1994).

In some areas the tubers are valued for their medicinal properties as are used in treatment of kidney and liver diseases, sores on the skin and head lice (Humanity development library 2.0). Traditionally, they are regarded as having anti-aphrodisiacal properties ~ reducing sexual desire. Clinical trials have indicated a reduction of up to 45 per cent in some male hormones when the tuber forms a considerable part of the diet but no loss in fertility has been observed.

Cultural Practices

Mashwa cultivation practices are the same as those described for oca (Hernando Bermejo and Leon, 1994). It is multiplied usually through its tubers. Mashwa is very hardy and can grow well on poor soils without use of fertilizers and pesticides, yet its yield can be almost double to that of potato. The growing cycle of this species varies from 220 to 245 days. Unlike oca and ullucu, mashwa produces a great quantity of viable seeds (Hernando Bermezo and Leon, 1994).

This crop requires cool and moist conditions and is tolerant to frost. Generally it is grown at 3,000m and above altitude, frequently in rotation with ullucu. It attains optimum growth only when it gets at least 12h day-length (Humanity development library 2.0).

Mashwa is usually cultivated in small beds on hillside terraces. Planting is usually done in rows 70-100cm apart with 40-70cm between the plants (Humanity development library 2.0). Though it can compete well with weeds but for good yield the beds should be earthed up and kept free of weeds.

In Colombia, it is planted as a companion crop to repel pests in potato fields. Its cultivation together with ullucu, oca and native potatoes could be accounted for by the nematicide and insecticide control properties that the plant has.

The tubers reach maturity in about 210 days. The tubers are dug by hand and are reported to keep better in the fresh state than the other Andean tuber crops ~ oca and ullucu, with a storage life at ambient temperatures of up to 6 months. Among the Andean tubers, mashua shows the highest yield (Hernando Bermejo and Leon, 1994), *i.e.* 20-30t/ha at 3,000m elevation though 70t/ha have been reported when grown under optimum conditions.

Pests and Diseases

Mashua is resistant to most of the insects, nematodes, fungi and other pathogens. This resistance is correlated to the content of glucosinolates present in the tuber (Hernando Bermejo and Leon, 1994).

References

1. Anon. 1989. Lost Crops of the Incas: Little-known Plants of the Andes with Promise for Worldwide Cultivation. National Research Council. National Academic Press, Andes.
2. Anon. 2008. Peace Diaries Workspace Archived 2008-01-23 at the Wayback Machine.
3. Grau, Ortega, Nieto and Hermann, 2003. Mashua (*Tropaeolum tuberosum* Ruiz and Pav.). Extensive monograph on mashua.http: //www.bioversityinternational.org/uploads/tx_news/Mashua__Tropaeolum_tuberosum_Ru per cent C3 per cent ADz__amp__Pav._880.pdf.
4. Hernando Bermejo J.E. and J. Leon (eds), 1994. Andean tubers (Arbizu, C. and M. Tapia). In: Plant Production and Protection Series No. 26. FAO, Rome (Italy), pp.149–63. https: //hort.purdue.edu/newcrop/1492/tubers.html.
5. Hernando Bermejo, J.E. and J. Leon (eds), 1994. Neglected Crops: 1492 from a Different Perspective. Plant Production and Protection Series No. 26 (FAO, Rome, Italy), pp. 149–63.
6. https: //en.wikipedia.org/wiki/Tropaeolum_tuberosum.
7. Humanity development library 2.0. Root crops: Anu (*Tropaeolum tuberosum*). nzdl.org.
8. Sanderson, Helen, Ghillean Prance and Mark Nesbitt (eds), 2005. The Cultural History of Plants. Routledge, p. 63.

22

Potato (*Solanum tuberosum* L.)

Family: Solanaceae (nightshade family).

Other common names: In India: Aaloo and Batata; elsewhere Ground apple, Potato, Spud, Tater and Tuber.

Synonyms: *Battata tuberosa* (L.) Hill; *Larnax sylvarum* subsp. *novogranatensis* N.W.Sawyer; *Lycopersicon tuberosum* (L.) Mill.; *Parmentiera edulis* Raf.; *Solanum andigenum* convar. *acutifolium* Lechn.; *S. a.* convar. *adpressipilosum* Lechn.; *S. a.* convar. *brachistylum* Lechn.; *S. a.* convar. *brevicalyces* Lechn.; *S. a.* convar. *brevipilosum* Lechn.; *S. a.* convar. *curtibaccatum* Lechn.; *S. a.* convar. *elongatibaccatum* Lechn.; *S. a.* convar. *longiacuminatum* Lechn.; *S. a.* convar. *macron* Lechn.; *S. a.* convar. *microstigmatum* Lechn.; *S. a.* convar. *nudiculum* Lechn.; *S. a.* convar. *obtusiacuminatum* Lechn.; *S. a.* convar. *puca-mata* Lechn.; *S. a.* convar. *stenon* Lechn.; *S. a.* convar. *titicacense* Lechn.; *S. a.* f. *alccai-huarmi* Bukasov and Lechn.; *S. a.* f. *ancacc-maquin* Bukasov and Lechn.; *S. a.* f. *arcuatum* Bukasov and Lechn.; *S. a.* f. *basiscopum* Bukasov and Lechn.; *S. a.* f. *bifidum* Bukasov and Lechn.; *S. a.* f. *caesium* Bukasov and Lechn.; *S. a.* f. *caiceda* Bukasov; *S. a.* f. *ccompetillo* Bukasov and Lechn.; *S. a.* f. *ccompis* Bukasov and Lechn.; *S. a.* f. *cevallosii* Bukasov and Lechn.; *S. a.* f. *chalcoense* Bukasov; *S. a.* f. *chimaco* Bukasov and Lechn.; *S. a.* f. *coeruleum* Lechn. ex Bukasov; *S. a.* f. *conicicolumnatum* Bukasov and Lechn.; *S. a.* f. *cryptostylum* Bukasov and Lechn.; *S. a.* f. *dilatatum* Bukasov and Lechn.; *S. a.* f. *discolour* Bukasov and Lechn.; *S. a.* f. *elongatipedicellatum* Lechn.; *S. a.* f. *globosum* Bukasov and Lechn.; *S. a.* f. *guatemalense* Bukasov; *S. a.* f. *huaca-layra* Bukasov and Lechn.; *S. a.* f. *huallata* Bukasov and Lechn.; *S. a.* f. *huaman-uma* Bukasov and Lechn.; *S. a.* f. *incrassatum* Bukasov and Lechn.;

S. a. f. *lanciacuminatum* Bukasov and Lechn.; S. a. f. *lapazense* Bukasov and Lechn.; S. a. f. *lecke-umo* Bukasov and Lechn.; S. a. f. *lilacinoflorum* Bukasov; S. a. f. *lisarassa* Bukasov; S. a. f. *llutuc-runtum* Lechn. ex Bukasov; S. a. f. *magnicorollatum* Bukasov and Lechn.; S. a. f. *microstigma* Bukasov and Lechn.; S. a. f. *nodosum* Bukasov; S. a. f. *ovatibaccatum* Bukasov and Lechn.; S. a. f. *pacus* Lechn. ex Bukasov; S. a. f. *pallidum* Bukasov and Lechn.; S. a. f. *pomacanchicum* Bukasov and Lechn.; S. a. f. *ppacc-nacha* Bukasov and Lechn.; S. a. f. *ppaqui* Bukasov and Lechn.; S. a. f. *sunchchu* Bukasov and Lechn.; S. a. f. *tenue* Bukasov and Lechn.; S. a. f. *tiahuanacense* Bukasov and Lechn.; S. a. f. *tocanum* Bukasov; S. a. f. *tolucanum* Bukasov; S. a. f. *uncuna* Bukasov and Lechn.; S. a. Juz. and Bukasov; S. a. subsp. *argentinicum* Lechn.; S. a. subsp. *australiperuvianum* Lechn.; S. a. subsp. *aya-papa* Bukasov and Lechn.; S. a. subsp. *bolivianum* Lechn.; S. a. subsp. *centraliperuvianum* Lechn.; S. a. subsp. *colombianum* (Bukasov) Lechn.; S. a. subsp. *ecuatorianum* Lechn.; S. a. subsp. *tarmense* Bukasov and Lechn.; S. a. var. *aymaranum* Bukasov and Lechn.; S. a. var. *bolivianum* Bukasov and Lechn.; S. a. var. *brevicalyx* Bukasov and Lechn.; S. a. var. *carhua* Vargas; S. a. var. *ccusi* Bukasov and Lechn.; S. a. var. *ckello-huaccoto* Bukasov and Lechn.; S. a. var. *colombianum* Bukasov; S. a. var. *cuzcoense* Bukasov and Lechn.; S. a. var. *digitotuberosum* Vargas; S. a. var. *grauense* Vargas; S. a. var. *hederiforme* Bukasov; S. a. var. *herrerae* Bukasov and Lechn.; S. a. var. *huairuru* Bukasov and Lechn.; S. a. var. *imilla* Bukasov and Lechn.; S. a. var. *juninum* Bukasov; S. a. var. *latius* Bukasov and Lechn.; S. a. var. *longibaccatum* Bukasov and Lechn.; S. a. var. *mexicanum* Bukasov; S. a. var. *platyantherum* Bukasov and Lechn.; S. a. var. *quechuanum* Bukasov and Lechn.; S. a. var. *sihuanum* Bukasov and Lechn.; S. a. var. *socco-huaccoto* Bukasov and Lechn.; S. a. var. *stenophyllum* Bukasov and Lechn.; *Solanum apurimacense* Vargas; *Solanum aracatscha* Besser; *Solanum aracc-papa* Juz. ex Rybin; *Solanum ascasabii* Hawkes; *Solanum boyacense* Juz. and Bukasov; *Solanum caniarense* Juz. and Bukasov; *Solanum cardenasii* Hawkes; *Solanum cayeuxi* Berthault; *Solanum chariense* A.Chev.; *Solanum chaucha* f. *purpureum* Hawkes; S. c. f. *roseum* Hawkes; S. c. Juz. and Bukasov; S. c. var. *ccoe-sulla* Ochoa; S. c. var. *ckati* Ochoa; S. c. var. *khoyllu* Ochoa; S. c. var. *puca-suitu* Ochoa; S. c. var. *surimana* Ochoa; *Solanum chiloense* (A.DC.) Berthault; *Solanum chilotanum* f. *magnicorollatum* Lechn.; S. c. f. *parvicorollatum* Lechn.; S. c. Hawkes; S. c. var. *angustifurcatum* Lechn.; S. c. var. *talukdarii* Lechn.; *Solanum chocclo* Bukasov and Lechn.; *Solanum churuspi* Hawkes; *Solanum coeruleiflorum* Hawkes; *Solanum cultum* (A.DC.) Berthault; *Solanum diemii* E.Brucher; *Solanum dubium* E.H.L.Krause; *Solanum erlansonii* Anon.; *Solanum esculentum* Neck.; *Solanum estradea* L.E.López; *Solanum goniocalyx* Juz. and Bukasov; S. g. var. *caeruleum* Vargas; *Solanum herrerae* Juz.; *Solanum hygrothermicum* Ochoa; *Solanum kesselbrenneri* Juz. and Bukasov; *Solanum leptostigma* Juz.; *Solanum leptostigma*

Juz. ex Bukasov; *Solanum macmillanii* Bukasov; *Solanum maglia* var. *chubutense* Bitter; *S. m.* var. *guaytecarum* Bitter; *Solanum mamilliferum* Juz. and Bukasov; *Solanum molinae* Juz.; *Solanum oceanicum* Brücher; *Solanum ochoanum* Lechn.; *Solanum paramoense* Bitter ex Pittier; *Solanum parmentieri* Molina ex Walp.; *Solanum parvicorollatum* Lechn.; *Solanum phureja* f. *orbiculatum* Ochoa; *S. p.* f. *sayhuanimayo* Ochoa; *S. p.* f. *timusi* Ochoa; *S. p.* f. *viuda* Ochoa; *S. p.* Juz. and Bukasov; *S. p.* subsp. *estradae* (L.E.López) Hawkes; *S. p.* subsp. *hygrothermicum* (Ochoa) Hawkes; *S. p.* var. *caeruleum* Ochoa; *S. p.* var. *erlansonii* (Bukasov and Lechnovitch) Ochoa; *S. p.* var. *flavum* Ochoa; *S. p.* var. *janck'o-phureja* Ochoa; *S. p.* var. *macmillanii* (Bukasov and Lechnovitch) Ochoa; *S. p.* var. *pujeri* Hawkes; *S. p.* var. *rubroroseum* Ochoa; *S. p.* var. *sanguineum* Ochoa; *Solanum riobambense* Juz. and Bukasov; *Solanum rybinii* Juz. and Bukasov; *S. r.* var. *bogotense* Hawkes; *S. r.* var. *boyacense* (Juz. and Bukasov) Hawkes; *S. r.* var. *pastoense* Hawkes; *S. r.* var. *popayanum* Hawkes; *Solanum sabinei* (A.DC.) Berthault; *Solanum sanmartinense* Brücher; *Solanum sendigena* Juz. and Bukasov; *Solanum sinense* Blanco; *Solanum stenotomum* f. *alcay-imilla* Hawkes; *S. s.* f. *canasense* Vargas; *S. s.* f. *canastilla* Hawkes; *S. s.* f. *catari-papa* Hawkes; *S. s.* f. *ccami* (Bukasov) Hawkes; *S. s.* f. *chilcas* Hawkes; *S. s.* f. *chincherae* Hawkes; *S. s.* f. *chojllu* Hawkes; *S. s.* f. *cochicallo* Hawkes; *S. s.* f. *cohuasa* Hawkes; *S. s.* f. *cuchipacon* Hawkes; *S. s.* f. *eucaliptae* Hawkes; *S. s.* f. *huallata-chinchi* Hawkes; *S. s.* f. *huamanpa-uman* Hawkes; *S. s.* f. *huanuchi* Hawkes; *S. s.* f. *kamara* Hawkes; *S. s.* f. *kantillero* Hawkes; *S. s.* f. *kehuillo* Hawkes; *S. s.* f. *koso-nahui* Hawkes; *S. s.* f. *negrum* Hawkes; *S. s.* f. *orcco-amajaya* Hawkes; *S. s.* f. *pallidum* Hawkes; *S. s.* f. *phinu* Hawkes; *S. s.* f. *phitu-huayacas* Hawkes; *S. s.* f. *piticana* Hawkes; *S. s.* f. *pitoca* Hawkes; *S. s.* f. *puca* Vargas; *S. s.* f. *roseum* Hawkes; *S. s.* f. *tiele* Hawkes; *S. s.* f. *yana-cculi* Hawkes; *S. s.* f. *yuracc* Vargas; *S. s.* Juz. and Bukasov; *S. s.* subsp. *goniocalyx* (Juz. and Bukasov) Hawkes; *S. s.* var. *ccami* Bukasov; *S. s.* var. *chapina* Hawkes; *S. s.* var. *cyaneum* Hawkes; *S. s.* var. *huicu* Hawkes; *S. s.* var. *keccrana* Hawkes; *S. s.* var. *megalocalyx* Hawkes; *S. s.* var. *peruanum* Hawkes; *S. s.* var. *pitiquilla* Hawkes; *S. s.* var. *poccoya* Vargas; *S. s.* var. *puca-lunca* Hawkes; *S. s.* var. *putis* Hawkes; *Solanum subandigenum* Hawkes; *Solanum sylvestre* Audib. ex Dunal; *Solanum tarmense* Bukasov; *Solanum tascalense* Brücher; *Solanum tenuifilamentum* Juz. and Bukasov; *Solanum tuberosum* f. *acuminatum* Bukasov and Lechn.; *S. t.* f. *alkka-imilla* Ochoa; *S. t.* f. *alkka-silla* Ochoa; *S. t.* f. *amajaya* Ochoa; *S. t.* f. *araucanum* Bukasov and Lechn.; *S. t.* f. *auriculatum* Bukasov and Lechn.; *S. t.* f. *azul-runa* Ochoa; *S. t.* f. *brachykalukon* Bukasov and Lechn.; *S. t.* f. *brevipapillosum* Bukasov and Lechn.; *S. t.* f. *camota* Bukasov and Lechn.; *S. t.* f. *chaped* Bukasov and Lechn.; *S. t.* f. *chiar-lelekkoya* Ochoa; *S. t.* f. *chiar-pala* Ochoa; *S. t.* f. *chojo-sajama* Ochoa; *S. t.* f. *conicum* Bukasov and Lechn.; *S. t.* f. *contortum* Bukasov and Lechn.; *S. t.* f. *coraila* Bukasov and Lechn.; *S. t.* f. *crassifilamentum* Bukasov and Lechn.; *S. t.* f.

elongatum Bukasov and Lechn.; *S. t.* f. *enode* Bukasov and Lechn.; *S. t.* f. *huacazapato* Ochoa; *S. t.* f. *huichinkka* Ochoa; *S. t.* f. *indianum* Lechn. ex Bukasov; *S. t.* f. *infectum* Bukasov and Lechn.; *S. t.* f. *isla-imilla* Ochoa; *S. t.* f. *janccko-kkoyllu* Ochoa; *S. t.* f. *jancko-chockella* Ochoa; *S. t.* f. *jancko-pala* Ochoa; *S. t.* f. *kunurana* Ochoa; *S. t.* f. *laram-lelekkoya* Ochoa; *S. t.* f. *latum* Bukasov and Lechn.; *S. t.* f. *mahuinhue* Bukasov and Lechn.; *S. t.* f. *milagro* Ochoa; *S. t.* f. *montticum* Bukasov and Lechn.; *S. t.* f. *nigrum* Ochoa; *S. t.* f. *oculosum* Bukasov and Lechn.; *S. t.* f. *ovatum* Bukasov and Lechn.; *S. t.* f. *overita* Ochoa; *S. t.* f. *pichuna* Bukasov and Lechn.; *S. t.* f. *pillicuma* Bukasov and Lechn.; *S. t.* f. *pulo* Ochoa; *S. t.* f. *sani-imilla* Ochoa; *S. t.* f. *sebastianum* Bukasov and Lechn.; *S. t.* f. *surico* Ochoa; *S. t.* f. *tenuipedicellatum* Bukasov and Lechn.; *S. t.* f. *thalassinum* Bukasov and Lechn.; *S. t.* f. *tinguipaya* Ochoa; *S. t.* f. *viride* Bukasov and Lechn.; *S. t.* f. *wila-huaycku* Ochoa; *S. t.* f. *wila-imilla* Ochoa; *S. t.* f. *wila-koyu* Ochoa; *S. t.* f. *wila-monda* Ochoa; *S. t.* f. *wila-pala* Ochoa; *S. t.* f. *yurac-taraco* Ochoa; *S. t.* subsp. *andigenum* (Juz. and Bukasov) Hawkes; *S. t.* subsp. *chiloense* (A.DC.) L.I.Kostina; *S. t.* var. *aethiopicum* Alef.; *S. t.* var. *alaudinum* Alef.; *S. t.* var. *album* Alef.; *S. t.* var. *anglicum* Alef.; *S. t.* var. *batatinum* Alef.; *S. t.* var. *bertuchii* Alef.; *S. t.* var. *borsdorfianum* Alef.; *S. t.* var. *brachyceras* Alef.; *S. t.* var. *brevipilosum* Bukasov and Lechn.; *S. t.* var. *bufoninum* Alef.; *S. t.* var. *californicum* Alef.; *S. t.* var. *cepinum* Alef.; *S. t.* var. *chiloense* A.DC.; *S. t.* var. *chilotanum* Bukasov and Lechn.; *S. t.* var. *chubutense* (Bitter) Hawkes; *S. t.* var. *conocarpum* Alef.; *S. t.* var. *cordiforme* Alef.; *S. t.* var. *corsicanum* Alef.; *S. t.* var. *crassipedicellatum* Bukasov and Lechn.; *S. t.* var. *cucumerinum* Alef.; *S. t.* var. *cultum*; *S. t.* var. *drakeanum* Alef.; *S. t.* var. *elegans* Bukasov and Lechn.; *S. t.* var. *elongatum* Alef.; *S. t.* var. *erythroceras* Alef.; *S. t.* var. *fragariinum* Alef.; *S. t.* var. *guaytecarum* (Bitter) Hawkes; *S. t.* var. *hassicum* Alef.; *S. t.* var. *helenanum* Alef.; *S. t.* var. *hispanicum* Alef.; *S. t.* var. *holsaticum* Alef.; *S. t.* var. *julianum* Alef.; *S. t.* var. *kaunitzii* Alef.; *S. t.* var. *laurentianum* Alef.; *S. t.* var. *lelekkoya* Ochoa; *S. t.* var. *leonhardianum* Alef.; *S. t.* var. *malcachu* Ochoa; *S. t.* var. *melanoceras* Alef.; *S. t.* var. *menapianum* Alef.; *S. t.* var. *merceri* Alef.; *S. t.* var. *multibaccatum* Bukasov and Lechn.; *S. t.* var. *murukewillu* Ochoa; *S. t.* var. *nobile* Alef.; *S. t.* var. *norfolcicum* Alef.; *S. t.* var. *nucinum* Alef.; *S. t.* var. *palatinatum* Alef.; *S. t.* var. *pecorum* Alef.; *S. t.* var. *peruvianum* Alef.; *S. t.* var. *platyceras* Alef.; *S. t.* var. *polemoniifolium* J.Rémy; *S. t.* var. *praecox* Alef.; *S. t.* var. *praedicandum* Alef.; *S. t.* var. *putscheanum* Alef.; *S. t.* var. *recurvatum* Bukasov and Lechn.; *S. t.* var. *reniforme* Alef.; *S. t.* var. *rockii* Alef.; *S. t.* var. *rossicum* Alef.; *S. t.* var. *rubrisuturatum* Bukasov and Lechn.; *S. t.* var. *rugiorum* Alef.; *S. t.* var. *runa* Ochoa; *S. t.* var. *sabinei* A.DC.; *S. t.* var. *saccharatum* Alef.; *S. t.* var. *salamandrinum* Alef.; *S. t.* var. *schnittspahnii* Alef.; *S. t.* var. *sesquimensale* Alef.; *S. t.* var. *sicha* Ochoa; *S. t.* var. *sipancachi* Ochoa; *S. t.* var. *strobilinum* Alef.; *S. t.* var. *taraco* Ochoa; *S. t.* var. *tener* Alef.; *S. t.* var. *tinctorium* Alef.; *S. t.* var. *ulmense* Alef.;

S. t. var. *versicolour* Alef.; *S. t.* var. *villaroella* Bukasov and Lechn.; *S. t.* var. *vuchefeldicum* Alef.; *S. t.* var. *vulgare* Hook.f.; *S. t.* var. *vulgare* Macloskie; *S. t.* var. *xanthoceras* Alef.; *S. t.* var. *yutuense* Bukasov and Lechn.; *Solanum utile* Klotzsch; *Solanum yabari* Hawkes; *S. y.* var. *cuzcoense* Hawkes; *S. y.* var. *pepino* Hawkes; and *Solanum zykinii* Lechn.

Introduction and Origin

Solanum ~ the largest genus of its family includes 3 economically important food crops: potato, tomato and brinjal (aubergine, eggplant). Potato is the most important and largest non-grain food crop occupying 4th rank in the world after wheat, rice and maize. It is a temperate crop grown under subtropical conditions in India, performing better where usually night temperature does not exceed 18°C in the season. The potato plant produces edible starchy stem tuber, *i.e.* enlarged stolons thicken to develop into storage organs called 'potato'. Today it is a staple food in many parts of the world and an integral part of much of the world's food supply. It is grown in more than 140 countries spread across both temperate and tropical regions; at elevations from sea level to 4,000m. China is the top potato producer in the world, followed by India, Russia, Ukraine and the United States. Looking into its importance, one entirely separate establishment in India named as ICAR-Central Potato Research Institute (CPRI), Shimla was started in 1949 to initiate systemic researches on potato. CPRI is established with certain of its Regional Stations in different climatic zones of the country. India in 1913 ranked third in potato area and its production in the world. In India, it is being grown in almost all the states though major states being Uttar Pradesh, West Bengal, Bihar, Madhya Pradesh, Punjab, Gujarat, Karnataka and Haryana.

Wild potato species can be found from southern United States to southern Chile (Hijmans and Spooner, 2001). The cultivated potato originated about 8,000 years ago near Lake Titicaca at 3,800m above sea level in the Andes mountain range of South America on the border between Bolivia and Peru. It is on the record that apart from Bolivia and Peru, it is native to Venezuela, Colombia, Equador, Argentina and central Mexico. Potatoes were introduced to Europe by the Spanish in the second half of the 16th century. European colonizers took potato to all corners of the globe. In India, potato was introduced in the early 17th century by colonial rulers (Portuguese/British) and since then it was primarily being grown only in the backyards of the British people till independence (Luthra *et al.*, 2006), however, Naik (2019) mentions that in India it is being grown since 1830. Major boost in potato production in India came after the establishment of CPRI in the year 1949. The crop is presently being grown in around 1.99 million hectares spread over 26 states of India, producing 44.3 million tonnes of potato tubers.

Potato production in India has undergone a revolution since independence and now India is the second largest producer of potato next only to China in the world. At present India is not only self-sufficient in potato production but is also exporting seed potatoes to certain countries as India has well-established seed production techniques to produce disease-free planting material. Ninety per cent of the potatoes in India come from the Indo-Gangetic planes. In recent years potato processing industry has grown rapidly in India due to rapid urbanization and changing life style of the people.

Etymology

The generic name was first used by Pliny the Elder (AD 23-79) for a plant also known as *Strychnos*, probably *S. nigrum*. The word may be from the Latin *sol*, meaning 'sun', referring to its status as a plant of the sun. The English word 'potato' is derived from Spanish *patata* (the name used in Spain). Although the name 'potato' originally referred to the sweet potato, the two plants are not closely related. The 16th-century English herbalist John Gerard referred to sweet potatoes as 'common potatoes' and used the terms 'bastard potatoes' and 'Virginia potatoes' for potatoes (Hijmans and Spooner, 2001). In the United States, potatoes are occasionally referred to as 'Irish or white potatoes' to distinguish them from sweet potatoes. The term potato translates roughly into English as 'earth or ground apple'.

Botany

Potatoes are much branched herbaceous dicotyledonous perennials though cultivated as an annual. The growth habit of potato varies between and within species. The plant has a rosette or semi-rosette habit. The tuber of potato is an enlarged, swollen and fleshy portion of an underground stem or stolon and has eyes (buds) from which next season's growth emerges. Eyes on tuber are concentrated near the apical end of the tuber with fewer near the stolon or basal end. Eye number and distribution are characteristic of the variety. The stem is erect in the early stage but later on becomes spreading and prostrate.

Potato plants

Flowers

Potatoes with different pigmentation

Depending on variety the stem grows to about 60cm high and the leaves die back after flowering, fruiting and tuber formation. The compound leaves are alternate and irregularly odd-pinnate. Depending on the cultivar potato inflorescence consists of 1-30 flowers that may be white, pink, red, blue or purple with yellow stamens. Star-shaped open flower has five petals and a pistil that generally protrudes above a cluster of 5 large bright yellow anthers. Opening of flower starts near the base of the inflorescence and proceeds upward at the rate of about 2-3 each day where at the peak bloom there are usually 5-10 open flowers. Flowers are open for only 2-4 days where receptivity of stigma and duration of pollen production are only about 2 days. The peak time of pollination is early morning. Although a substantial amount of self-fertilizing occurs in potatoes, they are mostly cross-pollinated by insects such as bumblebees which carry pollen from other potato plants. Tubers form in response to decreasing day length, although this tendency has been minimized in commercial varieties. Potato plants produce small green berry-fruits each containing about 300 seeds.

When aerial part of the plant dies the tubers survive underground over winter until spring. When these regenerate new shoots, these use the stored food present in the tuber to grow. As the main shoot develops from the tuber, the base of the shoot close to the tuber produces adventitious roots and lateral buds on the shoot. The shoot also produces stolons that are long etiolated stems. The stolon elongates during long days with the presence of high auxin levels that prevent root growth of the stolon. Before new tuber formation begins, the stolon should have attained a particular age which may be akin to every variety. The enzyme lipoxygenase produces a hormone jasmonic acid which is involved in the control of potato tuber development.

Potatoes, both *S. tuberosum* and most of its wild relatives are self-incompatible as they bear no useful fruit when self-pollinated. This trait is problematic for crop breeding as all sexually-produced plants must be hybrids. Self-compatibility has successfully been introduced both to diploid potatoes (including a special line of *S. tuberosum*) by CRISPR-Cas9.

Genetics

The basic chromosome number in genus *Solanum* is 12. *Solanum tuberosum* is a tetraploid with 2n=4x=48. The genus *Solanum* also includes four diploid species with 24 chromosomes (*S. stenotomum*, *S. phureja*, *S. goniocalyx* and *S. ajanhuiri*), two triploid species with 36 chromosomes (*S. chaucha* and *S. juzepczukii*) and one pentaploid cultivated species (*S. curtilobum*) with 60 chromosomes. The cultivated diploid species are found only in South America. The origin of tetraploidy is unclear. The cultivated *S. tuberosum* subsp. *tuberosum* can be either an auto-tetraploid (doubling of the chromosomes of a diploid species) or an allo-tetraploid (doubling of the chromosomes of a diploid hybrid between two related species) (Hawkes, 1990).

Subspecies

Only one wild species of potato, *S. chacoense* is present in India and that too near CPRI, Shimla. It has a diploid genome and as such cannot be crossed with the tetraploid *S. tuberosum* ssp. *andigena* or *S. tuberosum* ssp. *tuberosum*. *S. tuberosum* ssp. *tuberosum*, however, can cross with ssp. *andigena* and also with wild species only in germplasm collection maintained under field conditions. *Solanum tuberosum* has two major subspecies: *S.t.* subsp. *andigena* (or Andean) and *S.t.* subsp. *tuberosum* (or Chilean). These cultivated tetraploid potato species are photoperiodically long day plants and require long days and short nights for their flowering, but more than 90 per cent of the potato growing areas in India are in the plains and in *Rabi* season (October planting) when the day lengths are shorter. Therefore, usually the flowering does not take place in the plains. Potato is mainly propagated through tubers.

Distribution

In Indian tropical and sub-tropical plains potato is grown during autumn-winter-spring seasons while in temperate/semi-temperate hills during summer-autumn. The ideal time of planting in plains corresponds to maximum and minimum temperatures of 30-32°C and 10-20°C, respectively. In the hills, planting is done when the maximum temperature is about 20-22°C while minimum about 12-15°C so optimum being 15-25°C. Minimum night temperature is of great significance for tuberization and for this 17°C is the optimum one though above 18°C night temperatures tuberization suffers, *vis-a-vis* tuber yield. Certainly, at low temperatures the vegetative growth of the plant is restricted and at very low temperatures even plants die. In the Nilgiri hills, the temperatures remain mild throughout the year where three consecutive crops are grown during summer, autumn and winter. The potato crop thrives up to snow line. The hill areas offer good scope for seed potato industry because of almost virus-free environment. Himachal Pradesh, Uttarakhand and Sikkim provide the bulk of the 'hill seed potato' for the plains of India. More than 80 per cent potato in India is grown in sub-tropical regions during winter which differs than those prevalent in potato season of temperate countries. Long photoperiods promote aerial vegetative growth and delay tuberization though short photoperiods reduce vegetative growth but promote tuberization and crop maturity. **If crop gets long day conditions up to 40 days after transplanting but afterwards short days conditions tuberization occurs to the maximum level.**

The planting and harvesting time depends on temperature and cropping pattern followed in the area. On the basis of soil, prevailing climatic conditions and other agronomic factors, India has been divided into eight potato zones. These zones lie in two major potato-growing areas, *i.e.* North Indian hills and North Indian plains; though southern hills, North Bengal and Sikkim hills; and plateau are three special areas.

1. **North-western plains** (<300m asl: Punjab, Haryana and Rajasthan) have neutral to slightly alkaline deep alluvial soils so here potato is grown during early autumn, *i.e.* September-November, autumn (October-January/February) or spring (January-April/May). The climate is warm at planting of early autumn crop and frost in later stages. It is warm at tuberization in spring crop. The temperatures during winter is not suited for rapid multiplication of aphid-vectors therefore this area is suitable for seed potato production. Major characteristics of the varieties are short day adaptation, early bulking, tolerance to higher temperatures, slow rate of degeneration and late blight resistance. For the main crop, resistance to frost is an added advantage.
2. **North-central plains** (<300m asl; Western-central U.P. and North-western districts of Gujarat and M.P.) have neutral to slightly alkaline deep alluvial soils. Potato crop of 90-100 days duration is grown during winter under mild temperatures and short days (10h/day) from October–January/February. The region is generally free from major potato diseases. Late blight does not follow a definite cycle of appearance in the region. Major characteristics of the varieties are short day adaptation, early bulking, moderate resistance to late blight and slow rate of degeneration.

   2.1. Early autumn crop: September to November/December; main autumn crop: October to January/February; late autumn: November/December to March. The climate is warm at planting of early autumn crop and frost in later stages of main autumn crop, warm during later stages of late autumn crop.

3. **Eastern plains** (<300m asl; Asom, Bihar, Jharkhand, West Bengal, Odisha, eastern U.P., and NE and eastern districts of M.P. and Chhattisgarh) in combination with west-central plains is aptly called the '**potato bowl**' of the country and accounts for three-fourth of Indian potato production. Soils in this region are slightly acidic to slightly alkaline deep alluvial where crop is grown during short and mild winter (November-February/March). Some parts of eastern plains have a strong preference for red skin tubers. Important desired features in the varieties are adaptability to short days, early bulking/maturity and resistance to late blight.
4. **Plateau region** (500-1,000m asl; central and peninsular India covering Maharashtra, Karnataka and parts of Gujarat, Madhya Pradesh and Odisha) has red sandy and fine-textured black soils. Two crops are raised both during kharif (rain-fed crop, planted in June-July and harvested in September-October) and *Rabi* season (irrigated, planted in November and harvested in January-February). The varietal requirements in this region are early bulking, ability to tuberize under higher temperatures, and resistance to early blight, late blight, bacterial wilt, viruses and mites as well as potato tuber moth and a slow rate of degeneration. Minimum temperatures (13-16°C) during *Rabi* season favour a low accumulation of reducing sugars and higher dry matter as compared to *Kharif* crop.
5. **North-western and central hills** (Jammu and Kashmir, Himachal Pradesh and Uttarakhand) where potato is usually cultivated in summer during April to September.
   5.1. The very high hills of Himalayas (3,000- 3,500m asl) have loamy-textured acidic soils. In this region the crop is grown during summer (June to September). At early stages the frost occurs, moisture stress is during early growth period and snowfall near harvest. The most important disease appearing every year in epiphytotic proportion is late blight. Varieties are required to be adapted to long-days (14h/day) and resistant to late blight.
   5.2. The Himalayan high hills (1,800- 3,000m asl) have acidic soils of varying texture. In this region the crop is grown during summer (March/April to August/September; rainfed). The frost and hails occur in early stages, moisture stress during early growth, and excess moisture during and after tuberization. The varieties are required to be adapted to long-days (14h/day) and resistant to late blight.
   5.3. The mid-hills of Himalayas (1,000- 1,800m asl) have acidic soils of varying texture. In this region, the crop is grown during spring

(January/February to May/June) and autumn (August/September to November/December). Frost and hails occur after planting of spring crop and frost before lifting of autumn crop. Varietal requirements include resistance to early blight, late blight, bacterial wilt and viruses. Bacterial wilt is the problem in mid-hills, especially in the Uttarakhand.

5.4. The low hills (600-1,000m asl) have sandy and fine textured black soils. Winters are short and mild. Rainy season is warm with indifferent soil moisture conditions. In this region, the crop is grown during winter (November to February/March; irrigated) and kharif (July/August to September/October; rainfed). Varietal requirements include resistance to early blight, late blight, bacterial wilt, viruses and tuber rots.

6. In **North-eastern hills** (Meghalaya, Manipur, Mizoram, Tripura, Nagaland and Arunachal Pradesh) where potato is grown either as a spring crop (between January/February-May/June) or autumn crop (August-November/December). In this semi-temperate and high rainfall zone, late blight and bacterial wilt are the two major problems. The varieties are required to be adapted to long-days and resistant to late blight and bacterial wilt.

7. In **Southern hills** (1,000–2,000m asl; mainly the Nilgiris and Palani hills in T.N.) with acidic soils of varying textures where potatoes can be cultivated round the year depending upon the elevation. The main crop is grown during summer (March/April-August/September; rainfed), though congenial temperatures prevail throughout the year hence it is also possible to take even autumn (August/September-December/January) as well as spring crops (January/February-May/June). The moisture stress may occur at early stages in summer crop. Late blight and cyst nematodes are the two major problems. Essential varietal requirements are early bulking, adaptability to long days and resistance to late blight and cyst nematodes.

8. **Sikkim and North Bengal hills** (Sikkim and hills of West Bengal; 1,000–2,000m asl) have acidic soils of varying texture. Potatoes are grown in spring (January/February-July/August; rainfed) and autumn (September-December). There may be excess moisture during and after tuberization in spring crop. Resistance to late blight and immunity to wart are major requirements of varieties suited to the zone. There is a distinct preference for red skin tubers.

Varieties and Cultivars

Potato has about 200 wild species and subspecies and some 5,000 cultivated varieties worldwide though some 3,000 in the Andes alone mainly in Peru, Bolivia, Ecuador, Chile and Colombia. Certain species and subspecies can be cross-bred with cultivated varieties. Cross-breeding has been done repeatedly to transfer resistances to certain pests and diseases from the gene pool of wild species. Based on common characteristics, the potato varieties can be categorized into few main groups such as **russet potatoes** (rough brown skin), **red potatoes**, **white potatoes**, **yellow/Yukon potatoes** and **purple potatoes**. For culinary purposes, varieties are often differentiated by their waxiness. Floury or mealy **baking** potatoes have more starch (20–22 per cent) than waxy **boiling** potatoes (16–18 per cent). They can also be differentiated from variation in the comparative ratio of two different potato starch compounds ~ amylose (a long-chain molecule) and amylopectin (a highly branched molecule). When cooked in water amylose diffuses from the starch granule and lends itself to dishes where the potato is mashed. Varieties with slightly higher amylopectin content help the potatoes retaining their shape after being boiled in water (*recipetips.com*). Potatoes that are good for making potato chips or potato crisps are called 'chipping potatoes' which means they meet the basic requirements of similar varietal characteristics of being firm, fairly clean and well-shaped. Immature potatoes particularly valued for their taste, may be sold fresh from the field as 'creamer' or 'new' potatoes. They are typically small and tender with a loose skin but here flesh contains a lower level of starch than other potatoes.

Dozens of potato cultivars have been selectively bred specifically for their skin or flesh colour including gold, red and blue varieties that contain varying amounts of phytochemicals including carotenoids for gold/yellow or polyphenols for red or blue cultivars. Carotenoid compounds include provitamin A alpha-carotene and beta-carotene which during digestion are converted to the essential nutrient, vitamin A. Anthocyanins mainly responsible for red or blue pigmentation in potato cultivars do not have nutritional significance but are used for visual variety and consumer appeal. In 2010, potatoes were bioengineered specifically for these pigmentation traits.

Varieties of Potato Produced in India

Early potato introductions in India belong to *S. tuberosum* ssp. *andigena*. Further selections by Indian farmers resulted in several indigenous varieties of which later on some 16 were identified by CPRI. Among these, 'Phulwa', 'Darjeeling Red Round' and 'Gola' became most popular. These indigenous varieties could be grown in Indian plains but yield is the limiting factor and the seed stock is riddled with many viruses.

Potato Varieties Developed by CPRI

'Kufri Kisan' (1958), 'Kufri Kuber' (1958), 'Kufri Kumar' (1958), 'Kufri Kundan' (1958), 'Kufri Red' (1958), 'Kufri Safed' (1958), 'Kufri Neela' (1963), 'Kufri Sindhuri' (1967), 'Kufri Alankar' (1968), 'Kufri Chamatkar' (1968), 'Kufri Chandramukhi' (1968), 'Kufri Jeevan' (1968), 'Kufri Jyoti' (1968), 'Kufri Khasigaro' (1968), 'Kufri Naveen' (1968), 'Kufri Neelamani' (1968), 'Kufri Sheetman' (1968), 'Kufri Muthu' (1971), 'Kufri Lauvkar' (1972), 'Kufri Dewa' (1973), 'Kufri Badshah' (1979), 'Kufri Bahar' (1980), 'Kufri Lalima' (1982), 'Kufri Sherpa' (1983), 'Kufri Swarna' (1985). 'Kufri Megha' (1989), 'Kufri Jawahar' (1996), 'Kufri Sutlej' (1996), 'Kufri Ashoka' (1996), 'Kufri Pukhraj' (1998), 'Kufri Chipsona 1' (1998), 'Kufri Chipsona 2' (1998), 'Kufri Giriraj' (1998), 'Kufri Anand' (1999), 'Kufri Kanchan' (1999), 'Kufri Arun' (2005), 'Kufri Pushkar' (2005), 'Kufri Shailja' (2005), 'Kufri Surya' (2006), 'Kufri Chipsona 3' (2006), 'Kufri Himalini' (2006), 'Kufri Himsona' (2008), 'Kufri Sadabahar' (2008), 'Kufri Girdhari' (2008), 'Kufri Khyati' (2008), 'Kufri Frysona' (2009), 'Kufri Neelima' (2010), 'Kufri Chipsona 4' (2010), 'Kufri Garima' (2012), 'Kufri Gaurav' (2012), 'Kufri Lalit' (2014), 'TPS Population' (2007), *etc.* (Kumar *et. al.*, 2014; https://cpri.icar.gov.in).

List of some varieties released in India indicating their main climate change related attributes, disease resistance/sensitivity to Early Blight (EB), Late Blight (LB) and Potato Leaf Roll Virus (PLRV) and preferences to Indian farmers are:

Variety Name (Major Producers)	*Yield Potential (t/ha)*	*Maturity in Days*	*Heat Tolerance*	*Drought Tolerance*	*Resistance to LB, EB or PLRV*	*Description*
Most popular varieties						
Kufri Pukhraj (Bihar, Punjab, Haryana, U.P., Maharashtra, H.P., M.P., Orissa and W.B.)	35–40	70–90	Sensitive	High	Moderately resistant to LB and resistant to EB	Large, oval and slightly tapered in shape with fleet eyes and yellow flesh
Kufri Jyoti (Bihar, Maharashtra, Gujarat, Haryana, Punjab, U.P., Karnataka and W.B.)	25–30	90–100	Sensitive	Medium	Moderately resistant to EB and LB	Eyes are small and swift, and the flesh is white; suitable for processing

Variety Name (Major Producers)	*Yield Potential (t/ha)*	*Maturity in Days*	*Heat Tolerance*	*Drought Tolerance*	*Resistance to LB, EB or PLRV*	*Description*
Kufri Bahar (Haryana, U.P., H.P. and J and K)	30–35	90–100	Sensitive	Medium	Sensitive to LB	Large and round-oval in shape with medium deep eyes
Bhura Aloo	24	>110	Low	Sensitive	Sensitive to LB	
Kufri Chipsona (Bihar and U.P.)	30–35	90–100	Sensitive	Medium	Resistant to LB	Medium and round-oval in shape with fleet eyes and yellow flesh; suitable for preparing chips and French fries
Kufri Sindhuri (Bihar, Gujarat, Maharashtra, Punjab, U. P., Karnataka and J and K)	30–35	>110	High	Medium	Resistant to PLRV, moderately to EB and Sensitive to LB	Deep coloured; medium in size; suitable for processing
Kufri Chandramukhi (Bihar, Gujarat, Haryana, Punjab, M.P., H.P., U.P., Orissa and W.B.)	20–25	70–90	Sensitive	High	Sensitive to LB	Large, round, white, somewhat flattened; flesh: white and dull; suitable for instant flakes and chips
Kufri Khyati	25–30	70–90	Sensitive	High	Resistant to LB	
Most promising varieties						
Lady Rosetta	30	70–90	High	High	Sensitive to LB	
Kufri Megha	25–30	90–100	Sensitive	Medium	Resistant to LB	

Variety Name (Major Producers)	*Yield Potential (t/ha)*	*Maturity in Days*	*Heat Tolerance*	*Drought Tolerance*	*Resistance to LB, EB or PLRV*	*Description*
Kufri Kanchan	25–30	90–100	Sensitive	Medium	Moderately resistant to LB	
Kufri Arun	30–35	90–100	Sensitive	High	Resistant to LB	
Kufri Surya	25–30	70–90	High	Medium	Sensitive to LB	

'Kufri Pukhraj' (released in 1998), a high yielding and early-maturing variety is by far India's most dominating variety covering 521,375ha (*i.e.* 33 per cent of total potato area) in 2015. It is the most dominating variety in Punjab, Gujarat and Bihar while in Uttar Pradesh and West Bengal it is the second most dominating variety. In Karnataka, it is of no considerable importance. Producers prefer this variety because it is early-maturing and produces sizeable yield even if harvested pre-maturely after 60 days; however, it shows limited yield potential under early sowing condition and is sensitive to heat (Kumar *et al.,* 2014).

'Kufri Jyoti' (released in 1968) is the second most important variety in India covering 325,665ha (*i.e.* 21 per cent of total potato area) in 2015. It is the dominant variety in Karnataka and West Bengal, and second most important variety in Punjab. Despite erosion in its resistance to late blight susceptibility, to cracking and lower yield compared to 'Kufri Pukhraj' but it is still preferred in several states because of good shape *cum* size of tubers and storability, *vis-à-vis* slow degeneration rate (Kumar *et al.,* 2014).

'Kufri Bahar' (released in 1980) is the third most important variety which covers 272,642ha (*i.e.* 17 per cent of total potato area). It is most popular in Uttar Pradesh; however, it is susceptible to late blight with moderate yields.

'Bhura Aloo' is another important variety, particularly in Bihar. It is a native variety cultivated with low productivity and late blight susceptibility. Farmers' preferences for red skinned potatoes keep the demand high for this along with 'Kufri Sindhuri' and 'Lal Gulal'.

Notable varieties used for industrial purposes are 'Kufri Chipsona 1', 'Kufri Chipsona 3', and introduced ones such as 'Atlantic', 'Kennebec' and 'Lady Rosetta'. 'Kufri Chipsona 1' (released in 1998) is primarily found in Uttar Pradesh and is suitable for industrial use due to its low sugar content, high dry matter *cum* good tuber size and shape. However, it is susceptible to heat, moderately susceptible to drought but highly resistant to late blight.

Most Adopted Varieties in 2015

List of most adopted potato varieties in 2015 in six states which represents 75 per cent of total potato area in India:

- Bihar: 'Kufri Pukhraj', 'Bhura Aloo', 'Kufri Sindhuri'
- Gujarat: 'Kufri Pukhraj', 'Kufri Khyati', 'Kennebec'
- Karnataka: 'Kufri Jyoti', 'FL-1533'
- Punjab: 'Kufri Pukhraj', 'Kufri Jyoti', 'Lady Rosetta'
- Uttar Pradesh: 'Kufri Bahar', 'Kufri Pukhraj', 'Kufri Chipsona 1'
- West Bengal: 'Kufri Jyoti', 'Kufri Pukhraj', 'Kufri Chandramukhi'

Potato varieties presently recommended for cultivation in different agro-ecological zones of India:

Agro-ecological zones	*Duration**	*Recommended varieties*
North-western plains	Early	'Kufri Ashoka', 'Kufri Chandramukhi', 'Kufri Jawahar', 'Kufri Khyati', 'Kufri Pukhraj' and 'Kufri Surya' which mature in 70-90 days.
	Medium	'Kufri Anand', 'Kufri Arun', 'Kufri Badshah', 'Kufri Bahar', 'Kufri Chipsona-1', 'Kufri Chipsona-3', 'Kufri Garima', 'Kufri Gaurav', 'Kufri Jyoti', 'Kufri Pukhraj', 'Kufri Pushkar', 'Kufri Sadabahar' and 'Kufri Sutlej', maturing in 90-110 days.
West-central plains	Early	'Kufri Chandramukhi', 'Kufri Jawahar', 'Kufri Khyati', 'Kufri Pukhraj', 'Kufri Lauvkar' and 'Kufri Surya', maturing in 70-90 days.
	Medium	'Kufri Anand', 'Kufri Arun', 'Kufri Badshah', 'Kufri Bahar', 'Kufri Chipsona-1', 'Kufri Chipsona-3', 'Kufri Frysona', 'Kufri Garima', 'Kufri Gaurav', 'Kufri Jyoti', 'Kufri Pukhraj', 'Kufri Pushkar', 'Kufri Sadabahar' and 'Kufri Sutlej', maturing in 90-110 days.
	Late	'Kufri Sindhuri' which matures in >110 days.
North-eastern plains	Early	'Kufri Ashoka', 'Kufri Chandramukhi', 'Kufri Khyati', 'Kufri Pukhraj' and 'Kufri Surya', maturing in 70-90 days.
	Medium	'Kufri Arun', 'Kufri Bahar', 'Kufri Chipsona-1', 'Kufri Chipsona-3', 'Kufri Frysona', 'Kufri Garima', 'Kufri Gaurav', 'Kufri Jyoti', 'Kufri Kanchan', 'Kufri Lalima', 'Kufri Lalit', 'Kufri Pukhraj', 'Kufri Pushkar', 'Kufri Sutlej', and TPS Population 92-PT-27, maturing in 90-110 days.
	Late	'Kufri Sindhuri' maturing in >110 days.

Agro-ecological zones	*Duration**	*Recommended varieties*
Plateau region	Early	'Kufri Chandramukhi', 'Kufri Jawahar', 'Kufri Khyati', 'Kufri Lauvkar', 'Kufri Pukhraj' and 'Kufri Surya', maturing in 70-90 days.
	Medium	'Kufri Badshah', 'Kufri Garima', 'Kufri Jyoti', 'Kufri Lauvkar' and 'Kufri Pukhraj', maturing in 90-110 days.
North-western hills	Medium	'Kufri Girdhari', 'Kufri Giriraj', 'Kufri Himalini', 'Kufri Himsona', 'Kufri Jyoti', 'Kufri Megha' and Kufri Shailja', maturing in 110-120 days.
North-eastern hills	Medium	'Kufri Girdhari', 'Kufri Giriraj', 'Kufri Himalini', 'Kufri Jyoti', 'Kufri Megha' and 'Kufri Shailja', maturing in 110-120 days.
North-Bengal and sikkim hills	Medium	'Kufri Girdhari', 'Kufri Jyoti' and 'Kufri Kanchan', maturing in 110-120 days.
Southern hills	Medium	'Kufri Girdhari', 'Kufri Giriraj', 'Kufri Himalini', 'Kufri Jyoti', 'Kufri Neelima', 'Kufri Shailja' and 'Kufri Swarna', maturing in 110-120 days.

* Plains: Early (70-90 days), Medium (90-110 days) and Late (>110 days). Hills: Early (100-110 days), Medium (110-120 days) and Late (> 120days)

Source: Kumar *et al.*, 2014; https://cpri.icar.gov.in.

Propagation

Home gardeners often plant a piece of potato with two or three eyes in a hill of mounded soil. Commercial growers plant potatoes as a row-crop using seed tubers, young plants or microtubers and may mound the entire row. Potato is mainly reproduced vegetatively (clonally) by means of tubers but sometimes by botanical or sexual seeds, *i.e.* 'True Potato Seeds' (TPS). Vegetative reproduction ensures a uniform crop whereas a plant produced by TPS could totally be different from the mother plant and no uniformity may be observed in the field. Some of the advantages TPS have are that they don't carry any disease, *vis-à-vis* seed can be stored in small area. Sexual seed is becoming more popular in places where disease pressure is very high as maintaining disease-free seed is becoming a problem.

Vegetative Propagation by Seed-Tuber

Purity of the cultivars and healthy seed tubers are the primary requirements for a successful crop. The tuber seed should be disease-free and well-sprouted. When potatoes are cultivated the tubers are cut into pieces, each consisting of at least one or two eyes (buds) and planted much deeper into the soil. Tubers should be cut longitudinally through the crown eye with each piece having about 30-40g weight. Before cutting the tuber the knife should be first disinfected using potassium permanganate solution. Before planting the cut pieces should be treated with a fungicide. Planting the pieces deeper creates more area for the plants to generate the tubers and so their size also increases. The pieces sprout shoots that grow to the surface. With growing of the plants the stolons are produced around the soil surface from the nodes. The tubers form close to the soil surface and sometimes even on top of the ground. These shoots are rhizome-like and generate short stolons from the nodes while in the ground. When the shoots reach the soil surface they produce roots and shoots that grow into the green plant.

True Potato Seeds (TPS)

TPS is a botanical seed developed in the berry of the plant as a result of fertilization. The stolons are easily recognized when potato plants are grown from botanical seeds. Plants propagated from tubers are clones of the parent whereas those propagated from seeds produce a range of different varieties. New varieties grown from seed can be propagated vegetatively by planting tubers, pieces of tubers cut to include at least one or two eyes or cuttings ~ a practice used in greenhouses for the production of healthy seed tubers. Potato crop can be raised from TPS using seedling transplants or through seedling-tubers produced in preceding crop season. In the former method, the TPS seedlings raised in the nursery beds are transplanted in the field and grown to maturity. Though in latter case, the TPS seedlings are grown to maturity in nursery beds to obtain seedling-tubers that are used as a seed for raising normal potato crop in next season.

Tissue Culture

Under traditional propagation methods, usually one tuber yields approximately 8 daughter tubers in one growing whereas with tissue culture around 100,000 identical plantlets can be produced in eight months. Each plant has root system, terminal or growth points and leaves. An apical meristem is present at the apex of a potato stem and it also has lateral growth points. Meristems have no vascular system hence are less prone to viral, fungal and bacterial infections.

Under this technique disease-free plantlets are grown in test tubes on the nutrient media. Each plantlet is cut into 3-10 nodal sections after 18-60 days. Each new cutting is then planted in a new test-tube. This can be repeated until the desired number of plantlets is obtained. Plantlets are then removed from the tubes and grown in sterile soil where they complete their entire growth cycle. Tubers produced are collected and stored for selling. This process is also followed to obtain certified seeds. The first seed lot is called nuclear seed and then after harvesting the product of this seed the grower gets 'generation 1' (G1) and so forth.

Tissue culture plays an important role in the production, *vis-a-vis* international exchange of disease-free planting materials of potatoes in the form of plantlets. The aseptic steps taken in tissue culturing, produce almost disease-free planting materials though there is also a risk of re-infection once the seeds are transferred from the aseptic conditions of laboratories to fields where there is high disease pressure. Some other drawbacks of this technique include ~ high costs, the need for skilled personnel and specialized equipment, and more crop cycles are required to produce large quantities of suitable seed-size tubers for ware-yam production.

Techniques for Potato Seed Production

Seed production under '**Seed Plot Technique**' is growing of potato seed crop in the plains during low aphid period to get rid of virus infection, use of insecticides, periodic rouging of infected plants and finally dehaulming the crop before aphids reach the critical level of 20 aphids/100 compound leaves. This way the tubers produced are safe for planting. These tubers are then cold-stored during April-September, sorted out and dried under shade by spreading on the floor of a well-aerated chamber for planting in October under north-western plains. Though for planting all sizes of tubers may be used but the ideal ones are those having 35-40mm size with about 45-50g weight.

For '**breeder's seed production**' and distribution to the state and central seed producing agencies, ICAR-CPRI has a national mandate and so at 5 regional stations, *viz.* i) Kufri (H.P.) under the hills, and under Indo-Gangetic plains at ii) Jalandhar (Punjab), iii) Modipuram (U.P.), iv) Gwalior (M.P.), and v) Patna (Bihar) this task is given with a targeted production of 25,200q. For this a scientific approach is being followed for tuber indexing, *vis-a-vis* field multiplication at stage-I, II, III and IV. For tuber indexing, Modipuram indexes some 12,000, Gwalior 10,000, Jalandhar 4,000, Patna 2,000 and Kufri 4,000, out of which about 27,000 tubers of 19 commercial varieties are indexed through ELISA test against all the common viruses such as PVX, PVS, PVA, PVY, PVM and PLRV. At *Stage I multiplication* the healthy indexed tubers are planted individually at 1×1m distance and tested individually by ELISA against 5 potato viruses apart from visual inspection for mosaics or any other abnormality, and in the next stage only healthy ones and true to types are planted. Under *stage II clonal multiplication*, the healthy produce of single hill in stage-I is planted in a row as a clone and then again subjected to all the tests as at stage-I, and the healthy ones of stage-II are bulked varietywise and planted in next stage. Under *stage III bulk multiplication*, these are planted varietywise for bulk multiplication. Here these are randomly check-tested against PVX and PVS, *vis-à-vis* visually for mosaics. Healthy produce is bulked for next stage multiplication to produce breeder's seed. Under *stage IV multiplication* for '**breeder's seed production**', selected stage-III tubers are planted for check-testing and visual inspection and the abnormal ones are rogued out.

Through **micropropagation** also potatoes are being multiplied. In this method, virus-free potato plants are being produced, using meristem culture and then mini/micro-tubers are produced from these micropropagated plants and then healthy seed-crop is produced through use of these minitubers. Minitubers are then multiplied through aeroponic system vertically.

Next step is botanical seeds (true potato seeds) by providing extra light for 4-5 hours as disease transmission through botanical seed is negligible, and some 40-50g TPS is enough for planting one hectare of land. ICAR-CPRI Shimla has developed and recommended a high-yielding TPS population '92-PT-27' for commercial cultivation in the country.

Growth and Development

The growth of potato, *vis-a-vis* most other tuber crops can be divided into five phases: i) emeregence of sprouts from the seed potatoes and beginning of root growth; ii) photosynthesis begins as the plant develops leaves and branches aboveground, and stolons develop from lower leaf axils in the underground stem; iii) the tips of the stolons swell forming new tubers and the shoots continue to grow and then flowers typically develop soon after; iv) tuber-bulking occurs as the plant begins investing the majority of its resources in its newly formed tubers; and v) in the last the tubers mature, the leaves and stems senesce and the tuber skins harden. At this phase, several factors are critical to a good yield ~ optimal soil moisture and temperature, soil nutrient availability and balance, and resistance to pest attacks.

New tubers may start growing at the surface of the soil. These surface tubers should be covered since exposure of tubers to light leads to an undesirable greening of the skins and the development of solanine ($C_{45}H_{73}NO_{15}$) as a protection from the sun's rays. Commercial growers hill them up by piling additional soil around the base of the plant as it grows. Home gardeners and small-scale growers can cover the growing area with mulches such as straw or plastic sheets.

Factors Influencing Growth and Quality

To have the tubers disease-free the areas where seed potatoes are grown should be carefully selected. These locations are selected at a quite sunny site for good growth of plants and for protecting these from various diseases and insect-pests. Environmental factors such as light, temperature, moisture and soil type influence the growth and quality of potatoes. Other factors such as the cultivar, size of mother seed tubers, seed-piece cutting, seed-piece types, cut-seed size, planter operation, plant stand, stem population, irrigation, nutrition, pest management, planting date and harvest date can be controlled by the grower.

Many factors such as length of growing season, air *cum* soil temperatures, light intensity *cum* duration, humidity and wind that influence potato growth are largely uncontrollable. Potatoes are sensitive to heavy frost as this hampers tuber development. Even cold weather makes potatoes more susceptible to bruising and possibly later rotting.

Agro-climatic parameters for growing potatoes in sub-tropical countries or regions are given in the following table:

Growing Season	*Winter*
Temperature during planting and harvesting	Planting: 25-30°C; harvesting: 10-20°C
Crop duration	90-100 days
Photoperiod during growth	10.3 hours/day
Mid-day water stress	Prominent
Day/night temperatures	25-30°C/4-15°C
Frost	Common
Result	Low yield, less dry matter and more reducing sugars

Agro-climatic parameters for growing potatoes in temperate regions or countries are given in the following table:

Growing Season	*Summer*
Temperature during planting and harvesting	15-25°C at both the levels
Crop duration	150-180 days
Photoperiod during growth	14h/day
Mid-day water stress	Absent
Day/night temperatures	25°C/15°C
Frost	Absent
Result	High yield, high dry matter and low reducing sugars

Climate change is predicted to have significant effects on global potato production. Like many crops, potatoes are likely to be affected by changes in atmospheric CO_2, temperature and precipitation, as well as interactions between these factors. Climate change will also affect disease and pest incidences in potatoes.

The rate of development of sprouts from seed pieces depends on soil temperature. At 6°C, sprout elongation is very little. It is slow at 9°C and is maximized at about 18°C. The optimum soil temperature for initiating tubers is 16-19°C. Tuber development declines as soil temperatures rise above 20°C and tuber growth practically stops at soil temperatures above 30°C. The number of tubers set per plant is greater at lower temperatures than at higher temperatures whereas higher temperatures favour development of large tubers. Yields are highest when average daytime temperatures are about 21°C. Cool night temperatures are important because they affect the accumulation of carbohydrates and dry matter in the tubers. At lower night temperatures, respiration rate is slowed down which enhances storage of starch in the tubers (Pandey *et al.*, 2008).

The potato can be grown almost in any type of **soil** but not the saline and alkaline ones. Soils which are naturally loose are preferred as these offer least resistance to the enlargement of the tubers. A well-drained sandy-loam soil rich in organic matter and having pH range of 5.5-7.5, preferably 5.2-6.4 is most suitable for potato cultivation.

Three successive **ploughings** with associated harrowing and rolling and eliminating weed-roots are desirable before planting. The land is ploughed at a depth of 24-25cm and exposed to sun. The soils having a higher pore-space offer least resistance to tuber development. Any existent weed should be removed. It would be better if at first ploughing the well decomposed FYM at 25-30 t/ha is mixed in the soil.

When still the soil is sufficiently moist, furrows are opened at a distance of 50-60cm before planting. The whole or cut tubers are planted 15-20cm apart in the centre of the ridge at a depth of 5-7cm and covered with the soil. The seed rate of potato depends upon the season of planting, duration, seed size, spacing, *etc.* It is generally 1.5-1.8t/ha for round varieties and 2-2.5t/ha for oval varieties.

For a better yield, potatoes require a continuous supply of soil water along with adequate soil aeration. When soil moisture is maintained above 65 per cent of the available soil water (ASW) capacity, yields are greatest. The amount of water needed by potatoes varies with the soil type, humidity, air temperature, air movement, plant and stem populations, variety and cultural practices. Low or fluctuating moisture levels can contribute to common scab, early dying, hollow heart, knobby tubers, low dry matter, low tuber set and low yield. Excessive soil moisture results in waterlogging and poor soil aeration which ultimately causes reduction in tuber yield, and in extreme cases causes tuber rot. An excess of moisture may also lead to enlarged lenticels which are openings of the epidermis. This detracts from their appearance and allows entry of disease organisms, causing tuber rot in storage (Pandey *et al.*, 2008).

Since potato plant has shallow and sparse root system hence irrigation has a special significance in its production. First irrigation is often light and usually given 5-7 days after planting when sprouts are just peeping up or going to peep. Subsequent irrigations are given at 7-15 days intervals depending upon the soil type and prevailing climatic condition. Drip irrigation system is most economical giving highest productivity and saving almost 50 per cent water. It also enables application of fertilizers through irrigation water. The sprinkler system gives uniform distribution of water and reduces water losses by percolation and run off. Sprinkler irrigation is beneficial on frosty nights as it reduces frost damage in potatoes. It is recommended for areas experiencing water scarcity, having undulating topography and/or when soil is unusually sandy. Under such situations, the use of sprinkler systems increases water use efficiency by 40 per cent as compared to furrow irrigation (vikaspedia.in).

Manure and Fertilizers

Due to high nutrient requirement of potato crop, application of fertilizers and organic manures is considered essential to obtain high economic yield. Where organic manures are not easily available it is green manuring which can substitute it to some extent. Optimum fertilizer dose varies greatly depending upon the soil type, soil fertility, climate, crop rotation, variety, length of growing season and moisture supply.

A fertilizer dose of 180-240kg N, 60-90kg P_2O_5 and 85-130kg K_2O per hectare is recommended for alluvial soils of Indo-Gangetic plains. In the hilly areas, two applications of 100-150kg N, single application of 100-150kg P_2O_5 *vis-a-vis* 50-100kg K_2O per hectare is recommended. In black soils of plateau areas about 120-150kg N and 50kg each of P_2O_5 and K_2O is recommended. In the acidic soils of southern plateau 120kg N, 115kg P_2O_5 and 120kg K_2O per hectare are recommended (vikaspedia.in).

Two third of N and entire dose of P and K are given at the time of planting. Remaining dose of N is given at the time of earthing up. The fertilizers are applied by band placement 5cm away from the tubers. Ammonium sulphate and ammonium nitrate are usually the best fertilizers for potato followed by calcium ammonium nitrate, ammonium chloride and urea (vikaspedia.in).

Interculture

Potato crop develops canopy in about 4 weeks after planting. By this time weeds must be controlled mechanically or by herbicide application to get competitive advantage for the crop. Large weeds should be removed before starting earthing up. Since manual weeding is quite uneconomic therefore animal-drawn three-tine cultivator is used which can cover one hectare per day. Alternatively pre-emergence spraying of weedicides such as Fluchloralin (0.70-1kg a.i./ha) or Pendimethalin (0.5kg/ha) is recommended for controlling annual *cum* broad-leaved weeds.

To keep the soil loose and to destroy the weeds, 2-3 soil earthings are required at an interval of 15-20 days. First earthing-up should be done when the plants are about 15-25cm high and the second in next 20 days to cover the emerging tubers properly. Double mould-board plough ridger or 3 and 5 row tractor drawn culti-ridger can be used to accomplish the earthing.

A good crop rotation maintains soil fertility, reduces soil erosion *cum* depletion and suppresses insect and disease outbreaks. The cropping sequence recommended in states should be followed such as potato-mung-paddy or potato-mung-groundnut in Bihar; potato-wheat-maize, potato-wheat-paddy or potato-wheat-green manure crop in Punjab; potato-mung-transplanted paddy in Asom; potato-bajra-groundnut in Gujarat and U.P.; and potato-okra-soybean in M.P.

Being a fast-growing *cum* short-duration crop, potato is ideal for intercropping with other crops. It can be successfully intercropped with sugarcane as the cultural operations and resources used in both the crops are mutually complimentary. The potato-fennel and potato-onion intercropping in Haryana; potato-mustard and potato-linseed in U. P.; and potato-wheat intercropping in Bihar are some of the profitable crop combinations.

Insect Pests, Diseases and Physiological Disorders

Potato crop is infested by various insects and nematodes. Some pests attack the foliage while others the tubers. Many of the diseases in potato are soil/seed borne and so once infected it is very difficult to control them. Similarly viruses affect the seed quality and so has a chain of implications. Not all pests and diseases are present everywhere; some are more widespread than the other while some are very localized, *e.g.* 'wart' is confined to the Darjeeling hills while golden nematode is confined to the Nilgiri hills. The proper diagnosis and control of pests and diseases is of utmost importance.

Cutworms (*Agrotis ipsilon, A. segetum, A. flammatra, A. spinifera*): These cosmopolitan, polyphagous and most destructive caterpillars are present worldwide. In India, cutworms are more serious in northern region. *A. segetum* is commonly found on hills and *A. ipsilon* in the plains. Its female lays 15-35 eggs either at the base of the plants or in the moist soils near the base of potato plants. Its young larvae feed on leaves but mature larvae cut the stems just near the ground and even make irregular holes in the tubers during night. After tuber formation, they start feeding on tubers and roots which results in a variety of holes and so severely reducing tuber yield as well as drastically its market value (Sharma *et al.*, 2013). They remain very active in the plains of northern India from October to March-April though in temperate areas during summer. To minimize its population, as a precaution, only well-rotten FYM should be used at the time of field preparation, *vis-à-vis* any pesticidal granules such as 5 per cent Aldrin or Heptachlor dust at 45kg/ha should be incorporated in the soil at the time of planting which would together control even white grubs. At the time of hilling, the hills or plant bases should be sprayed with 0.05 per cent Chlorpyriphos 20EC or Chlorfevininphos and as a safety major the spray should be repeated after 20-30 days. At 10^9 spores per ml of biopesticide Bt can also be together used on bunds and on the crops for controlling this pest.

White grubs (*Holotrichia longipennis,* syn. *Lachnosterna longipennis*; and *L. coracea*): These cosmopolitan *cum* polyphagous grubs are white-coloured with dark brown heads which become 'C-shaped' when disturbed. They are distributed in Asia, Africa, Americas and European countries. Female beetle lays its eggs in the soil near the plants. These eggs hatch in May coinciding with onset of rainfall, and even otherwise in the year when rainfall is more their damage is also more and in the temperate areas its damage is to a tune of 85 per cent. These newly hatched larvae (grubs) start feeding immediately on developing tubers on the hills thereby deteriorating their market value drastically. These also feed on potato roots. The female in its whole life cycle lays 4 to 41 tiny eggs. It is very difficult to control, however, the pesticidal granules such as 5 per cent Aldrin or Heptachlor dust at 45kg/ha should be incorporated in grub-infested soils at the time of planting or at the time of hilling.

Leaf-eating semiloopers {Bihar hairy caterpillar (*Spilosoma obliqua*), cabbage green semilooper (*Plusia ni*), semilooper (*Plusia chalcites*), fruit borer (*Helicoverpa armigera*), ragi cutworm (*Spodoptera exigua*), tobacco caterpillar (*Spodoptera litura*), *etc.*}, all being polyphagus in nature and their females lay eggs usually in clusters on leaves which on hatching start feeding on foliage gregariously, mostly defoliating the plants. To kill them the plants are sprayed with 0.05 per cent monocrotophos or 0.2 per cent carbaryl. Further, *Helicoverpa armigera* is controlled through release of egg-parasitoid (*Trichoderma chilonis*) @ 2,50,000/ha/week.

***Epilachna* or Hadda beetle** (*Epilachna vigintioctopunctata* and *E. occellata*): It is a polyphagus pest as the adults and grubs feed voraciously on the leaves skeletonising the leaves leaving the leaf-veins intact so afterwards such leaves dry out. The affected plants should be sprayed with 0.05 per cent monocrotophos or 0.2 per cent carbaryl.

Aphids: There are more than 12 polyphagous species of aphids that infest potato crops. These include green peach aphid (*Myzus persicae*) and cotton aphid (*Aphis gossypii*). Depending on weather conditions, their life cycle is of 25-45 days. They lie on the underside of tender leaves and shoots and continue sucking the sap from phloem of potato stems, leaves and roots using the stylets. Affected plants become weak whereas the leaves yellow and curl downwards. Aphid secretes honeydew which gives rise to sooty mould and other fungal diseases. These are controlled by spraying Malathion (0.05 per cent) or oxy-demeton methyl (0.025 per cent).

Leafhoppers/Jassids (*Alebroroides dravidans, Empoasca fabae, Orosius albicinctus, Seriana equata*): Adults and nymphs of jassids are sap sucking insects. Excessive feeding causes paling, drastic yield reduction and ultimate death of potato plants. Jassids are also responsible for transmitting various viruses and MLOs to plants. *A. dravidans* transmits infection of 'purple top roll' (PTR) and 'witche's broom' (WB); *S. equata* transmits MF (marginal flavescence) and PTR; and *O. albicinctus* PTR, WB and MF. These diseases can be controlled by controlling jassids and aphids as elaborated under aphids.

Whiteflies (*Bemisia tabaci*): These milky-white insects suck sap from leaf-underside and other tender parts during which virus acquisition and transmission take place. While controlling aphids, these tiny insects will automatically be controlled.

Leaf hoppers (*Amrasca biguttula*): It is a polyphagous pest distributed worldwide causing major loss to potatoes. In India, it is distributed in all potato growing regions of Indo-Gangetic plains. These suck the cell-sap from lower side of the leaves causing extensive damage by direct feeding on the plants. Severely infested field gives a burnt look, and this appears in a circular ring commonly known as 'hopper burn' (Sharma *et al.*, 2013).

Thrips (*Thrips palmi*): It is a predominant species and most efficient vector of 'stem necrosis disease' caused by 'groundnut bud necrosis virus' belonging to the Tospovirus group. Pale yellow or whitish adults feed on young leaves. Bean-shaped and colourless to pale-white eggs are deposited in leaf tissue in a slit cut by the female. Their nymphs feed in groups particularly along the leaf midribs and veins and usually on older leaves causing 15-30 per cent loss in northern Gujarat, parts of M. P. and Rajasthan (Sharma *et al.*, 2013).

Leaf eating caterpillars (*Spilosoma obliqua, Spodoptera exigua*): The caterpillars of both the species cause damage by feeding on potato leaves. These are controlled by spraying the crop with deltamethrin (0.0025 per cent).

Potato tuber moth (*Phthorimea operculella*): Though this caterpillar is mainly a pest of stored tubers but sometimes has been observed infesting even the field tubers in all tropical and subtropical potato producing countries in Asia, Africa and NC and S America. Wings of these brownish grey moths are folded to form a roof-like structure. Their freshly laid white eggs are deposited singly or in small clutches resembling strings of beads. The larvae destroy the crop by injuring the leaves and boring into petioles and terminal shoots causing wilting of plants. It then enters the tubers and feeds on them causing mines therein (Sharma *et al.*, 2013). Its population is drastically reduced when at the time of field preparation only well-rotten farmyard manure has been used. Earthing up should be done carefully so that the tubers are not exposed in the field to ovipositing female moths. Infested crop should be sprayed with 0.1 per cent carbaryl. This treatment is also required before keeping the potato tubers in cold storage.

Mite [(*Polyphagotarsonemus latus*), commonly known as yellow mites or broad mites]: These 8-legged creatures are highly polyphagous in nature and cause serious damage to potato crop during *Kharif* season in Maharashtra and Karnataka. It also affects early planted potato crop during *Rabi* season in Punjab and western UP. Usually found on the upper part of the plant, they feed on apical shoots and on abaxial surface of young leaves by sucking the cell sap. The typical symptoms are bronzing, curling and discolouration of leaves (Sharma *et al.*, 2013).

Aphids feeding on leaf

Whiteflies feeding on potato leaf

Leaf hopper

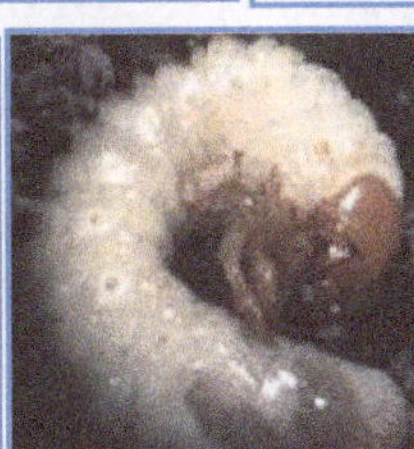
White grub

Cutworm larvae

Root knot nematode (*Meloidogyne incognita*): These obligate parasites of the roots and tubers of potato plants are distributed worldwide. Nematode-infested root system of potato plants is poorly developed. Heavily infested plants remain severely stunted with dull and unhealthy foliage. Later on the lower leaves turn yellow and brown and wither leaving only the young leaves at the top. The browning and withering of the foliage gradually extend and ultimately cause premature death of the plant with significant reduction in tuber yield. Their infestation on roots forms small galls or knots (warty pimple-like outgrowths) but in case of heavy infestation the plants are stunted and exhibit early maturity leading to significant reduction in tuber number, size and yield (Sharma *et al.*, 2013) especially in hill-grown potatoes though of not much significane in the plains because on the hills these are grown during summer when the nematodes remain quite active while in the plains these are grown during winter when these remain quite inactive. To get rid of root-knot nematodes the field should be deeply ploughed and in summer solarized for two months. Moreover, the burning of plant debris in the field also minimizes its infestation. Planting of potato in the second week of October and in the beginning of January minimizes its attack. Alternate planting of marigold is also quite effective in minimizing nematode population. Moreover, the cropping of maize, wheat, bajra, beans, *etc.* is also quite useful in minimizing its attack. Application of *neem* or *Pongamia* cake @ 20 quintals per hectare at the time of field preparation will also minimize nematode population drastically. Variety 'H.C. 294' is resistant to root-knot nematode infestation. Active ingredients of Carbofuran @ 3kg/ha or Aldacarb 2kg/ha or Ethoprop dusting twice @ 10kg/ha, once at planting and other with hilling are quite effective.

Potato/golden cyst nematodes (*Globodera rostochiensis, G. pallida*): In India it was first recorded in 1961 in Nilgiris (T.N.) though initially it was recorded in Andes Mountains. Globally, *G. rostochiensis* has five pathotypes namely Ro1, Ro2, Ro3, Ro4 and Ro5 and four races (R1A, R1B, R2A and R3A) whereas *G. pallida* has three pathotypes (Pa1, Pa2 and Pa3) and seven races (P1A, P1B, P2A, P3A, P4A, P5A and P6A). In India, pathotypes Ro1 of *G. rostochiensis* and Pa2 of *G. pallida* are the most prevalent forms in the Nilgiris hills. The other prevalent pathotypes are Ro2 and Ro5 of the former and Pa1 and Pa3 of the latter. Due to attack of these nematodes leaves become dull and start drying under direct sun and the plants become weak requiring comparatively frequent watering. In this case, affected plants die during mid- to late-summer on the hills with leaves yellowing and drying up from the bottom of the stems upwards. On digging such plants white, yellow or brown cysts of spherical nematode up to 1mm diameter are found on the roots. This causes tuber yield reduction to a tune of 65-71 per cent (Khurana and Pandey, 2008). Cyst nematode resistant exotic varieties are

'Accent', 'Alhambra', 'Cara', 'Javelin', 'Kingston', 'Maris Piper', 'Montana', 'Nadine', 'Pentland', 'Rocker', 'Sante', 'Stemster', 'Ukama'; and indigenous one is 'Kufri Swarna' (adopted for growing in 40 per cent of potato growing areas of Nilgiris). Since the eggs of potato cyst nematode remain viable inside the cysts for many years hence its control is quite difficult. For its effective control, all sorts of control measures should be applied together such as use of effective nematicide at right time together with proper cropping scheme and growing of nematode-resistant varieties.

Late blight (*Phytophthora infestans*): The infection appears mostly at the tips or edges of the lower leaves with circular or irregular water-soaked pale-green lesions which rapidly grow into large brown to purplish-black necrotic spots followed by withering and death of leaves. A white mildew consisting of sporangia and spores of the pathogen can be seen on lower surface of the infected leaves especially around the edges of the necrotic lesions (Sharma *et al.*, 2013). Cloudy and humid weather favours quick spread of the disease. Later the disease may spread as slightly sunken *cum* dark patches on tubers with a reddish-brown discolouration beneath. This discolouration spreads into the flesh causing tuber rotting. As a precaution only healthy (disease-free) seed-tubers should be planted. Excess of nitrogen and irrigation should be avoided. The crop can be sprayed with Bordeaux mixture, copper oxychloride and mancozeb @ 0.25 per cent well in advance. To reduce chances of infection developing on tubers, the tubers require to be earthed up deeply. Late blight resistant cultivars such as 'Cara', 'Estima', 'Kondor', 'Maris Peer', 'Pentland Crown', 'Record' and 'Romeo' are recommended for growing in England.

Early blight (*Alternaria solani*): The infection appears on lower leaves causing small dark brown to black oval or angular, dry and papery necrotic spots on leaves. Concentric rings of raised and depressed tissues within the leaf spot give it a bull's eye or target board appearance. Leaf tissues around the spots become yellow and chlorotic. High humidity *cum* low temperature is favourable for spread of this fungus. This pathogen is carried over from year to year on infected plant debris. Crop rotation, spraying the crop with mancozeb @ 0.2 per cent 30-35 days after planting, and growing tolerant varieties such as 'Kufri Jeevan' are the ways to combat this problem.

Leaf spot disease/potato gangrene (*Phoma exigua, P. sorghina*): Slightly sunken spots with clearly defined round-edged *cum* slightly wrinkled areas develop on the tubers shortly after lifting but there is no any visible fungal growth. Affected tubers turn pale-pink, darken and rot. *F. exigua* infection also causes large (1-1.25cm), round, light and dark slaty blotches on leaves. *P. sorghina* infection causes pin-head leaf spots up to 4mm diameter and on tubers up to 1.7cm dark slaty spots. After planting, such tubers may not produce a crop or may produce a poor gappy crop. While lifting, bruised and injured tubers should be separated and consumed immediately and only healthy ones should be sent to cold storage after proper curing. Any tuber showing *Phoma* infection should immediately be disposed off. To get rid of these maladies, only healthy tubers should be used for planting. In the field, nitrogen fertilization should be in a quite balanced quantity. Urea at 1 per cent should be sprayed at 45 days of planting and repeated again after 10 days. In temperate regions, 1 per cent Bordeaux mixture and 0.3 per cent copper oxychloride sprayings are quite effective against gangrene. Potato field should be quite away from any other solanaceous crop.

Potato silver scurf (*Helminthosporium solani*): It is pale and inconspicuous silvery markings or lines developing on the tuber skin but in no way the size, shape, flesh and eating quality of the tuber is affected except the skin markings. These markings may turn black if tubers are stored under too humid conditions as there fungus may produce spores.

Cercospora leaf blight (*Cercospora solani tuberosi, C. concors*): The spread of this disease is favoured by low-temperature and high-humidity as is found on the hills. The disease occurs on leaves and stems, initially on lower leaves but afterwards progressing to the upper part of the plant, forming 2-5mm round brown spots with white centre which after coalescing engulf most part of the leaves. Such leaves dry and fall down soon. On stems, its infection is initially grey but soon becoming black. In case of infection of *Cercospora concors*, under-surface yellowing and falling of leaves start from lower to upper part of the plant though still upper leaf surface may remain light green. Usually the spots may be up to 5mm in diameter though in highly susceptible varieties the spot may engulf entire leaf. The disease does not spread during dry season of summers on the hills. Spread of disease is favoured in nitrogen-deficient soil, therefore balanced dose of fertilizers is desired to minimize this problem. To check further spread of disease the stems of diseased plants should be removed and burnt. Spraying the crop with 5:5:50 Bordeaux mixture, mancozeb @ 2kg/ha or Difolatan 4F will control the disease to maximum extent.

Black scurf disease (*Rhizoctonia solani*): This is a serious disease of potato worldwide. Its infection makes the tubers misshapen and hampers the yield and quality of the tubers, *vis-à-vis* reduction in tuber yield to a tune of up to 35 per cent. Infected seed-tubers have blighted sprouts, later on causing rot of the stem *vis-à-vis* stolons so the field looks gappy with poor standing. Various types of symptoms appear on tubers, sprouts, stems and stolons. Irregular, black and lumpy encrustations coupled with raised black-scurfy spots starting from pin-head to pea-seed size are formed on the tubers. Its infection may also appear as pitting, deforming and cracking of tubers, rotting of the lower part of stem, *vis-à-vis* formation of aerial tubers on stems. Due to its infection, entire or a part of aerial stem exhibits injury-type orangish-grey symptoms which make the upper part of stem looking rosy to violet with upcurling of leaves. The toxin generated due to disease infection is partially responsible for tuber-deformity. The mycelium and spores present on tubers or in the soil persist for long but germinate at 8-30°C temperature range, optimum being 18-23°C. Only disease-free tubers and that too sprayed with 3 per cent boric acid should be planted in solarized soil. Afterwards in the field crop benomyl, thiabendazole or carboxin should be sprayed at 0.2 per cent when danger of infection is anticipated. *Trichoderma viride, T. harzianum* or *Bacillus subtilis* may also be used in its control.

Charcoal rot (*Macrophomina phaseolina*): It is a serious soil- and tuber-borne fairly widespread potato disease causing mostly wet or dry charcoal-coloured tuber rot in storage or stem blight *cum* wet charcoal rot during dry and warm (28-35°C) weather. In stem blight, there is sudden yellowing and then withering of plants. Tuber rot is rather more serious where symptoms appear as small dark light grey to black, soft and water-soaked lesions around eyes (buds), lenticels and stolon-end. Subsequently, the lesions become filled with black mycelium and sclerotia of the pathogen. Just below the skin, fleshy tissues get blackened to a depth of 2-5mm and ultimately the tubers are reduced to foul-smelling wet-rotten mass. The lesions may shrink and develop symptoms similar to dry rot under low moisture conditions. This disease can be kept at bay by using disease-free hill-grown seed tubers and that too as early cropping so that the crop may not experience 28°C temperature. Moreover, light but frequent irrigations are provided so that soil temperature is lowered.

Wart (*Synchytrium endobioticum*): It is a serious soil-borne disease but in India its infection is confined only to the Darjeeling hills (W. B.) where it was first recorded in 1953. Affected plants show warty growth-protuberances on stems, stolons and tubers though at all the stem bases at soil level and roots are not affected. Infected tubers initially exhibit proliferation akin to cauliflower heads or cock's-comb which of the green-yellow colouration in its small form is sometimes also observed on stems as well as on buds at soil level, together with stolons and tuber but never on roots. Under a very high humid condition, intermittent rains and at 10-20°C temperature the upper portion of plants such as floral buds, lateral buds as well as on leaves its malformation may be observed. These warts or lumps are initially white or cream but as these turn old the colour changes to black from grey. The warts consist of distorted, proliferated and branched structures grown together into a mass of hypertrophied tissue. In the field as well as in storage badly infected tubers show watery-rot. If the pathogen once enters a field, the same field can not be used for next 25-30 years for potato cultivation even if it has been kept fallow. Disease-free seed-tubers should be used for planting. Soil treatment with 5 per cent Formalin is also effective.

Potato dry rot (*Fusarium* spp.): Out of some 14 *Fusarium* species responsible for causing dry rot of potato tubers in storage, two are most common, *viz. Fusarium oxysporum* and *F.* solani. *Fusarium* infection in storage causes wrinkling on tuber skin at one end, followed by rapid shrinking inwards, therefore concentric rings of wrinkles occur along with discolouration. The fungus may infect the tubers when still they are in the ground or as they are being lifted. The pathogen enters through the lenticels, eyes or wounds caused due to scab infection, pest attack or due to abrasions. If infected tubers are planted, the pathogen causes poor growth. Affected tissues become black, grey or brown. It appears as small brown lesions on surface of the affected tubers which enlarge subsequently and become dark, sunken and wrinkled producing white, pink or blue pustules of fungal spores. In later stages a cavity develops in the centre of the concentric ring and whitish, pinkish or dark brown growth of fungal mycelium may become visible. Harvesting of tubers should be only when these are fully mature. The handling of tubers from lifting to storage should be done quite carefully so that in no way these are injured and there may not be any visible injury mark of any kind. After lifting, the tubers should be cured in shade at 13-18°C and at well ventilated place for about one fortnight and then stored. Storage should be at a cool, frost-free and dry place. Only resistant varieties should be planted. Treating the seed tubers for 30 seconds in 1 per cent formalin solution before planting is quite effective. Thiabendazole at 1200ppm or benomyl at 0.2 per cent spraying on tubers before planting is also advisable. Avoid division of seed tubers for planting but in case one has to do this it would be better to treat the cut-tubers with Dithiocarbamate. Use of *Bacillus subtilis, Pseudomonas fluorensis* or *Trichoderma viride* is also effective in this case.

Wilt: *Verticillium albo-atrum, Ozonium taxanum parasiticum, Sclerotinia sclerotiorum, Phytophthora erythroseptica, Colletotrichum atromenterium, Spondylocadium atrovirens, Penicillium verucosum, Gilmaniela humicola, Phomopsis tuberivora, Roselinia, Pestalotia versicolor, Aspergillus fumigatus, Rhizopus arrhizus, Pythium ultium, Jivotrichum candidum, Coleofomiae empetryh, etc.* cause various types of potato wilt in different regions of the country though are not of significant importance. However, while growing potatoes, phytosanitary practices should carefully be applied, tubers should be treated before planting *vis-à-vis* after lifting as described above and then stored.

Powdery scab (*Spongospora subterranea*): This potato disease is confined to the hills and valleys only. Its symptoms initially appear as small, smooth, circular and light brown sunken spots on potato tubers but later becoming scab-like raised as pimples, however, unlike common scab the lesions of powdery scab are round, raised, filled with powdery mass of spores and surrounded by ruptured remains of epidermal layers so commercial value of the tuber is highly reduced. Ultimately, these pustules break open to release the powdery mass. This disease can be minimized by planting only healthy stock and treating the seed tubers with 0.3 per cent Agallol-3 for 10 minutes or with 0.1 per cent mercuric chloride for 1.5h before planting.

Leaves infected with late blight

Leaves infected with early blight

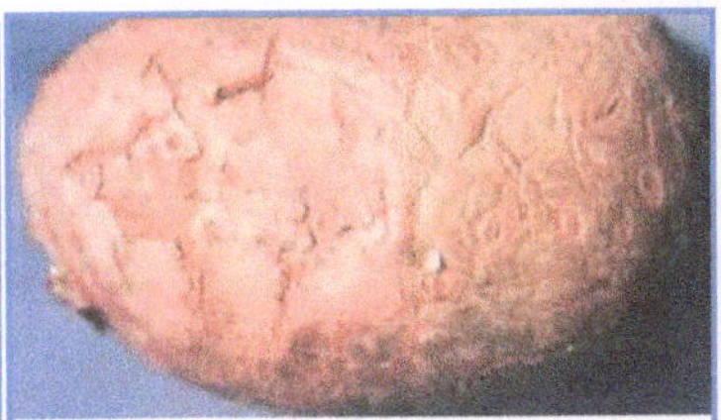

Powdery scab lesions surrounded by ruptured remains of epidermis

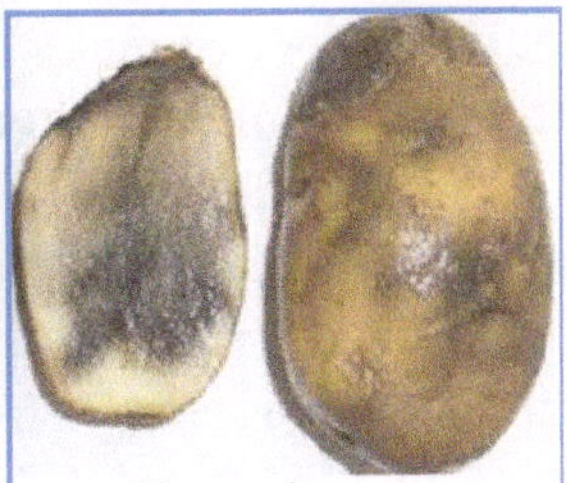

Charcoal rot

Brown bacterial rot or wilt (*Ralstonia solanacearum*): This pathogen on potato in India was first recorded in 1892, then with the name of *Pseudomonas solanacearum* though now it is renamed as *Ralstonia solanacearum*. Its infection initially causes a slight wilting of youngest leaves during hot sunny days though afterwards showing recovery when sun is weak but again next day these plants show the same symptoms with no recovery at all. Due to loss of plant turgidity, initially the leaves droop down, followed by total wilt, especially at mid-day. In advanced stages, bacterial mass may ooze out milky exude from tuber-eyes (Virupaksh *et al.*, 2012). Infection causes tuber pitting and on cutting this oozes out a thick liquid though such symptoms do not appear in varieties which do not exhibit leaf-wilting. Cross section of stems of such plants shows vascular discolouration. The vascular tissues of a transversely cut tuber show water-soaked brown circles where dirty white sticky drops appear in 2-3 minutes though brown rot symptoms on tubers may appear at later stage of disease development. Planting of infected tubers exhibit such symptoms within a month therefore seed-potatoes should be collected only from certified agencies. Apart from seed-potatoes the disease also spreads through agricultural implements and irrigation water. High

RH, high soil moisture and a temperature range of 28-30°C are quite favourable for quick spread of this disease. To get rid of this problem the best solution is the use of disease-free planting material, ploughing of the field in winter along with use of 12kg stable bleaching powder per hectare of land. However, proper cropping system is also quite effective in getting rid of this malady and the best for this is intercropping with maize or beans. Also when isolated plants show wilting, the soil in the radius of 10-15cm around the infected plant should be treated with 5 per cent Formalin or 1 per cent copper sulphate solution up to a depth of 2.5cm just after the harvest to kill bacterial inoculum.

Soft bacterial rot (*Erwinia carotovora, Bacillus* spp. and *Pseudomonas* spp.): This problem is observed in the field as well as in storage. In field it may cause up to 10 per cent loss but in storage to a tune of 80 per cent. In this case the plants usually remain stunted with grey stems, coupled with leaf curling and yellowing of plants. Brown-black or jet-black slimy areas just above the ground level on stems appear which is known as 'black leg'. Affected tissue on the tuber surface show discolouration, occasional wrinkling *cum* depressions, *vis-a-vis* internal tissues turning soft and mushy therefore entire tuber disintegrates. Potato var. 'Kufri Deva' shows resistance to this problem to a greater extent, and the tubers of this variety can be stored at ordinary room condition in the plains for more than 7 months. However, the tubers must be dipped for 30 minutes in 3 per cent boric acid before planting. Prior to planting or with 1st and last watering, field application of 17kg/ha of stable bleaching powder is recommended as a safeguard against soft rot of tuber in storage. Nitrogen application in the field should never be more than 150kg/ha and tuber lifting below 19°C temperature are the measures that minimize the incidence of this disease. Before planting the seed tubers (whole or cut) may also be dipped in 0.01 per cent streptocycline or in 0.01 per cent mercuric chloride solution for up to 2.5h followed by keeping these at a well-ventilated place for up to 4 days for formation of corky layer.

Common scab (*Streptomyces scabies*): This disease has now become prevalent in some isolated pockets of the plains. Its symptoms appear as small brownish or reddish superficial rough spots along the tuber surface during juvenile period of tubers. These lesions later become corky due to elongation and division of invaded cells. The symptoms range from a few superficial or raised brown spots on the skin to dark corky pits (1-3mm deep) into the potato tuber resulting in the reduced market value of the produce. Green manuring lessens the incidence of infection by this bacterium. To get rid of this problem the seed tubers require to be dipped in 2.5g/l of Agallol-6 or Emisan-6 for 30 minutes before planting. From 5th week of planting to the 9th week the field should be kept well-irrigated. A 4-year crop rotation avoiding sugarbeet and carrot is quite effective against this disease.

Soft rot and **black leg** (*Pectobacterium* species) infection at initial stage causes a small area of tuber tissue around lenticels or stolon attachment point becoming water-soaked and developing soft lesions. Under high humidity these lesions may enlarge and spread to larger area. In advanced stages of decay, tubers are usually invaded by other organisms and the decaying tissue becomes slimy with brown liquid ooze *cum* foul smell. Black leg symptoms appear in cooler regions. Affected haulms become black at collar region just above the ground. The lesions first enlarge into stripes on stem and petioles, turning black and then invade the affected parts causing soft rot and toppling of the stem and leaves. Shoot bases develop a blackened *cum* shrivelled cortex with stunted plant growth. Leaflets develop reddish tip colouration. Branches become stiffened and more upright than normal. Affected haulms at soil level become jet black. Such tubers develop watery rot, emitting offensive sulphurous odour. Only disease-free healthy tubers should be planted. Too early planting should be avoided. Affected plants should be uprooted with soil intact and destroyed. The tubers should be washed with chlorinated water before storage.

Symptoms of bacterial wilt

Brown discoloration of stem tissues

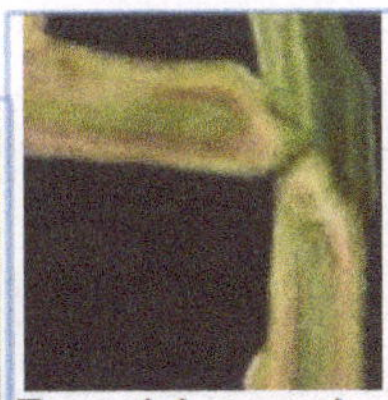

Bacterial streaming in clear water from stem cut section of potato

Brown rot symptoms on tubers

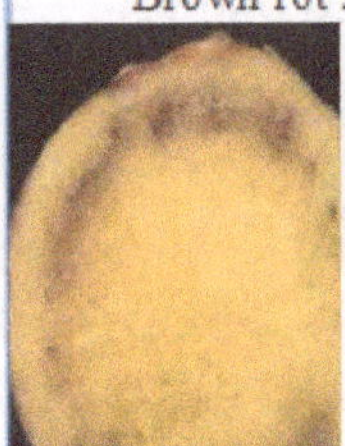

Vascular browning

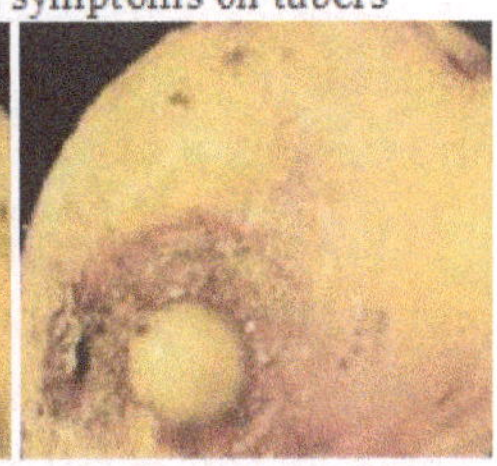

Oozing of bacterial mass through an eye of tuber

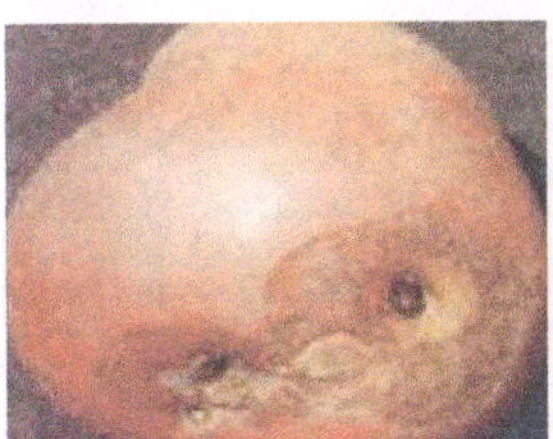

Soft rot

Black leg

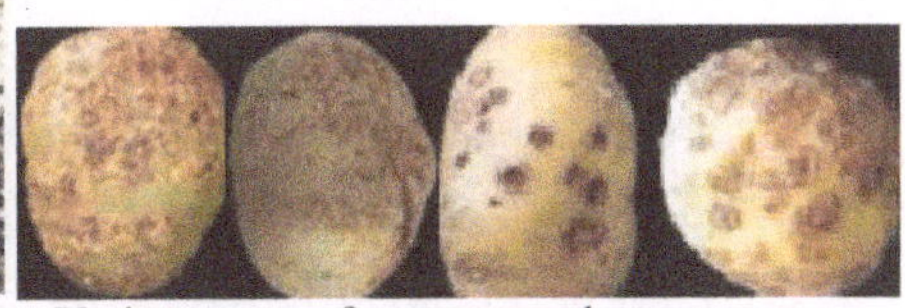

Various types of common scab symptoms

Potato virus A (PVA) belongs to the genus *Potyvirus* in the family Potyviridae which too is transmitted by aphids. It has single stranded positive sense RNA and causes mild mosaic symptoms somewhat similar to those caused by Potato virus X. However, unlike PVX the mottles caused by PVA appear on the veins and the infected leaves look shiny. Infected plants show an open habit. In combination with PVX, it may produce severe leaf crinkle.

Potato virus M (PVM) is a member of the genus *Carlavirus* of the family Flexviridae. PVM is also transmitted by aphids and causes mottle, mosaic, crinkling, spoon-shaped rolling of leaves and stunting of shoots. Severity of symptoms depends on the combination of potato cultivars and PVM isolates.

Potato virus X (PVX) type-species of the genus *Potexvirus* belongs to the family Alphaflexiviridae and the order Tymovirales. This virus is largely confined to members of the family Solanaceae. It is transmitted mechanically and initially the infected plants often remain symptomless. When symptoms are expressed, the leaflets express patterning of light and dark greening. Small, lighter and irregular blotches occur between the veins. PVX can interact with PVY and PVA to cause more severe symptoms such as rugose mosaic (PVX+PVY) and crinkle mosaic (PVX+PVA).

Potato virus Y (PVY) is a type member of the genus *Potyvirus* of the family Potyviridae. PVY is a monopartite, single stranded RNA virus that is transmitted by aphids. There are multiple strains of the virus including the common strain, PVYO. The symptoms may vary with the strains of the virus and variety of potato. Generally, the affected plants are stunted with mild or severe mosaic and veinal necrosis. Original strain of PVYO causes mosaic, necrotic spots, mild rugosity and drying of leaves.

Potato spraing virus occurs due to infection of 'tobacco rattle virus' (TRV) and rarely due to 'potato mop top virus'. These viruses are carried through free-living nematodes in the soil. The infection may occur due to planting of infected tubers. In this case the infected tuber is characterized by tan- or red-brown rings or arc of discolouration in the flesh, *vis-à-vis* development of some corky tissue. Its infection may sometimes distort the tubers along with yellow mottling of stems and foliage. Since TRV has a wide host range so effective weed control measures should be adopted, resistant varieties should be planted, and affected plants should immediately be uprooted and destroyed. There is no any effective chemical control measure for this malady.

Potato leafroll virus (PLRV) is transmitted by aphids. PLRV is the type species of the genus *Polerovirus* in the family Luteoviridae. Primary symptoms are confined to top young leaves which usually stand upright, roll and turn slightly pale in certain cultivars. Secondary symptoms, rather prominent in older leaves, develop when the plants are grown from infected seed tubers. Infected plants have characteristic dwarfed, upright and pale appearance with rolling of lower leaves that turn yellow, brittle and leathery. In some cultivars, a reddish or purple discolouration develops on the margins and underside of leaves.

Apical leaf curl disease on potato plants

Potato virus S (PVS) is a member of the genus *Carlavirus* of the family Betaflexiviridae transmitted mainly by contact and by aphids. PVS is symptomless on the majority of potato cultivars with occasional mild leaf symptoms of rugosity, vein deepening and bronzing. However, symptoms such as rugosity, mosaic, stunting, bronzing, vein deepening, early drop of leaves, and fine necrotic spots on lower leaves have also been reported occasionally in some cultivars.

Tomato leaf curl New Delhi virus-potato is a variant of 'tomato leaf curl New Delhi virus' (ToLCNDV), which is a bipartite geminivirus transmitted by whitefly. It causes the apical leaf curl disease of potato and affected plants show curling/crinkling of apical leaves with a conspicuous mosaic symptom. In case of seed-borne infection, the entire plant shows the symptoms of leaf distortion with mosaic accompanied with severe stunting of the plant. It can be easily identified when the concentration of virus is very high.

Stunted and mosaic symptoms on potato plants

Rolling and crinkling symptoms on potato plants

Marginal flavescence is caused by mycoplasma-like organisms which is prevalent in the NW hills and is transmitted by leaf hoppers (*Orosius albicinctus, Seriana equate, Empoasca fabae, etc.*). Affected plants exhibit intense chlorosis on the top leaves so plant growth is stunted. Tuber yield due to infection of this disease is badly affected and such tubers produce hairy sprouts.

Witch's broom disease is confined to Deccan plateau. It is caused by mycoplasma-like organisms. This disease perpetuates through infected tubers and is transmitted by leaf hoppers. The infected plants show stunting and spreading type of growth with numerous filamentous and hyaline stems bearing aerial tubers *cum* small and simple leaves. Infected tubers produce thin hair-like sprouts.

In case of all the virus and virus-like diseases, the control measures are:

i) Use of disease-free tubers for planting
ii) Regular roguing of infected plants from field and burning them
iii) Before and after use, all the tools and implements should thoroughly be washed with 3 per cent trisodium phosphate solution

iv) In case of seed tuber production, before outbreak of aphid infestation, the plants should be dehaulmed

v) To control the aphids and leaf hoppers as vectors, the plants should be sprayed up to 5 times at fortnightly intervals with systemic insecticides such as 0.03 per cent dimethoate or phosphamidon, or 0.05 per cent methyl demeton

Physiological Disorders

Potato hollow heart is caused by rapid growth of tubers which become oversized and form cavity in the centre with the death of the small area of pith cells. This results in adjacent cracks and hollowness as the centre expands during potato growth. This disorder can be managed by maintaining optimal level of soil moisture conditions and by avoiding over-fertilization, especially nitrogen. Some varieties are less prone to this defect. To overcome this problem, potatoes should be grown under uniform stands and by ensuring regular and adequate but not over-watering especially during dry weather.

Internal potato rust spot is a latent disorder of tubers and plants but when cut open the tubers are found marked with many irregular small rusty brown spots. It is probably due to unsuitable soil conditions as usually this disorder occurs on sandy and clay soils or in the soils having poor organic matter coupled with low potassium and lime levels. To avoid such problems, plenty of humus in the soil should be applied along with proper irrigation as the plants grow. However, the soils should also be limed well before planting.

Knobbiness occurs due to uneven growth of tuber cells/tissues. This may be due to uneven watering conditions that lead to an obstruction in tuber growth. Heavy irrigation after a long dry spell leads to growth of some cells very fast resulting in knobbiness. Knobbiness can be prevented by frequent and optimum irrigation.

Black heart is caused by sub-oxidation conditions forming under storage of potato tubers by piling as the air does not get into the centre. It occurs due to higher temperature and excessive moisture resulting into blackening of tissues in the centre. Though there is no decay, the appearance of the tuber may lower the sale. This condition can be controlled by providing proper ventilation and by keeping potato tubers in layers, *i.e.* tubers should not be stored in the heap.

Greening of tubers is caused by various factors such as mechanical injury, premature harvest, excessive fertilizer application, *vis-a-vis* exposure of tubers to sunlight which increases the glycoalkaloid contents, and this may lead to solanin production which in nature is slightly poisonous though on boiling this disappears. This condition can be prevented by proper earthing up of tubers in the field when tuberization is taking place. To avoid such conditions, the tubers after lifting should be cured and stored in dark.

Cracking is due to boron deficiency or uneven water supply. Its remedy is frequent and optimum water supply and application of Borax @ 20kg/ha.

Sun scalding occurs generally in the autumn crop when the sunshine is more with high temperature (>30°C). Emergence of sprout and leaflets is drastically affected at that time, *i.e.* tip-burn. Water should be passed through the furrows to lower the soil temperature.

Translucent end is related to environmental stress due to drought and heat. It is generally found at the proximal end of the tubers that show glossy appearance and irregularity in shape. These glossy areas are high in sugar and low in total soluble solids. It may also result into decay under storage. At least 50 per cent moisture should be maintained in the field and excessive nitrogen supply should be avoided.

Black spot or **internal browning** of potato tubers occurs in vascular tissues within 3 days of mechanical injury due to certain phenolic compounds. It is quite related to genetic make up of the plant, *vis-à-vis* storage and growing conditions.

Freezing injury occurs due to exposure of tubers to freezing temperature during or after harvest. Usually it takes place at -1.5°C or below temperature. In the form of a ring there is discolouration of tissues as it affects the vascular tissues so it is also called the ring necrosis. When fine elements or cells of vascular ring are affected then it is called 'net necrosis'. Tubers show more damage towards proximal end. This condition can be prevented by avoiding exposure of tubers to freezing temperature during storage or harvest.

Sprouting is often a serious problem in storage and can be inhibited by spraying maleic hydrazide @ 1000-6000ppm about 2-3 weeks before harvesting. Chemicals like Chloro IPC (N-tetra chloro isopropyl carbonate)@ 0.5 per cent and/or nomyl/amyl alcohol @ 0.05-0.12mg/ha also helps in inhibiting sprouting.

Swollen lenticel disorder is caused by exposure of the tuber to very wet conditions in the field or in storage due to oxygen deprivation. Besides the tuber giving an unmarketable appearance, it also gives an opportunity for entering of certain pathogenic organisms and therefore bacterial soft rot or pink rot may occur. It can be prevented by creating proper drainage facilities in the field and by avoiding over-watering. Store should be well-ventilated and devoid of condensation.

Harvest

The time of harvest is very important in potato. The development of tuber continues till vines die. The main crop is ready for harvest when majority of the leaves turn yellow-brown. This could be between 70-120 days of planting depending upon the area, soil type and variety planted. In the hills, the crop should be normally harvested when the soil is not very wet. Tubers lifted during monsoon have a poor keeping quality and also develop various types of rots. When the leaves turn yellow-brown the tops are cut near the ground level. The potatoes are dug out from the field by ploughing after 8-10 days. These potatoes are manually picked from the field and stored in shade. Manual harvesting of potatoes is quite labour-intensive, time consuming and a lot of damage occurs to the tubers. Various low cost bullock-drawn and tractor-drawn potato diggers have been developed at CPRI, Shimla which exposes 80 per cent of the tubers and covers 1-3ha of area per day.

Commercial Harvesting

Gardeners usually lift the potatoes with a long-handled, three-prong 'grape' or 'graip', *i.e.* a spading fork or a potato hook which is similar to the graip but with tines at a 90° angle to the handle. In larger plots, the plough is the fastest implement for digging potatoes. Commercial harvesting is typically done with large potato harvesters which scoop up the plant and surrounding earth. This is transported up an apron chain consisting of steel links several metres wide which separates some of the dirt. The chain deposits into an area where further separation occurs. Different designs use different systems at this point. The most complex designs use vine choppers and shakers, along with a blower system to separate the potatoes from the plant. The workers thus continue to sort out plant material, stones and rotten potatoes before the potatoes are continuously delivered to a wagon or truck. Further inspection and separation occurs when the potatoes are unloaded from the field vehicles and put into storage.

Postharvest

Curing

The harvested potatoes are surface-dried and kept in heaps for 10-15 days in shade for curing of skin. Potatoes are usually cured after harvest to improve skin-set. This is the process by which the skin of the potato becomes resistant to skinning damage. Potato tubers may be susceptible to skinning damage at harvest, *vis-a-vis* at handling operations. Curing allows the skin to fully set and any wound to heal. Wound-healing prevents infection and water-loss from the tubers during storage. Curing is normally done at relatively warm temperatures

(10-16°C) with high humidity and good gas-exchange if at all possible. The tubers should not be exposed to direct sunlight as they become green. All damaged and rotten tubers should be removed. The produce should be kept in a cool place before sending to the market.

Storage

Storage facilities need to be carefully designed to keep the potatoes alive and for slowing down the natural process of sprouting which involves the breakdown of starch. It is crucial that the storage area be dark, well aerated, is able to maintain 4°C constant temperature and may sustain long-term storage. For short-term storage, the temperatures of 7-10°C are preferable. Temperature below 4°C converts the potato starch into sugar which alters the taste and cooking qualities and leads to higher acrylamide levels in the cooked product especially in deep-fried dishes. The discovery of acrylamides in starchy foods in 2002 has led to international health concerns. They are believed to be probable carcinogens and their occurrence in cooked foods is being studied for potentially influencing health problems. Chemicals are used to suppress sprouting of tubers during storage. Chlorpropham (CIPC) is the main chemical used but toxicity concerns have led to it being banned in the EU. Alternatives are application of MH to the standing crop or using ethylene, spearmint, orange oils and 1,4-dimethylnaphthalene.

Under optimum conditions in commercial warehouses, potatoes can be stored for 10–12 months. The commercial storage and retrieval of potatoes involves several phases such as i) drying of surface moisture; ii) wound-healing at 85-95 per cent RH and 25°C temperature; iii) a staged cooling phase; iv) a holding phase; and v) a reconditioning phase during which the tubers are slowly warmed. Mechanical ventilation is used at various points during the process to prevent condensation and accumulation of CO_2.

Nutrition and Toxicity

Potatoes are second only to soybean for amount of protein these contain, with the major storage protein being patatin ~ one of the most nutritionally balanced known plant proteins (Liedl *et al.*, 1987). According to U.S. Department of Agriculture, a typical raw potato is 79 per cent water, 17 per cent carbohydrates (88 per cent starch), 2 per cent protein but negligible amount of fat. A single 150g tuber provides up to 45 per cent of recommended daily allowance (RDA) for vitamin C, 10 per cent vitamin B6, 8 per cent niacin, 6 per cent folate as well as significant amounts of other essential mineral nutrients required; however it lacks many other essential nutrients such as vitamin B12, biotin, thiamine, riboflavin, alfa carotene, lycopene, retinol and alfa tocopherol, *vis-à-vis* important minerals such as copper, iodine and molybdenum. Raw potato starch is poorly digested by humans, hence, it is rarely eaten raw. The contents of vitamin B6 and vitamin C decline notably when a potato is baked, though there is little significant change in the amount of other nutrients. The glycemic index (GI) of potatoes can vary significantly depending on the cultivar, growing conditions and storage, preparation and cooking methods (mashed, cubed or consumed whole; boiled, baked or cooked with gravy; eaten hot or cold; *etc.*), and accompanying foods consumed (especially the addition of various high-fat or high-protein toppings). Reheated or pre-cooked and cooled potatoes may yield a lower GI effect due to formation of resistant starch.

Potato contains glycoalkaloids alpha-solanine and alpha-chaconine (Maga, 1980) that are mainly concentrated in the flowers and sprouts (200 to 500mg/100g). The concentration of the glycoalkaloids is usually less than 10mg/100g in healthy potato tubers and this is normally reduced by peeling (Wood and Young, 1974; Bushway *et al.*, 1983). In bitter varieties the alkaloid concentration can go up to 80mg/100g in the tuber as a whole and up to 150-220mg/100g in the peel. At higher concentrations solanine and other potato glycoalkaloids are toxic and cause a burning and persistent irritation similar to hot pepper. They are not destroyed during normal cooking because the decomposition temperature of solanine is about 243°C. Levels of glycoalkaloids may build up in potatoes which are exposed to bright light for long periods or from wounding during harvest or during post-harvest handling and storage, especially at temperatures below 10°C (Jadhav and Salunkhe, 1975). Solanine poisoning has been known to cause severe illness but it is rarely fatal (Jadhav and Salunkhe, 1975). Potato also contains proteinase inhibitors which act as an effective defence against insects and micro-organisms but are not problematic to humans because they are destroyed by heat. Lectins or haemogglutenins are also present in potato. These toxins are capable of agglutinating the erythrocytes of several mammalian species including humans (Goldstein and Hayes, 1978) but this is of minimal nutritional significance as haemogglutenins are also destroyed by heat, and potatoes are certainly cooked before they are eaten.

Uses

Culinary Use

Potatoes are cooked in many ways ~ alone or with other vegetables; skin-on or peeled; whole, cut or mashed; and with seasonings or without. The only requirement involves cooking to swell the starch granules. It can be cooked like other vegetables, boiled, baked and fried (shallow or deep). Most potato dishes are served hot but some are first cooked but served cold such as potato salad, crispy potato chips and *bhujiya, etc.* In India, the most popular potato dishes are potato-vegetable (with or without gravy), and several fast food items such as *batata vada, samosa* (a spicy mashed potato mixed with a small amount of vegetable stuffed in conical dough and deep fried), *aloo chat/aloo tikki* (boiled, mashed and deep-fried potato served with chutney), *aloo pakoda* (oil-fried raw or boiled potato mixed with gram flour), *etc.* Potato as a cooked vegetable is prepared mixed with almost all other vegetables, *etc.* In northern India, *aludum* (spicy curry of boiled potato) and *alu paratha* (stuffed chapatti prepared with butter or mustard oil) are the favourite part of the diet. A dish called *masala dosa* from South India is very notable all over India. It is a thin pancake of rice and pulse rolling over the spicy smashed potato and eaten with sambhar (cooked spicy *cum* sour pulse) and chutney. *Poori* in south India is a favourite dish which is liked with boiled *cum* smashed spicy potatoes. *Vada pav* is a popular vegetarian fast food dish in Mumbai and other regions in the Maharashtra in India. *Aloo posto* (a curry with potatoes and poppy seeds) is immensely popular in East India, especially Bengal. During fasting, potato is eaten in various forms of dishes, *e.g.* boiled, fried, in the form of chips and French fries, boiled and mixed with milk or curd, *etc.*

Industrial Use

Potatoes are used to brew alcoholic beverages such as vodka, poitín or akvavit. Potato starch is used in the food industry as a thickener and binder for soups and sauces, in the textile industry as an adhesive, and for manufacturing of paper and boards.

Other Uses

When harvesting the potatoes, too small and/or blemished potatoes along with whole plant are used as fodder to feed the livestock. Quite small tubers are stored in bins as animal feed and served to animals from time to time. Some farmers prefer to steam them before feeding the animals.

References

1. Bushway, R.J., J.L. Bureau and D.F. McGunn, 1983. Alpha chaconine and alpha solanine content of potato peels and potato peel products. *J. Food Sci.*, **48**: 84-86.
2. Goldstein, I.J. and C.E. Hayes, 1978. The lectins: carbohydrate binding proteins of plants and animals. *Adv. Carbohydr. Chem. Biochem.*, **35**: 127-340.
3. Hawkes, J.G. 1990. The Potato: Evolution, Biodiversity and Genetic Resources. Belhaven Press, London and Smithsonian Institute Press, Washington, D.C., 259 pp.

4. Hijmans, R.J. and D.M. Spooner, 2001. Geographic distribution of wild potato species. *Amer. J. Bot.*, **88** (11): 2101–12. doi: 10.2307/3558435. JSTOR 3558435. PMID 21669641.
5. https: //cpri.icar.gov.in/AddLink/getIMG/6165.
6. Jadhav, S.J. and D.K. Salunkhe, 1975. Formation and control of chlorophyll and glycoalkaloids in tubers of *Solanum tuberosum* L. and evaluation of glycoalkaloid toxicity. *Adv. Food Res.*, **21**: 307-354.
7. Khurana, S.M.Paul and S.K. Pandey, 2008. Aaloo ki Kheti (in Hindi). ICAR, New Delhi, 291p.
8. Kumar, V., S.K. Luthra, V. Bhardwaj and B.P. Singh, 2014. Indian Potato Varieties and Their Salient Features. Technical Bulletin No. 78 (revised). ICAR-Central Potato Research Institute, Shimla (H.P.).
9. Liedl, B.E., T. Kosier and S.L. Sesbarought, 1987. HPLC isolation and nutritional value of a major potato tuber protein. *Amer. Potato J.*, **31**: 1177-82.
10. Luthra, S.K., S.K. Pandey, B.P. Singh, G.S. Kang, S.V. Singh and P.C. Pandey, 2006. Potato Breeding in India. CPRI Tecnical Bulliten No. 74.
11. Maga, J.A. 1980. Potato glycoalkaloids. *CRC Crit. Rev. Food Sci. Nutr.*, **12**: 371-405.
12. Naik, P.S. 2019. Potato. In: Handbook of Horticulture, vol. 1, pp. 593-9. ICAR, New Delhi.
13. Pandey S.K., J.P. Singh and Jai Gopal, 2008. Potato varieties and cropping systems in India. *Potato J.*, **35**(3-4): 103-10.
14. recipetips.com. Creamer Potato. Retrieved 18 July 2008.
15. Sharma, S., Meena Thakur, V. K. Chandla, B. P. Singh, S. K. Chakrabarti and Shashi Rawat, 2013. Integrated Management of Potato Pests: A Field Mannual. CPRI, E Book series available on www.cpri.ernet.in.
16. vikaspedia.in/agriculture/crop-production/package-of-practices/vegetables-1/potato#: ~: text=Potatoes per cent 20are per cent 20a per cent 20rich per cent 20source,leucine per cent 2C per cent 20tryptophane per cent 20and per cent 20isoleucine per cent 20etc.
17. Virupaksh, U. Patil, Jai Gopal and B.P. Singh, 2012. Improvement for bacterial wilt resistance in potato by conventional and biotechnological appraoches. *Agric. Res.*, **1**(4): 299-316.
18. Wood, F.A., and D.A. Young, 1974. TGA in potatoes. *Agric. Can. Publ.*, No. 1533, pp. 1-3.

23

Soh-phlong/Sohphlang (*Flemingia vestita* Benth. ex Baker, 1876)

Family: Fabaceae. **Subfamily**: Faboideae.

Synonyms: *Flemingia procumbens* Roxb.; *Moghania vestita* (Benth.) ex Baker Kuntze; *Moghania procumbens* (Roxb.) Mukerjee

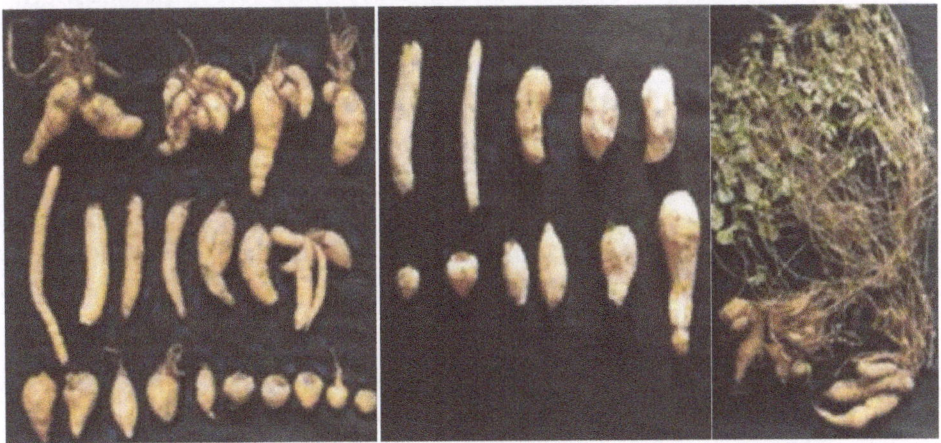

Root tubers (unpeeled) Peeled tubers Harvested tubers with dried aerial parts

***Source*: Sivaraman *et al.*, 2019 (cropped image)**

Introduction and Origin

The name Soh-phlong is a combination of two words of Khasi language, "Soh" (meaning fruit) and "Phlong" (meaning grass); the term used for juicy and refreshing tubers valued as fresh fruit which are dug from field totally covered with its trailing stems and branches (Sivaraman *et al.*, 2019). Soh-phlong is distributed in the western Himalaya extending eastwards upto Khasi Hills, where it is locally used for its medicinal importance (Edison *et al.*, 2006; Jaiswal, 2010; Divya and Pushpa Bharati, 2017).

This leguminous tuber crop grows in the wild along the mountain slopes of Himalayas and is distributed in some part of China, Nepal, Khasi Hills and Jaintia Hills of Meghalaya, and sparsely found in Laos, Philippines and Vietnam (Van Valkenburg and Bunyapraphatsara, 2001; National Research Council, 2002). However, in spite of its several culinary and medicinal uses, soh-phlong has gained very little attention and its commercial cultivation is conned to only Khasi and Jaintia hills of Meghalaya (Jyrwa *et al.*, 2017) where it is grown as a cash crop and sold as peeled raw tubers at Rs. 100-300/kg in the markets of Meghalaya. Tubers are available in the local village markets and in Shillong markets by October till May (Sivaraman *et al.*, 2019).

Botany

Soh-phlong is a perennial herb having shapeless edible tuberous roots with delicate, dark to pale yellow and easy to peel skin and cream-coloured juicy flesh that has a sweet, nut-like flavour. Specific varieties are not known, however, there is a wide variation in tubers with respect to shape (cylindrical, fusiform, napiform, round and oval), size (2.5-8.6cm in length and 1.5-3cm in width) and weight (5-40g). The prostrate but weak stem is 60-80cm long, highly branched and densely pubescent (Talang *et al.*, 2019). Its chromosome number is 2n=22 (Bir and Sidhu, 1967).

The pinnately compound leaves are digitately 3-foliolate and pubescent having obovate-cuneate leaves. Terminal leaflets are oblong or nearly obovate and 1-5×0.5-2.5cm, though lateral leaflets are obliquely elliptic and slightly smaller. Petiole is 1-2cm long, wingless and pubescent (Sivaraman *et al.*, 2019; Talang *et al.*, 2019).

Inflorescence about 2–10cm long is densely pubescent having axillary or terminal raceme with lanceolate bracts. Flowers are bright-red and usually appear after the monsoon season, *i.e.* August and September but sometimes during autumn which suggests that day length sensitivity could be an issue if seed production is required (Chaudhri, 2005; Ren and Gilbert, 2010; Sivaraman *et al.*, 2019). Calyx is 5-lobed, lobes are linear-lanceolate, lower one being longest and longer than the tube. Corolla is elliptical and slightly longer than calyx. Fruits are hairy sub-cylindrical pods. Seed is globose in shape and brown or black in colour.

Nutritional Value

Soh-phlong tuber is particularly rich in phosphorus and proteins (en.wikipedia.org/wiki/Flemingia_vestita).

Chemical Composition of Sohphlong Flour Samples (Marboh and Mahanta, 2020)

Parameters (per cent)	*Flour of*	
	Cultivated Sohphlang (CS)	*Market Sohphlang (MS)*
Moisture	11.22±0.25[a]	10.58±0.23[b]
Ash	2.12±0.15[a]	1.51±0.13[b]
Protein	7.86±0.44[a]	7.35±0.32[a]
Fat	0.71±0.26[a]	0.78±0.13[a]
Carbohydrate	78.09±0.39[a]	79.78±0.16[b]
Dietary fibre	7.07±0.36[a]	7.39±0.22[a]
Starch	67.95±0.15[a]	64.61±0.31[b]
Amylose	18.90±0.17[a]	16.75±0.16[b]
Reducing sugar	3.47±0.83[a]	7.89±0.65[b]
Acidity	0.15±0.08[a]	0.19±0.11[a]
Ascorbic acid	17.58±1.12[a]	14.33±0.92[b]
Cyanide	ND	ND

Values are expressed as mean±standard deviation of triplicate assays. Means with different superscripts [a,b] within a row are significantly different (p<0.05). All values were reported in per cent dry basis (db) except moisture which was reported as per cent wet basis (wb); ND not detected.

Mineral Composition of Sohphlong Flour Samples (Marboh and Mahanta, 2020)

Mineral	*CS Flour (mg/100g)*	*MS Flour (mg/100g)*
Macro-mineral		
Potassium	872.63±0.82[a]	856.56±0.65[b]
Sodium	31.45±0.92[a]	26.84±0.65[b]
Magnesium	171.65±0.50[a]	152.23±0.47[b]
Calcium	31.15±0.22[a]	25.96±0.24[b]
Phosphorus	23.33±0.47[a]	19.98±0.15[b]
Na/K	0.036	0.030

Mineral	*CS Flour (mg/100g)*	*MS Flour (mg/100g)*
Ca/P	1.33	1.34
Micro-mineral		
Iron	6.72±0.27[a]	4.56±0.52[b]
Copper	9.38±0.29[a]	6.96±0.42[b]
Manganese	7.97±0.24[a]	4.93±0.35[b]
Zinc	3.61±0.16[a]	2.70±0.17[b]

Values are reported as mean±standard deviation (n=3). For each macro-mineral and micro-mineral, means with different superscripts [a,b] within a row are significantly different (p<0.05). All values were reported in per cent dry basis (db).

Source: www.ncbi.nlm.nih.gov/pmc/articles/PMC7447699/

Phytochemicals

Alkaloids, carbohydrate and flavonoids were found in aqueous extract of rhizomes of soh-phlong, and the petroleum ether extracts revealed the presence of alkaloids, carbohydrate, flavonoids and phytosterols. Carbohydrate, phenol, tannic acid, flavonoid and phytosterol were present in chloroform extracts, and methanolic extracts were found to contain alkaloids, carbohydrate, phenol, tannic acid and flavonoids (Langstang *et al.*, 2019).

Soh-phlong is rich in bioactive isoflavones such as genistein, daidzein, formononetin and pseudobaptigenin (Rao and Reddy, 1991) and they are reported to be important in prevention of osteoporosis, cancer and vascular pathologies (Dabkeviciene *et al.*, 2012).

The flour extracts of the conventional-assisted-extraction (CAE), microwave-assisted-extraction (MAE) and ultrasound-assisted-extraction (UAE) were analysed for total phenolic content (TPC), total flavonoid content (TFC) and antioxidant activities (AA). CS flour exhibited good nutritional properties, TPC, TFC and AA than MS flour. Furthermore, MAE gave better extraction efficiency than UAE and CAE. Genistein as the predominant compound among the phenolic compounds identified is comparable with soybean (Marboh and Mahanta, 2020).

Isoflavones particularly genistein have been reported to confer health promoting properties by acting as anti-oxidant, anti-diabetes and anti-cancer (Penalvo *et al.*, 2004).

Uses

The tubers of soh-phlong that are usually eaten in raw form can serve as a versatile source of food and nutritional security to the people. The tuber also offers

Identification and Quantification of Identified Compounds in Sohphlong Flour Extract Samples (Marboh and Mahanta, 2020)

Peak No.	Compounds	Retention Time (min)	UAE (μg/g)		MAE (μg/g)		CAE (μg/g)	
			CS Flour	MS Flour	CS Flour	MS Flour	CS Flour	MS Flour
1	Ascorbic acid*	4.41	52.17 ± 0.11^{Aa}	50.67 ± 0.23^{Bd}	63.73 ± 0.19^{Cb}	63.33 ± 0.08^{De}	83.66 ± 0.34^{Ec}	19.33 ± 0.16^{Ff}
	Phenolics							
2	Daidzin	17.96	12.20 ± 0.09^{Aa}	8.05 ± 0.22^{Bd}	13.81 ± 0.13^{Cb}	9.06 ± 0.41^{De}	14.22 ± 0.50^{Ec}	4.05 ± 0.26^{Ff}
3	Coumaric acid	19.05	5.66 ± 0.35^{Aa}	51.09 ± 0.07^{Bd}	24.36 ± 0.12^{Cb}	71.15 ± 0.18^{De}	86.70 ± 0.39^{Ec}	3.17 ± 0.27^{Ff}
4	Rutinhydrate	19.90	23.53 ± 0.13^{Aa}	7.61 ± 0.25^{Bd}	27.85 ± 0.16^{Cb}	5.13 ± 0.07^{De}	15.70 ± 0.17^{Ec}	4.01 ± 0.26^{Ff}
5	Quercetin	22.66	25.89 ± 0.04^{Aa}	21.04 ± 0.12^{Bd}	29.27 ± 0.14^{Cb}	23.72 ± 0.07^{De}	24.67 ± 0.23^{Ec}	22.93 ± 0.10^{Ff}
6	Genistein	23.77	63.36 ± 0.41^{Aa}	34.12 ± 0.16^{Bd}	96.47 ± 0.06^{Cb}	79.40 ± 0.17^{De}	77.80 ± 0.12^{Ec}	60.44 ± 0.07^{Ff}
7	Formononetin	25.79	50.60 ± 0.15^{Aa}	27.52 ± 0.08^{Bd}	56.21 ± 0.19^{Cb}	28.93 ± 0.07^{De}	22.77 ± 0.22^{Ec}	36.18 ± 0.26^{Ff}
8	Biochanin-A	27.82	34.94 ± 0.05^{Aa}	36.66 ± 0.10^{Bd}	76.16 ± 0.11^{Cb}	44.06 ± 0.08^{De}	16.54 ± 0.19^{Ec}	24.58 ± 0.13^{Ff}
	Isoflavones#							
1	Daidzin	12.62	160.11 ± 0.12^{Aa}	135.25 ± 0.45^{Bd}	485.40 ± 0.27^{Cb}	165.14 ± 0.07^{De}	325.07 ± 0.16^{Ec}	108.75 ± 0.11^{Ff}
2	Formononetin	25.53	99.11 ± 0.12^{Aa}	60.91 ± 0.33^{Bd}	313.37 ± 0.11^{Cb}	84.90 ± 0.06^{De}	86.64 ± 0.09^{Ec}	49.51 ± 0.29^{Ff}
3	Biochanin-A	26.48	258.14 ± 0.17^{Aa}	211.35 ± 0.10^{Bd}	602.38 ± 0.20^{Cb}	360.05 ± 0.09^{De}	200.92 ± 0.12^{Ec}	190.81 ± 0.15^{Ff}
4	Genistein	29.86	1983.40 ± 0.14^{Aa}	1261.11 ± 0.22^{Bd}	2113.40 ± 0.43^{Cb}	1569.58 ± 0.29^{De}	1643.47 ± 0.38^{Ec}	1252.92 ± 0.21^{Ff}

Values are reported as mean±standard deviation (n=3). Means with different capital superscripts $^{A, B; C, D}$ and $^{E, F}$ within a row showed the significant difference (p<0.05) between CS flour and MS flour of UAE, MAE and CAE, respectively. Means with different small superscripts $^{a, b, c}$ and $^{d, e, f}$ within a row showed the significant difference (p<0.05) between the treatments of CS and MS flours, respectively*Vitamin C#Dilution factor of 1.5 was used for isoflavones extracts of MS flour sample.

scope for exploiting its health benefitting functional ingredients (Marboh and Mahanta, 2020). Easy to peel, crispy white or creamy tubers have a sweet nutty flavour and are usually eaten raw (Talang *et al.*, 2019). It is generally consumed after peeling off the outer yellowish skin along with ground *Perilla* seeds (locally known as 'Neileih'), salt and chilli. The tubers have anthelmintic and vermifugal properties and are administered to children for deworming, stomach upset and for cooling (Singh and Arora, 1973; Das *et al.*, 2004; Sunita, 2014; Tandon *et al.*, 1997; Tandon and Saha, 2004; Tandon and Das, 2007).

Ready to Eat tubers of Sohphlang

By LEMS1010 - Own work, CC BY-SA 3.0,
https://commons.wikimedia.org/w/index.php?curid=29509515

The tuber has been traditionally used as an indigenous vermifuge among the Khasis. Raw tuber or the root peel is directly consumed for the treatment of soft-bodied intestinal worms (Hrckova and Velebny, 2013; Rao and Reddy, 1991). Genistein, a polyphenolic isoflavone isolated from the tuber extracts, proved to contain anti-helmintic principle, highly potent against nematodes and cestodes (Rao and Reddy, 1991; Toner *et al.*, 2008). This compound was found to be effective against human tapeworms such as *Echinococcus multilocularis* and *E. granulosus metacestodes* (Naguleswaran *et al.*, 2006), and the sheep liver fluke *Fasciola hepatica* (Toner *et al.*, 2008).

Being a leguminous crop, soh-phlong is known to improve soil quality by fixing atmospheric nitrogen, hence mixed cropping with *F. vestita* was found to give better economic returns with a net gain in nitrogen of up to 250kg/ha/year (Gangwar and Ramakrishnan, 1989).

Cultivation

Soh-phlong is generally propagated vegetatively through small tubers that are usually planted in a virgin soil during March-April for one year. After one crop, the land is usually left fallow or utilized for plantation of other crops. Farmers use to grow the crop after the gap of minimum of five years in rotation. Some farmers grow it along with turmeric, ginger, taro and potato (Sivaraman *et al.*, 2020; Awasthi *et al.*, 1981).

The crop is planted in *jhum* land, on bund at planting distance of 30-45cm in rows that are 35-40cm apart. Earthing-up and weedings are required as soon as weeds start emerging but preferably when plants attain a height of about 8-10cm during June-July (Sivaraman *et al.*, 2020; Awasthi *et al.*, 1981).

Fowering starts from August onwards and the crop takes seven months to reach maturity. Harvesting starts in mid-November. The tubers are manually harvested by digging up with a spade and stored in a pit covered with soil and they are taken out as per market demand (Sivaraman *et al.*, 2020; Awasthi *et al.*, 1981).

Freshly harvested tubers are graded on the basis of large, medium and small sizes. Large-medium sized tubers fetch better price in market. However, an average yield of 3,000kg/ha has been reported in Meghalaya (Singh and Arora, 1973). Soh-phlong is grown as a cash crop and sold as peeled raw tubers at Rs. 100-300/kg in the markets of Meghalaya. Tubers are available for sale in local village and Shillong markets by October till May (Sivaraman *et al.*, 2020; Talang *et al.*, 2019).

After grading, freshly harvested small tubers are kept for self consumption and as propagating material for next season growing. For propagation purpose tubers are stored for the next planting season (5-6 months) inside a pit dug near the field and undisturbed till planting season. The size of pit could be approximately 40-50×30 cm or depending on the quantity of tubers to be stored. For market sale, tubers are stored underground intermittently and delivered periodically based on market demand. For marketing usually round to fusiform medium-large sized tubers with good taste are preferred (Sivaraman *et al.*, 2020; Talang *et al.*, 2019).

References

1. Awasthi, R.P., S. Kari and J.S. Grewal, 1981. Effect of soil burning on the growth and yield of potato in Khasi Hills Meghalaya. *Indian J. agr. Sci.*, **51**(5): 315.
2. Bir, S.S. and S. Sidhu, 1967. Cytological observations on the north Indian members of family Leguminosae. *Nucleus*, **10**(1): 47–63.

3. Chaudhri, A.B. 2005. Forests Plants of Eastern India. *Ashish*, pp. 205–206. ISBN 978-8170245377.
4. Dabkeviciene, G., B. Butkute, N. Lemeziene, V. Jakstas, E. Vilcinskas and V. Janulis, 2012. Distribution of formononetin, daidzein and genistein in *Trifolium* species and their aerial plant parts. *Chemija.*, **23**(4): 306–11.
5. Das, B., V. Tandon and N. Saha, 2004. Effects of phytochemicals of *Flemingia vestita* (Fabaceae) on glucose 6-phosphate dehydrogenase and enzymes of gluconeogenesis in a cestode *Raillietina echinobothrida. Comp. Biochem. Physiol. C. Toxicol. Pharmacol.*, **139**(1-3): 141-6.
6. Divya, P.M. and Pushpa Bharati, 2017. Utilisation of minor tuber crops grown in Western Ghats of Karnataka. *J. Farm Sci.*, **30**(3): 400-403.
7. Edison, S., M. Unnikrishnan, B. Vimala, S.V. Pillai, M.N. Sheela, M.T. Sreekumari and K. Abraham, 2006. Biodiversity of Tropical Tuber Crops in India, 60p.. National Biodiversity Authority, Chennai, Tamil Nadu, India.
8. Gangwar, A.K. and P.S. Ramakrishnan, 1989. Cultivation and use of lesser-known plants of food value by tribals in north-east India. *Agric., Ecosys. and Environ.*, **25**(2-3): 253-67.
9. Hrckova, G. and S. Velebny, 2013. Pharmacological Potential of Selected Natural Compounds in the Control of Parasitic Diseases, 64p. Springer, Holland (ISBN 978-3-7091-1324-0).
10. Jaiswal, V. 2010. Culture and ethnobotany of Jaintia tribal community of Meghalaya, Northeast India: a mini review. *Indian J. Trad. Know.*, **9**(1): 38-44.
11. Jyrwa, R.B., B. Sangita and B. Bharadwaj, 2017. *Flemingia vestita* – a review on the unexplored plant from the North-east India. *WJPLS*, **3**(1): 650-58.
12. Langstang, W., M. Kharsahnoh, T. Khongwir and N. Venugopal, 2019. Pharmacognostic studies of *Flemengia vestita. Spectrum Sc. and Tech.*, **6**:21-8 (ISSN: 2349 2937). (https://app.anthonys.ac.in/documents/documents/journal/article/1/j_OqDWUxP.pdf).
13. Marboh, V. and C.L. Mahanta, 2020. Characterisation and antioxidant activity of sohphlong (*Flemingia vestita*), a tuberous crop. *J. Food Sci. Technol.*, 57(10):3533–44. (PMCID: PMC7447699. PMID: 32903822. doi: 10.1007/s13197-020-04344-2).
14. Naguleswaran, A., M. Spicher, N. Vonlaufen, L.M. Ortega-Mora, P. Torgerson, B. Gottstein and A. Hemphill, 2006. *In vitro* metacestodicidal activities of genistein and other isoflavones against *Echinococcus multilocularis* and *Echinococcus granulosus. Antimicrobial Agents Chemoth.*, **50**(11): 3770–78.

15. National Research Council, 2002. Tropical Legumes: Resources for the Future. Books for Business/ The Minerva Group, Inc., pp. 37–38 (ISBN 978-0894991929).

16. Penalvo, J.L., T. Nurmi, H. Adlercreutz H. 2004. A simplified HPLC method for total isoflavones in soy products. *Food Chem.*, **87**(2):297–305. (doi: 10.1016/j.foodchem.2003.12.028).

17. Rao, H.S. and K.S. Reddy, 1991. Isoflavones from *Flemingia vestita. Fitoterapia*, **62**(5): 458.

18. Ren, S. and M.G. Gilbert, 2010. *Flemingia* Roxburgh ex W. T. Aiton, *Hort. Kew.* (ed. 2): **4**: 349. 1812, nom. cons., not Roxburgh ex Rottler (1803) (PDF). *Flora of China*, **10**(3): 232–7. (ISSN 1043-4534).

19. Singh, H.B. and R.K. Arora, 1973. Soh-phlong, *Moghania vestita* - a leguminous root crop of India. *Eco. Bot.*, **27**(3): 332-8.

20. Sivaraman, Nivedhitha; Anjula Pandey; Subarna Hajong; Hammylliende Talang; Sudhir Ahlawat and A. Mishra, 2019. Exploration, Germplasm Collection and Variability Study on an Underutilized Root Tuber Crop "Soh-phlong" (*Flemingia procumbens* Roxb.) from Meghalaya, India. *Indian J. Pl. Gen. Res.*, **32**(3):347-53. (DOI 10.5958/0976-1926.2019.00037.8).

21. Sunita, S. 2014. Chromatographic evaluation of a phytoestrogen genistein from *Flemingia vestita* (Benth): an endemic plant of northeast India. Pharmacognosy Commun., **4**(4): 2-9.

22. Talang, H.D., H. Rymbai, M. Bilashini Devi, A.K. Jha and P. Chaudhuri, 2019. Sohphlong–a potential indigenous leguminous tuber crop of Meghalaya. *Intern. J. Minor Fruits, Med. and Arom. Plants*, **5** (2): 53-6.

23. Tandon, V. and B. Das, 2007. *In vitro* testing of anthelmintic efcacy of *Flemingia vestita* (Fabaceae) on carbohydrate metabolism in *Rallietina echinobothrida. Methods*, **42**(4): 330-38.

24. Tandon, V. and N. Saha, 2004. Anthelmintic efcacy of *Flemingia vestita* (Fabaceae): Alterations in glucose metabolism of the cestode. *Parasit. Intern.*, **53**(4): 345-50.

25. Tandon, V., P. Pal, B. Roy, H.S.P. Rao and K.S. Reddy, 1997. *In vitro* anthelmintic activity of root-tuber extract of *Flemingia vestita*, an indigenous plant in Shillong, India. *Parasitol. Res.*, **83**: 492-8.

26. Toner, E., G.P. Brennan, K. Wells, J.G. McGeon and I. Fairweather, 2008. Physiological and morphological effects of genistein against the liver fluke, *Fasciola hepatica. Parasitology*, **135** (10): 1189–1203.

27. van Valkenburg, J.L. and N. Bunyapraphatsara, 2001. Plant Resources of South-East Asia No. 12(2): Medicinal and Poisonous Plants 2. Backhuys Publisher, Leiden, The Netherlands, pp. 267-70.

28. www.ncbi.nlm.nih.gov/pmc/articles/PMC7447699/

24

Swamp Taro [*Cyrtosperma merkusii* (Hassk.) Schott]

Family: Araceae.

Common names: Gallan and Giant swamp taro. **Other names**: Baba (in the Marianas Islands), Babai (in Kiribati), Birah hutan, Geli-geli and Keladi pari (in Malaysia), Brak (in Palau), Galiang (in Bikol, Philippines), Iaraj (in the Marshall Islands), Kakake (in Solomon Islands, Melanesia), Kli (in Malay-Pattani, Thailand), Lak [in Yap (Federated States of Micronesia)], Lovo (in Fiji), Mane (in Nepal), Mwahng (in Pohnpei), Navia (in Vanuatu), Palau/Palauan/Palawan (in Bisaya, Philippines), Pasruk (in Kosrae), Pula'a (in Samoa, via Kana), Pulaka (in Lau, Tokelau and Tuvalu), Puraka (in Cook Islands), Simiden (in Chuuk) and Swamp taro (in Papua New Guinea) (https://en.wikipedia.org; Anon., 2016).

Synonyms: *Arisacontis chamissonis* Schott, *Cyrtosperma bantamense* Koord., *C. chamissonis* (Schott) Merr., *C. cuspidilobum* Schott, *C. dubium* Schott, *C. edule* Schott, *C. ferox* N.E.Br. and Linden, *C. intermedium* Schott, *C. lasioides* Griff., *C. merkusii* var. *giganteum* Nadeaud, *C.m.* var. *intermedium* (Schott) Engl., *C. nadeaudianum* J.W.Moore and *Lasia merkusii* Hassk (Anon., 2013).

Introduction and Origin

Giant swamp taro believed to have been originated in Indonesia is a riverine and 'swamp crop' similar to taro but with larger leaves, *vis-a-vis* larger *cum* coarser roots. It has been introduced into the Philippines, Papua New Guinea and the Pacific Islands in pre-European times and is now grown throughout Oceania and South and SE Asia. It is an important staple food and source of carbohydrate in the harsh atoll and mountainous islands of Pacific, especially in Tuvalu and Kiribati of the Central Pacific where its cultivation is an important cultural and culinary tradition but is now under threat from rising sea level and displacement from the growing use of imported food products (https://en.wikipedia.org). The cultivation is quite difficult and both the cultivation as well as cooking are very time-consuming. Although the rhizomes are rich in nutrients, especially calcium, they need to be cooked for hours to reduce toxicity (Gorisek, 2010).

Babai cultivation in Butaritari, Kiribati

Cyrtosperma merkusii is probably the largest of the root crop plants known collectively as 'taro'. It is considered a minor crop that is usually produced and traded only locally by smallholders. In SE Asia, it is primarily considered an emergency or famine food although in some regions in the Philippines it is still a main staple food. At present, it is the dominant aroid crop in Micronesia. *Colocasia esculenta*, the most widely cultivated taro is the major aroid crop in Polynesia and Melanesia whereas *Cyrtosperma merkusii* is kept in reserve as emergency food when taro and other crops have failed or are when in short supply.

Giant swamp taro can produce food under conditions such as droughts, floods and fresh and brackish water swamps where other crops fail, thus making it a very useful plant in Oceania. It is promising for marginal lowland areas as well and deserves much more attention.

Botany

Swamp taro, outwardly similar to *Colocasia esculenta*, is a giant herbaceous perennial reaching to a height of up to 3-4m (rarely up to 6m) and grows solitarily or in clumps, with 6-8 huge sagittate (rarely hastate) leaves arising from a short subterranean stem. Its diploid chromosome number is 2n=26. The arrow-shaped, usually 0.3-1.3m × 0.8m leaf blades (1-2m long as stated by Kay, 1987) are borne on stout petioles that are terete, 0.4-3m long (oftenly 1-2m) and about 10cm in diameter (Anon., 2016). Petioles are usually heavily to very heavily armed with stout, conical, straight or upturned spines towards the base but less spiny or even unarmed in some cultivated forms, with a prominent solid pulvinus at the apex and inside with longitudinal air ducts (Anon., 2016). The blades are sometimes armed abaxially, held horizontally to vertically with the posterior lobes down and usually somewhat longer than the anterior one. Primary venations of the anterior lobe are curvinerved in small lobes to pinnate in large lobes (Anon., 2016).

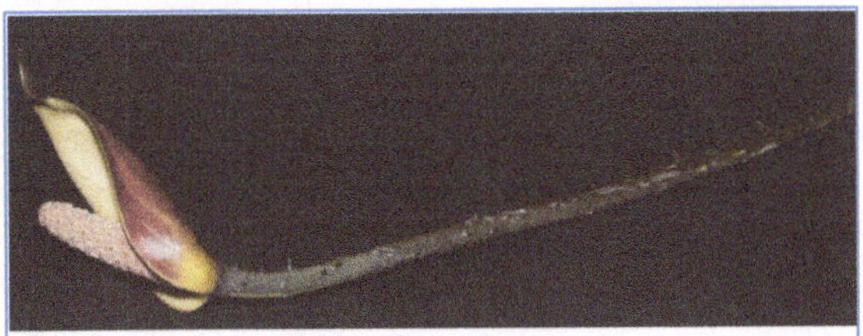
Flowering stem

Flowering spathe

The inflorescence with a long and thick yellowish spathe and a purplish spadix is usually solitarily formed on a peduncle which is similar to but usually shorter than the petiole. Spathe is thick, fleshy and very variable in size, colour and shape ~ 2.5-30cm long, white, yellow, green or purple, erect to deflexed to rolling back; spadix is sessile or stipitate, 2-24cm long, as long as or more than half the length of the spathe. The bisexual flowers are hexamerous; ovary unilocular, (1-)2-ovulate, receptive stigma wet; and stamens exserted from the tepals at male anthesis. Reddish-orange, sessile and 1(-2)-seeded fruit is a berry. Seed is campylotropous, almost circular in profile to shallowly kidney-shaped, 5-11mm long, brown, with about 3 raised, warty, longitudinal crests or sparsely and more or less irregularly warty and faintly crested (Anon., 2016).

The giant swamp taro is believed to be the largest plant in the world which produces edible rhizomes. Rhizomes of giant swamp taro vary greatly in shape, size and colour; they can be short and slender to tuberous and very large, cylindrical to conical or almost spherical, up to 1-2m long and 0.6-1m in diameter (Anon., 2016; https://en.wikipedia.org). The rhizome sizes vary with cultivar and age. These are usually 15-25kg in three years but it can weigh up to 80-90 kg or more in a 8-10 year old plant though a 10 year old plant becomes fibrous and may not be suitable for eating (Humanity development library 2.0; Kay, 1987). Rhizome offsets (often mistakenly called cormels) that are produced as side shoots on the parent rhizome after about three years, send up leaves and develop into suckers. The fleshy and starchy rhizomes have a taste similar to sweet potato or taro though it is drier in texture.

Cultivars

Swamp taro is morphologically quite variable as is also expressed by its various names. There are no registered or formally classified cultivars of the cultivated forms of this plant though several locally identified cultivars have been recognized depending on leaf shape, colour and spininess as well as size of the rhizomes (Anon., 2016).

Other *Cyrtosperma* Species

Previously, the genus *Cyrtosperma* was believed to be widespread from Asia to Africa and South America but in 1980s the African and South American species were subsequently moved into separate genera as a result of which it is now considered to be native only to SE Asia and some Pacific islands (Govaerts and Frodin, 2002). The genus is now known to be most prominent in New Guinea. It is the only known big genus in Araceae that is known to be found east of Wallace's line (Brown, 2000; Hay, 1988).

Cyrtosperma johnstonii
(Source: en.wikipedia.com)

Species in the genus include *Cyrtosperma beccarianum* A.Hay (New Guinea), *C. bougainvillense* A.Hay (Solomon Islands), *C. brassii* A.Hay (Louisiade Archipelago), *C. carrii* A.Hay (Papua New Guinea), *C. cuspidispathum* Alderw. (New Guinea), *C. giganteum* Engl. (New Guinea), *C. gressittiorum* A.Hay (Louisiade Archipelago), *C. hambalii* A.Dearden and A.Hay (Western New Guinea), *C. johnstonii* (N.E.Br.) N.E.Br. (Solomon Islands), *C. kokodense* A.Hay (Papua New Guinea), *C. macrotum* Becc. ex Engl. (New Guinea), and *C. merkusii* (Hassk.) Schott (New Guinea, Solomon Islands, Borneo, Java, Sumatra, Philippines, Fiji, Samoa, Kiribati, Santa Cruz Islands, Cook Islands, Marquesas, Caroline Islands, Mariana Islands; naturalized on various other islands) (https://en.wikipedia.org).

All species except *C. merkusii* are confined to New Guinea and surrounding islands. *C. johnstonii* (W. Bull) N.E. Brown is known only from cultivation, mainly as a gigantic ornamental, possibly also producing edible tubers.

Nutritional Value

Although there are many *Cyrtosperma* cultivars, only few have been analyzed for nutrient content. Lois *et al.* (2008) reported that the samples collected in the Federated States of Micronesia (Pohnpei, Chuuk and Yap) and in the Republic of Palau were analyzed for carotenoids (β- and α-carotene, β-cryptoxanthin, lutein, zeaxanthin, and lycopene) and minerals (including iron, zinc and calcium). Of the 34 cultivars analyzed, β-carotene concentrations varied from 50 to 4486μg/100g.

Yellow-fleshed cultivars generally contained higher carotenoid concentrations. Substantial concentrations of zinc (5.4–46.1mg/100g), iron (0.3–0.8mg/100g) and calcium (121–305mg/100g) were found in 10 cultivars that were analyzed for mineral content (wet weight basis). All the cultivars were acceptable for taste and production factors (Lois *et al.*, 2008).

Cyrtosperma merkusii rhizomes per 100g of edible portion contain approximately 63-68g water, 0.8-0.9g protein, 0.1-0.2g fat, 29-34g carbohydrates, 1.5g fibre, 1g ash, 28-55mg P, 330-570mg Ca, 1mg Fe, 0.03-0.05mg thiamine, 0.07-0.10mg riboflavin, 0.9-1.2mg niacin and 0.1-3mg ascorbic acid. The energy value ranges between 500-550 kJ/100g. In the Philippines, the starch content of the tubers ranges between 7 to 23 per cent and the starch grains are rounded to angular-rounded, and (4-)11(-18)μm in size (Anon., 2016).

Yellow (YP) and white (WP) sections of giant swamp taro yielded 40g/kg and 51.5g/kg (dry weight), respectively of pure mucilage made up of D-glucose (44.95-78.85 per cent), D-galactose (8.70-25.35 per cent), D-mannose (3.20-10.45 per cent), D-arabinose (2.45-5.20 per cent) and small amounts of glucoronic acid and rhamnose (Richard *et al.*, 2014).

Toxicity in Aroids

The rhizomes and leaves may also contain irritating calcium oxalate raphides and some hydrocyanic acids, both of which can be removed and made inactive by drying, cooking or by steeping in water.

Uses

Mature starchy rhizomes of *C. merkusii* are usually produced in 1.5-3 years and are consumed only after thorough cooking. They are peeled, cut into sections and boiled, steamed or baked and also sometimes grated or crushed and mixed with other foods such as coconut milk (Anon., 2016; en.wikipedia.org). They can be mashed into dough, which is kneaded and left to ferment before being baked in an oven (Facciola, 1998). They may be sliced, fried and eaten with sugar.

Leaves and inflorescence can also be cooked as vegetable (Tindall, 1983). The leaves are also used as packing material for food, as mulch and sometimes medicinally (Anon., 2016; en.wikipedia.org). A decoction of the flower cluster is used as an emmenagogue and embolic (Uphof, 1959; Anon., 2016; en.wikipedia.org).

In Nepal where it is called *Mane*, all plant parts (leaf, stem, rhizomes) are enjoyed after being thoroughly boiled and roasted. The stem requires prolonged boiling and the water is replaced once to remove irritating chemicals. If cooked carefully, the rhizomes taste like taro and the leaves like spinach. But without careful washing, the food causes an unpleasant tingling or scratchy sensation (Limbu, 2012).

C. merkusii also has ornamental value and in certain parts of Micronesia it is highly valued as prestigious ornamental (Anon., 2016; en.wikipedia.org).

Processing and Trade

The rhizomes can be peeled, sliced and scalded, and then dried in the sun to preserve it for several months. Until recently the plant was grown solely for home consumption, but in a few islands it is now sold in the local market though it lacks international trading.

Propagation

Rhizome offsets which send up leaves and develop into suckers are produced as side shoots on the parent rhizome after about three years. These suckers are the most common planting material, although sometimes the top of the main rhizome of the harvested plant is used. When using the cut setts, they should be the upper 3-5cm top of the rhizome and the lower 30-50cm of the petioles. In each case one or two of the youngest leaves in the shoot are retained. The larger pieces are usually considered stronger and more vigorous. These propagules can be planted in moist or flooded soil which is kept loose around the base of the plant so as to ensure a nicely shaped rhizome.

Growth and Development

Swamp taro is a slow growing crop and some cultivars can take up to 15 years to mature. The rhizomes of several cultivars become edible after about 18-36 months of growth though they can be left in the ground for up to 10 years or even more. The rhizomes continue to increase in size and maintain their edibility for many years (Anon., 2016). Usually rhizomes are harvested when 3-6 years old, with a minimum of 1 year and a maximum of 10 years. In general, a rhizome is considered mature when flowering starts (Anon., 2016). Older rhizomes begin to harden and become increasingly woody and fibrous so becoming completely inedible. Although faster-growing cultivars also exist, their rhizomes are not considered to be as tasty as those from slow-growing types and the quick growers appear to be more susceptible to adverse conditions (Anon., 2016).

In the harsh environments of Tuvalu and Kiribati where the plant has deep cultural significance, the cultivation is increasingly threatened by rising sea levels caused by global warming. The plant does not thrive in brackish water (Knox, 2002) as roots start rotting, the leaves turning yellow and the plant is stunted (Gorisek, 2010).

Factors Influencing Growth and Quality

Although environmental factors such as temperature, moisture, light and soil type influence the growth and quality, other factors such as varieties, quality of planting material, planter operation, irrigation, nutrition, pest management, planting date and harvest date, *etc.* are equally important which require to be controlled appropriately by the grower. Also, a good ground preparation, harrowing and ploughing and proper planting method are always required. The most profitable yields can be attained only when all the factors are at optimum levels.

C. merkusii is a tropical species adapted to grow within fresh water and coastal swamps at sea-level, and usually occurs between latitudes 18°N (Mariana Islands) and 20°S (Cook Islands), in lowland swamp forests, open swamps or in man-made swamp pits dug to below the level of the freshwater lens (Anon., 2016; en.wikipedia.org). Ideal growing conditions are natural swamplands about 0.5-2m deep, rich in humus and having provision of slow-flowing irrigation water. A more or less continuous supply of water is essential but the plant cannot be grown in streams where water is running swiftly or in a fresh marine soil but not on a slope where soil is frequently washed away by heavy rains. It requires high temperatures and suffers chilling injury. It does not tolerate temperatures below 4°C (Anon., 2016). The plant is able to tolerate drought (on atolls often because of the porosity of the soil), salt sea-water sprays, mild brackish water and hurricanes or storms much better than most other aroid crops. Though not suitable for growing in upland or rainfed conditions, it can be grown under year-round high rainfall as a rainfed crop up to 150m altitude.

The plant is tolerant of a wide range of soil types and acidity and can be grown in areas of moderate rainfall if the soil is deep and swampy, and when it is at least partially shaded and sheltered from wind.

Season

Swamp taro can be planted year-round under satisfactory growing conditions. In Nepal *mane* grows in the tropical and sub-tropical forests along stream banks where these are gathered in January–February.

Land Preparation and Planting Method

The only sure supply of fresh water for the islanders on coral atolls in the pacific is the hydrostatic lens which floats at variable depth on the salt water that permeates the lower levels of the coral, sometimes several feet below the surface of the land, and planting procedures have been developed to suit these particular circumstances. In atolls, pits extending to 10x20m across are dug deep enough to reach the fresh water layer which may be 0.5-3m or more below the surface. Then the individual holes are dug for each plant at a distance of 40-100 cm for the larger types whereas the smaller cultivars may be as close as 30x30 cm or 50x50 cm. These holes are then filled with a mixture of soil, sand and green manure compost to create a suitable growing medium and the suckers or setts are planted in the sand so that its upper roots are at the water level. Planting depth varies but 10-15cm is fairly standard (Anon., 2016). Additional leaf mulch may be added as the plants grow or each single plant may be surrounded by a bottomless basket woven *in situ* from *Pandanus* or coconut leaves, and the enclosed area filled with a mixture of chopped leaves for composting and soil. This type of basket cultivation gives the largest rhizomes though with a slow growth rate.

In other areas, the planting method is similar to those employed for wet land cultivation of taro (*Colocasia*) but great attention is paid to mulching and shading, and if possible natural shade such as overhanging trees, bushes, *etc.* is provided until the plants are 1-2m high. For non-puddled or firmer soils, deep planting holes or furrows are prepared (15-100cm deep) and after the setts or suckers are placed in position the furrows are partially filled with soil and/or compost up to 10-15cm above the base of the sucker or setts.

In wet land cultivation, *C. merkusii* is often interplanted with *Colocasia esculenta*. Being a slow grower, *C. merkusii* requires 3-6 years to mature, hence 2-3 crops of *Colocasia esculenta* can be intercropped while *C. merkusii* is growing. The *Colocasia esculenta* is planted at a distance of 1-1.5 m, and *C. merkusii* is interplanted between it.

Harvest

The cultivars are considered ready for harvesting when flowering starts. However, harvesting is generally done whenever the food is needed because the rhizomes can remain in the soil without loss of quality for about 10 years. The tubers are dug by hand and normally eaten as soon as harvested.

Postharvest

As per requirement, continuous harvesting and replanting is done as per the normal procedure. Although storage is not usually practiced the rhizomes are sometimes buried in a damp place where they may be kept for up to 6 months. If left with the top portion, harvested tubers can be stored well under cool conditions for few months. Sometimes tubers are peeled, sliced and dried in the sun and the dried ones can be stored for several months.

Yield

Depending on the cultivar, growing condions and regions, individual tubers vary greatly in size, from an average of 2kg in Truk and 4.5kg in Yap, to as much as 20-50kg or more for the giant types. The largest recorded rhizome weighed about 180kg in Ponape Island. With moderate soil fertility, annual rhizome yield averages 10t/ha though up to 35t/ha has been obtained in the Philippines.

Pests and Diseases

C. merkusii is probably more resistant to pest and diseases as compared to *Colocasia esculenta*. The most serious pest is reported to be the taro beetle (*Papuana huebneri*) which tunnels into the rhizome. Lindane or primiphos methyl can be used against these insects in mid-summer. Other minor pests include *Aphis gossypii*, which has been reported to transmit virus diseases. This can be controlled by spraying tobacco decoction and through jet-sprays of water. On some islands such as Mortlock Islands rats cause serious crop losses which can be controlled by using zinc phosphide pellets. In case of *Pythium* tuber rot, the infected plants should immediately be removed along with adhering soil. Before planting, the rhizomes should be dipped in Bordeaux mixture or copper oxychloride. In case of infection of 'dasheen mosaic virus' (DMV) on the leaves, the plants should be uprooted and burnt.

References

1. Anon. 2016. *Cyrtosperma merkusii*. Plant Resources of South-East Asia (PROSEA). *PlantUse English*. Retrieved 04: 55, September 30, 2022 from https: //uses.plantnet-project.org/e/index.php?title=Cyrtosperma_merkusii_(PROSEA) and oldid=219000.
2. Anon. 2013. The Plant List (version 1.1). Published on the Internet; http: // www.theplantlist.org/(accessed 1st January).
3. Brown, Deni, 2000. Aroids: Plants of the Arum Family. Timber Press, USA.
4. Facciola, S. 1998. Cornucopia II. Kampong Publications, California.
5. Gorisek, Sasa, 2010. Tuvalu Food Security (PDF). United Nations.
6. Govaerts, R. and D.G. Frodin, 2002. World Checklist and Bibliography of Araceae (and Acoraceae): 1-560. The Board of Trustees of the Royal Botanic Gardens, Kew.
7. Hay, A. 1988. *Cyrtosperma* (Araceae) and its Old World allies. *Blumea*, **33**: 427-69.
8. https: //en.wikipedia.org/wiki/Cyrtosperma_merkusii

9. Humanity development library 2.0. Root crops: Swamp taro (*Cyrtosperma chamissonis*). nzdl.org.

10. Kay. D.E. (revised by B. Gooding), 1987. Root Crops. Tropical Development and Crop Research, London.

11. Knox, Angie, 2002. Sinking feeling in Tuvalu. BBC News (retrieved 10 May, 2010).

12. Limbu, Prakash, 2012. Nepal Food Culture (PDF). ICIMOD, Kathmandu, Nepal.

13. Lois Englberger, Joseph Schierle, Klaus Kraemer, William Aalbersberg, Usaia Dolodolotawake, Julia Humphries, Robin Graham, Anne P. Reid, Adelino Lorens, Kiped Albert, Amy Levendusky, Eliaser Johnson, Yumiko Paul and Fernando Sengebau, 2008. Carotenoid and mineral content of Micronesian giant swamp taro (*Cyrtosperma*) cultivars. *J. Food Compos. and Anal.*, **21**(2): 93-106. https: //doi.org/10.1016/j.jfca.2007.09.007. (https: //www.sciencedirect.com/science/article/pii/S0889157507001329).

14. Richard Marcel Nguimbou, Thaddee Boudjeko, Nicolas Yanou Njintang, Makhlouf Himeda, Joel Scher and Carl M.F. Mbofung, 2014. Mucilage chemical profile and antioxidant properties of giant swamp taro tubers. *J Food Sci. Technol.*, **51**(12): 3559–67.

15. Tindall, H.D. 1983. Vegetables in the Tropics. MacMillan, Oxford, England.

16. Uphof, J.C. Th. 1959. Dictionary of Economic Plants. Weinheim, Germany.

25

Sweet Potato [*Ipomoea batatas* (L.) Lam.]

Family: Convolvulaceae (bindweed or morning glory family).

Synonyms: *Batatas batatas* H.Karst., *B. edulis* var. *porphyrorhiza* (Griseb.) Ram.Goyena, *Convolvulus apiculata* M.Martens and Galeotti, *C. batatas* L., *C. esculentus* Salisb., *Ipomoea batatas* var. *edulis* (Thunb.) Makino, *I. batatas* var. *lobata* Gagnep. and Courchet, *I. chrysorrhiza* Hook.f., and *I. edulis* Makino and *I. wallii* Hemsl.

Other common names: Sweetpotato (for plant as well as its edible tuberous roots). **Hindi**: Ganji, Meetha aloo or Shakarkand/Shakarkandi.

Introduction and Origin

Ipomoea is the largest genus in its family with over 600 species but *I. batatas* is the only crop which is of major importance as a food plant. Its thickened roots (enlarged storage root) are widely used as food whether raw, boiled or fire-roasted in the tropical and subtropical countries such as India, Pakistan, Myanmar, Bangladesh, United States of America, Indonesia, China, Japan, Philippines, Thailand, Vietnam, Nigeria and many other countries. However, *I. aquatica* spreading on the water surface with roots in the soil on the margins of the water bodies is another crop which is also used as leafy vegetable when fried with potatoes. Many other species are even poisonous. Sweet potato is a staple food crop in many of the developing countries, and also serves as animal feed and raw material for many industrial products. According to a study by the United Nations Food and Agriculture Organization, sweet potatoes are the most efficient staple food to grow. Sweet potato is recognized as the third most important tuber crop after potatoes and cassava with a world production of 103.11 million tonnes. These root tubers are eaten raw, boiled, roasted and as cooked vegetables. They are also used for canning, dehydrating, flour making, as a source of starch, for making glucose syrup and for alcohol distillation. It is second most important root tuber and the 7th most important food crop of the world. It ranks as the foremost root crop in respect of calorie value and dry matter production. Its smooth-skinned root tubers may be of various surface colours such as white, yellow, orange, red and purple. Sweet potatoes in their storage roots combine the properties of cereals, fruits and vegetables owing to its content respectively of starch, pectin and vitamins. Sweet potato yields high amount of energy per unit area per unit time and is expected to bridge the food shortages and malnutrition. Studies have shown that in sweet potato storage roots the carbohydrate and ethanol yield are roughly three times that of corn (Rodgers *et al.*, 2007) though quite cultivar-dependent.

Sweet potato is a short maturity crop tolerant to a wide range of growing conditions. The crop has relatively few natural enemies and grows well in poor soils. As the rapidly growing vines shade out weeds, it requires relatively little labour, fertilizer or harmful pesticides. In India, it is largely grown in north and eastern parts of India comprising of Odisha, Bihar, Chhattisgarh, Jharkhand, West Bengal, Madhya Pradesh and Uttar Pradesh which collectively contribute to 90 per cent of the total Indian production. Out of these, leading states are Odisha, Uttar Pradesh and West Bengal. In South India, it is grown at several places and is traditionally popular among a large portion of India's population. In India, its production is estimated to be 11,32,360 metric tonnes with a total area of 111,760ha. India occupies 12th, 8th and 5th rank globally in terms of area, production and productivity, respectively while 5th rank in Asia in terms of area, production and productivity. India exports sweet potatoes to United Arab Emirates, Nepal, Maldives and other Asian countries. Being rich source of dietary fibre, natural sugar, complex carbohydrates, protein, carotenoids, vitamin C, iron and calcium the sweet potatoes have been ranked number one in nutrition out of all the vegetables, by the Centre for Science in the Public Interest, USA. A per day regular intake of 100g of orange-fleshed sweet potato tubers can provide the daily recommended amount of vitamin A for children and adults. To overcome oxidative stress, the purple-fleshed tubers of sweet potato can provide good amounts of bio-available antioxidant to overcome oxidative stress as its flesh is rich in anthocyanin (0.9-1g/100g).

Till now, sweet potato was believed to have been originated and domesticated in either Central or South America. However, a new research conducted by a paleobotanist of Indiana University has revealed that *shakarkandi* (sweet potato) originated in Asia and was present much earlier than was believed till now. The fossil leaves found in Meghalaya, northeast India, have the hallmarks of morning glory ~ a vast family of flowering plants that includes sweet potatoes. The research team including Indiana University Prof. Emeritus David Dilcher and some of his Indian colleagues found 57 million years old fossils of leaves of the sweet potato family (morning glory family) in Eastern India. The family of plants is now said to have originated in the East Gondwana landmass during the late Paleocene epoch. The part of the Gondwana land where the plant first grew, later became a part of Asia.

Etymology

The generic name *Ipomoea* is derived from the Greek word *ipos* meaning 'woodworm' or 'bind weed' and *homoios* is for 'resembling', owing to its resemblance to 'bindweed'. Sweet potato has a twining habit similar to bindweed. The species name *batatas* has come originally from the Taino name for sweet potato.

Although the sweet potato is not botanically closely related to the common potato but they have a shared etymology. Also the soft and orange sweet potato is often called a 'yam' in parts of North America though it is quite distinct from the botanical yam (*Dioscorea*) which belongs to the monocot family Dioscoreaceae.

Botany

Sweet potato vine is a dicotyledonous herbaceous perennial with alternate triangle-shaped or palmately lobed leaves screwed along the stems, and medium-sized sympetalous flowers. The stems usually crawl on the ground forming adventitious roots at the nodes. The leaf stalk is 13 to 50cm and the variable heart or kidney to egg-shaped, rounded or triangular and spear-shaped leaf blades are 5-13cm long with entire, toothed or often 3-7 times lobed, cut or divided leaf margins. The leaves are usually green in colour though due to accumulation of anthocyanins especially along the leaf veins, they can be purple also. Leaf surfaces are usually bare, rarely hairy, and the tip is rounded or pointed. Depending on the variety, the total stem length can be 0.5-4m though there are certain varieties where stems may run even up to 16m.

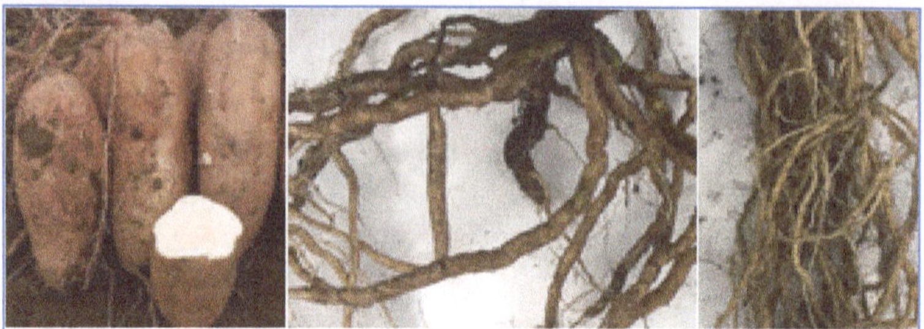

Roots of *Ipomoea batatas*, Tiazhong6 cultivar | Roots of *I. trifida* | Roots of *I. triloba*

It produces short-stalked and 5-fold bisexual flowers which open only for a few hours before sunrise but somewhere after midnight starting from 2AM but getting closed in the following morning before withering. Upright standing inflorescences arising from the leaf axils bear solitary to a few-clustered flowers though there are varieties never producing a flower. Elongated, point-tapering, spiky and usually finely hairy or ciliate sepals are only 0.7-1.5cm long where inner three are a little longer. Overgrown, 4-7cm long and funnel-shaped folded crown with a shorter hem are lavender to purple-lavender, usually with darker throat though there may also occur white crowns. The enclosed stamens are of unequal length with glandular filaments. Two-chambered ovary has relatively short styles. Seeds are produced from cross-pollination. The long and tapered tuberous root has a smooth skin with skin colour of near-white, yellow, orange, red, brown, purple and beige depending upon the variety.

Genetics

Cultivated sweet potato is a highly heterozygous crop and has a chromosome number of 2n=6x=90 ~ a natural hexaploid though the structure of the hexaploid genome has yet not been defined. The progenitor of sweet potato is also still under debate. Out of about 600 *Ipomoea* species, only sweet potato is cultivated as the food crop though 13 wild species are closely related to it (Austin, 1988). Of these, hexaploid *I. trifida* (H.B.K.) Don. has been considered the most likely progenitor. *I. trifida* was generated from the hybridization of the diploid *I. leucantha* Jacq. and tetraploid *I. littoralis* Blume. The species *I. trifida* contains different cytotypes such as diploids (2n=30), tetraploids (2n=60) and hexaploids (2n=90).

Varieties

Varieties play a very significant role in yield improvement of any crop and on this line various R and D organizations are working on sweet potato. Major breeding methods in sweet potato include introduction of germplasm, its evaluation and clonal selection, *vis-a-vis* intervarietal *cum* interspecific hybridization and selection. Majority of the improved varieties have been developed from the progenies from open-pollinated seeds. Several hundreds of sweet potato varieties are available in different parts of the world. In India, 1,100 accessions of sweet potato germplasm are maintained under *in vitro* condition at CTCRI field gene bank. Out of these 407 are exotic collections and remaining 693 indigenous.

Storage tubers of different varieties of sweet potatoes

(www.seattletimes.com; www.hgtv.com; www.jessicagavin.com)

Sweet potatoes have great potential as a feedstock for bio-ethanol production. Monday (2009) reported that varieties 'X-1617' and 'Beauregard' produce highest tuber and ethanol yield. Higher the dry matter and starch the greater is the ethanol yield. Nyiawung *et al.* (2010) reported that ideal dry matter, starch and amylase/amylo-pectin content *cum* ratio that a sweet potato variety should have for maximum ethanol yield are 30-33 per cent, 20-24 per cent, 23-30 per cent, 70-77 per cent and 0.27-0.46 per cent, respectively. Sweet potato varieties 'WS 14905' and 'TU0002' produce 66g/l of ethanol by flour-based fermentation and 34.4g/l from fresh tubers of var. 'Beauregard' (Nyiawung *et al.*, 2010). Promising varieties of sweet potato along with improved agrotechniques can boost root production. Adoption of improved agronomic practices alone can increase the yield of sweet potato significantly. Cultivars of the sweet potato have been bred to bear tubers with flesh and skin of various colours. Sweet potato cultivars with white or pale yellow flesh are comparatively less sweeter and moister than those with red, pink or orange flesh.

The orange-fleshed sweet potato contains carotene which is a precursor of vitamin A. Sweet potato tubers contain carotene up to 20mg/100g fresh weight (Sakamoto *et al.*, 1987; Yamakawa, 1997). The high beta carotene rich orange flesh sweet potato varieties cultivated in India are 'Kamala Sundari', 'Gouri', 'Sankar', 'Sree Kanaka', 'ST-14', *etc.* The carotene content in 'ST-14' is 13.83mg/100g of fresh root (Vimala *et al.*, 2006). High anthocyanin rich purple-fleshed sweet potatoes have purple skin and flesh due to the accumulation of anthocyanins (Philpott *et al.*, 2004; Terahara *et al.*, 2004) that are natural soluble food pigments and such varieties are 'ST-13'. Red and purple pigmentation in various parts of sweet potato such as stem, leaf and storage root is caused by presence of acylated anthocyanins (Fan *et al.*, 2008). Sweet potato anthocyanins are comparable to that of red cabbage in terms of their quality as natural food colourants. In Japan, 'Ayamurasaki' ~ a high anthocyanin sweet potato variety was developed by the Kyushu National Agricultural Experiment Station (Yamakawa, 1997). CTCRI Regional Centre, Bhubaneswar has developed a high anthocyanin line 'ST-13' (Anon., 2003). The variety 'PIS 189-257' from Thailand is also rich in anthocyanin (Narin and Reungmaneepaitoon, 2005). The high extractable starch rich (21 per cent) variety 'ST-10' is popular in the country for food nutrition and processing. Some of the high yielding improved varieties cultivated in India are 'Bidhan Jagannath', 'BCSP-5', 'Birsa Sakarakand-1', 'Co-1', 'Co-2', 'Co-3', 'Co CIP 1', Gautam', 'Gouri', 'H-41', 'H-42', 'H-268', 'Indira Sakarakand-1', 'Kalinga', 'Kamala Sundari', 'Kanjanghad', 'Kiran', 'Kishan', 'Pusa Safed', 'Rajendra Sakarakand-5', 'S-1221', 'Samrat', 'Sankar', 'Sourin', 'Sree Arun', 'Sree Bhadra', 'Sree Kanaka', 'Sree Nandini', 'Sree Rethna', 'Sree Vardhini', 'Sree Varun', 'Tripti', 'Varsha' and 'VL Sakarkand-6'. Most of these varieties are developed by CTCRI, Kerala. Certain local cultivars such as 'SP 4', 'SP 13', 'SP 18', 'SP local' and 'Musiri Thandel' are also suited for cultivation in Tamil Nadu and adjoining areas.

Salient features of a few promising varieties of sweet potato in India

- ☆ '**Bidhan Jagannath**' is semi-spreading variety recommended for growing in West Bengal. This matures in 80-110 days with 20-22t/ha of tuber yield. This bears ovate tubers with skin and rind being deep purplish and flesh light orange. It contains β-carotene at 3.5mg/100g.
- ☆ '**CIP-440038**' variety is recommended for cultivation in Tamil Nadu which matures in 130-140 days with an average tuber yield of 25-30t/ha where dry matter recovery is 18.27 per cent and β-carotene 6mg/100g. It bears spreading type vines with white tuber skin and orange flesh. It has good cooking quality.
- ☆ '**CIP-440127**' is a sweet potato variety recommended for Odisha, which matures in 105-110 days with 22-24t/ha of tuber yield. Its tubers are elliptic, skin and rind are yellow and flesh is orange. Its tubers contain 25-26.6 per cent dry matter content. It contains β-carotene 6.5mg/100g. It tolerates salt-stress, mid-season drought and weevil attack.
- ☆ '**CIPSWA-2**' is a semi-compact variety for Odisha, maturing in 100-110 days with 20-22t/ha of tuber yield. It contains 5.5mg/100g of β-carotene.

Its tubers are round-elliptic with pink skin and orange flesh. It is tolerant to salt stress.

- 'COCIP-1' is a spreading type variety recommended for growing in Tamil Nadu which matures in 95-100 days with tuber yield of 25-28t/ha. This bears elliptic-ovate tubers with pink skin and yellow flesh. It is suitable for chip making, mashing and for beverage.
- '**Co 1**' is a variety suitable for cultivation in Tamil Nadu which matures in 120-135 days with 20-25t/ha of tuber yield and is tolerant to weevil. Its vines are moderately vigorious and the tuber is fusiform having light pink skin and white flesh.
- '**Co 2**' is a variety suitable for Tamil Nadu, maturing in 110-115 days with 20-25t/ha of tuber yield. Its vines are semi-spreading with round elliptic tuber, white flesh which cooks well and tastes sweet.
- '**Co 3**' is a variety suitable for cultivation in Tamil Nadu, maturing in 105-115 days with 20-28t/ha tuber yield, containing 9.8mg/100g of β-carotene. Its plants growth is bushy, tuber is fusiform, tuber skin is red and the flesh is orange.
- '**Co 3-4**' is recommended for growing in Kerala and Odisha. Its maturity period is 90-95 days and tuber yield 25-28t/ha with dry matter content of 33.5 per cent. It is semi-spreading type bearing short fusiform tuber having pink skin, purple rind and white flesh. It is tolerant to weevil.
- '**Gouri**' is a variety suitable for growing in Odisha, which matures in 110-120 days, with average tuber yield of 18-25t/ha. Its vines are semi-erect, tuber round elliptic, skin red, flesh orange, dry matter 22-27 per cent and β-carotene 4.5-5.5mg/100g. It is tolerant to moisture stress.
- **'Goutam' (Pol. 21-1)** is a spreading type and is recommended for growing in Odisha. It matures in 105-110 days with an average tuber yield of 18.9-21t/ha having 29.9-31 per cent dry matter content. Its tubers are round-ovate where skin *cum* flesh colour being creamy-white. It has a quite good cooking quality and is tolerant to weevil.
- **'H-41'** and **'H-42' are semi-spreading type** hybrids taking around 120 days to mature with an average yield of about 20-25t/ha.
- **'IGSP-14'** is recommended for growing in Tamil Nadu. It matures in 119-120 days with 25-28t/ha of tuber yield, having 23.28 per cent dry matter and 7mg/100g of carotene. Its growth is compact bearing round-elliptic tubers with pink skin and white flesh.
- **'Indira Madhur'** is a semi-spreading type sweet potato recommended for growing in Chhattisgarh region. It matures in 110-120 days with

21-25t/ha of tuber yield. Its tubers are long elliptic with red skin and orange flesh. It contains 4.5mg/100g of β-carotene. It is field tolerant to weevil.

- **'Indira Nandini'** is a semi-spreading type, maturing in 110-120 days with 25-30t/ha of tuber yield. It is recommended for growing in Chhattisgarh. Its tubers are elliptical-round with red skin and creamy flesh. Its tubers are suitable for culinary purpose. It is field-tolerant to weevil.
- **'Indira Narangi' (IGSP. C-15)** is an erect growing variety suitable for growing in Chhattisgarh and Andhra Pradesh. It matures in 90-100 days with an average tuber yield of 23-25t/ha. Its tubers possess 25.17 per cent dry matter and 5.72mg/100g of carotene. It produces fusiform tubers with purple-red skin, pink rind and orange flesh. It is tolerant to weevil.
- **'Indira Naveen'** is semi-spreading variety recommended for growing in Chhattisgarh region. It matures in 90-100 days with 26-30t/ha of tuber yield. Its tubers contain 25.5 per cent dry matter content. Its tubers are elliptical-round with pinkish-red skin and creamy flesh. It is suitable for culinary purpose and is field tolerant to weevil.
- **'Indira Priya' (IGSP-14)** is an erect growing sweet potato variety suitable for growing in Chhattisgarh and Tamil Nadu regions. Its tubers mature in 100-110 days with average tuber yield of 25-30t/ha. Its tubers are fusiform with pink skin *cum* rind and pale-yellow flesh, and its tubers contain 24.97 per cent dry matter. Tubers are tolerant to weevil.
- **'Kalinga' (90/704)** is recommended for growing in Odisha, Bihar, Jharkhand and West Bengal. This matures in 105-110 days with average tuber yield of 17.2-20t/ha. It is a spreading type of variety bearing round elliptic tuber, purple-red skin, white flesh and 29.6 per cent dry matter content. It has excellent cooking quality.
- **'Kamala Sundari'** is a sweet potato variety developed for West Bengal and Odisha regions, maturing in 110-120 days with average tuber yield of 21-24t/ha. Tubers contain 26.47 per cent dry matter and 8.2mg/100g β-carotene. Its spreading plants contain long elliptical tubers with orange skin *cum* rind and orange flesh. It has good cooking quality.
- **'Kiran'** is a variety recommended for growing in Andhra Pradesh. It matures in 110-130 days with an average yield of 20-22t/ha having 3-4mg/100g of β-carotene. It has spreading vine, tubers elliptic, skin red, flesh orange, and dry matter content 31-35 per cent. It is a weevil escaping variety.

- **'Kishan' (Pol. 13-4)** is a semi-spreading type recommended for growing in Odisha which matures in 110-120 days with an average tuber yield of 17-20.8t/ha and 32-34 per cent of dry matter content. It bears elliptical tubers with reddish-purple skin and creamy-white flesh. Its flesh is mealy and sweet with good cooking quality. It is tolerant to weevil.
- **'Konkan Ashwini'**, developed to be cultivated in Konkan region of Maharashtra, is a short duration crop (maturing in 105-110 days) giving high yield (15-20t/ha). It is a semi-spreading type bearing elliptical-round tubers with pink skin and creamy-white flesh. It has good cooking quality.
- **'Narendra Shakarkand-9'** is a variety with semi-erect growth for growing in Uttar Pradesh. Its tubers mature in 120-130 days with 16-20t/ha of average tuber yield having 30-34 per cent dry matter content. Its tubers are elliptic, skin cream and flesh whitish-cream. It has excellent cooking quality.
- **'Rajendra Shakarkand-5 (RS-5)'** is recommended for cultivation in Bihar and Andhra Pradesh, maturing in 105-120 days with an average yield of 18-25t/ha. Its vines are spreading type, tubers cylindrical, skin light brown, flesh white and is tolerant to waterlogging, weevil, hairy caterpillar and *Cercospora* disease.
- **'Rajendra Shakarkand-35 (RS-35)'** is recommended for growing in Asom, Karnataka, Bihar, West Bengal and Maharashtra. It matures in 105-120 days with an average yield of 25-30t/ha. Its vines are medium spreading, tubers cylindrical, skin light brown and flesh white. Its tubers have good cooking quality.
- **'Rajendra Shakarkand-43 (RS-43)'** is recommended for cultivation in Asom, Karnataka, Bihar and West Bengal. Its tubers mature in 110-120 days with an average yield of 20-23t/ha. Its moderately spreading vines are used as green fodder. Its tubers are cylindrical, skin off-white and flesh white. It is moderately tolerant to *Cercospora* leaf spot and weevil.
- **'Rajendra Shakarkand-47 (RS-47)'** is recommended for growing in the plains of Bihar. It matures in 120-130 days with an average tuber yield of 25-32t/ha. Its vines are spreading type, tuber skin is red and the flesh is white. It is tolerant to *Cercospora* leaf spot and sweet potato weevil.
- **'Rajendra Shakarkand-92' (RS-92)** is a spreading variety for the plains of Bihar. It matures in 75-120 days with 20-25t/ha of average tuber yield. Its tubers are cylindrical with red skin and yellowish-white flesh. It is moderately tolerant to weevil, flood and low temperature.

- **'Samrat' (S-30)** is recommended for growing in Andhra Pradesh. It matures in 90-120 days with an average tuber yield of 18-22t/ha. Its vines are medium spreading, tuber fusiform, skin white, flesh white and with 35 per cent dry matter. It is moderately resistant to drought.
- **'Sankar'**, a medium duration variety with good yield and excellent cooking quality, is released for Odisha. It matures in 120 days with an average yield of 13.73t/ha.
- **'ST-10'** is a semi-compact sweet potato for growing in Odisha, which matures in 105-110 days with average tuber yield of 24-26t/ha. This bears round tubers with yellowish-white skin and white rind *cum* flesh. Its tubers contain extractable starch 21 per cent and dry matter content of 27.4-29.7 per cent. It has excellent cooking quality and is tolerant to mid-season drought and weevil.
- **'ST-13'** is a semi-compact variety maturing in 110-120 days with average tuber yield of 18-22t/ha, which is recommended for growing in Odisha. Its tubers contain 90mg/100g anthocyanin, and extractable starch 18 per cent. This bears long elliptic tubers with dark purple skin and flesh. It is highly tolerant to salt stress and weevil.
- **'ST-14'** is a spreading type variety recommended for cultivation in Odisha, which matures in 105-110 days with average tuber yield of 20-24t/ha, containing 27-28 per cent dry matter and 11.5-12.5mg/100g of β-carotene. Its round-elliptic tubers have yellow skin and dark orange flesh. It has excellent cooking quality and is ideal for food security and processing industry.
- **'Sourin' (Pol. 4-9)** is a spreading type maturing in 105-110 days with 16.2-18.5t/ha tuber yield and 26.9-27.5 per cent dry matter content. It can withstand mid-season drought. It is tolerant to weevil. Its red-skinned tubers are round-elliptic with white rind and creamy-white flesh.
- **'Sree Arun'** is an early maturing variety with an average yield of 22.5-31.5t/ha, having pink skin and creamy flesh which was developed by CTCRI, Sreekariyam.
- **'Sree Bhadra' (S-1010)** is a semi-spreading very early variety (maturing in 90-105 days) with broad leaves which are dark purple when emerging, recommended for growing in Kerala, Maharashtra, Bihar and Madhya Pradesh with an average tuber yield of 20-22t/ha. Its short tubers are spherical, light pink-skinned, pink rind and cream flesh. It contains 30-33 per cent dry matter, 20 per cent starch, 2.9 per cent total sugar

and 972IU/100g carotene content. It is used as a trap crop for root-knot nematode.

Sree Vardhini

Varsha

Sree Bhadra

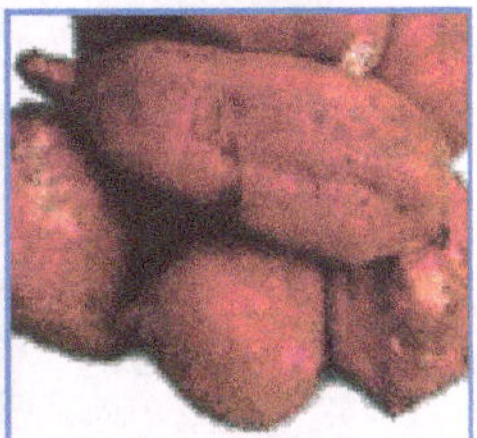
Sree Retna

- **'Sree Kanaka'** has creamy skin and dark orange flesh. It gives an average yield of 11-17t/ha.
- **'Sree Nandini' (76-OP-217)** is a drought tolerant white sweet potato variety recommended for growing in Kerala which matures in 105-120 days with 20-25t/ha of tuber yield having white flesh and 25-26 per cent dry matter. It has good cooking quality and is tolerant to weevil. It is a spreading type where tubers are fusiform.
- **'Sree Rethna' (X-108-2)**, a short duration (90-105 days) spreading type variety which bears 5-lobed and medium sized purple leaves. Tubers are rounded/spherical in shape with purple skin and orange flesh. Tuber contains 31 per cent dry matter, 23 per cent starch, 2 per cent sugar and 3500 IU/100g carotene content. It is recommended for growing in Kerala.
- **'Sree Vardhini" (76-OP-219)**, an open pollinated seed progeny, is a semi-spreading and early maturing (100-105 days) dual purpose variety recommended for cultivation in Kerala with an average yield of 20-25t/ha and carotene content of 1200IU/100g. Its tubers are fusiform, skin is pink and the flesh is light orange. It contains 23-25 per cent dry matter.
- **'Varsha' (H-268)**, a semi-spreading type variety released for Konkan region of Maharashtra, which matures in 120-130 days with average tuber yield of 17-22t/ha. Its vines are semi-spreading type where tubers are fusiform with red skin and white flesh and contains 23-25 per cent dry matter content. This is drought tolerant with good cooking quality and sweet in taste.
- **'Sree Varun'** bears creamy skin of tubers. It is an early maturing variety (90-100 days) with an average yield of 21.5-30t/ha.

'Sree Vardhini'	*'H-42'*

Sree Vardhini is a high yielding variety of sweet potato. It is a selection from open pollinated seed progeny of S-13. Semi-spreading variety. Early maturing (100-105 days) variety with dual purpose for human consumption and animal feed (foliage). It is rich in carotene (1200 IU/100g) and average yield is 20-25 T/Ha.	H-42 is a high yielding hybrid variety of sweet potato with sweet and low-fibre tubers. Maturity time is 120 days and average yield is 22-25t/ha.
'H-41'	*'Varsha'*

H-41 is a high yielding hybrid variety of sweet potato with sweet, low fibre and excellent cooking quality tubers. It is drought tolerant. Duration is 120 days and average yield is 20-25t/ha.	*V*arsha is a high yielding hybrid variety of sweet potato with non- branching tall (2-2.5m) plants. Carotene content in tubers is 466 IU/100g and starch content in fresh tubers 25-27 per cent. Duration 10 months and average yield is 23-25t/ha. Released for Konkan region of Maharashtra.

'Gouri'	*'Sankar'*
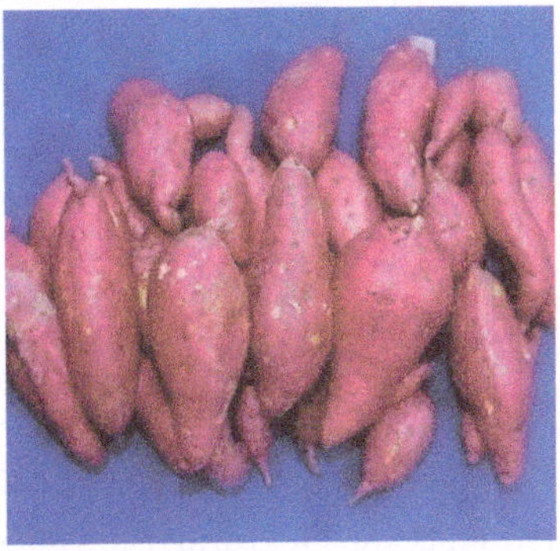	
Gouri is a medium duration variety with high carotene content (4.5-5.5mg/100g). It matures in 110-120 days and the average yield is 30t/ha. This variety tolerates mid-season moisture and suitable for Kharif and Rabi.	Sankar is a medium duration variety with good yield and excellent cooking quality, released for Odisha. It matures in 120 days and the average yield is 13.73t/ha.
'Sree Bhadra'	*'Sree Rethna'*

Sree Bhadra is a semi-spreading, broad-leaved, short duration (90-105 days) and high yielding sweet potato variety. Emerging leaves are dark purple on greenish-brown vines. Tubers are short, fusiform, pink skinned and with cream flesh colour, flesh containing 33 per cent dry matter, 20 per cent starch, 2.9 per cent total sugar and 972 IU/100g carotene content.	Sree Rethna is a spreading type short-duration (90-105 days) but high yielding variety. Medium-sized emerging leaves are purple. Rounded to spherical tubers have purple skin and orange flesh containing 31 per cent dry matter, 23 per cent starch, 2 per cent sugar and 3500 IU/100g carotene content.

'Sree Kanaka'	*'Sree Nandhini'*

Early maturing (75-85 days) sweet potato variety having cylindrical tubers with dark orange flesh and a very high beta carotene content.	It is a drought tolerant early maturing (100-105 days) variety suitable for paddy fallows as a catch crop.
'Sree Arun'	*'Sree Varun'*

It is a early maturing (90 days) sweet potato with good cooking quality. Its skin is pink but flesh is cream.	It is an early maturing (90 days) variety with good cooking quality. It has pink skin and cream flesh.
'Kalinga'	*'Goutam'*
	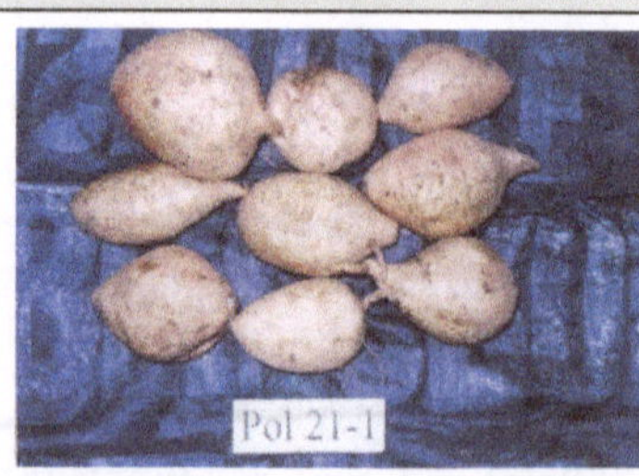
It is a medium duration (105-110 days) variety with excellent cooking quality having high starch content (28 per cent). It can be used as food, fodder and for starch extraction. Its skin is purple-red and flesh cream.	It is a medium duration (105-110 days) variety with good cooking quality. It is suitable for upland and hilly area cultivation and suitable for Kharif and Rabi season.Its skin is white and flesh is cream.

'Sourin'	*'Kishan'*
	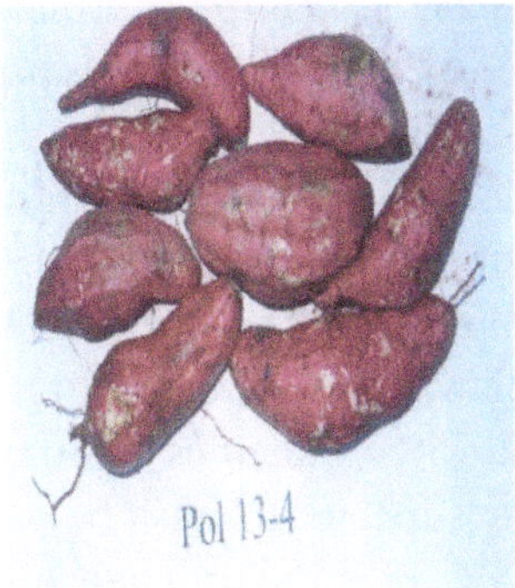
Sourin is a medium duration (105-110 days) variety with good cooking quality. It is suitable for rainfed and irrigated conditions and are good for sandy loam soils. It withstands mid-season drought. Its skin is red and flesh creamy white.	It is a medium duration (110-120 days) variety with high starch content (29-30 per cent), suitable for growing in sandy loam and black sandy soils for Kharif and Rabi seasons. Its skin is reddish-purple and flesh white.

Propagation

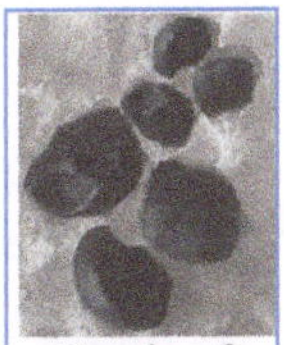
Seeds of Sweet potato

Sweet potatoes rarely flower when the daylight is longer than 11 hours as is normal outside of the tropics. True seeds and tissue culture are used for breeding only. Tissue culture is growing of small plant parts in test tubes with a nutrient medium in laboratory to regenerate the complete plant (plantlets). Farmers or seed producers mostly propagate sweet potatoes by stem (vine) or by adventitious shoots called 'slips' that grow out from the tuberous roots during storage. Sweet potatoes are relatively easy to plant because they are planted by vine cuttings rather than seeds.

Vine cuttings for planting in the main field can be obtained either from freshly harvested plants or from nursery raised either from stored tubers or from vines of the freshly harvested crop. In colder climates where vines do not develop well, producers plant roots. Recurrent use of vines can cause increased weevil infestation even though there is less chance of root yield reduction (Ray *et al.*, 1983; Nair 2006). Vines obtained from nursery should be healthy and vigorous for maximum root production. Cuttings of sweet potato vine, either edible or ornamental cultivars, having the vine length of 20-40cm with 3-5 nodes are best for tuber production. The apical portion of the

Sprouted storage root

vine is best for getting higher root yield as compared to mid-stem cuttings (Choudhary *et al.*, 1986). Thick and woody basal portions should be avoided as sometimes these fail to establish and there is also greater chance of weevil incidence due to closeness with the crown portion where sweet potato weevil multiplies (Nair, 2006). Cut vines with intact leaves are usually stored under shade for 2-4 days prior to planting in the main field to promote better root initiation, easy establishment of vines and higher root yield (Ravindran and Mohankumar, 1989; Biswal, 2008). To reduce the bulkiness, the leaves of the vines can be removed when the vines are to be transported to distant places. Stored vines are generally superior to fresh ones in respect of leaf area index (LAI), crop growth rate (CGR), root bulking rate, number of roots per plant and root and vine yields. Storing of vines for a long time may cause these getting dried, *vis-a-vis* failure of establishment in the field.

Preparing Nursery

The nursery preparation must be started 2.5-3 months prior to planting in the main field. About 100m^2 of primary nursery area and about 100kg of medium-sized seed roots (125-150g each) are required for planting one hectare of land. At the time of nursery bed preparation, around 100kg of farmyard manure or compost should be applied. The roots must be planted at a gap of 20cm on ridges that should be 60cm apart. For quick growth of vines they are usually top-dressed with 1.5kg urea per 100m^2 at 15 days after planting. For the first 10 days the nursery should be irrigated every alternate day and thereafter twice a week. After 40-45 days the vines will be ready for planting in the main field or they can be cut to a length of 20-30cm for further multiplication in the secondary nursery (Anon., 1987).

Vines obtained from the primary nursery can be further multiplied in the secondary nursery to an extent of 500m^2 to produce enough planting material for planting in the large field. At the time of preparation of nursery bed, around 500kg of FYM or compost should be applied and ridges (around 60cm apart) should be made. Vines obtained from the primary nursery or from freshly harvested crop are planted in the secondary nursery at a spacing of 20cm within ridges. To ensure enough vegetative growth 5kg of urea is usually applied in two splits on 15 and 30 days after planting. Now these are regularly watered until ready for transplanting. The vines will be ready for transplanting in the main field after 40-45 days (Anon., 1987).

Plug Transplanting

In recent decades, there is a remarkable progress in environmental control technology to achieve rapid and uniform growth of high quality transplants that can be directly transplanted into the main fields. Sweet potato plug plants can be produced using virus-free stock plants in closed systems. Use of plug transplants reduces post transplanting stresses and labour cost. Mechanical transplanters can be used to plant the plug plants into the main field. Both single node cuttings and shoot tip cuttings with one unfolded leaf of sweet potato can be used to produce plug plants in closed systems in multicell trays with recommended cell volume of around 55ml. Much higher efficiency of sweet potato cultivation can be achieved using plug transplants because of the labour saving, mechanical transplanting by the transplanters and quick production.

Growth and Development

Throughout the world, the sweet potato is cultivated from 40°N to 32°S under contrasting system of agriculture (Mukherjee, 2019). Depending on the cultivar *cum* soil and environmental conditions tuberous roots mature in 2-9 months. Pesticides are rarely needed as sweet potatoes have few natural enemies. They grow well in many farming conditions. With care, early-maturing cultivars can be grown as an annual summer crop in temperate areas. The agronomic practices for sweet potato vary from place to place depending upon prevailing environmental conditions and variety. Promising varieties of sweet potato along with improved agrotechniques can boost up tuber production.

To be disease-free, the areas where seed vines are grown should be carefully selected. Environmental factors such as light, temperature, moisture and soil type influence the growth and quality of sweet potatoes. Other factors such as ground preparation, planting methods, variety of sweet potato, vine-cutting health, irrigation, nutrition, planting time, pest management and time of harvest can be controlled by growers appropriately. When all the factors at optimum levels have been met with, quality root tubers with most profitable yields can be attained. In several sweet potato growing countries, the agronomical advancement has increased the tuber yield of 10-30 per cent.

Many factors such as length of growing season, air and soil temperatures, light intensity and duration, humidity, and wind that influence the growth of sweet potato are largely uncontrollable. Sweet potatoes are cultivated throughout tropical and warm temperate regions wherever there is sufficient water to support their growth. Sweet potato requires a moderately warm climate (21-27°C, optimum being 24°C) with abundant sunshine and warm nights. Shade causes drastic yield reduction. Annual rainfall of 750–1,000mm is considered most suitable with a minimum of 500mm in the growing season. As it is a photosensitive crop, sunny days and cool nights are favourable for better tuber development. The plant does not tolerate frost and is also sensitive to drought at the tuber initiation stage, *i.e.* 50–60 days after planting. It is not tolerant to waterlogging and if aeration is poor it may cause tuber rots and reduced growth of storage roots. Heavy rainfall, high temperature and excess cloudiness encourage vegetative growth.

In India, sweet potato is grown throughout the country utilizing monsoon rain during *Kharif* (June-August) and with supplemental irrigation during *Rabi* (October-December). For higher storage root yield, the rainfed crops should be planted immediately after onset of monsoon (Nedunchezhiyan and Reddy, 2006) whereas dry season crops should be planted in October-November (Nath *et al.*, 2006). The major area under sweet potato is planted during *Kharif* season, though *Rabi* crop enjoys warm sunny days and cool nights with moderate rainfall enhancing higher root yield as compared to *Kharif* crop. Planting in the main field should be done during June-July and in the nursery beds during March to May.

When there is no any other source of irrigation this is the only possible method. Traditionally, the sweet potato vines are planted in the plains of north India during August-September and the crop is harvested in February-March. In low-lying areas, vines are planted during January-February after the second crop of rice or after the harvest of tobacco. This helps to utilize the residual moisture after the second crop of rice and the organic matter and compost left by tobacco. In north India these are planted when wheat sowing starts. In Kerala, it is planted as upland crop in June as rainy season crop while in October-November with irrigation facilities. In Karnataka, Andhra Pradesh and Maharashtra, this is planted during *Kharif* and *Rabi* but with irrigation. In Tamil Nadu these are planted from May to September but with irrigation. Under upland of Odisha and Madhya Praesh, these are planted as *Kharif* crop from June to July. In Diara land of Bihar and Uttar Pradesh these are planted in November after recession of flood. This is planted in August-September with supplemental irrigation in the midland areas of Odisha, Bihar, Jharkhand, Chhattisgarh and Madhya Pradesh. In Eastern Uttar Pradesh, West Bengal, Asom and NEH region, the sweet potato is planted from second fortnight of September to first fortnight of October as autumn crop with supplemental irrigation; in the upland as rainy season crop while it is planted in February as summer crop but with supplemental irrigation.

Sweet potatoes can grow on a variety of soil types but it should be well-drained. Sandy-loams are most suitable though clay loam soils are also good provided these have been mixed thoroughly with farmyard manure @ 250 quintals per hectare at the time of field preparation. However, fertile sandy loam with clay subsoil is also ideal. With little fertilizers, it can also be grown in poor soils. Heavy clayey soils restrict the storage root development due to compactness whereas sandy soils encourage formation of long cylindrical pencil-like roots

(Nedunchezhiyan and Ray, 2010). Sweet potato is mostly grown in acidic soils, however soil pH of 5.5-6.7 is found to be optimum. High soil pH invites pox and scurf diseases in sweet potato whereas at low pH sweet potato suffers from aluminium toxicity (Nedunchezhiyan and Ray, 2010) to which these are very sensitive and will die about six weeks after planting if lime is not applied along the planting in such soils. Sweet potato is also sensitive to saline and alkaline conditions (Dasgupta *et al.*, 2006). Best growth of sweet potato is obtained when soil is having bulk density of 1.3-1.5g/cm^3. Salt-tolerant beta carotene rich (14mg/100g) orange-flesh 'ST-14' and anthocyanin rich (0.9mg/100g) purple flesh 'ST-13' varieties are developed by CTCRI for cultivation in coastal belts of Odisha and certain other areas of the country.

Land Preparation and Planting Method

The ploughing or digging to a depth of at least 20cm with associated harrowing and eliminating weed-roots are desirable operations before planting. To bring the soil to the fine tilth before planting, land must be ploughed 3-4 times followed by planking every time. Any existent weed should be removed. Various methods such as mound method, ridge and furrow method, bed method and flat method are practiced in sweet potato cultivation in different localities. In areas experiencing problems of drainage, planting sweet potatoes on mounds is preferable. Mounds permit better soil aeration with fewer tendencies for soil compaction. The ridge and furrow system is recommended in sloppy lands for control of soil erosion. Under Indian conditions, the highest root yield is realized when planted on mounds (Ravindran and Mohankumar, 1985). In Vietnam, planting is done on beds formed at a width of 1-1.4m and to a height of 0.4-0.5m (Ngoan, 2006). In Bangladesh, higher yields are reported with trench planting followed by ridge and flat method of planting in alluvial soils under irrigated conditions (Bhuiyan *et al.*, 2006).

Sweet potato cuttings are normally planted with middle portion buried in the soil and both ends exposed. These are also planted in an inclined position with half of its length buried in the soil, or planted horizontally to the soil surface with 5 or 6 nodes. Though horizontal planting is laborious but results in higher transplant survival and better development of the root system than other methods (Nair, 2000). CTCRI has recommended to plant the cuttings in the soil with both the ends exposed and the middle portion buried (Nair, 2006).

The ideal depth of planting is 7-10cm. In sweet potato, close spacing is generally recommended to achieve maximum root yield. Row distance of 30-60cm and a distance of 15-20cm between plants are recommended to achieve maximum root yield. However, no specific spacing is followed when sweet potato is planted on mounds. The vines are planted on mounds by accommodating 3-6 slips or vines per mound. A general spacing of 60×20 cm has been recommended by CTCRI to accommodate 83,000 plants per hectare. In Bihar, the mixed culture of two sweet potato varieties, 'Cross-4' and 'RS-5' in alternate rows with 30×15cm spacing recorded maximum root yield (Mishra *et al.*, 1985).

Irrigation

Sweet potato vines are succulent and fragile and immediately after planting if sufficient moisture is not available in the soil these dry up. Therefore, for proper sprouting and establishment sufficient soil moisture at the time of planting is to be ensured. In the early part of the growing season sweet potatoes generally require, on an average, 2mm of water per day which gradually increases to 5-6mm per day prior to harvest. Sweet potato is mostly grown under rainfed conditions. Hence, planting is carried out on rainy day or immediately after rain. It is also grown in dry season under protective irrigation. Under such conditions if sufficient moisture is not available after planting, irrigation is needed to be provided on alternate days initially for the first fortnight and thereafter once in 7-10 days. Moisture stress during crop growth significantly affects the storage root yield (Ravi and Indira, 1996).

Sweet potato is not tolerant to water-logging, and if aeration is poor, it may cause tuber rots and reduced growth of storage roots. Biswal (2008) found that when soil moisture was high or in compact soil sweet potato had luxuriant vegetative growth with little or no tuberization of root. The combined effect of a long wet period during root initiation and a drought during root bulking may depress the yield significantly.

Manure and Fertilizers

In soils that are low in organic matter considerable amount of organic manure (at the rate of 5-10t/ha, depending on the soil quality) is recommended at the time of field preparation to maintain soil productivity. Usually, farmyard manure, compost or green manure (legume) is used as organic manure for sweet potato.

Fertilizers recommended are 50:25:50kg/ha of NPK. Phophorous fertilizer may be applied at the time of planting as full dose while N and K may be applied in two equal splits coinciding with interculture and earthing up operations. A moderate dose of 50-75kg N/ha is optimum for root production in sweet potato (Nair *et al.*, 1996; Sebastiani *et al.*, 2006; Biswal, 2008). Higher levels of N sometimes decrease the root yield (Hartemink, 2003) and encourage vine growth. Continuous use of N fertilizer under certain situation may have detrimental effects on root quality. Therefore, use of organic source of N is essential to improve quality characters. Sweet potato response to phosphorus is very low as a dose of

25-50kg P_2O_5/ha is considered optimum (Mohanty *et al.*, 2005; Akinrinde, 2006; Sebastiani *et al.*, 2006). Relative efficiency of rock phosphate as source of P to sweet potato was equal to single superphosphate in direct effect but superior to it in residual value (Kabeerathumma *et al.*, 1986).

Potassium is a major key element essential in the synthesis and translocation of carbohydrates from the tops to the roots (Byju and Nedunchezhiyan, 2004). A moderate dose of 75-100kg K_2O is recommended for sweet potato (Nair *et al.*, 1996). The quality characters such as starch and protein contents were found to increase with increased K levels (Biswal, 2008).

Intercultural Operations

After 10 days of sweet potato planting, it establishes in the soil. It starts vigorous vegetative growth after third week. Bulking of roots starts 4-6 weeks after transplanting, hence growing vines should be turned over once a week to avoid anchorage to soil at nodes which will hamper root production by forming undesirable small tubers there. Over-grown vines are regularly pruned leaving only 30cm to increase tuber yield. Though rapidly growing vines of sweet potato cover the soil quickly, weeding is necessary particularly at early stages of the crop growth when grown with wider spacing. Herbicides such as DCPA (Dacthal) can be used to rid the soil with weeds that may interfere with crop growth.

Earthing up of the soil also brings about weed control besides improving the physical condition of the soil. To keep the soil loose and to destroy the weeds, at least two weedings followed by earthing up have to be given between 15-30 and 35-60 days after planting (Nedunchezhiyan and Ray, 2010).

Home gardeners and smaller-scale growers can cover the growing area with **mulches** such as straw or plastic sheets. Mulching is a common practice in rainfed ecosystem in small-holder farming. Mulching conserves soil moisture and reduces soil temperature as well as augments rhizosphere microflora (Kundu *et al.*, 2006).

When temperature and rainfall are favourable, sweet potato grows rampantly producing large quantity of vines at the cost of storage roots (Nedunchezhiyan and Byju, 2005). Under such situations, part of the vine preferably the top portion can be removed and used as leafy vegetable or forage or planting material. Too much pruning may result in significant decrease in total and marketable root yield.

A good **crop rotation** maintains soil fertility, reduces soil erosion, depletes and suppresses insect and disease outbreaks. Rotating with other crops on a 3 or 4 year cycle is the standard procedure to prevent pest attack and disease infections. Corn and squash are the beneficial preceding and succeeding crops for sweet potato.

Sweet potato is **intercropped** with other seasonal crops such as pigeonpea, maize, *etc.* (Nedunchezhiyan *et al.*, 2010) and also grown as intercrop under plantation/orchard crops with the aim of crop intensification and profit maximization (Nedunchezhiyan *et al.*, 2007).

Pests and Diseases

Although sweet potato is able to produce high yield even under relatively adverse conditions, a number of pests, diseases and nutritional disorders may affect the crop. Wild pigs, rats and other herbivores also cause damage and loss in some countries (Ray and Ravi, 2005). Sweet potato weevils (*Cylas formicarius*) and virus diseases probably contribute the most to yield losses, although leaf-feeding insects such as sweet potato butterfly (*Acraea acerata*) may also cause significant losses during outbreaks. Nutritional disorders can cause slight to complete yield loss. These are the main limiting factors mostly in unfertilized crops.

Insect infestation and pathogen infections can be prevented through integrated crop management (ICM), which includes adequate cultural practices and the conservation of natural enemies to manage sweet potato pests, diseases and nutritional disorders. The cultural practices include the selection of healthy planting material from well-adapted varieties, rotation, proper sanitation practices of field and maintenance of soil fertility. Conservation of natural enemies involves avoidance of pesticide use; enhancement, *vis-a-vis* introduction of necessary natural enemies, and favourable cultivation practices.

Sweet potato weevil (wikipedia.com)

A major pest in most sweet potato growing countries is **sweet potato weevil** (*Cylas formicarius* Fab.) (Ray and Ravi, 2005) whose larvae and adult both feed on the root tubers causing extensive damage in field as well as in storage. In sweet potato weevil endemic areas, early harvest of roots (Mohanty *et al.*, 2005) and destroying the crop residues are recommended to avoid weevil damage. It would be better to remove previous sweet potato crop residues and alternate host, *i.e.*, *Ipomoea* sp. and destroy them. The integrated pest management (IPM) strategy developed by CTCRI, Kerala can effectively manage the weevil by following their recommended guidelines as furnished below. IPM practice may reduce 50-60 per cent weevil infested storage roots thus increasing more than 20 per cent storage root yield (Sethi *et al.*, 2003).

1. Before planting, dip the vine cuttings in fenthion or fentrothion 0.05 per cent solution for 10 minutes.

2. Re-ridge the crop two months after planting.
3. Install synthetic sex pheromone traps @ 1 trap/100 m^2 area to collect and kill the male weevils.
4. Destroy the crop residues after harvest by burning.
5. Control strategies recommended by CTCRI include: i) field sanitation to destroy alternate hosts; ii) selection of uninfested planting material; iii) moderation of cultural methods such as earthing up, regular irrigation, crop rotation, *etc.*; iv) spraying the crop at monthly intervals with 0.05 per cent of fenitrothion, fenthion or monocrotophos; v) soil drenching with fenthion, endosulfan, quinalphos, monocrotophos or fenitrothion 0.05 per cent at 50 and 70 days after planting.

Larvae and adults of **West Indian sweet potato weevil** [*Euscepes postfasciatus* (Fairm.)] that feed on roots and vines of the sweet potato is a serious pest, especially in the drier parts of South America, the Caribbean and the Pacific (Yasuda, 1998). The larvae tunnel the roots similar to sweet potato weevil and cause the roots producing bitter toxic terpenoid compounds making them inedible. The **sweet potato vine borer** (*Omphisa anastomosalis, Metasia coniotalis*) has been reported in India, Malaysia and China. It is considered to be as destructive as the sweet potato weevil (Rajamma and Premkumar, 1994). Its light-purple and/or yellowish-white larvae bore into the stem leading to the storage roots which ultimately causes yellowing, dying and wilting of the plant. The borers can easily be identified by presence of faecal matter on the soil surface and holes on the stem. Its management includes field sanitation including weeding especially of other *Ipomoea* spp. If infestation is severe pheromone traps are used to monitor and control the insect. However, at planting, care should be taken that planting material is quite healthy. Further such field should be left uncultivated for a few seasons. Defoliators (*Diacrisia obliqua, Enchromia polymena, etc.*), sap feeders (*Halticeis minutus*) and vine borer (*Omphisa anastamosalis*) are problematic pests in India which can be controlled through crop hygiene, crop rotation and if infestration is quite heavy then even crop may be sprayed with 0.05 per cent fenthion or fenitrothion. Hilling up is also effective as this seals the holes. However, earwigs and ants also feed on the weevil larvae developing within sweet potato vines. **White grubs** (*Phyllophaga ephilida*) are the larvae of scarab beetles (May and June beetles) feeding on main stem, roots and on tubers by making tunnels. Infested plants wilt and die. These are managed by adopting improved cultural practices including deep summer ploughing (solarization) to expose the grubs and pupae present in the soil, proper drainage to avoid waterlogging, and crop rotation with soybean which reduces grub population. Application of biocontrol agents such as *Bacillus popilliae* and *B. lentimorbus* bacteria may also kill the grubs.

Nematodes

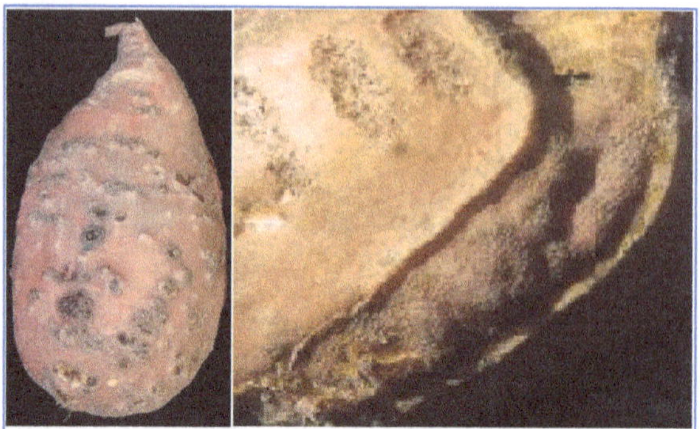
Galls and cracks caused by root knot nematodes

Storage roots and vines of sweet potato are attacked by various nematodes. Two major known nematodes in the tropics that attack fibres as well as fleshy roots of sweet potatoes and reduce yield and quality are *Meloidogyne* spp. (**root-knot**) and *Rotylenchulus reniformis* (Mohandas, 2006). In their infestation the plants become stunted, foliage turning yellowing and flowers appear abnormal. Reniform nematode (*Rotylenchulus reniformis*) when attacks fleshy roots, the roots develop cracks which enlarge as the tubers grow. Rotation is the best way to keep its population reduced. Other pathogens also penetrate through the wounds inflicted by nematodes. Nematodes can be kept under check through applying neem cake in the soil @500kg/ha at the time of last ploughing and before ridge *cum* furrow making. 'Sree Bhadra' variety from India is found to be resistant to root-knot nematode (Mohandas, 2006).

Diseases

Though diseases in sweet potato field are observed but incidence is sporadic. Fungal diseases are usually not very serious in the tropics under field conditions (Woolfe, 1992). **Stem rot** (*Fusarium oxysporum* Schlect. f.sp. *batatas*) is reported destructive in United States (Woolfe, 1992). It can be managed by selecting only disease-free root tubers for seed, using cut transplants rather than slips, practicing crop rotation and by treating the seed roots with an appropriate fungicide prior to planting. *Fusarium lateritium* causes chlorotic leaf distortion. *Cercospora batatae* and *C. ipomoeae* cause leaf spots on sweet potato. Leaf spot diseases can be kept under check by spraying the crop with dipotassium hydrogen phosphate or disodium hydrogen phosphate at 0.001 per cent sprayings 3-4 times at fortnightly intervals. **Black rot** (*Ceratocystis fimbriata*) is another fungus that may result in stunted growth, wilting and yellowing of plants, and dropping of leaves *cum* subsequent plant death. Circular brown-black patches of rot appear on tubers. Cultural practices to control or manage black rot include disease-free seed material for planting, transplanting material should be of above-ground portion and treatment of seed material should be with an appropriate fungicide prior to planting. **Leaf and stem scab** (*Sphaceloma batatas*), common in Asia and Australia, is most severe in regions with frequent fog, rain or dew accumulation. It causes small brown lesions on leaf veins which become corky in texture. The veins get shrinked and cause leaves to curl. On stem, lesions are slightly raised and have purple to brown centres with light brown margins. Scabby lesions get formed on stems after these coalesce. Cultural practices to manage it include avoiding overhead irrigation, crop rotation, use of only disease-free planting material and destroying sweet potato crop residue immediately after harvest. Application of appropriate fungicides can help to control the disease.

More than 20 viral diseases have been identified on sweet potato crop. Out of all the viruses infecting sweet potato, important ones are 'sweet potato feathery mottle virus' (SPFMV), 'sweet potato sunken vein virus' (SPSVV), 'sweet potato virus disease' (SPVD) and 'sweet potato mild mottle virus' (SPMMV). Other viruses include 'sweet potato chlorotic leaf spot virus', 'sweet potato chlorotic stunt virus', 'C-6 virus', 'sweet potato caulimo virus', 'sweet potato vein clearing virus', 'sweet potato vein mosaic virus', 'sweet potato leaf curl virus', and 'sweet potato mild mosaic virus'. Viruses are important in parts of East and West Africa but not in the rest of the world. The virus diseases can be managed through field tolerant varieties, use of virus-free planting materials as well as through meristem-cultured plants (Prasanth *et al.*, 2006). Sweet potato virus disease (SPVD) is a disease complex caused by two viruses ~ sweet potato chlorotic stunt virus (SPCSV) and SPFMV. SPFMV is transmitted by a wide range of aphid species whereas SPCSV is transmitted by white flies (*Bemisia tabaci*). The symptoms include severe stunting of infected plants and distorted *cum* chlorotic mottle or vein clearing of the leaves. SPCSV enhances the accumulation of SPFMV. The symptoms caused by SPCSV alone are negligible though symptom caused by SPFMV is mild and won't cause significant damage to the plant. Common symptoms include appearance of feathery and purple patterns on the leaves. The cultural practices to manage and control SPVD include using of disease-free planting material, crop rotation and removing *cum* burning of the infected plants. Beside these, mycoplasma may also cause little leaf where pale-yellowing and stunting of plants and leaves occur due to the career agent leafhoppers. Virus or MLO infected plants should immediately be uprooted and burnt.

Bacterial soft rot is caused by *Erwinia chrysanthemi* bacterium. Its symptoms include brown to black water-soaked lesions on stems and petioles, expanding rapidly causing large areas of soft rot on the stem, stem may collapse causing several vines to wilt and the plant may die, storage roots may develop areas of soft rots which are initially colourless but turns brown with a black margin. It can be prevented by avoiding wounds to storage roots at all stages of growth by planting only disease-free seed materials and by growing disease resistant varieties. Any storage root which becomes infected must be discarded. Also only the vines above the soil surface should be used for transplanting. **Bacterial wilt** (*Ralstonia solanacearum*) causes wilting of new sprouts and is symptomized by water-soaked bases which turn yellow-brown to dark brown in colour. Vascular system of the sprouts is discoloured brown. Infection of healthy transplants causes the lower portions of the stems to become water-soaked and turn a similar colour to infected sprouts. Yellow-brown streaks may develop inside storage roots and if infection is severe, gray-brown water-soaked lesions may be present on the root surface. The bacterium causing the disease is an important pathogen of other crops but in sweet potato it is only severe in certain regions of China where it can cause 70-80 per cent yield reduction. Quarantine procedures have been put in place in regions of China where the disease is severe. Only disease-free storage roots should be used for planting and planting should only be done in sites free of the disease. Rotating sweet potato with a flood crop such as paddy rice or a non-host such as corn or wheat can be beneficial. Growing sweet potato during cooler periods of the year allows some avoidance of the disease. **Pox** is caused by *Streptomyces ipomoeae* bacterium that favours light sandy soils and may result in plant's poor growth and reduced yield. Its symptoms are circular dark brown and corky lesions on tubers which are V-shaped in cross section, cracked and distorted tubers resembling dumbbells and rotting feeder roots. Use of resistant varieties is the best way to keep this disease under check. If resistant varieties are not available, soil should be maintained at a low pH which is unfavourable to the pathogen, and crop rotation should be practiced.

Harvest

Sweet potato is harvested from 90 to 150 days after planting depending upon the location, season and variety. With improved crop management practices, on an average, it yields 20-25t/ha of root tubers. The maturity of the roots can be determined by cutting the roots as cut surface of the immature roots exhibit a dark greenish colouration while in mature roots the cut ends dry clearly. The field is irrigated 2-3 days prior to harvesting to facilitate easy lifting of the roots. After removing the vines the roots are dug out without causing any injury (Nair, 2000). In the tropics, sweet potato is harvested usually by manual labours. Sweet potato takes about 5-6 months for maturity in North India, while only 4 months in the South. Staggered harvesting facilitates marketing to realize reasonable price for the produce. In temperate regions, sweet potatoes are most often grown on larger farms and are harvested before first frosts. To some extent the yield per hectare usually gets increased when crop remains in the field longer but danger of weevil damage increases and rots become more noticeable with age.

Postharvest

After harvest, several commercial growers cure the sweet potatoes tubers to improve storage, flavour and nutrition. Curing also allows wounds on the periderm of the harvested root to heal. Curing process generally requires drying of the freshly dug tubers on the ground in the shade but at a well aerated place from a few hours to a few days and only then tubers are stored at 29–32°C with 90-95 per cent relative humidity from 5 to 15 days. Tubers stored at 13–15°C with >90 per cent relative humidity may last well up to 13 months, however, more colder temperatures injure the roots.

Nutritive Value

Sweet potatoes have a medium to high glycemic index (GI), varying from 44 to 96. GI is a measure of speed of rise of blood sugar levels after a meal. Boiled sweet potatoes have lower GI values as compared to baked, fried or roasted. Sweet potatoes have about 80 per cent of rapidly digested starch that is quickly broken down and absorbed, increasing the GI value. Tuber consumption in large amount in a single meal may be harmful for people with type 2 diabetes due to this having high GI value. They have around 9 per cent of slowly digested starch that breaks down slowly and causes a smaller rise in blood sugar levels. They also have 11 per cent of resistant starch that is not digestible and acts as fibre ~ feeding the friendly gut bacteria. The amount of resistant starch may be slightly increased by cooling the sweet potatoes after cooking.

Cooked or baked sweet potato has around 76 per cent water, 21 per cent carbohydrates, 2 per cent protein, and almost negligible fat. Baked sweet potato in a 100 gram reference amount provides 90 calories and a very rich contents (approximately 20-25 per cent 'daily value', DV) of vitamins A, B_6 and C, manganese and potassium. Though it has higher nutrient density than cereals but provides less edible energy and protein per unit weight than cereals. Sweet potatoes are important source of carbohydrates and the starch makes up around 53 per cent of the carbohydrate content out of which simple sugars such as glucose, fructose, sucrose and maltose comprise around 32 per cent of the total carbohydrate content. Cooked sweet potatoes are relatively high in fibre as a medium-sized sweet potato tuber contains it to a tune of 3.8g. The soluble fibre in the form of pectin is around 15-23 per cent while insoluble fibre in the form of cellulose, hemicelluloses and lignin is around 77-85 per cent. Soluble fibre such as pectin may increase belly fullness hence decreases food intake and reduces blood sugar spikes by slowing down the digestion of sugars and starches. A high intake of insoluble fibres is associated with health benefits such as a reduced risk of diabetes and improved gut health. Sweet potatoes contain only

2 per cent protein out of which some 80 per cent is unique sporamins that are produced to facilitate healing whenever the plant is subjected to physical damage. Despite being relatively low in protein, sweet potatoes are an important source of this macronutrient in many developing countries. Certain varieties with dark orange flesh have more beta-carotene (a precursor of vitamin A) as compared to varieties with light coloured flesh. Beta-carotene is converted to a higher vitamin A content once digested.

Culinary Use

In Asia, sweet potatoes are mainly produced for human consumption, animal feed and to a limited extend as a raw material for industrial purposes such as starch and alcohol production. The large, starchy, sweet-tasting tuberous roots are used as a root vegetable. Though the tubers can be eaten raw but then it upsets the stomach therefore these are often cooked before consumption as this increases its nutrition and digestibility. The carbohydrate rich storage root is used as a subsidiary food after boiling, baking, roasting or frying. In some regions of India, sweet potato is roasted slowly over kitchen coals or boiled and eaten with some dressing. The boiled tubers are peeled and cut into pieces and seasoned for a vegetable dish as part of the meal. The boiled tubers can also be consumed as evening snack. In some parts of India the fresh sweet potato tuber is chipped, dried and then ground into flour that is then mixed with wheat flour and baked into *chapattis* (bread). Some Indian communities prepare sweet potato pickles and snack chips. The tuber flesh can be cut and fried similar to French fries or can be candied with brown sugar to prepare sweets.

The young shoots, leaves and vine tips are edible as a green vegetable. Older growths may be used as animal fodder. Young sweet potato leaves are also used as baby food, particularly in SE Asia and East Asia. Mashed sweet potato tubers are used similarly throughout the world.

Industrial Use

Varieties having high dry matter, starch, carotene and anthocyanin have been developed that are used in industrial applications apart from traditional usage as food and feed. Sweet potato roots are used in various countries to produce industrial products such as starch, liquid glucose, citric acid, monosodium glutamate and ethanol. It has great potential as a feedstock for bio-ethanol production. However, to exploit it as bio-fuel crop, varieties with high yields and dry matter content termed as industrial sweet potatoes (ISP) have to be developed. North Carolina State University (NCSU) has developed varieties which have as high as $67.8 gl^{-1}$ of ethanol for flour-based fermentation and $34.9 gl^{-1}$ for fresh sweet potato sugar fermentation (Santa-Maria *et al.*, 2006).

Ornamental Use

Ornamental sweet potatoes are popular landscape, container and bedding plants. These ornamental cultivars are not poisonous. Cultivars are available in many colours such as green, yellow and purple. Some ornamental varieties such as 'Blackie' flower more than others. Although the leaves are edible, the tubers do not have a good taste. Sweet potato vine can also be used in home aquaria, trailing out of the water with its roots submerged. It is a sort of aquaponic system where the waste product of aquatic life, *i.e.* toxic ammonia and nitrates are absorbed by the submerged roots as nutrients thus purifying the water and improving the living conditions for fish or other aquatic animals, *vis-a-vis* providing refuge in their extensive root systems. The vines rapidly form roots in water and can grow in it indefinitely in good lighting with a steady supply of nutrients.

Other Uses

The juice of red sweet potatoes can be combined with lime juice to make a cloth-dye. By varying the proportions of its juices, every shade from pink to black can be obtained. Purple-fleshed sweet potatoes contain natural soluble food pigments called anthocyanin. Anthocyanins have applications in pharmaceutical and cosmetic industries due to its bright colour, non-poisonous nature, rich nutrition *cum* safe and health care function (Rice-Evans and Packer, 1998). Anthocyanins from purple sweet potatoes have high heat and light stability (Cuevas Montilla *et al.*, 2011). Sweet potato anthocyanins are effective natural food colourants for preparation of beverages. These are used in juices, alcoholic beverages, jams, confectioneries, bread, snacks and noodles. Recent findings of the radical scavenging, antimutagenicity and efficacy against liver disease of sweet potato anthocyanins (Terahara *et al.*, 2004) indicated that purple-flesh sweet potato may contribute to maintaining good health of human beings. The anthocyanins were proved to be the active component in memory enhancement (Cho *et al.*, 2003). Recently, Japanese markets have coupled anthocyanin extracts from purple-fleshed sweet potato with cordyceps mushroom extracts as a memory enhancing dietary supplement.

References

1. Akinrinde, E.A. 2006. Phosphorus fertilization effect on dry matter production and biomass partitioning in sweet potato (*Ipomoea batatas*) grown on an acidic loamy-sand Alfisol. *J. Food Agri. Environ.*, **4**(3-4): 99-104.
2. Anon. 1987. Cultural practices of sweet potato, 8pp. *Extension Bulletin.* Central Tuber Crops Research Institute, Thiruvananthapuram.

3. Anon. 2003. Annual Report 2002-2003, 92pp. Central Tuber Crops Research Institute, Thiruvananthapuram.

4. Austin, D.F. 1988. The taxonomy, evolution and genetic diversity of sweet potatoes and related wild species. In: Exploration, Maintenance, and Utilization of Sweet Potato Genetic Resources (ed. Gregory, P.). First Sweet Potato Planning Conference, 1987. International Potato Center, Lima, Peru, pp. 27–60.

5. Bhuiyan, M.K.R., M. Hossain, M.A.E. Hussain and S. Attaluri, 2006. Research and development activities of sweet potato in Bangaladesh, pp. 1-17. In: *14th Triennial Symp. Intern. Soc. Trop. Root Crops*, held on 20-26 Nov. 2006 at CTCRI, Thiruvananthapuram (India).

6. Biswal, S. 2008. Response of Sweet Potato (*Ipomoea batatas* L.) to Irrigation and Fertility Levels. Ph.D. Thesis, Orissa Agricultural University and Technology, Bhubaneswar, India

7. Byju, G. and M. Nedunchezhiyan, 2004. Potassium: A key nutrient for higher tropical tuber crops production. *Fert. News*, **49**(3): 39-44.

8. Cho. J., J.S. Kang, P.H. Long, J. Jing, Y. Back and K.S. Chung, 2003. Antioxidant and memory enhancing effects of purple sweet potato anthocyanin and cordyceps mushroom extract. *Arch. Pharm. Res.*, **26**: 821-5.

9. Choudhary, S.H., S.U. Ahmed and A.F.M. Sharafuddin, 1986. Effect of number of nodes in different types of vine cuttings on the growth and yield of sweet potatoes. *Bangladesh Hort.*, **14**(1): 29-33.

10. Cuevas Montilla, E., S. Hillebrand and P. Winterhalter, 2011. Anthocyanins in purple sweet potato (*Ipomoea batatas* L.) varieties. *Fruit, Vegetable and Cereal Science and Biotechn.*, **5** (Special Issue 2): 19-24.

11. Dasgupta, M., P.C. Kole, M.R. Sahoo and A. Mukherjee, 2006. Screening of sweet potato genotypes for salinity stress. In: *14th Triennial Symp. Inter. Soc. Trop. Root Crops*, pp. 166-7. CTCRI, Thiruvananthapuram (India).

12. Fan, G., Y. Han, Z. Gu and D. Chen, 2008. Optimizing conditions for anthocyanins extracted from purple sweet potato using response surface methodology (RSM). *LWT Food Sci. Tech.*, **41**: 155-60.

13. Hartemink, A.E. 2003. Integrated nutrient management research with sweet potato in Papua New Guinea. *Outlook on Agriculture*, **32**(3): 173-82.

14. Kabeerathumma, S, B. Mohankumar and V.P. Potty, 1986. Efficacy of rock phosphate as a source of phosphorus to sweet potato. *J. Indian Soc. Soil Sci.*, **34**(4): 866-9.

15. Kundu, D.K., R. Singh and S. Roy Chowdhury, 2006. Effect of rice straw mulch and irrigation on nutrient availability in soil and tuber yield of sweet potato (*Ipomoea batatas* L.) in coastal Orissa. In: Root and Tuber Crops: Nutrition, Food Security and Sustainable Environment (eds Naskar, S.K., M. Nedunchezhiyan, K. Rajasekhara Rao, P.S. Sivakumar, R.C. Ray, R.S. Misra and A. Mukherjee), pp. 117-22. Regional Centre of Central Tuber Crops Research Institute, Bhubaneswar.

16. Mishra, S., S.S. Mishra, P.K. Ray and S.J. Singh, 1985. Comparison between single and double harvesting of sweet potato. In: *Proc. Natl Symp. Prod. Util. Trop. Root Crops*, pp. 135-7; held on Nov. 27-28 at Trivandrum (India).

17. Mohandas, C. 2006. Nematode problems in tuber crops. In: *14th Triennial Symposium of International Society of Tropical Root Crops*, pp. 150-51, held Nov. 20-26, at CTCRI, Thiruvananthapuram (India).

18. Mohanty, A.K., K. Sethi, S. Samal, S.K. Naskar and M. Nedunchezhiyan, 2005. Relationship of optimum stage of harvest with weevil incidence in sweet potato under different agroclimatic conditions of Orissa. *Orissa J. Hort.*, **33**(1): 43-5.

19. Monday, T. 2009. Industrial Sweet Potato a Viable Biofuel Crop for Alabama. M.Sc. Thesis, Aub. University, Alabama (Canada).

20. Mukherjee, Archana. 2019. Sweet Potato. In: Handbook of Horticulture, vol. 1, pp. 667-75. ICAR, New Delhi.

21. Nair, G.M. 2000. Cultural and manurial requirements of sweet potato. In: *Production Technology of Tuber Crops* (eds Mohankumar, C.R., G.M. Nair, James George, C.S. Ravindran and V. Ravi). CTCRI, Thiruvananthapuram, Kerala (India), pp 44-64.

22. Nair, G.M. 2006. Agro-techniques and planting material production in sweet potato. *Quality Planting Material Production in Tropical Tuber Crops* (ed. Byju. G.), CTCRI, Thiruvananthapuram (India), pp 55-8.

23. Nair, G.M., V.M. Nair and C. Sreedharan, 1996. Response of sweet potato to phasic stress irrigation in summer rice fallows. *J. Root Crops*, **22**(1): 45-9.

24. Narin, P. and S. Reungmaneepaitoon, 2005. Breeding of sweet potato for health-promoting attributes in Thailand. In: *Concise Papers Second Int. Symp. Sweet Potato and Cassava*, held 14-17 June 2005 at Kuala Lumpur, Malaysia, pp 69-70.

25. Nath, R., C.K. Kundu, A. Majumder, S. Gunri, A. Chattopadhyay and H. Sen, 2006. Productivity of sweet potato as influenced by cultivar, season and staggered harvesting in laterite ecosystem of West Bengal. In: *14th Triennial*

Symp. Intern. Soc. Trop. Root Crops, held on 20-26 Nov. 2006, CTCRI, Thiruvananthapuram (India), pp. 213.

26. Nedunchezhiyan, M. and G. Byju 2005. Effect of planting season on growth and yield of sweet potato (*Ipomoea batatas* L.) varieties. *J. Root Crops*, **31** (2): 111-4.
27. Nedunchezhiyan, M., G. Byju and S.K. Naskar, 2007. Sweet potato (*Ipomoea batatas* L.) as an intercrop in a coconut plantation: growth, yield and quality. *J. Root Crops*, **33**(1): 26-9.
28. Nedunchezhiyan, M, K. Rajasekhara Rao, K. Laxminarayana and B.S. Satapathy, 2010. Effect of strip intercropping involving sweet potato (*Ipomoea batatas* L.) on soil moisture conservation, weevil infestation and crop productivity. *J. Root Crops*, **36**(1): 53-8.
29. Nedunchezhiyan, M. and R.C. Ray, 2010. Sweet potato growth, development, production and utilization: overview. In: Sweet Potato: Post Harvest Aspects in Food (eds Ray, R.C. and K.I. Tomlins,), pp. 1-26. Nova Science Publishers Inc., New York (USA).
30. Nedunchezhiyan, M. and D.S. Reddy, 2006. Effect of time of planting and varieties on growth and yield of sweet potato (*Ipomoea batatas* L.) under rainfed conditions. In: *Root and Tuber Crops: Nutrition, Food Security and Sustainable Environment* (eds Naskar, S.K., M. Nedunchezhiyan, K. Rajasekhara Rao, P.S. Sivakumar, R.C. Ray, R.S. Misra and A. Mukherjee), pp. 123-7. Regional Centre of CTCRI, Bhubaneswar (India).
31. Ngoan, T.N. 2006. Status of root crop production, utilization and marketing in Vietnam. In: *14th Triennial Symp. Int. Soc. Trop. Root Crops*, 9pp, held on 20-26 Nov. 2006, CTCRI, Thiruvananthapuram (Kerala), India.
32. Nyiawung, K.Z., D.G. Mortley, A. Issah, C.K. Bonsi, W.A. Hill and B.T. Vaughan, 2010. Evaluation of ten sweet potato cultivars as potential feedstock for biofuel production. *J. Root Crops*, **36**(2): 242-9.
33. Philpott, M, K.S. Gould, C. Lim and L.R. Ferguson, 2004. *In situ* and *in vitro* antioxidant activity of sweet potato anthocyanins. *J. agri. Food Chem.*, **52**: 1511-3.
34. Prasanth, G., V. Hegde, T. Makeshkumar, M.L. Jeeva and S. Edison, 2006. Development of diagnostics for sweet potato feathery mottle virus. In: *14th Triennial Symposium of International Society of Tropical Root Crops*, 145 pp, held on 20-26 Nov. 2006, CTCRI, Thiruvananthapuram (Kerala), India.

35. Rajamma, P. and T. Premkumar, 1994. Pests of sweet potato. In: Advances in Horticulture: Tuber Crops, vol. 8 (eds Chadha, K.L. and G.G. Nayar), pp. 457-84. Malhotra Publishing House, New Delhi.

36. Ravi, V. and P. Indira, 1996. Anatomical studies on tuberization in sweet potato under water deficit stress and stress-free conditions. *J. Root Crops*, **22:** 105-111.

37. Ravindran, C.S. and C.R. Mohankumar, 1985. Effect of method of land preparation and depth of planting on the yield of sweet potato. In: *Proc. Nat. Symp. Production and Utilization of Tropical Tuber Crops*, pp. 129-33, held on Nov. 27-28 at CTCRI, Trivandrum (India).

38. Ravindran, C.S. and C.R. Mohankumar, 1989. Effect of storage life of vines with and without leaves on the establishment and root yield of sweet potato. *J. Root Crops*, **15**(2): 145-6.

39. Ray, P.K., S. Mishra and S.S. Mishra, 1983. Sweet potato productivity as affected by recurrent use of vine as planting material. *South Indian Hort.*, **31** (6): 280-83.

40. Ray, R.C. and V. Ravi, 2005. Post harvest spoilage of sweet potato and its control measures. *Crit. Rev. Food Sci. Nutr.*, **35:** 623-44.

41. Rice-Evans, C. and I. Packer, 1998. Flavonoids in Health and Disease, 504p. Marcel Dekker, New York.

42. Rodgers, J.A., M.K. Bomford, B.A. Geier and A.F. Silvernail, 2007. Evaluation of Alternative Bioethanol Feedstock Crops. Kentucky Academy of Science, Louisville, KY, USA.

43. Sakamoto, S., S. Marumine, Y. Ide and O. Yamakawa, 1987. 'Benihayato': A new sweet potato cultivar registered. *Bull. Nat. agric. Res. Cent. Kyushu Okinawa Region*, **24**: 255-77.

44. Santa-Maria, M.C., C.G. Yencho, W.F. Thompson, R. Kelly and B. Sosinski, 2006. Genetic engineering of sweet potato for endogenous starch bioprocessing. *Am. Soc. Pl. Biol.* Available online: http: //abstracts.aspb.org/pb2006/public/P46/P46020.html.

45. Sebastiani, S.K., A. Mgonja, F. Urio and T. Ndondi, 2006. Response of sweet potato to application of nitrogen and phosphorus fertilizer, agronomic and economic benefits in the Northern highlands of Tanzania. In: *14th Triennial Symposium of International Society of Tropical Root Crops*, 205pp., held on 20-26 Nov. 2006 at CTCRI, Thiruvananthapuram (India).

46. Sethi, K., A.K. Mohanty, S.K. Naskar and M. Nedunchezhiyan, 2003. Efficacy of IPM technology over farmers/traditional practices in management of sweet potato weevil (SPW) in tribal areas of Orissa. *The Orissa J. Hort.*, **31**(2): 65-8.

47. Terahara, N., I. Konczak, H. Ono, M. Yoshimoto and O. Yamakawa, 2004. Characteristics of acylated anthocyanin in callus induced from storage root of purple-fleshed sweet potato, *Ipomoea batatas* L. *J. Biomed. Biotechn.*, **5**: 279-86.

48. Vimala, B., R. Thushara, B. Nambisan and J. Sreekumar, 2006. Evaluation of orange-fleshed sweet potatoes for higher yield and carotene content. In: *14th Triennial Symp. Intern. Soc. Trop. Root Crops*, 55pp., held on 20-26 Nov. 2006, at CTCRI, Thiruvananthapuram (Kerala), India.

49. Woolfe, Jennifer A. 1992. Sweet Potato: An Untapped Food Resource. Cambridge University Press, Cambridge, UK, 643 pp.

50. Yamakawa, O. 1997. Development of new cultivation and utilization system for sweet potato toward the 21st century. In: Proc. Intern. Workshop on Sweet potato Production System toward the 21st Century, pp. 1-8, held on December 9-10, 1997, at Miyakonojo, Miyazaki, Japan.

51. Yasuda, K. 1998. Integrated control of sweet potato weevils *Euscepes postfasciatus* and *Cylas formicarius* (Coleoptera: Curculionidae). In: *Proc. intern. Workshop on Sweet potato Production System toward the 21st Century*, pp. 317-22, held on Dec. 9-10, 1997, at Miyakonojo, Miyazaki, Japan.

26

Tannia [*Xanthosoma sagittifolium* (L.) Schott]

Family: Araceae.

Other common names: American taro, Arrowleaf elephant ear, Cocoyam, Malanga, New cocoyam, Tanier and Yautia.

Synonyms: *Alocasia talihan* Elmer ex Merr., *Arum sagittifolium* L., *A. xanthorrhizon* Jacq., *Caladium edule* G.Mey., *C. mafaffa* Engl., *C. sagittifolium* (L.) Vent., *C. utile* Engl., *C. xanthorrhizon* (Jacq.), *Xanthosoma appendiculatum* Schott, *X. atrovirens* K.Koch and C.D.Bouche, *X. blandum* Schott, *X. edule* (G.Mey.) Schott, *X. ianthinum* K.Koch and C.D.Bouche, *X. jacquinii* Schott, *X. mafaffa* Schott, *X. nigrum* Stellfeld, *X. peregrinum* Griseb., *X. poeppigii* var. *mafaffa* (Schott) J.F.Macbr., *X. roseum* Schott, *X. utile* K.Koch and C.D.Bouche, *X. violaceum* Schott and *X. xantharrhizon* (Jacq.) K.Koch.

Introduction and Origin

The name *Xanthosoma* derives from Greek words *xanthos* meaning 'yellow' and *soma* meaning 'body', referring to the stigma or the inner yellow tissues of some species (Quattrocchi, 2000). Specific epithet means arrow-leaved. In parts of Africa, both tannia and taro (*Colocasia esculentum*) are known as cocoyam, and in a few other places both are listed as taro. In West Africa, *Colocasia* spp. is called 'old cocoyam/yam' and *Xanthosoma* spp. is called 'new cocoyam/yam'. The common name 'new cocoyam' reflects the late introduction of this species into areas where *Colocasia* (cocoyam) was previously established. Elephant-ear is also used for some other large-leaved genera in the Araceae such as *Alocasia*, *Caladium* and *Colocasia*. In Manipur (India), it is popularly known as 'Yendem Amubi'. It grows best in humid tropical rainforest climates and can be found naturalized along stream banks and in moist shady areas (Manner, 2011).

Xanthosoma sagittifolium, often confused with taro (*Colocasia esculenta*), is a tropical aroid that possibly originated in the wet valleys of NE South America though naturalized and widely cultivated in other tropical regions. It became an important food crop in most of the Caribbean islands, Florida (USA), western Africa, parts of Egypt, India and Oceania (O'Hair and Maynard, 2003). This pantropical species has been domesticated in most communities in Oceania, Africa and Asia (Ramanatha, *et al.*, 2010). Africa is the major producer with W and C Africa notably Nigeria, Ghana and Cameroon contributing to over 60 per cent of the total African production. In India, *Xanthosoma* is grown in the states of West Bengal, Odisha, Kerala, Andhra Pradesh, Tamil Nadu, Maharashtra, Karnataka, Asom and NE states including Manipur. In NE states it is a major edible crop.

Several *Xanthosoma* species, especially *X. sagittifolium*, are grown for their starchy rhizomes and/or leaves *cum* tender shoots. Most other species such as *X. roseum* are used as ornamental plants though some species such as *X. brasiliense* are occasionally grown solely for their edible leaves that can be cooked as spinach. Being a very important staple food crop in the tropics, tannia is the second most widely cultivated edible species in the family Araceae after taro and eddo (*Colocasia esculentum*) and is probably one of the world's oldest crops. It is the sixth most consumed tuber crop in the world after potatoes, cassava, sweet potatoes, yams (*Dioscorea*) and taro (*Colocasia*). Though the young leaves, petioles and inflorescence are also edible, tannia is most valued for its edible starchy rhizome or rhizome offsets that can be cooked and eaten similar to potatoes.

X. sagittifolium, though widely cultivated for underground stems, is also included in the 'Global Compendium of Weeds' (Randall, 2012). It is currently listed as invasive in Florida, Puerto Rico, Costa Rica, the Galapagos Islands, Micronesia and French Polynesia (Anon., 2008; Acevedo-Rodríguez and Strong, 2012; Chacon and Saborio, 2012; Anon., 2013).

In recent years, tannia's production has increased significantly particularly in parts of West Africa due to its greater resistance to *Phytophthora* blight compared to taro and also as it requires no exacting conditions for its cultivation. Tannias being shade-tolerant are of particular importance for intercropping with plantation crops such as cocoa, rubber, banana/plantain, oil palm or coconut, or in the homesteads mainly in association with banana/plantains, cassava, sweet potato, yams and other local vegetables (Susan John, 2019). Its future probably depends upon whether costs of production can be reduced by improved cultural techniques such as mechanisation, use of herbicides, *vis-a-vis* fertilisers. High degree of genetic variability suggests that there is considerable potential for improvement of this crop through selection and breeding.

Botany

Tannia is a fast-growing plant spreading through slender rhizomes and exuding milky sap when cut. This species can form mature plants from rhizomes within 14-20 weeks. Once established, mature plants can produce large amount of foliage in the first 6-9 months and may also produce up to or more than 10 rhizomes within 10 months (Langeland *et al.*, 2008).

Tannia is a stout, robust and glabrous herbaceous perennial reaching up to or more than 1.5-2.5m height. These plants are acaulescent when young, though mature plants have up to 1m long, thick, erect and fleshy stem with numerous leaf scars and sometimes with aerial roots; and the enlarged *cum* ovoid base which produces laterally subterranean elongated edible rhizomes. Several large, long-stalked, sagittate or hastate (narrow-triangular) and elephant-ear-like leaves are about 40-75cm long (sometimes up to 1.2m long and 40-90cm wide) having a prominent marginal vein. Stiff, ribbed, erect, green and 1-1.5m long petioles are invaginate on lower $2/3^{rd}$ part with straight, wavy or sometimes involute margins. The tannia leaf differs from those of *Colocasia* in that the leaf stalk joins the blade at the margin between the lobes (not into the surface of the blade), and the tips of the lobes are pointed and not rounded though in *Colocasia* the petiole attachment is peltate or more to the middle-leaf (Manner, 2011).

Infrequently produced flower is pale green to greenish-white and up to 13-15cm or sometimes even with longer spathe, enclosing a white spadix which is slightly shorter than the spathe. Fruit is a small yellow berry. Some cultivars never flower and seed is rarely produced. A starchy rhizome produced at the base of the plant bears usually 10 or more 10-25cm long offsets, weighing up to 1kg each. It has chromosome number 2n=26.

Offset rhizomes of different varieties of *Xanthosoma sagittifolium*

Peeled rhizomes of *X. sagittifolium*

Source: U.S. National Library of Medicine; https://www.ncbi.nlm.nih.gov.

Like many species in the family Araceae, this plant has the ability to generate heat in the spadix as the pollen becomes ready for fertilization. This heat greatly increases the strength of the aroma released by the plant thus attracting more pollinating insects. It can also have the effect of making the insects more active thus increasing the level of fertilization (Bown, 2000).

Varieties

Most varieties currently cultivated are believed to be landraces that are generally distinguished by the flesh colour of the rhizome offsets (red, purple to pink, white and yellow). Varieties with corresponding names in their local dialects are being used for their identification (Doku, 1966; Onokpise *et al.*, 1999). Of the few tannia varieties, most have white flesh. Some authors sometimes classify the less common pink and yellow-fleshed types as separate species ~ *X. violaceum* Schott and *X. atrovirens* C. Koch, respectively. The same clone may have more than one common names. Common **white-fleshed varieties** in the English speaking Caribbean are 'South Dade White', 'Bruce' and 'Rabess' whereas **purple- to pink-fleshed varieties** are 'Vinola' and 'Jamaique'. Although the main mother rhizome of the **yellow-fleshed varieties** is edible, it is not the edible part of white- and purple to pink-fleshed varieties and it is occasionally boiled for animal feed. The edible portions of the white- and purple to pink-fleshed types are the underground club-shaped offsets (*i.e.* lateral rhizomes) of the main rhizome. Young leaves and tender shoots are occasionally cooked as leafy green vegetables. Cultivars with purple stems or leaves are also variously called **blue taro**, **purple taro**, **purple-stem tannia** and **purple elephant's ear** among others. 'Konkan Haritpani' from Maharashtra is the only properly released variety of tannia in India which is well adopted in Konkan and adjoining areas as a rainfed crop maturing in 190-210 days.

Nutritional Value

Rhizomes or lateral offsets of cocoyam comprise mostly of starch which is highest among all the edible aroids. It is postulated to have superior nutritional value over other major root and tuber staples of West Africa, especially in terms of their protein digestibility and mineral composition (Ca, P and Mg) (Chukwu *et al.*, 2008; Ekwe *et al.*, 2009; Lim, 2015, 2016; Opara, 2003). Rhizome is an appreciable source of dietary energy, proteins and vitamins (Opara, 2003) and is high in potassium, zinc and nicotinic acid as well as a low inhibitor of trypsin compared to other edible aroids (Bradbury and Holloway, 1988). Cocoyam leaves are known to have appreciable levels of antioxidants, vitamins and dietary fibre (Ekwe *et al.*, 2009; Lebot, 2009) but these potential health benefits have barely been explored in the food and pharmaceutical industries.

Average Composition (per 100g of raw product) of Leaves of Edible Aroid Tannia and Taro Leaves

Constituent	*Tannia*	*Taro*
Moisture (g)	90	90
Carbohydrate (g)	5.3	5.7
Protein (g)	2.5	2.4
Fat (g)	1.0	0.6
Fiber (g)	2.1	1.5
Ash (g)	1.3	1.3
Calcium (mg)	95	98
Phosphorus (mg)	388	49
Iron (mg)	2.0	2.0
Vitamin A (IU)	3300	180
Ascorbic acid (mg)	37	11

Source: Leung, Burtrum and Chang (1972).

Toxicity in Aroids

All parts of the plant contain antinutrients, predominantly calcium oxalate crystals which are toxic when fresh (Ramanatha *et al.*, 2010; Sefa-Dedeh and

Nutritional Profile of Cocoyam Varieties from different Geographical Areas

Parameter/ Source	*Mean Values of 37 Samples from Tonga and Papua New Guinea (rhizome offsets) (Bradbury and Holloway, 1988)*	*Mean Values from E. Africa (Tanzania and Uganda (Ndabikunze et al., 2011)*	*Values of Indigenous Varieties from Nigeria (Rhizome offsets) (Odebunmi et al., 2010)*	*Mean Values of different Portions (Apical, middle and distal) of Rhizome-offsets from Red and White Varieties of Cocoyam in Ghana (Sefa-Dedeh and Agyir-Sackey, 2004)*		*Mean Values from Cameroonian Accessions (Opara, 2003)*		
	Proximate (per cent)	*Proximate (g/100g)*	*Proximate (per cent)*	*Proximate (g/100g)*		*Rhizome-offsets*	*Leaves*	*Shoots*
				Red	*White*			
Moisture	67.1	68.41	80.99	67.83	63.33			
Energy kJ 100 per g	521					133	34	33
Protein	1.55	4.75	5.47	4	5.12	2.0	2.5	3.1
Carbohydrate		20.95				31	5	5
Starch	27.6		11.03	22.83	27.2			
Sugar	0.42							
Crude fibre	0.99	1.96	1.28	1.54	1.39	1	2.1	3.2
Fat	0.11	0.43	0.2	0.59	0.43	0.3	1.6	0.6
Ash	1.04	3.51	1.03	3.2	2.55			
Minerals (mg/100g)								
Calcium	8.5	110.17		13.51	12.15	20	95	49
Phosphorous	53	207.50		53.23	45.7	47	388	80
Magnesium	27	90.62		73.1	63.57			

Contd.

Sodium	6.6	23.98		22.13	36.1			
Potassium	530	908.25		1248.33	1112			
Sulfur	7.9							
Iron	0.4	4.54		3.05	3.53	1	2	0.3
Copper	0.19	0.63						
Zinc	0.52	2.72		32.6	41.9			
Manganese	0.17	1.95						
Aluminum	0.53							
Boron	0.09							
Vitamins (mg/100g)								
Vitamin A (ret. + -car./6)	0.005					Trace	3.3	
Thiamin	0.024					0.1		
Riboflavin	0.032					0.03		
Nicotinic acid	0.8					0.5		
Ascorbic acid						10	37	82
Pot. Nic. Acid = Trp/60	0.33							
Total vitamin C (AA + DAA)	14							

Blank space indicates data not provided by author(s).

Source: U.S. National Library of Medicine; https://www.ncbi.nlm.nih.gov.

Agyir-Sackey, 2004) though calcium oxalate is easily broken down either by thoroughly cooking the plant or by fully drying it, making the plant safe for consumption. People with a tendency to rheumatism, arthritis, gout, kidney stones and hyperacidity should take special precaution if including this plant in their diet (Brown, 1995).

Uses

Although all parts of tannia plant are edible, none of the parts are consumed raw because of its acridity. Rhizomes, especially offsets can be boiled, baked, fried and cooked before consuming similar to those of potatoes and taro. The mother rhizome of yellow-fleshed varieties is edible whereas white and purple-to pink-fleshed varieties are mostly used to feed animals after boiling or as with other offset-rhizome of tannia are peeled, dried and ground to produce flour. Smaller offsets can be boiled with skin and served with sauce or chutney where before eating the skin is removed. They can be boiled, mashed and added to soups. They are mainly used as food in cultivated areas, and the plant parts are also used as fodder/feed and medicine, including its use as antipoisonous agents against tarantula, scorpion and snake bites. Young leaves and petioles can be cooked and eaten similar to spinach. In Manipur (India), stem of tannia is used to prepare stew and soup mainly for women in post-natal care to prevent anaemia. *X. sagittifolium* possessess antimicrobial, antioxidant, anti-diabetic and hypolipidaemic activities.

Industrial Processing

Tannia leaves having size of 20-30 × 20-30cm are usually made into rolls. The bulkiness *cum* short shelf life of tannia rhizomes are major challenges to utilization. Moreover, processing of tannia is currently limited to freezing of peeled offsets and bagging them in consumer packs in the form of a ready to cook frozen food though there are many potential uses similar to those described for taro. Industrial processing of rhizomes into flour may be a viable economic outlet for storage and further use (Lebot, 2009; Ramanatha *et al.*, 2010). The flour could be used in the snack and complementary (baby) food sector to develop ready to eat healthy snacks and complementary foods. These could also serve the general purpose market as improved soup/sauce/gravy thickeners (Falade and Okafor, 2014). Rhizomes can also be processed into ready to prepare chips and crisps similar to potatoes. The rhizomes also have the potential to be used in the beverage industry as adjuncts (Eneh, 2013).

The self life of leaves can be enhanced by processing them into purées or flours to be used in preparation of soups, sauces and similar products.

Starch is occasionally made from the grated offset-rhizomes. The starch grains of tannia are relatively large with average diameter of 17-20 microns though less easily digested than those of *Colocasia*. Some authorities have reported good properties of starches from West African cocoyam varieties (Falade and Okafor, 2013; Sefa Dedeh and Agyir Sackey, 2002) though economically it may not be viable to process for industrial use because this cannot compete with starch yield of cassava which is more abundant staple with higher starch yield for industrial application.

Propagation

Xanthosoma rarely flowers so in cultivation seeds are rarely produced. However, when flowering also occurs, for evolving new varieties, it requires hand pollination for seed production. Formal breeding programmes in West Africa have focused on developing root rot resistant genotypes. Tannia can be vegetatively propagated from small rhizomes (offsets), 5cm section top cut setts from larger/mother rhizome with few buds, and by separating new plantlets formed at the rhizome tips (offshoots). Healthy, unbruised and disease-free uniform-sized planting material should be selected for producing new plants. In the home gardens, usually it is grown as ratoon crop for a period of 5-6 years where every year the rhizomelets are separated for home consuption. In parts of Maharashtra and Gujarat only the leaves are harvested every year and the rhizome and rhizomelets are left as such in the field for further perpetuation so these become there as community plants. In Tamil Nadu, they grow it under irrigated condition both as sole crop and as intercrop in coconut garden for a period of 10-12 months. Usually they use the planting materials as sprouts emerging from rhizomes after 1-3 months of harvest and these are kept under shade with irrigation until planting. In the main field these are transplanted at a spacing of 60×60cm.

Cultural Practices

Climate and Soil

It is a crop of 270-300 day's duration in the field though most cultivars mature usually within 190-210 days. Tannias can be grown on a wide variety of soils except hard clays and pure sandy soils, having pH range of 5.5 to 6.5 though can tolerate 4.5-7.8 pH range. For optimum yield these require a well-drained, deep and fertile soil with a pH of about 4.8 and 20 per cent exchangeable Al^{3+} provided the soil is rich in organic matter (Tindall, 1983; Susan John, 2019). Waterlogged soils are at all not suitable for tannia cultivation. The crop grows best at 1,350-1,500m a.s.l. in humid and tropical conditions, *i.e.* high rainfall areas with an evenly distributed annual rainfall in the range of 1,000-2,000mm, however, it can easily tolerate 750-6,000mm. Tannias can also be grown as an upland crop under irrigation though certain early-maturing cultivars can even be grown without irrigation in comparatively dry situations such as exposed steep slopes. It can easily be found naturalized along stream banks and in moist shady areas (Manner, 2011). Tannia is often restricted to lowland areas as here climatic conditions favour rapid growth of plants. It grows best at elevations below 800m, though successful crops have been produced at an elevation up to 1,900m (Tindall, 1983) provided a suitable cultivar is selected. Its preferred mean annual temperature is within the range of 20 to 28°C though it can tolerate the range of 10 to 35°C. It feels always comfortable in full sun (Tindall, 1983) though certain varieties adapt well to grow in shaded conditions under coconut, cocoa, banana or coffee plantations (Manner, 2011).

Though environmental factors such as temperature, moisture, light and soil type influence the growth and quality, other factors such as soil type, preparation of land, varieties, quality of planting material *cum* planting method, planter operation, planting cum harvesting season, irrigation, nutrition, pest management, *etc.* are equally responsible and can be controlled by the grower. Most profitable yields can be attained only when all the factors are at optimum levels.

Soils, Land Preparation, Manures-*cum*-Fertilizers and Planting

Soil preparation and fertilizer use for tannia are almost same as for taro. Susan John (2019) mentions that dolomite (soil amendment), cattle manure and NPK @ 1t/ha, 25t/ha and 80:50:150kg/ha, respectively is recommended for tannia cultivation. Dolomite is applied at the time of land preparation, and just before planting the soil is thoroughly mixed with cattle manure and full dose of phosphorus. Of the total nitrogenous and potassic fertilizers, each of 1/3rd should be applied at 60, 120 and 180 days after transplanting. In the interspaces, together with tannia planting cowpea as green manure @ 20kg/ha should be sown. To bring the soil to the fine tilth before planting, land must be ploughed 2-3 times to a depth of 20-40 cm, followed by harrowing and planking. Any existent weed should be removed. Field should be levelled and beds as per the convenience should be made. Planting is often done on ridges. If the planting materials are offsets they are planted 7.5-10cm deep with the growth buds pointing downwards, however, if pieces of main rootstock are used, about 2.5cm is left above the ground.

Aroids such as tannia and taro (*Colocasia esculenta*) can be grown throughout the year mostly in subtropical and tropical regions though in drier areas (100cm rainfall) it is usually planted just before or at the start of the rainy season as these require abundant of water. Usually it is grown from sea level up to 1,350-1,500m with a wide temperature range (13-29°C), however, the mean soil temperature being above 20°C. It is less tolerable to drought, *vis-à-vis* it does not like waterlogging at any time. Though not much but it can tolerate light shading. Required photoperiod for tuberization is less than 12h. Though the interplant spacing is variable, the most commonly adopted one is around 100×100cm. However, in practice it is from 60×60cm to 1.8×1.8m. The wider spacing gives a higher yield per plant though 1.5×1.5m has been reported giving maximum rhizome yield per hectare. Wider spacings involve increased maintenance due to more land requirement, *vis-a-vis* it takes longer time for leaves to cover the area between plants and rows so weeds become menacing and their removal involves more labours. Narrow-spaced plants cover entire bed area within a short time through their growth so weeds do not get a chance to grow so vehemently and flourish. These spacing distances refer only to tannia grown in monoculture as there in no any general practice for mixed cultivations.

Root formation and its rapid growth takes place immediately after planting, followed by rapid shoot growth. Rhizomes usually mature in 9-12 months after planting though the crop can sometimes be obtained after 6 months of planting, however, delayed harvesting assures higher yield. Rhizomes can remain viable underground and survive through unfavourable environmental conditions such as drought. Rhizomes can also be stored for more than 18 weeks by storing under dry conditions though unplanted offsets can sprout within a few weeks under hot *cum* humid conditions (Langeland *et al.*, 2008; Manner, 2011). Tannia is often grown there where soil moisture is too low for successful cultivation of *Colocasia* but production is most successful in areas of high air humidity.

Mulching is quite profitable in this crop and 20-40t/ha of FYM is recommended for taking a good yield. There is little knowledge about mineral fertilising both in Puerto Rico and in the Pacific Islands. **Nitrogen** and **potassium** both at 100kg/ha in split applications have given good results.

Weed control is extremely important, especially for young plantings that are especially susceptible to weed competition. If the planting is done on level ground, sometimes earthing up is needed. The application of the pre-emergence herbicide diuron at 1.6 kg/ha has been recommended; other effective herbicides are atrazine, prometryn and ametryn.

Depending on the areas and adapted farming methods, tannia can be grown in monoculture, though it is usually grown in crop **rotation system**, oftenly as the first crop in shifting agricultural systems. It can also be intercropped with plantation crops such as cocoa, rubber, bananas and coconuts.

Harvest

When the crop is grown for greens, leaf harvesting usually starts from third month onwards at fortnightly intervals till the younger leaves remain succulent and green. Some 40-50 leaves are harvested from one plant during its entire growth period. Per hectare some 3-4t of greens (leaves along with petioles) are harvested as petioles are also edible. For rhizomes, the lateral offsets form the crop whereas the central or mother rhizome is usually not eaten and often is not harvested except for the yellow variety of tannia. The harvest time depends greatly on the cultivar and growth conditions. Delays in harvesting do not seriously affect tuber quality or yield. When the older leaves of mature plants start yellowing and drying, *i.e.* usually 9-11 months of planting, the tubers are harvested. The mature rhizomes do not deteriorate if left in the ground and can be harvested as required. The whole plant may be dug up often by hand, and the offset rhizomes are separated from the main rhizome. Sometimes the soil is dug away from the plant and after harvesting the exposed offsets of the parent plant are covered with soil and left to produce a new crop. In this way the plant may continue to crop for several years, usually at least up to 6 years, before there is a noticeable decline in quality of the rhizomes. Average rhizome yield per hectare is 12-25 tonnes and almost the same of the non-edible central rhizome which is used as planting stock. Harvesting should be done carefully to avoid bruising the rhizomes otherwise they may develop serious rots after storage. In some countries mechanical harvesters are used for harvesting of tannia rhizomes. In case of ratoon cropping, the soil around the clump should be dug out to remove the mature side rhizomes. After harvesting, the central as well as side-rhizomes are sun-dried for 4-5 days so that these are fully cured. Now these are stored through embedding in dry soil or sand for up to 4-5 months under ventilated, dry and semi-dark conditions in case these are not immediately sold out. Mature rhizomes if are not harvested timely, these may start sprouting when rain occurs and then these become watery and less suitable for consumption. Usually per plant yield of rhizome is 3-4kg with market price of Rs. 20-30 per kg.

Yield

The average yield is about 6-12t/ha though in Trinidad, several experiments have given rhizome yields of 30-32.5t/ha and in Puerto Rico 25-37t/ha. Average yield of offset rhizome in the South Pacific is 20t/ha when grown under monoculture though in the farmer's field in mixed cultivation commonly it gives 5-7t/ha of rhizome yield.

Postharvest

Tannias can be stored for periods of up to 6 months under dry and well-ventilated conditions, though the loss of eating quality after 8 weeks was recorded when rhizomes were stored at ambient temperatures. The quality can be maintained for 18 weeks or more if the rhizomes are stored at 6-7°C and 80 per cent RH. In Cameroon, traditional storage in pits in a confined atmosphere has been found more satisfactory than storing on trays in well-ventilated huts.

Offset rhizomes have a natural dormancy period of 1–3 months. They are planted year-round in moist production areas and harvested as per requirement. Offset rhizomes can be removed individually or the entire plant can be uprooted and the offsets separated. In commercial settings, offset rhizomes are washed, surface-dried and packaged in waxed boxes prior to shipment. The storage period of the offset rhizomes can be extended by several months by holding them at 7°C and 80 per cent relative humidity (RH). Although not common but offset rhizomes can be dried and stored for extended periods.

Pests and Diseases

In general, tannias are relatively free from serious attack of insect-pests and diseases which may probably be due to the fact that the crop is grown mainly on small scale and not as an extensive monoculture. In Trinidad, Venezuela and Guyana, a dynastic beetle (*Ligyrus ebenus*) has been reported to attack the crop which is controlled effectively by spraying 0.25 per cent Malathion. Other pests which sometimes attack tannias include *Aphis gossypii* in Surinam; *Aphis gossypii, Aspidiotus destructor, Corythucha gossypii, Pentalonia nigronervosa* and *Tetraleurodes ursorum* in the Antilles; *Aspidiotus destructor* in Polynesia; *Euetheola bidentata* in Surinam; and *Graphocephala propior*, *Quinta cannae* and *Cacographis ortholatis* in Venezuela. One can get rid of all these pests through clean cultivation, jet water spraying and through the use of tobacco decoction. Since the plant is edible hence restricted use of insecticide is permitted and that too before about one month of crop harvesting. Malathion at 0.2 per cent spraying will control all these pests effectively.

Several root rots can affect tannias with varying degrees of severity. During 1930s, the root rot was particularly serious in Ghana for which investigations suggested that the causal agent could have been a fungal or bacterial infection or a combination of both, *vis-à-vis* the nutritional status of the crop was also a factor influencing the severity of attack. Tannias are susceptible to various *Pythium* rots. In Puerto Rico, a soft rot known as 'currutaca' was observed due to *Pythium ultimum* infection while in New Caledonia *P. irregulare* is reported occasionally to cause serious losses. In Venezuela, the crop is sometimes infected with *Cercospora chevalier*, *C. verruculosa* and *Punctellina solteroi*. When these rots start appearing, the crop should be sprayed with 0.2 per cent benomyl alternate with 0.2 per cent Captafol after 15 days. Dasheen mosaic virus as a secondary

Infection has been observed especially in case of Mg deficiency so care should be taken that soil is rich in minerals in case of its cultivation. Virus infected plants should immediately be uprooted and burnt.

When the crop is grown in laterite soils of Kerala, lower leaves start showing interveinal chlorosis characteristic to Mg-deficiency prevalent from 3rd month of growth which devastates the entire crop so dolomite @ 1 ton per hectare at the time of land preparation should be added in the soil as a precautionary measure. The third or fourth leaf at four month of planting should have N, P, K and Ca as 3.2, 0.5, 2.3, 0.9 and 1.3 per cent, respectively (Susan John, 2019).

Integrated Pest Control

Integrated pest control recommended for tannias to ensure healthy and resistant plants include:

- ✰ Good soil preparation along with good drainage
- ✰ Use of healthy, unblemished and unbruished planting materials which are free from diseases, dipping of planting materials for half an hour before planting in fungicide solution (Captan + benomyl), wide spacings, *vis-à-vis* proper weed control, field sanitation and fertilising.
- ✰ Disease- and pest-resistant cultivars should be used where possible.
- ✰ In monoculture practice, different cultivars should be planted together or in adjacent plots to observe various degrees of susceptibility within the planted areas.
- ✰ Regular inspections should be made so that appropriate pest control may be applied as early as possible.
- ✰ Natural biological control is important and chemicals should be used only when very necessary.
- ✰ After harvest all plant residues should be removed.
- ✰ Continuous planting of adjacent areas should be avoided as this gives the opportunity for pests to have uninterrupted access to plants at their most susceptible age.

References

1. Acevedo-Rodríguez P, and M.T. Strong, 2012. Catalogue of the Seed Plants of the West Indies. *Smithsonian Contributions to Botany*, **98**: 1192 pp. Smithsonian Institution, Washington DC, USA.
2. Anon., 2008. Database inventory of introduced plant species in the rural and urban zones of Galapagos. Charles Darwin Foundation, Galapagos, Ecuador.

3. Anon., 2013. Pacific Islands Ecosystems at Risk. University of Hawaii. Honolulu, Hawaii, USA. HEAR, http: //www.hear.org/pier/index.html.

4. Bown, D. 2000. Aroids: Plants of the Arum Family. Timber Press, Portland (Oregon), USA.

5. Bradbury, D. H. and W.D. Holloway, 1988. Chemistry of Tropical Root Crops: Significance for Nutrition and Agriculture in the Pacific. Australian Centre for International Agricultural Research, Canberra, Australia.

6. Brown. D. 1995. Encyclopaedia of Herbs and their Uses. Dorling Kindersley, London.

7. Chacon, E and G. Saborio, 2012. Red Interamericana de Información de Especies Invasoras. San Jose, Costa Rica: Asociacion para la Conservación y el Estudio de la Biodiversidad (http: //invasoras.acebio.org).

8. Chukwu, G.O., K.C. Ekwe and S. Anyaeche, 2008. Cocoyam Production and Usage in Nigeria. National Root Crops Research Institute (NRCRI) News Bulletin, No. 1, pp 2.

9. Doku, E. V. 1966. Root crops in Ghana. *Ghana J. Sci.*, **6:** 15–36.

10. Ekwe, K.C., K.I. Nwosu, C.C. Ekwe and I. Nwachukwu, 2009. Examining the underexploited values of cocoyam (*Colocasia* and *Xanthosoma* spp.) for enhanced household food security, nutrition and economy in Nigeria. Proceedings of the International Symposium on Underutilized Plants for Food Security, Income and Sustainable Development (eds. Jaenicka, H., J. Ganry, I. Zeledon Hoeschle and R. Kahara). *Acta Hort.* No. 86, pp. 71–8.

11. Eneh, O. C. 2013. Towards food security and improved nutrition in Nigeria: Taro (*Colocacia antiquorum*) grit as carbohydrate supplement in energy food drink. *African J. Food Sci.*, **7:** 355–60.

12. Falade, K.O. and C.A. Okafor, 2013. Physicochemical properties of five cocoyam (*Colocasia esculenta* and *Xanthosoma sagittifolium*) starches. *Food Hydrocolloids*, **30:** 173–81.

13. Falade, K.O. and C.A. Okafor, 2014. Physical, functional and pasting properties of flours from corms of two cocoyam (*Colocasia esculenta* and *Xanthosoma sagittifolium*) cultivars. *J. Food Sci. Techn.*, **52**: 3440–48.

14. Langeland, K.A., H.M. Cherry, C.M. McCormick and K.A. Craddock Burks, 2008. Identification and Biology of Non-native Plants in Florida's Natural Areas. University of Florida IFAS Extension Bulletin. Gainesville, Florida, USA.

15. Lebot, V. 2009. Tropical Root and Tuber Crops. Cassava, Sweet potato, Yams and Aroids.CABI Publishing, UK.

16. Leung, W.T.W., R.R.B. Burtrum and F.H. Chang, 1972. Food Composition Table for Use in East Asia. Food and Agriculture Organization (with permission). Rome, Italy.

17. Lim, T.K. 2015. *Xanthosoma sagittifolium*. Edible Medicinal and Non-Medicinal Plants, pp. 498–509. Springer. New York and London.

18. Lim, T.K. 2016. Edible Medicinal and Non-Medicinal Plants (1st edn). Springer, New York and London.

19. Manner, H.I. 2011. Farm and forestry production and marketing profile for Tannia (*Xanthosoma* spp). In: Specialty Crops for Pacific Island Agroforestry (ed. C.R. Elevitch), pp. 1-16. Permanent Agriculture Resources (PAR), USA.

20. Ndabikunze, B.K., H.A.L. Talwana, R.J. Mongi, A. Issa-Zacharia, A.K. Serem, V. Palapala and J.O.M. Nandi, 2011. Proximate and mineral composition of cocoyam (*Colocasia esculenta* L. and *Xanthosoma sagittifolium* L.) grown along the Lake Victoria Basin in Tanzania and Uganda. *African J. Food Sci.*, **5**: 248–54.

21. Odebunmi, E.O., O.O. Oluwaniyi and M.O. Bashiru, 2010. Comparative proximate analysis of some food condiments. *J. Appl. Sci. Res.*, **6**: 272–4.

22. O'Hair, S.K. and D.N. Maynard, 2003. Encyclopedia of Food Sciences and Nutrition (2nd edn). England.

23. Onokpise, O.U., J.G. Wutoh, X. Ndzana, J.T. Tambong, M.M. Meboka, A.E. Sama and M. Bruns, 1999. Evaluation of Macao Cocoyam Germplasm in Cameroon, pp. 394-6. In: Perspectives on New Crops and New Uses (ed. Janick, J.). ASHS Press, Alexandria, VA.

24. Opara, L.U. 2003. Edible Aroids: Post Harvest Operations. Food and Agriculture Organization of the United Nations. Rome, Italy.

25. Quattrocchi, U. 2000. CRC World Dictionary of Plant Names, vol. IV(R-Z), 2849p. Taylor and Francis, USA.

26. Ramanatha, R.V., P.J. Matthews, P.B. Eyzaguirre and D. Hunter, 2010. The Global Diversity of Taro: Ethnobotany and Conservation. Biodiversity International, Rome, Italy.

27. Randall, R.P. 2012. A Global Compendium of Weeds. Department of Agriculture and Food Western Australia, 1124 p. Perth, Australia.

28. Sefa Dedeh, S. and E.K. Agyir Sackey, 2002. Starch structure and some properties of cocoyam (*Xanthosoma sagittifolium* and *Colocasia esculenta*) starch and raphides. *Food Chem.*, **79:** 435–44.

29. Sefa-Dedeh, S. and E.K. Agyir-Sackey, 2004. Chemical composition and the effect of processing on oxalate content of cocoyam *Xanthosoma sagittifolium* and *Colocasia esculenta* cormels. *Food Chem.*, **85:** 479–87.

30. Susan John, 2019. Tannia. In: Handbook of Horticulture, vol. 1, pp. 676-7. ICAR, New Delhi.

31. Tindall, H.D. 1983. Vegetables in the Tropics. MacMillan, Oxford. U.K.

27

Taro [*Colocasia esculenta* (L) Schott]

Family: Araceae.

Other common names: Elephant ears. **Hindi**: Arbi/Arvi/Taro (eddoe/eddo group, bearing small or medium rhizomes in large numbers; *Colocasia esculenta* var. *antiquorum*); Bunda (dasheen group, bearing a few but large rhizomes; *Colocasia esculenta* var. *esculenta*); Ghuiyan; and Kachalu.

Synonyms: *Alocasia dussii* Dammer; *A. illustris* W.Bull; *Aron colocasium* (L.) St.-Lag.; *Arum chinense* L.; *A. colocasia* L.; *A. colocasioides* Desf.; *A. esculentum* L.; *A. lividum* Salisb.; *A. nymphaeifolium* (Vent.) Roxb.; *A. peltatum* Lam.; *Caladium acre* R.Br.; *C. colocasia* (L.) W.Wight nom. illeg.; *C. colocasioides* (Desf.) Brongn.; *C. esculentum* (L.) Vent.; *C. glycyrrhizum*. Fraser; *C. nymphaeifolium* Vent.; *C. violaceum* Desf.; *C. violaceum* Engl.; *Calla gaby* Blanco; *C. virosa* Roxb.; *Colocasia acris* (R.Br.) Schott; *C. aegyptiaca* Samp.; *C. colocasia* (L.) Huth nom. inval.; *C. euchlora* K.Koch and Linden; *C. fonstanesii* Schott; *C. gracilis* Engl.; *C. himalensis* Royle; *C. neocaledonica* Van Houtte; *C. nymphaeifolia* (Vent.) Kunth; *C. peltata* (Lam.) Samp.; *C. vera* Hassk.; *C. violacea* (Desf.) auct.; *C. virosa* (Roxb.) Kunth; *C. vulgaris* Raf.; *Leucocasia esculenta* (L.) Nakai; *Steudnera virosa* (Roxb.) Prain; and *Zantedeschia virosa* (Roxb.) K.Koch. **Synonyms of *Colocasia antiquorum***: *Caladium antiquorum* (Schott) Andre, *Colocasia fontanesii* Schott, *C. gaoligongensis* H.Li and C.L.Long, *C. gongii* C.L.Long and H.Li and *C. lihengiae* C.L.Long and K.M.Liu.

Introduction and Origin

Colocasia is derived from the Greek word *kolokasi* which itself is derived from Ancient Greek word *kolokasion*, literally meaning 'lotus root'. The specific epithet *esculenta* in Latin means 'edible'. English term 'taro' may have been borrowed from New Zealand's Maori language. The names elephant-ear and cocoyam are also used for some other large-leaved genera in the Araceae, notably *Xanthosoma* and *Caladium*.

Colocasia esculenta, one of the oldest tuber crops grown in India, originated probably in the Indomalayan realm (*i.e.* East India, Nepal and Bangladesh) which is being grown throughout the tropics for its edible starchy rhizome and often as an ornamental in temperate areas for its large glossy leaves. Its rhizomes, leaves and petioles are used as vegetables. Similar to yams, the rhizomes are used as staple food in African, Oceanic and South Asian cultures. Taro is believed to be native of Southern India and SE Asia though widely naturalized across the globe in tropical and subtropical areas. Africa ranks first in area and production of colocasia, followed by Asia.

Another tropical vegetable eddoe (eddo), often considered identifiable as the species *Colocasia antiquorum* Schott, is closely related to taro and is also used mainly for its thickened rhizome. Young leaves can also be cooked and eaten but unlike taro these have a somewhat acrid taste so require careful cooking (Purseglove, 1972). Originally Linnaeus described *Colocasia esculenta* and *Colocasia antiquorum* as two different species but later many botanists considered both as the members of a very variable single species *Colocasia esculenta* (Hill, 1939). *Colocasia* is the 5th most consumed root crop in the world after potatoes, cassava, sweet potatoes and yams (*Dioscorea*). It is most widely cultivated edible species in the family Araceae and is probably one of world's oldest crops. This important edible aroid is found abundantly in NE India. Some similar taro plants include giant taro (*Alocasia macrorrhizos*), swamp taro (*Cyrtosperma merkusii*), and arrowleaf elephant's ear (*Xanthosoma sagittifolium*). In India, this crop is cultivated in Kerala, Tamil Nadu, Andhra Pradesh, Odisha, West Bengal, Asom, Bihar, Jharkhand, Chhattisgarh, Uttar Pradesh, Rajasthan, NE States, Himachal Pradesh and Andaman and Nicobar Islands.

Botany

Colocasia esculenta is a tender monocotyledonous herbaceous perennial growing 0.3-2m high bearing edible rhizomes of different shapes and sizes. Also there are types which either produce long slender stolons (aboveground runners) or not-forming starchy rhizomes. The lumpy rhizomes with rough ridges have a brown skin and a white or pink interior. Large, entire, ovate to sagittate leaves are held on the end of erect, long, thick *cum* succulent petioles coming directly from underground rhizome. Petiole size may range 0.8 to 1.2m. Petiole stands erect whereupon centre of the lower surface of peltate leaf is joined with the leaves facing upward or outward. Unlike *Colocasia,* the leaves of *Alocasia* and *Xanthosoma* are usually not peltate so are held more upright. Smooth, waxy, triangular-ovate, sub-rounded and apex-mucronate leaves are long-petioled, up to 40cm long, 25cm wide, usually dark green above and light green beneath with the tip of the basal lobes rounded or sub-rounded. In their native habitat the leaves usually grow up more than 90cm long and 60cm wide (depending on the species) but they tend to be much smaller when grown elsewhere. Flat sagittate leaves often having prominent veins are shaped like a large ear or shield hence the common name 'elephant's-ear'. It produces shallow, fibrous and adventitious roots, *vis-a-vis* stolons or suckers. The lateral buds present on the rhizomes arise as new lateral rhizomes, stolons and suckers.

Colocasia corms

C. esculenta flowers and leaves

Leaf colours in some ornate cultivars may range from black or brown to chartreuse gold. In addition, the leaves will vary in glossiness from a matte finish (common) to a high gloss-shine (rare in *Colocasia*). Quite commonly the leaves contain thick-mid veins and major side veins of a contrasting colour. Major vein colour may be green, purple, creamy ivory, yellow or pink. The colouration may only occur at the 'piko' (the 'belly button' on the top of the leaf opposite to the point of attachment to the petiole) or may run through all the major veins of the leaf. In addition, as the season progresses, mid-vein colouration often leaks into the surrounding lamina tissue and may form an oval shape as with *C. esculentum* 'Nancy's Revenge'. This trait appears to be triggered at the same time as flowering and may be under the same genetic control. It is not uncommon for the bottom of the leaf to have a different lamina or vein colouration than the top of the leaf which provides a nice flash of colour on windy days. Newly emerged leaves may have a different colour than mature ones.

Though flowering in taro is quite uncommon, especially in the Midwest, but old plantings can sometimes bloom producing typical aroid-type inflorescences bearing a white to yellow or light green spathe surrounding the spadix. The inflorescence that arises at leaf axils in cluster of 2–5 is normally hidden underneath the foliage consisting of spathe, spadix, staminate flowers, pistillate flowers, neutral flowers and sterile appendage (Thirugnanavel *et al.*, 2015). Staminate and pistillate flowers are small, sessile and monoecious. The stigma becomes receptive one day before the anther dehiscence and remains receptive for upto two days. The pistillate flowers are covered by spathe. Self-incompatibility, natural constriction by spathe and different flower maturity times prevent self pollination and in turn enable the cross pollination. The flowers can be large, fragrant and attractive. Pollination is carried out by insects ~ fruit flies being the major pollinating agents in *colocasia*. In absence of pollinators, natural seed set may occur due to entry of rain washed pollens into the stigma. The fruit is a globular green or yellow berry containing 1–10 hard and ovate seeds that germinate in 7–14 days (Thirugnanavel *et al.*, 2015).

Genetics

The basic chromosome number of *Colocasia esculenta* is 14 and similar to several other species of the genus it too has two cytotypes ~ diploid with 2n=2x=28 chromosomes and triploid with 2n (3x)=42 (Coates *et al.*, 1988). The diploid cultivars are usually fertile but have little cold tolerance while triploids are usually sterile but with more cold-tolerance. An additional set of chromosomes provide triploids with increased levels of adaptability and hardiness at high elevation and latitude (Ochiai *et al.*, 2001). In NE India where diploids and triploids both are found, the triploids seem to have evolved in response to the climatic conditions on the hills (Kuruvilla and Singh, 1981). Somatic chromosome numbers in some of the varieties are 28 in 'Muktakesi', 'Sree Kiran', 'Sree Pallavi', 'Mothan' and 'Sunajhili' while 42 in 'Banky', 'Duradin' and 'Telia'.

Colocasia species rarely cross with other species of the genus and supposedly with *Alocasia* though a few intergeneric and interspecific hybrids have been discovered or hybridized including *C. esculenta* × *A. brisbanensis*, *C. esculenta* × *C. gigantea* and *C. gigantea* × *A. macrorhizos* (Hunt *et al.*, 2013).

Cultivars and Varieties

The starchy, compact and thick rhizome has brown coloured scars and scales on outer surface and ground starchy parenchyma. Based on the rhizome morphology, the plant is botanically classified into two groups, *viz.* **dasheen** [banda (bunda) in Hindi; *Colocasia esculenta* var. *esculenta*; 2n=3x=42] and **eddoe** [arvi (ghuiyaan) in Hindi; *Colocasia esculenta* var. *antiquorum*; 2n=2x=28] (Thirugnanavel *et al.*, 2015). Dasheen type is characterized by large mother rhizome with very few small rising cormel-like lateral rhizomes, and the eddoe type is characterized by small to medium-sized inedible mother rhizome with equal size of numerous edible lateral rhizomes (Thirugnanavel *et al.*, 2015). These two types of colocasias are commonly cultivated throughout India.

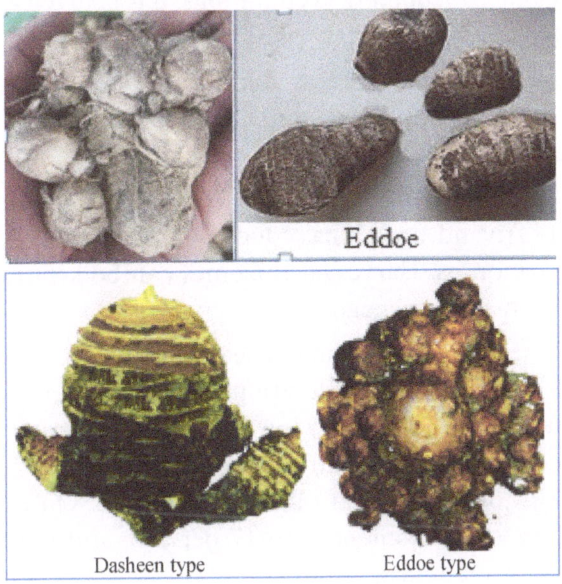

With wide variability, *Colocasia* has a large number of local cultivars grown in different parts of India as given here under:

- ☆ **'Gaderi'**, a much sought after taro variety from Kumaon, especially from Lobanj (distt. Bageshwar).
- ☆ **'Muktakeshi'**, released by CTCRI, is an erect, medium-tillering variety having round-cormous rhizome with cylindrical lateral rhizomes. At field condition, these are resistant to 'taro leaf blight', 'dasheen mosaic', aphids, cutworms and scale insects. This variety is suitable for uplands and lowlands during summer and rainy seasons.

Sree Pallavi

- ☆ **'Pani Saru-1'**, released by CTCRI, is a variety of taro with good cooking quality. Lateral rhizomes have elliptical shape. This can be grown in upland and low/submerged conditions. It is field tolerant to leaf blight.
- ☆ **'Pani Saru-2'**, released by CTCRI, is a variety of taro with good cooking quality. Lateral rhizomes are round to elliptical in shape. This can be grown in upland and low land/waterlogged/swampy conditions. It is field tolerant to leaf blight.
- ☆ **'Sree Kiran'** is the first hybrid taro variety in India released from CTCRI (Trivandrum). It has good cooking quality with crop duration of 6.5–7 months. Harvested rhizomes can be kept for long duration (farmextensionmanager.com).
- ☆ **'Sree Pallavi'**, a high yielding variety released by CTCRI (Trivandrum) is 1-1.5m tall having crop duration of usually 6.5–7 months and a yield of about 15–18t/ha. This is a low tillering-type variety which per plant produces some 20-25 club-shaped, small-sized tuberous-rhizomes. It is field tolerant to leaf blight and mosaic. On the basis of fresh weight, its starch content is 16-17 per cent and protein 2-3 per cent (farmextensionmanager.com).
- ☆ **'Sree Rashmi'**, a high yielding variety released from CTCRI (Trivandrum) is a 1-1.5m tall with 7-8 months crop duration having an average yield of about 15–20t/ha under low input levels. All the plant parts are non-acrid. Lateral rhizomes have good cooking quality and excellent taste. On the basis of fresh weight, its starch content is 15 per cent and protein 2.5 per cent (farmextensionmanager.com). The edible leaves, mother rhizome and attached small cormel-like rhizomes are acrid-free.

Some other popular cultivars include 'Banky', 'Duradin', 'Megha Taro-1', 'Mothan', 'Rangma', 'Sree Kiran', 'Sunajhili', *etc.*, however, 'Muktakeshi', 'Pani Saru-I', 'Pani Saru-II', 'Sree Kiran', 'Sree Pallavi', 'Sree Rashmi', *etc.*, were released by

CTCRI. 'Bidhan Chaithanya', 'Jhankhri', 'Muktakeshi', *etc.* are resistant cultivars to taro leaf blight.

C. *esculenta* 'Black Coral'

i) 'Ahina Local', 'Bidhan Chaithanya', 'Birsa Arvi-1', 'Jhankhri', 'Kadma', 'Kakakachu', 'Muktakeshi', 'Nadia Local', 'NDC 1', 'NDC 2', 'NDC 3', 'Panchamukhi', 'Rajendra Arvi-1', 'Saharshamukhi', 'Satamukhi', 'Telia', 'White Gauriya', *etc.* are promising varieties for **eastern zone.**

ii) 'Saharshamukhi' and 'Satamukhi' for **western zone**.

iii) 'Indira Arvi-1', 'Muktakeshi', 'Narendra Arvi-1', 'Narendra Arvi-2', 'Narendra Arvi-3', 'Rajendra Arvi-1' and 'Satamukhi' for **central zone.**

iv) 'Co-1', 'Muktakeshi', 'Satamukhi', 'Sree Pallavi' and 'Sree Rashmi' are recommended for **southern zone**.

v) Though in NE States, local land races such as 'Chigi', 'Rangma', 'Tamachokham' and 'Tamatim' are very popular but for this region even 'Megha Taro-1, 'Muktakeshi' and 'Rajendra Arvi-1' are also quite promising.

Colocasia varieties can be divided into **clumpers** and **runners**, although some cultivars may put a root or two over the dividing lines. The traditional edible variety grown for food is *C.e.* var. *aquatilis*. It produces very long stolons (aboveground runners) forming a large mass of plants in a short time. Though it is good for large plantation, it is less desirable in a garden of limited space. Many modern 'elephant ears' bred with these early plants still have aboveground runners although the degree of running is somewhat less. The underground runners of *Colocasia* varieties are *C.e.* **var.** *antiquorum* **'Black Beauty'**, *C.e.* '**Coal Miner**' and *C.e.* '**Illustris**'. The clumping varieties have an attractive vase-shape in the garden, and many of the cultivars available today are very slow or are non-existent runners.

Constituents

Flavonoids and triterpenoids are the two pharmacologically active compounds mainly present in the *Colocasia* leaf extracts. The isolated flavonoids contain vicenin-2, isovitexin, iso-vitexin 3'-O-glucoside, vitexin X'-O-glucoside, iso-orientin, orientin, orientin 7-O-glucoside and leteolin 7-O-glucoside. The leaves of the plant also contain fibres, calcium oxalate, minerals, starch, vitamin A, B, C, *etc.* (Pawar *et al.*, 2018). Phytochemical investigations on *C. esculenta* leaf exhibit flavones, apigenin, luteolin and anthocyanins (Khare, 2007). *Colocasia* leaves contain phytochemicals such as anthraquinones, apigenin, catechins, cinnamic acid derivatives, vitexin and isovitexin (Gupta *et al.*, 2019).

C. esculenta rhizomes contain globulins accounting for 80 per cent of the total tuber proteins. On dry weight basis, taro rhizome has been reported to have 70–80 per cent starch with quite small granules. High level of carbohydrate content observed in raw taro, taro powder and total amino acids recorded in the tubers are in the range of 1,380-2,397mg/100g (Krishnapriya and Suganthi, 2017).

Toxicity in Aroids

Out of all the edible aroids ~ elephant foot yam (*Amorphophallus paeonifolius*), giant swamp taro (*Cyrtosperma chamissonis*), giant taro (*Alocasia macrorrhizos*), tannia (*Xanthosoma violaceum*) and taro (*Colocasia esculenta*) have been shown to contain varying amounts of cyanide in their leaves and stems as found through a screening assay and a more accurate acid hydrolysis method (Bradbury *et al.*, 1995). The amount of cyanide present in the leaves is 0-3mg HCN/100g fresh weight and in the stems 0—0.3mg HCN/100g fresh weight of *Colocasia* and *Xanthosoma* though only about 1-5 per cent of that is found in cassava leaves and tubers which is not a cause of concern for human nutrition (Bradbury *et al.*, 1995).

Like several other plants in the family of aroids the stems, rhizomes, leaves and petioles of taro contain numerous needle-like and stinging crystals of calcium oxalate (raphides) that can cause serious reactions if ingested raw so they are required to be thoroughly cooked before consumption.

Nutritional Value and Health benefits

The rhizome is free from gluten and rich in starch, minerals and energy. One of the finest sources of dietary fibre (4.1g/100g), it is also rich in potassium and magnesium and has good amount of vitamin B complex and other vitamins such as vitamin C, D, *etc. Colocasia* has several health benefits as well as it helps in lowering the blood pressure and weight (Thirugnanavel *et al.*, 2015). It is one of the anti-ageing vegetable due to the presence of essential amino acids and omega-3 fatty acids thus helps in regeneration of cells (Thirugnanavel *et al.*, 2015). Being a low glycemic index food, properly boiled rhizomes will reduce the glucose level in the body thereby becoming quite beneficial as food for diabetic patients (Thirugnanavel *et al.*, 2015). Overconsumption of rhizome without proper cooking may lead to kidney stones (Thirugnanavel *et al.*, 2015). Proper cooking is essential to nullify effect of calcium oxalate. The nutrient availability/100g present in raw tuber includes moisture 70.3 per cent, fat 0.1 per cent, protein 3.2 per cent, starch 21.2 per cent, energy 97 Kcal, vitamin B_1 0.09mg, vitamin B_2 0.03mg, calcium 31mg, beta carotene 34µg, phosphorous 68mg, magnesium 106mg, sodium 1.6mg, potassium 356mg, sulphur 7.4mg, iron 0.63mg, copper 0.2mg, zinc 3.6mg, manganese 0.34mg and boron 0.09 mg.

Taro Cooked (without salt)

Nutritional Value per 100g

Energy	594 kJ (142 kcal)
Carbohydrates	34.6g
Sugars	0.49g
Dietary fibre	5.1g
Fat	0.11g
Protein	0.52g
Vitamins	***Quantity Per cent DV†***
Thiamine (B1)	9 per cent 0.107 mg
Riboflavin (B2)	2 per cent 0.028mg
Niacin (B3)	3 per cent 0.51mg
Pantothenic acid (B5)	7 per cent 0.336mg
Vitamin B6	25 per cent 0.331mg
Folate (B9)	5 per cent 19µg
Vitamin C	6 per cent 5mg
Vitamin D	8 per cent 1.2 µg
Vitamin E	20 per cent 2.93mg
Minerals	***Quantity Per cent DV†***
Calcium	2 per cent 18mg
Iron	6 per cent 0.72mg
Magnesium	8 per cent 30mg
Manganese	21 per cent 0.449mg
Phosphorus	11 per cent 76mg
Potassium	10 per cent 484mg
Zinc	3 per cent 0.27mg

IU = International units

†Percentages are roughly approximated using US recommendations for adults.

Source: USDA Food Data Central

Taro Leaves (Raw)

Nutritional Value per 100g

Energy	177 kJ (42 kcal)
Carbohydrates	6.7g
Sugars	3g
Dietary fiber	3.7g
Fat	0.74g
Protein	5g
Vitamins	***Quantity Per cent DV†***
Vitamin A equiv. beta-Carotene lutein zeaxanthin	30 per cent 241μg 27 per cent 2895μg 1932μg
Thiamine (B1)	18 per cent 0.209mg
Riboflavin (B2)	38 per cent 0.456 mg
Niacin (B3)	10 per cent 1.513mg
Vitamin B6	11 per cent 0.146mg
Folate (B9)	32 per cent 126μg
Vitamin C	63 per cent 52mg
Vitamin E	13 per cent 2.02mg
Vitamin K	103 per cent 108.6μg
Minerals	***Quantity Per cent DV†***
Calcium	11 per cent 107mg
Iron	17 per cent 2.25mg
Magnesium	13 per cent 45mg
Manganese	34 per cent 0.714mg
Phosphorus	9 per cent 60mg
Potassium	14 per cent 648mg
Zinc	4 per cent 0.41mg

IU = International units

†Percentages are roughly approximated using US recommendations for adults. Source: USDA Food Data Central

Source: https://en.wikipedia.org/wiki/Taro#Distribution_and_habitat

Health Benefits

The health benefits of taro include its ability to improve digestion, lowering blood sugar levels, preventing certain types of cancers, protecting the skin, boosting vision health, increasing circulation, aiding the immune system and preventing heart disease apart from supporting muscle and nerve health. Taro rhizome has much more dietary fibre as compared to potatoes; hence including it in diet improves digestive function and can relieve issues like diarrhoea, constipation, acid reflux and stomach ulcers. It also aids in healthy weight loss because fibre moves slowly through the digestive system and makes to feel fuller between meals. The carbohydrate content in taro rhizome is a resistant starch thus stabilizing blood sugar and helping with weight management and reduced risk of diabetes. These starches are also suitable for low-carb and keto diets. Taro rhizome has significant level of potassium that helps to control high blood pressure by breaking down excess salt, thus reducing stress on cardiovascular system and helping to prevent development of chronic heart problems.

Taro root contains various antioxidants including beta-carotene and cryptoxanthin. These antioxidants can help to improve vision by preventing the free radicals from attacking ocular cells and causing macular degeneration or cataracts. Free radicals are the dangerous byproducts of cellular metabolism that can cause healthy cells to mutate and turn into cancerous cells. Cryptoxanthin eliminates these free radicals thus lowering chance of developing both lung and oral cancers.

Uses

Rhizomes, lateral rhizomes, leaves, petioles and flowers are edible (Bose *et al.*, 2003) but must be first thoroughly cooked (usually boiled, baked or roasted) prior to eating in order to remove calcium oxalates which render uncooked rhizomes and stems inedible. Taro rhizome is a vegetable used in a variety of cuisines around the world. It has a mild *cum* nutty taste, starchy texture, *vis-a-vis* nutrition benefits that make it a healthier alternative to other root vegetables like potatoes. Larger varieties usually have a stronger flavour while smaller roots add more moisture. A ready-to-eat rhizome is firm, unblemished and feels heavy for its size. The rhizomes and leaves are rich in nutrients and have even medicinal values (Thirugnanavel *et al.*, 2015). These can be prepared in a variety of ways such as deep-frying in oil to be eaten on the side with rice or cooking in a tangy tamarind sauce with spices, onion and tomato. Taro rhizomes can be thinly sliced and fried to make chips.

In India, taro or eddoe is a common dish that can be served in many ways. The boiled or roasted rhizomes can be eaten like potatoes, as vegetables (with or without gravy) or sliced and fried to prepare chips. The green leaves are rolled and steamed after applying gram flour-paste, salt, turmeric and red chilli powder from inside of the leaves. The steamed preparation is then cut into pieces and deep-fried to prepare *pakora*-like dish that can be eaten as snacks with chutney or sauce. The leaf-*pakora* can also be prepared by cutting the leaves into small pieces, and after mixing with gram flour and spices, it is deep fried. Young leaves can be cooked and eaten like spinach.

Propagation

Though taro is always propagated asexually in farmers' fields, it is an allogamous and highly heterozygous species where natural pollinations do occur between flowering diploids. New genotypes germinating spontaneously can be clonally selected by farmers but these can also capture somaclonal variants when these appear sufficiently different from mother plants to be considered as new cultivars and renamed.

Although the flowering in taro is very uncommon but seed production is possible whenever flowering occurs. Natural breeding and population spread occurs in wild taro. This can be propagated by **seeds** that are usually collected about 30 days after fertilization and are surface-sown soon after drying. Germination takes place in 20-22 days.

Like most other root and tuber crops, taro too is usually propagated **vegetatively**. Cultivars are propagated through rhizomes and lateral offsets while in the wild vegetative propagation occurs through stolons. The planting material includes the small rhizomes/offsets, cut setts of larger/main rhizomes with few buds and by separating new plantlets formed at the rhizome tips on the running types. The long runners form nodes along their length and new growth-buds exist at each of these nodes. Lateral rhizomes (offsets of main rhizome) weighing about 20-25g are good planting material. About 800-1,000kg of cut setts as seed material are required for one hectare of land (Thirugnanavel *et al.*, 2015). Healthy, unbruised and uniform-sized planting material should be selected and stored in a cool place at least for 3 months before planting.

Factors Influencing Growth and Quality

Although environmental factors such as temperature, moisture, light and soil type influence the growth and quality, other factors such as varieties, quality of planting material, planting method, planting *cum* harvesting date, irrigation, nutrition, pest management, *etc.* are also equally important provided there had been proper ground preparation through harrowing and ploughing. The most profitable yields can be attained only when all factors are at optimum levels.

The warm and moist **climate** with mean temperature of 21–27°C and well-distributed rainfall of 1,000mm during growth period is highly favourable for its growth as well as development. In areas where rainfall is less a good amount of supplementary irrigation is required for successful crop cultivation. It can be grown up to an altitude of 1,500m though in some places it can be grown up to 2,400m elevations (Thirugnanavel *et al.*, 2015). Most *Colocasia* prefer bright sunlight unless grown in extremely hot and low humidity climates where some light shade will be necessary.

Colocasia can be grown in a wide range of **soils** and climatic conditions. It grows best in a well-drained fertile sandy-loam soil with a pH range of 4.2-7.5 though ideal pH range is 5.5-7. Being a heavy feeder, *Colocasia* requires rich **organic soil** with plenty of composted material, *i.e.* 10-15t/ha of farmyard manure which should thoroughly be mixed in the soil at the time of land preparation. CTCRI, Trivandrum recommends a fertilizer dose of 80kg N, 25kg P_2O_5 and 100kg K_2O/ha. Nitrogenous and potassic fertilizers are recommended in 2-3 split doses at monthly intervals. For most part of India and especially in Andhra Pradesh, Asom and Odisha the economical dose of N:P:K in practice is usually 80:60:80kg/ha where full dose of phosphorus and half dose of nitrogen and potassium is applied within a week after sprouting and remaining half dose of nitrogen and potassium after one month of first application along with weeding and earthing up (Thirugnanavel *et al.*, 2015). In Uttar Pradesh it is 80:80:120kg/ha, in West Bengal 100:80:100kg/ha while for Maharashtra it is 120:60:120kg/ha (Sunitha, 2019).

The ideal **planting time** is rainy season whereas February-March is for irrigated areas. Ideal planting time in M.P., Chhattisgarh, Bihar, Odisha and U.P. is usually February and June; in T.N. and Kerala it is April-June; in Andhra Pradesh is February; and in Asom it is April. On the hills it can be planted during April and lifted by October. **Planting** is done on ridges or on flat-beds. The field must be ploughed 2-3 times to a fine tilth to a depth of 20cm, all the weed-rootstocks and foreign materials should be taken out with fork, followed by levelling and then making of ridges and furrows. In sandy loam soil, pit method of planting is followed. Planting should be done at a depth of 2.5 to 7.5cm at 45cm spacing on ridges made at 60cm apart. Flat bed method can be adopted under upland conditions having good drainage. Planting in small pits is good in flat bed planting. In the field, the lateral rhizomes or seed setts should be planted slightly deeper than they grow in the pot or when planting dormant rhizomes. However, at planting there in the soil should be sufficient moisture to sustain the sprouting of the crop. Mulching the soil after planting helps the crop conserving soil moisture and saving the crop from abnoxious weeds which promotes rhizome yield.

Colocasia grows well under abundant soil moisture level, therefore at planting time in the soil there should be sufficient moisture for proper sprouting of rhizomes. The field should be **irrigated** at 8-15 days intervals depending on weather conditions *cum* soil moisture status and soil types as clayey-loam soils require shallow but delayed irrigation though sandy-loam soils frequent and deep. The irrigation should be stopped 3-4 weeks before harvest of rhizomes. *Colocasia* can successfully be cultivated as a rain-fed crop where distribution of rainfall is good but protective irrigation should be given when rainfall is not regular (Thirugnanavel *et al.*, 2015).

Weeding, light **hoeing** and **earthing** up should be carried out at 30-45 days and 60-75 days after planting. Deep cultivation should be avoided (Thirugnanavel *et al.*, 2015) as this may injure the developing rhizomes and roots. Inefficient small suckers from the mother plants have to be removed along with second weeding. Only 3 suckers per plant should be retained. Periodically the dead leaves should be severed off carefully without injuring the growing plants. Earthing up should be to cover the developing rhizomes and hoeing for soil aeration.

In NE region, the *Colocasia* is cultivated widely as a **mixed crop** along with paddy in *jhum* fields and as a mixed crop along with *Xanthosoma*, maize, leafy vegetables, chillies, *etc.* in homestead gardens. It is cultivated as mono-crop in some of the isolated areas (Thirugnanavel *et al.*, 2015).

Harvest and Storage

The rhizomes are ready for **harvesting** when leaves start yellowing and withering, *i.e.* from August to December depending upon maturity of the crop and existing season. The harvesting can be delayed for some weeks in dry weather. A fortnight before lifting the field should be lightly watered but lifting should be carried out during dry weather (Thirugnanavel *et al.*, 2015). Normally the plants are lifted by hand using a long stick with a spade. The harvested rhizomes are initially kept in the field hut and then transferred to home where the rhizomes are cleaned (Thirugnanavel *et al.*, 2015). The damaged rhizomes should be separated from marketing lot and consumed within a few days. Healthy rhizomes should be spread on the ground for surface drying. These should not be packed in air tight containers but in jute bags or basket which will prevent rotting during storage.

The yield is highly variable depending on the variety planted. Usually the rhizome yield is 8–10t/ha after 6–8 months (Thirugnanavel *et al.*, 2015).

Pests and Diseases

Though taro is susceptible to attack by a number of diseases and pests, only a few cause serious reduction in growth and production. 'Taro leaf blight', '*Pythium* root and rhizome rot', 'softrot', and 'dasheen mosaic virus' are the major diseases. Rhizome borers, whiteflies, hoppers, thrips, and mites are the major insect-pests infesting the crop. Although most of the insect-pests and diseases are not serious, 'taro leaf blight' and 'rhizome borer' are serious problems and cause large economic damage to the crop worldwide, especially in NE India (Thirugnanavel *et al.*, 2015).

Corm borer

Rhizome borer (*Aplosonyx chalybaeus*) is the most damaging pest of *Colocasia*, causing up to 30 per cent damage to the leaves and 80–90 per cent to the rhizomes. This shiny metallic blue or pink beetle was first reported from Meghalaya as its infestation is noticed in May in NE India with more activity being seen during June to August. The female lays eggs inside leaf sheath where hatching takes place in 4–5 days. The grubs feed mainly on the rhizomes by making holes and tunneling into it, causing rhizome rot. Their infestation makes the rhizome unfit for marketing and as animal feed. Severely affected plants show signs of wilting. The adult beetles that hide in the cracks and crevices in the soil and in sheaths remain quite active during daytime, feeding on the leaves making several circular holes. The infestation is high in mono-cropping than in the mixed cropping system (Thirugnanavel *et al.*, 2015). Regular watering will minimize their population otherwise their control is very difficult as at this time no insecticide may be recommended. If in the soil the borer infestation of the crop is anticipated, at the time of bed making 5 per cent Heptachlor dust @ 45kg/ha should be applied and thoroughly mixed in the soil which will protect the plants from attack of all the underground insects.

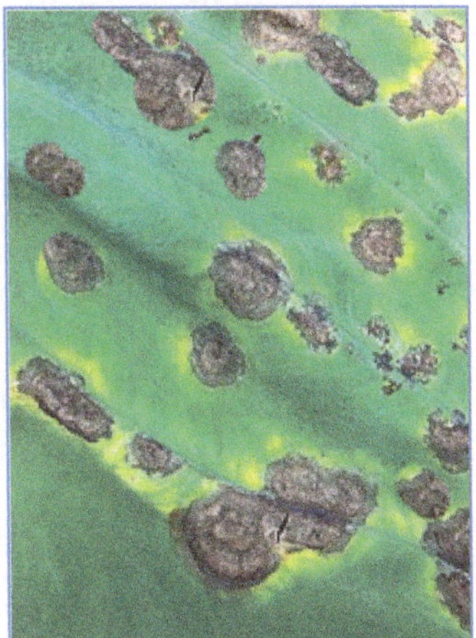

TLB disease - Brown lesions

Taro leaf blight (TLB) is caused by *Phytophthora colocasiae*. Though it affects mainly the leaves but the infection can extend to petioles. In rhizomes, the fungus causes rhizome rot. Heavy incidence causes up to 50 per cent crop loss. Its symptoms include oval or irregular purplish or brownish necrotic lesions with water-soaked periphery on the leaves. The affected portion becomes papery and sometimes produces 'shot-hole' appearance. In severe cases, the entire leaf lamina and the petioles are affected giving a blighted appearance and collapse of the plant. A prominent sign of *P. colocasiae* is the white ring of sporangia around the edge of lesions which grow up rapidly and produce reddish-orange exudates. The TLB spreads quickly at 20–28°C temperatures coupled with more than 85 per cent RH. Sporangia and zoospores are spread by rain-splash and wind-blown rain between plants or within the same plant. Infected planting material spreads the disease over long distances. The zoospores can survive in the soil for up to 3 months. The leftover rhizomes in the field after harvest can also serve as inoculums (Thirugnanavel *et al.*, 2015). The disease can be managed by using disease resistant varieties *cum* healthy planting materials, hot water disinfections before planting, early planting to avoid heavy monsoon rains, proper field sanitation, crop rotation and spray with fungicides [Mancozeb (0.2 per cent) or Ridomil MZ 72 @ 2g/l of water] and treating the seed rhizomes with biocontrol agents such as *Trichoderma viride.*

CTCRI recommended either of Folpet or Dithane-M45 at 0.2 per cent spraying at fortnightly intervals from first appearance of the late blight disease. For this, four sprays may be required at weekly intervals. Unhealthy rhizomes and plant debris should be collected and burnt. For planting, only healthy materials *cum* resistant/tolerant cultivars such as 'Jankhari', 'Muktakesi', 'Nadia Local' or 'Topi' should only be selected. To escape the disease infection, early maturing varieties should be planted.

Multiple **viruses** have been found to infect taro including the four main viruses such as 'dasheen mosaic virus' (DsMV), 'taro bacilliform virus' (TaBV), '*Colocasia* bobone disease virus' (CBDV) and 'taro vein chlorosis virus' (TaVCV) (Sivaprasad *et al.*, 2011; Yang *et al.*, 2003; Pearson *et al.*, 1998; 1999; Yeturu *et al.*, 2013; Matthews *et al.*, 1996), and a new virus 'taro bacilliform CH virus' (TaBCHV) (Wang *et al.*, 2018; Kazmi *et al.*, 2015; Kidanemariam *et al.*, 2018). Virus infected plants should be gently removed from the field and burnt to check further spread. Since many viruses spread through insect-pests such as aphids, leafhoppers, plant hoppers, thrips, whiteflies, beetles, scale insects, mites and nematodes so while planting the soil should be mixed with 45kg/ha of heptachlor and further in case of their attack the crop should be sprayed with 0.2 per cent methyl parathion so that field is free of these insect-pests which will ultimately check spread of viruses too.

Water stagnation in the field may cause the rhizome becoming hard to cook. This occurs both in dasheen and eddoe varieties. Therefore, proper drainage in the field at all the times should be maintained.

References

1. Bose, T.K., J. Kabir, T.K. Maity, V.A. Parthasarathy and M.G. Som, 2003. Vegetable Crops, Vol. 2, pp. 413–42. Naya Udyog Publishers, Kolkata.
2. Bradbury J. H., S.V. Egan and P.J. Matthews, 1995. *Phytochem. Analysis*, **6**(5): 268 – 71.

3. Coates, D.J., D.E. Yen and P.M. Gaffey, 1988. Chromosome variation in taro, *Colocasia esculenta*, implications for origin in the Pacific. *Cytologia*, **53**(3): 551–60.

4. Food Security and Nutrition. Oxford and IBH Publishing Co. Pvt. Ltd., New Delhi.

5. Hill, Albert F. 1939. The Nomenclature of the Taro and its Varieties, *Botanical Museum Leaflets*. Harvard University, **7**(7): 113–8.

6. http: //farmextensionmanager.com/colocasia-variety/

7. Hunt, H.V., H.M. Moots and P.J. Matthews, 2013. Genetic data confirms field evidence for natural breeding in a wild taro population (*Colocasia esculenta*) in northern Queensland, Australia. *Gen. Resour. Crop Evol.*, **60**(5): 1695–1707.

8. Kazmi S.A., Z. Yang, N. Hong, G. Wang and Y. Wang, 2015. Characterization by Small RNA Sequencing of Taro Bacilliform CH Virus (TaBCHV), a Novel Badnavirus. *PLoS ONE*, **10**: 1–13.

9. Khare, C.P. 2007. Indian Medicinal Plants. 1, Springer, New York, 900p.

10. Kidanemariam D.B., A.C. Sukal, A.D. Abraham, F. Stomeo, J.L. Dale, A.P. James and R.M. Harding, 2018. Identification and molecular characterisation of Taro bacilliform virus and Taro bacilliform CH virus from East Africa. *Plant Pathol.*, **67**: 1977–86.

11. Krishnapriya, T.V. and A. Suganthi, 2017. Biochemical and phytochemical analysis of *Colocasia esculenta* (L.) Schott tubers. *Intern. J. Res. Pharmacy and Pharmacy Sci.*, **2**(3): 21-5.

12. Kuruvilla, K.M. and A. Singh, 1981. Karyotypic and electrophoretic studies on taro and its origin. *Euphytica*. **30**(2): 405-13.

13. Matthews, C.G., K.S. Milne, R.L.S. Forster and H.F. Neilson, 1996. Comparison of four potyvirus isolates infecting aroid species. *Acta Hort.*, No. 432, pp.354–63.

14. Ochiai, T., V.X. Nguyen, M. Tahara and H. Yoshino, 2001. Geographical differentiation of Asian taro, *Colacasia esculenta* (L.) Schott, detected by RAPD and isozyme analyses. *Euphytica*, **122**(2): 219–34.

15. Pawar, H.A., P.D. Choudhary and S.R. Kamat, 2018. An Overview of Traditionally Used Herb, *Colocasia esculenta* as a Phytomedicine. *Medicinal Aromat Plants*, **7**(4): 1-7.

16. Pearson, M.N., W.T. Bussell and J.J.C. Scheffer, 1998. New plant disease record in New Zealand: Dasheen mosaic potyvirus infecting taro (*Colocasia esculenta* Schott). *N. Z. J. Crop. hortic. Sci.*, **26**: 69–70.
17. Pearson, M.N., G.V.H. Jackson, J. Saelea and S.G. Morar, 1999. Evidence for two rhabdoviruses in taro (*Colocasia esculenta*) in the Pacific region. *Australas. Pl. Path.*, **28**: 248–53.
18. Purseglove, J.W. 1972. Tropical Crops: Monocotyledons. Longman and John Wiley, Harlow and New York.
19. Ramanatha, R.V., P.J. Matthews, P.B. Eyzaguirre and D. Hunter, 2010. The Global Diversity of Taro: Ethnobotany and Conservation (eds Ramanatha Rao V., M.P.J. Eyzaguirre, B. Pablo and D. Hunter). *Bioversity International,* Rome (Italy), 202p.
20. Sivaprasad, Y., B.V. Bhaskara Reddy, C.V.M. Naresh Kumar, K. Raja Reddy and D.V.R. Sai Gopal, 2011. First report of groundnut bud necrosis virus infecting taro (*Colocasia esculenta*). *Australas. Pl. Path.*, **6**: 30–32.
21. Sunitha, S. 2019. Colocasia. In: Handbook of Horticulture, vol. 1, pp. 661-2. ICAR, New Delhi.
22. Thirugnanavel, A., Bidyut C. Deka, Lily Rangnamei, Megokhona Meyase and J.K. Patel, 2015. *Colocasia* and its genetic diversity in Northeast India. *Ann. Rep.*, ICAR Research Complex for NEH Region, Umiam (Meghalaya).
23. Wang, Y., W.B. Borth, J.C. Green, L. Hamim, K. Cao, J.S. Hu and M.J. Melzer, 2018. Genome characterization and distribution of Taro bacilliform CH virus on taro in Hawaii, USA. *Eur. J. Plant Pathol.*, **150**: 1107–11.
24. Yang, I.C., G.J. Hafner, J.L. Dale and R.M. Harding, 2003. Genomic characterisation of taro bacilliform virus. *Arch. Virol.*, **148**: 937–49.
25. Yeturu, S., B.V.B. Reddy, S. Asadhi and D.V.R.S. Gopal, 2013. Sequence diversity of the nucleoprotein gene of peanut bud necrosis virus isolates from South India. *J. Phytopathol.*, **162**: 542–7.

28

Ullucu (*Ullucus tuberosus* Caldas)

Family: Basellaceae.

Synonyms: *Basella tuberosa* Kunth, *Chenopodium tuberosum* Ruiz, *Gandola tuberosa* Moq., *Melloca peruviana* Moq., *M. tuberosa* (Kunth) Lindl., *Ullucus aborigineus* Brucher, *U. kunthii* Moq., *U. tuberosus f. albiflorus* Kuntze, *U. tuberosus f. rubriflorus* Kuntze and *U. tuberosus* subsp. *aborigineus* (Brucher) Sperling.

Other common names: Ulluco. **Regional names**: Illaco, Ulluma or Ullucu (in Aymara) (Anon., 1989); Melloco (in Ecuador); Chungua or Ruba (in Colombia); Ulluku or Milluku (in Quechua); Olluco or Papalisa (in Bolivia and Peru); and Ulluma (in Argentina) (Busch and Savage, 2000; Arbizu *et al.*, 1997).

Introduction and Origin

The generic name *Ullucus*, as well as the common name 'ullucu' have been derived from the Quechua word *ulluku. Ullucus tuberosus*, probably originated in the high Andes of Peru and Bolivia, which is the only species in its genus and one of the most widely grown and economically important root crops in the Andean region of South America, second only to the potato (Herman and Heller, 1997). There is no clear wild progenitor of ullucu in the Andes, so perhaps it developed elsewhere and since went extinct in its original range. Ullucu is found primarily in Peru, Bolivia, Ecuador, and to a lesser extent in Colombia, Venezuela, Argentina and Chile. This crop is challenging to grow in the Pacific Northwest and very difficult in the rest of the country. Although a popular crop in its native region, it is little grown outside the Andes. In New Zealand, it has been cultivated for about 20 years and through a mutation breeding programme there, it has produced some new varieties. It was introduced into Sri Lanka early in the present century, but its cultivation is still virtually confined to the Andes (Humanity development library 2.0).

Nine heirloom ulluco varieties

Source: https://www.cultivariable.com

Ullucu is one of the more popular members of the Andean root and tuber crops. It is a staple carbohydrate foodstuff in parts of the Andes where it is cooked and eaten in a manner similar to potatoes. It is primarily grown for its edible tubers but unlike potato the edible leaves are also used similar to spinach. The tubers of ullucu have been eaten in the Andean populations since the time immemorial as this is an important source of protein, carbohydrate and vitamin C to people living in the high altitude (around 2,500 to 4,000m a.s.l.) mountainous regions of South America (Busch and Savage, 2000).

Subspecies and Cultivars

Ullucus tuberosus has a wild subspecies, *U.t.* subsp. *aborigineus*, which is usually triploid, but sometimes diploid forms are also available. The diploid wild ullucu may be either the progenitor of domesticated ulluco or merely a feral form. The subspecies appears to set invariably small pink tubers. Plants are decumbent with vines ranging from 30cm to 3m in length. The tubers are somewhat bitter and more mucilaginous than the domesticated varieties. This wild ullucu may give perennial leaf vegetable along with the edible tubers. Domesticated ullucus are generally diploid but sometimes triploid or tetraploid forms are also available.

In Colombia, there is a unique group of seemingly intermediate ullucus cultivated in the NE Andes that appear similar to wild ullucu with long stems and smaller tubers but colour and flavour are characteristic to domesticated ulluco (Parra-Quijano *et al.*, 2012).

There are hundreds of different, incredibly diverse varieties, but apparently with almost no sexual reproduction, usually the main generator of genetic diversity. The ullucu collection of the International Potato Center includes 360 varieties. Some high yielding varieties are 'Pica de Pulga', 'Purple' and 'Cusco Market'.

Plants bearing red tubers are the most frost-resistant though yellow-tuberous ones are consumed by most people as this type is flooded in Ecuador markets (Pietila and Jokela, 1988). The attractiveness of the colour of cultivars varies among countries and regions. In the New Zealand market, the preferred skin colour is red over plain yellow and mixtures of yellow, green and red (Busch and Savage, 2000).

In New Zealand, they have successfully produced new varieties with new colours and low geosmin content by inducing mutations using gamma irradiation (Pathirana *et al.*, 2011). Octoploid ullucus have been produced by treatment with oryzalin (Viehmannova *et al.*, 2012).

Botany

Ullucu plants are similar to potato in size though with more sprawling nature. It is a scrambling herbaceous perennial but with succulent leaves and stems, growing to a height of about 38cm before sprawling. The green to red stems are angular in cross section, and often have different colouration at the edges. The alternate leaves are broadly oval to cordate, 5-20cm long, 5-12cm broad, somewhat fleshy, variable in colour according to cultivar, ranging from dull green to bright green with red spots and purplish-yellow borders (Humanity development library 2.0).

Flowering is rare in ullucu though typically it flowers about two months after planting and most varieties flower profusely. Flowering and seed set continue in most varieties until the onset of tuber production in the fall. The flowers produce a single dry fruit which contains a single seed (Rousi *et al.*, 1988). Ullucu is an obligate outbreeder or nearly so (Pietila, 1995). Majority of accessions are diploids (2n=24), but rarely triploids (2n=36), and tetraploids (2n=48) are also recorded (Mendez *et al.*, 1994).

Smooth, starchy and spherical or elongated to slightly curved tubers formed on underground stolons are similar in size to potato. An ullucu plant forms a large number of tubers, usually more than 100, but at least

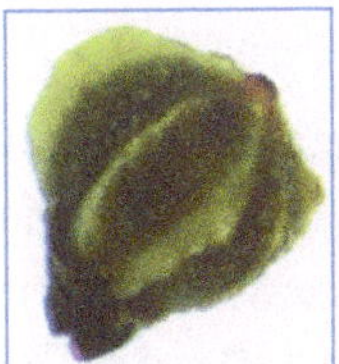

Ulluco seed

Ulluco seeds

half of them are often too small to eat, though per plant average yield is 30 tubers. Remaining tubers that are rather longer than broader measure 2.5-7.5cm though these have been known to grow up to 15cm (en.wikipedia.org/wiki/ullucus; Humanity development library 2.0). Tuber skin is thin and soft with inconspicuous buds. The flesh is low in fibre content. Brightly coloured waxy skin of the tubers also varies with the cultivar ~ being white, yellow, green, orange, red, purple and sometimes with spots, and the flesh is normally white or yellow. Over 70 cultivars have been recognised (Humanity development library 2.0). These colourful tubers of *Ullucus* varieties are regarded as one of the most striking foods in the Andean markets (Scheffer *et al.*, 2002).

Production

Production of ullucu is mostly limited to the Andean region. The commercial trade of this crop is in larger South American cities. More than 15,000ha of land is under its cultivation in Peru and production in mid-1960s was estimated to be about 14,000t/ha, of which roughly 10,000t was for human consumption, 2,000t for seed, and losses due to spoilage were estimated at about 1,000t (Humanity development library 2.0).

By Nzfauna - Own work, CC BY-SA 3.0, https://commons.wikimedia.org/w/index.php?curid=16996227

Nutritive Value

Ullucu tubers contain high levels of protein, calcium and carotene. Fresh tubers of ullucu are a valuable source of carbohydrates, comparable to one of the most world spread root crop ~ the potato. It also contains high fibre levels, moderate protein and only little fat (<2 per cent). Regarding the vitamin content, ullucu tubers contain a significant value of vitamin C (11.5 mg/100g), which is higher than commonly used vegetables such as carrots (6mg/100g) and celery but lower than yams (17.1mg/100g) or potato (19.7mg/100g) (en.wikipedia.org/wiki/ullucus). Dietary value variability is pronounced between cultivars.

Betalains Pigment

Ullucu has unique betalains pigments in the base form of betacyanins and acid form of betaxanthins (en.wikipedia.org/wiki/ullucus; Lim, 2015). Betalains act as antioxidants and may be nutritionally important. Out of 32 types of reported betalains in ullucu, 20 are in the form of betaxanthins and the remaining

Nutritional Values of Ulluco (Lim, 2015; Reyes *et al.*, 2017; Busch and Savage, 2000) grown in South American countries in comparison to nutrient content of major staple root crops (Anon., 2016) per 100g portion

	Ulluco	*Potato*	*Cassava*	*Sweet Potato*	*Yam*	*Oca*
Nutrients						
Energy (kJ)	311 (74.4 kcal)	322	670	360	494	255 (61 kcal)
Water (g)	83.7 - 87.6	79	60	77	70	84.1
Carbohydrates (g)	14.4 - 15.3	17	38	20	28	13.3
Dietary fibre (g)	0.9 - 4.9	2.2	1.8	3	4.1	-
Fat (g)	0.1 - 1.4	0.09	0.28	0.05	0.17	-
Protein (g)	1.1 - 2.6	2.0	1.4	1.6	1.5	1.0
Sugar (g)	-	0.78	1.7	4.18	0.5	-
Vitamins						
Retinol (A) (μg)	5	-	-	-	-	1
Thiamine (B1) (mg)	0.05	0.08	0.09	0.08	0.11	-
Riboflavin (B2) (mg)	0.03	0.03	0.05	0.06	0.03	0.13
Niacin (B3) (mg)	0.2	1.05	0.85	0.56	0.55	0.43
Vitamin C (mg)	11.5	19.7	20.6	2.4	17.1	38.4
Minerals						
Calcium (mg)	3	12	16	30	17	2
Iron (mg)	1.1	0.78	0.27	0.61	0.54	1.6
Phosphorus (mg)	28	57	27	47	55	36

Source: en.wikipedia.org/wiki/Ullucus and en.wikipedia.org/wiki/Oxalis_tuberosa

12 in the form of betacyanins (Svenson *et al.*, 2008). A high concentration of betacyanins is responsible for red or purple tuber varieties, whereas, a high concentration of betaxanthins is responsible for the yellow or orange colouring of the tubers. Inspite of having significant amount of betalains pigments, the antioxidant capacity of the ullucu is low in comparison to three other Andean tuber crops ~ native potato, oca and mashua because of the absence of flavonoids, carotenoids and anthocyanin pigments which are much more abundant sources of antioxidant compounds than betalains (Campos *et al.*, 2006). The stability of the betalains pigments makes ullucu a promising industrial crop of natural pigments (Manrique *et al.*, 2017).

Sometimes people after eating a lot of ullucu experience a change in urine or faeces colour. This is known as beeturia ~ a condition where body cannot break down the betalain pigments in ullucu. People most commonly experience this when eating beets and ullucu is also coloured by the same group of pigments. Most medical information indicates that this is a harmless condition that occurs in about 10 per cent of people but it may indicate either an iron deficiency or excessive oxalate in some cases.

Uses

At all stages of growth, all the parts of ullucu plant are edible. Though the tubers and leaves both can be eaten raw or cooked, most commonly cooked; tubers being the primary part used as food; and the leaves can be used similar to spinach. The tubers have a much firmer texture than potatoes and taste like a combination of beet and potato. They are typically cooked whole with the skin on and require no preparation other than washing. Some varieties produce smaller, bite-sized tubers with higher tuber counts per plant and are suited for soups as they only need to be rinsed and added into the soup pot where they will thicken the broth and remain intact even with long boiling. These are also good for eating raw in salads. Some other varieties produce a smaller number of larger tubers that substitute well for potatoes and beets in many recipes and are best roasted either alone or with other vegetables and meats. They are commonly boiled whole or cut and fried; though they don't mash very well on their own but make a nice addition to mashed potatoes.

Similar to several other vegetables, ullucu too loses colour when cooked though colour loss can be minimized by acidifying cooking water and by minimizing cooking time. Also, adding ullucus near the end of cooking will preserve much of the colour. Yellow varieties tend to retain their colour better when cooked than red varieties.

Propagation

True seeds are rare in this crop and are used only for breeding. Ullucu produces sexual seeds only under very narrow environmental conditions. Moreover, the seed has a low rate of embryo formation and is very difficult to germinate. Ullucu is planted almost exclusively from seed tubers. Whole or cut tubers are planted at about the same time as potatoes, although harvest takes place later than for most potatoes. Although the tubers can be directly planted in the main field but a few people would like to pre-sprout them before planting. To prepare presprouted seedlings, tubers are planted in nursery beds or in pots at room temperature. Sprouted tubers can wait to be planted for a month or more. Once they have sprouted, it is best to store the tubers where they can get some light because sprouts that grow in darkness will become spindly and fragile. It is recommended to pot the tubers in some barely damp soil with the tip of the sprout exposed and then these are transplanted after the last frost. It is very tolerant about transplanting. It is not necessary to pot sprouted tubers individually; they can be put a bunch in a single container and then divided during transplantation.

Ulluco seedling

Ullucu can also be propagated using stem cuttings. Sections of ullucu stem root easily and quickly in water or wet soil. They may even form tubers while rooting in water if the lighting period is 12 hours or less. Ullucu roots so easily that it may be possible to simply chop up stems and cover them lightly with soil, as long as the conditions are wet enough.

Growth and Development

The timing of planting and harvesting, climate, photoperiod, cultural practices, and pests and diseases will vary considerably by location. In the Andes, some varieties may produce a crop in as little as 5 or as long as 9 months but the average is 6 to 8 months (Anon., 1989). If the growing season is shorter than 7 months, plants can be started in pots; but the thing which is important is to keep them alive until at least the middle of November in a short season climate. Ullucu can be grown much like potatoes but it tolerates a narrower range of climate. Whole or cut tubers are planted in spring. Planting occurs at about the same time as for main crop potatoes, although harvest takes place later than those of potatoes.

In the Andes, once the day length falls to about 12.5 hours in the autumn, very thin stolon-forming on the stems can be seen, which drops to the ground forming tubers. The same process is occurring below the ground, wherever the stems touch, and those will be the source of earliest and largest tubers. At this stage, a little of fine FYM or compost should be carefully added close to each plant to increase the yield.

Ullucu thrives under very narrow climatic conditions and grows poorly where summer temperatures exceed 24°C. The plant grows best in cool moist conditions under short day-lengths of about 12 hours (the production of stolons and stolon-borne tubers is stimulated by even shorter days of 10 hours). Ullucu has considerable resistance to frost and thus it is well suited to the Andean altiplano or high valley conditions. It is cultivated at elevations between 1,500 and 4,000m where it gives higher yields than the Andean cultivars of potato. Tubers will become green when exposed to light but this in any way does not alter flavour or become toxic as in case of potatoes (Humanity development library 2.0).

Cultivation

The three Andean tubers (oca, mashwa and ullucu) are grown in the same agro-ecological zone and their soil requirements and cultivation practices are almost akin to potato (https://hort.purdue.edu/newcrop/1492/tubers.html). For this reason they are dealt with together. In the northern area of the Peruvian sierra, the traditional form of cultivation is on Melgas where after potato harvesting the land is divided into 3-5 plots each of which is planted with one of the Andean tubers (Arbizu and Tapia, 1994).

On the high plateau of Puno and in the agro-ecological zone of the semi-humid Puna a mixture of tubers is sown. By contrast, in the Quechua agro-ecological zone, the oca and ullucu are planted together with maize. These crops show a high response to agricultural work such as fertilization, earthing up, hoeing and above all the control of pests and diseases. Their production increases to levels of 40 to 50 tonnes per hectare being comparable to the highest potato yields (Arbizu and Tapia, 1994).

Ullucu thrives well even in relatively poor soils though loose and light soils are preferably better to facilitate harvesting. Before planting, the field should be ploughed or dug 2-3 times thoroughly and all the remnants of weeds should be taken out. Compost at the rate of 100-150 quintals per hectare should thoroughly be mixed in the soil before final field preparation. Organic manure is always better than addition of inorganic fertilizers. Ullucu doesn't expect much as fertilizers may harm the roots. Cordova (2017) found earthworm castings as a better proposition for ullucu than goat manure. The tubers or presprouts are planted in September-October in furrows in well-cultivated soil. Highest yield of usable tubers has been observed with furrows 80-90cm apart and plants at 30cm along the rows (Humanity development library 2.0). The planting depth should be roughly equal to the tuber diameter. Vining types (wild ullucu) are best grown on mounds, planted in the centre, allowing the vines to trail down the sides though bush types on flat ground. Plants require 2-3 soil-earthings, first at 30 days of planting and then after 45 days from first earthing. If still the tuber formation is visible after some time the third earthing may be effected after another 45 days of second earthing. In no case the forming tubers should be visible on the bunds from outside. This aids in smooth formation of tubers.

The crop normally requires weeding only at eary stages. It grows as a thick ground cover and is able to suppress weeds very effectively when densely planted. However, it will take a while to spread the canopy effectively so by then it requires at least 2 weedings, one after about one month of planting and the other after 20-30 days. Mulching in this crop is not recommended as the stems drop descending stolons to the ground which then will not be able to penetrate the soil which consequently will hamper tuber production.

Ullucu is frequently intercropped with oca (*Oxalis tuberosa*), garlic or onions with reasonably good results. These plants will be harvested before the ullucu plants grow really large in the fall.

Harvest and Yield

The tubers are ready for harvest in 6-8 months. In North America where climate is warm *cum* frost-free, *i.e.* around 16-20°C, earliest harvest of ullucu can be done after the middle of November, achieving about 50 per cent of the possible yield, roughly 227-453g per plant. Harvesting in the first week of December provides around 80 per cent of the possible yield though harvesting from late December to early January provides 100 per cent tuber yield. At full maturity, most varieties form more than 100 tubers per plant but only about a third of them are of marketable size (roughly 2.5cm or more in diameter). Usually a proportion of the smaller tubers from the harvest are preserved for use as seed tubers the following year. The tubers are lifted usually by hand and often eaten immediately (Humanity development library 2.0). At the time of harvest, if the aboveground part of the plants is not killed by frost then after digging the tubers the foliage can be buried back in the holes to get a secondary crop from ullucu.

Postharvest

Under careful storage conditions of 1-3°C temperatures, 90 per cent humidity and full darkness, ullucu has been known to store for two years. It is pretty easy to get a year out of them. Small tubers can suffer dehydration so these should be checked regularly. If these tubers are becoming soft, these should be shifted in some barely damp soil or in the crisper of refrigerator. Even badly desiccated ullucu tubers that start to look like raisins will usually recover if introduced to some damp soil.

Pests and Diseases

The most important pest is stated to be the Andean weevil (*Premnotrypes solani*). Nematodes of *Globodera* spp. attack the plant. Reported fungi include *Aecidium cantense* and *Rhizoctonia solani*. A number of viruses have been found in virtually all ullucu material examined. These viruses include an 'ullucu strain of tobacco mosaic virus' (TMV/U) and 'papaya mosaic virus' (PMV/U), as well as three new viruses designated 'ullucu virus A, B and C', respectively. Apart from slight leaf mottling, these viruses appear to be symptomless but laboratory experiments using virus-free material prepared by meristem culture suggest that growth and yield are considerably retarded by their presence. No vector is known to be involved in transmitting these viruses. Grubs of the weevil can be controlled by hand picking and killing or by use of lindane and primiphos-methyl. Nematodes are kept at bay by mixing carbofuran in the soil at the time of field preparation. Fungi attack can be minimized through spraying the crop with Captafol 0.2 per cent alternate with Bavistin 0.1 per cent at fortnightly intervals twice. Any plant showing visible virus symptoms should be uprooted and burnt. In any case if attack is mild no pesticide should be used and instead the affected portion should be removed and destroyed.

References

1. Anon., 1989. Lost Crops of the Incas: Little-known Plants of the Andes with Promise for Worldwide Cultivation. National Research Council. National Academic Press, USA.
2. Anon. 2016. Nutrient data laboratory. USDA (Retrieved 10 August 2016).
3. Arbizu, C., Z. Huaman and A. Golmirzaie, 1997. Other Andean Roots and Tubers. In: Biodiversity in Trust: Conservation and Use of Plant Genetic Resources in CGIAR Centres (eds: Fuccillo, D., L. Sears and P. Stapleton), pp. 39-56. Cambridge University Press, Cambridge, England.
4. Arbizu, C. and M. Tapia, 1994. Andean Tubers, pp. 149-63. In: Plant Production and Protection Series No. 26 (eds Hernando Bermejo J.E. and J. Leon). FAO, Rome, Italy (CIP. Lima, Peru). https: //hort.purdue.edu/newcrop/1492/tubers.html.
5. Busch, J. and G.P. Savage, 2000. Nutritional composition of ulluco (*Ullucus tuberosus*) tubers. *Proc. Nutr. Soc. New Zealand*, pp. 55-65.
6. Campos D, G. Noratto, R. Chirinos, C. Arbizu, W. Roca and L Cisneros-Zevallos, 2006. Antioxidant capacity and secondary metabolites in four species of Andean tuber crops: Native potato (*Solanum* sp.), mashua (*Tropaeolum tuberosum* Ruiz and Pavon), oca (*Oxalis tuberosa* Molina) and ulluco (*Ullucus tuberosus* Caldas). *J. Sci. Food Agric.*, **86**(10): 1481-8.
7. Cordova, P. A. A. 2017. La fertilizacion organica del melloco (*Ullucus tuberosus* l.) Y su rentabilidad en la comunidad de tuncarta. *INNOVA Res. J.*, **1**(12): 132-49.
8. Hermann, M. and J. Heller, 1997. Andean roots and tubers: Ahipa, arracacha, maca and yacon. *Intern. Pl. Gen. Res. Inst.*, Rome (Italy).
9. https: //en.wikipedia.org/wiki/Oxalis_tuberosa.
10. https: //en.wikipedia.org/wiki/Ullucus.

11. Humanity development library 2.0. Root crops: Ullucu (*Ullucus tuberosus*). nzdl.org.
12. Lim, T.K. 2015. Edible, Medicinal and Non-medicinal Plants (Vol. 9): Modified Stems, Roots and Bulbs. Springer Science+Business Media Dordrecht, pp. 741–5.
13. Manrique, I., C. Arbizu, F. Vivanco, R. Gonzales, C. Ramirez, O. Chavez, D. Tay and D. Ellis, 2017. *Ullucus tuberosus* Caldas: coleccion de germoplasma de ulluco conservada en el Centro Internacional de la Papa (CIP) (in Spanish). Centro Internacional de la Papa (CIP), Lima (Peru). hdl: 10568/81224.
14. Mendez, M., C. Arbizu and M Orrillo, 1994. *Niveles de ploidia de los Ullucus cultivados y silvestres* (in Spanish). Valdivia (Chile): Universidad Austral de Chile. Resúmenes de trabajos presentados al VIII Congreso Internacional de Sistemas Agropecuarios y su proyeccion al tercer milenio, p. 12.
15. Parra-Quijano, M., S. Panda, N. Rodrýìguez and E. Torres, 2012. Diversity of *Ullucus tuberosus* (Basellaceae) in the Colombian Andes and notes on ulluco domestication based on morphological and molecular data. *Genet. Resour. Crop Evol.*, **59**: 49–66.
16. Pathirana, R., C. Sansom, B. Smallfield, N. Perry, T. Frew, G. Timmerman-Vaughan and J. Heyes, 2011. *In vitro* mutagenesis to adapt ulluco (*Ullucus tuberosus*) to New Zealand conditions. In the report of 'The New Zealand Institute for Plant and Food Research Limited'. New Zealand.
17. Pietila, L. 1995. Pollination requirements for seed set in ulluco (*Ullucus tuberosus*). *Euphytica*, **84**(2): 127-31.
18. Pietila, L. and P. Jokela, 1988. Cultivation of minor tuber crops in Peru and Bolivia. *J. agric. Sc. Finland*, **60:** 87–92.
19. Reyes Garcia, Maria, Gomez-Sanchez Prieto, Ivan and Espinoza Barriento Cecilia, 2017. *Tablas peruanas de composicion de alimentos* (in Spanish). Ministerio de Salud, Instituto Nacional de Salud, Lima, pp. 66–67.
20. Rousi, A., M. Yli-Rekola, P. Jokela, R. Kalliola, L. Pietila and J. Salo, 1988. The fruit of *Ullucus* (Basellaceae) - an old enigma. *Taxon*, pp. 71-75.
21. Scheffer, J.J.C., J.A. Douglas, R.J. Martin, C.M. Triggs, S. Hallow and B. Deo, 2002. Agronomic requirements of ulluco (*Ullucus tuberosus*) – a South American tuber. *Agronomy* (N.Z.), **33**: 41-7.
22. Svenson J., B.M. Smallfield, N.I. Joyce, C. Sansom and N. Perry, 2008. Betalains in red and yellow varieties of the Andean tuber crop ulluco (*Ullucus tuberosus*). *J. Agr. and Food Chem.*, **56**(17): 7730-37.

23. Viehmannova, I., M. Travnièkova, E. Špatenkova, M. Èerna, and P. Travnicek, 2012. Induced polyploidization and its influence on yield, morphological and qualitative characteristics of microtubers in *Ullucus tuberosus*. *Plant Cell, Tissue and Org. Cult.*, **109**(1): 83-90.

29

Wild Arum (*Arum maculatum* L.)

Family: Araceae.

Other common names: Adam and Eve, Adder's meat, Jack in the pulpit, and Lamb-in-a-pulpit (Gordon, 1891); Adder's root, Arum, Arum lily, Bobbins, Cows and Bulls, Cuckoopint, Cuckoo plant, Devils and Angels, Friar's cow, Lords-and-Ladies, Naked boys, Portland arrowroot, Starch Root, Wake Robin and Wild Arum (Prakash *et al.*, 2018).

Synonyms: *Arisarum maculatum* (L.) Raf.; *Arum heldreichii* Orph. ex Boiss.; *A. immaculatum* (Rchb.) Rchb.; *A. malyi* Schott; *A. pyrenaeum* Dufour; *A. trapezuntinum* Schott ex Engl.; *A. vernale* Salisb.; *A. vulgare* Lam. and *A. zeleborii* Schott

Introduction and Origin

Arum in Greek stands for 'poisonous'. *Arum maculatum* has native range from Europe to West Turkey and the West Caucasus (Govaerts and Frodin, 2002; en.wikipedia.org). It is widespread across temperate northern Europe. This shade-loving flowering plant spreads quickly by self-seeding and is usually found growing in woodland areas and riversides where it can occasionally grow as a weed in partly shaded spots.

All parts of this plant if eaten raw are poisonous to humans and animals. The plant can cause skin irritation and illness. The fruits are harmless to birds. They eat the berries and disperse the seeds.

Botany

Arum maculatum with chromosomr number of 2n=56 has tuberous root system. It is an upright herbaceous flowering perennial growing to 20-60cm high (gardenershq.com), bearing 7-20cm long (Stace Clive, 2005) mid deep green, arrow-shaped glossy leaves blotched purplish-black (plants.ces.ncsu.edu). The leaves appear in the spring (April–May) and are followed by flowers borne on an erect poker-shaped inflorescence called a spadix that is deep purple to pale yellow and is partially enclosed in a pale green spathe or leaf-like hood (en.wikipedia.org; plants.ces.ncsu.edu) with a greenish-white interior having a purple band (plants.ces.ncsu.edu). The spathe can be up to 25cm high and the fruiting spike which follows later in the season may be up to 5cm (Stace Clive, 2005). The flowers are hidden from sight and clustered at the base of the spadix with a ring of female flowers at the bottom and a ring of male flowers above them (en.wikipedia.org).

There is a ring of hairs above the male flowers that works as an insect trap. The inflorescence has the ability to heat itself above the ambient air temperature (up to 15°C warmer) to such a degree that it is quite noticeable to the touch (pfaf.org; Hemming *et al.*, 1963). The foul and urinous smell of spadix and its warmer temperature attracts the insects, especially owl-midges, *Psychoda phalaenoides* (Lack and Diaz, 1991) which get trapped beneath the ring of hairs and are dusted with pollen by the male flowers before escaping. Insects carry the pollen to the spadices of other plants where they pollinate the female flowers. The foul

Arum maculatum
CC.BY.SA.3.0,.https://commons.wikimedia.org/w/index.php?curid=68710

Berries of *Arum maculatum*
By.Frank.Vincentz.-.Own work, CC.BY-SA.3.0,.https://commons.wikimedia.org/w/index.php?curid=3195268

smell disappears once the flower has been pollinated. In autumn the lower ring of flowers (female) forms a cluster of extremely poisonous but attractive bright red to orange berries that are up to 5cm long (Stace Clive, 2005). These berries remain even after the leaves and spathes have withered away.

Arum has deep-rooted but lateral rhizome that multiplies each year and is difficult to remove. This stores the starch and may be quite large. In mature specimens, the rhizomes may be as much as 400mm below ground level. These brownish and oblong fleshy tubers resemble somewhat those of potato having white flesh when fresh (healthbenefitstimes.com). These yield a milky insipid juice but soon producing a burning and pricking sensation when touched or tasted. The acridity is lost during the process of drying and by application of heat when the substance of the tuber is left as starch (healthbenefitstimes.com). When thoroughly baked the tubers become edible due to nutritious starch (healthbenefitstimes.com).

Constituents

Active substances present in roots of *A. masculatum* are gums, proteins, carbohydrates, starch, acids and essential oil which contain the poisonous substance of 'aroine' (mediplantepirus.med.uoi.gr). All parts of arum plants are considered poisonous and contain calcium oxalate crystals and soluble oxalates (plants.ces.ncsu.edu).

Toxicity

Like many other aroids, cuckoopint too contains a bitter poisonous sap (britannica.com). The attractive berries are extremely poisonous to many animals including humans but harmless to birds which eat them and propagate the seeds (en.wikipedia.org). These needle-shaped crystals of oxalates irritate the skin, mouth, tongue and throat which result in swelling of throat, difficulty in breathing, burning pain and upsetting the stomach (ncbi.nlm.nih.gov). There is no known antidote to *A. maculatum* poisoning. Airway management may reduce the mortality and aggressive fluid administration may prevent renal injury (Prakash *et al.*, 2018).

Uses

When roasted or thoroughly boiled, the tuberous root of the cuckoo-pint may be edible (en.wikipedia.org). The leaves of *Arum maculatum* are also used to make soup in the Andýrýn region of Turkey where they use it with yogurt and boiled for long hours to eliminate the toxicity. This process results in a sour soup which is called 'Tirºik' (en.wikipedia.org).

The grounded roots were once traded under the name of Portland sago. It was used like salep (orchid flour) to make saloop, a drink popular before the introduction of tea or coffee. It was also used as a substitute for arrowroot. It can be highly toxic if not prepared correctly.

The whole plant is used as antirheumatic, expectorant and laxative. The root and the leaves have astringent properties. It is also administered for the treatment of arthritis, insomnia, cough, running nose, bronchitis and pneumonia. In homoeopathy it is recommended in the form of poultice which is applied externally to the skin against bleeding and flushes. The plant is administered as antirheumatic, expectorant and laxative in the form of powder from the dried root dissolved in water (daily dose from 5 to 20g). Its roots and leaves are applied externally to the skin in the form of poultice against rheumatic pain, sprains, boils and hematomas (mediplantepirus.med.uoi.gr).

The arum lily may be used as an ornamental in mild climates in shade gardens, borders and woodlands. The roots are a traditional source of starch for stiffening clothes.

Propagation and Cultivation

The cuckoopint plant can be propagated both by seeds and tubers. It is best to sow the seeds outdoor in the autumn at a depth of 3cm. If the grower intends to start seedlings off inside then seeds should be prepared 12 weeks before they are to be planted out (in the spring or autumn). If growing from tubers then they should be buried at a depth of 7-8cm directly in the field.

The plants of the genus are not hardy enough below freezing temperatures. Cuckoopint and related plants can take up to six months to germinate. Ideal temperature for germination is 13-18°C. Seedlings should be transplanted at a distance of 30 to 45cm apart either in sunny or partially shaded areas. In hot summers, the plant may need to be shaded in the afternoon for protection. The soil should be cool, moist and rich in humus (britannica.com).

References

1. mediplantepirus.med.uoi.gr/pharmacology_en/plant_details.php?id=111
2. britannica.com/plant/cuckoopint
3. en.wikipedia.org/wiki/Arum_maculatum
4. gardenershq.com/Arum-Cuckoopint.php
5. Gordon, W.J. 1891. Our Country's Flowers and How to Know Them. Simkin, Kent and Co., Marshall, Hamilton, London.

6. Govaerts, R. and D.G. Frodin, 2002. World Checklist and Bibliography of Araceae (and Acoraceae): 1-560. The Board of Trustees of the Royal Botanic Gardens, Kew, England.
7. healthbenefitstimes.com/cuckoo-pint-arum/
8. Hemming, F.W., R.A. Morton and J.F. Pennock, 1963. Constituents of the unsaponifiable lipid fraction from the spadix of *Arum maculatum*. *Proc. Royal Society B: Biological Sciences*, **158** (972): 291–310. Bibcode: 1963RSPSB.158.291H. doi: 10.1098/rspb.1963.0049. JSTOR 90493.
9. Lack, A.J. and A. Diaz, 1991. The pollination of *Arum maculatum* L.- a historical review and new observations (PDF). *Watsonia*, **18**: 333–42. Archived from the original (PDF) on 2009-01-09. Retrieved 2012-02-15.
10. ncbi.nlm.nih.gov/pmc/articles/PMC6018264/#: ~: text=Arum per cent 20maculatum per cent 2C per cent 20commonly per cent 20known per cent 20as,this per cent 20plant per cent 20are per cent 20considered per cent 20toxic.
11. pfaf.org/user/Plant.aspx?LatinName=Arum+maculatum
12. plants.ces.ncsu.edu/plants/arum-maculatum/
13. Prakash Raju, K.N.J., K. Goel, D. Anandhi, V.R. Pandit, R. Surendar and M. Sasikumar, 2018. Wild tuber poisoning: *Arum maculatum* - A rare case report. *Int. J. Crit. Illn. Inj. Sci.*, **8** (2): 111–4. doi: 10.4103/IJCIIS.IJCIIS_9_18 (inactive 31 July 2022). PMC 6018264. PMID 29963416.
14. Stace Clive, 2005. New Flora of the British Isles (2nd ed.), 778p. Cambridge University Press, England.

30

Winged Bean/Asparagus Bean/ Dragon Bean [*Psophocarpus tetragonolobus* (L.) D.C.]

Family: Fabaceae. **Subfamily**: Faboideae.

Other common names: Asparagus bean, Cigarillas, Dragon bean, Four-angled bean, Four-cornered bean, Goa bean, Manila bean, Mauntius bean, Princess bean (en.wikipedia.org/wiki/Winged_bean).

Introduction and Origin

Generic name *Psophocarpus* means 'noisy fruit' as the pods make a popping noise when dehiscing to release the seeds, and the specific name *tetragonolobus* refers to the four angles of the fruit (Rajan, 2018). Winged bean's historical origin is uncertain. Many authors place the origin in the Papua New Guinea and Indonesian island regions where many genetic strains of Winged Bean exist (Echo community, 2023). The species is not found growing wild although it has been noted growing as an escape in Myanmar and the Philippines. The native range of this crop, as found in different literatures over internet, is Tanzania (POWO, 2023), New Guinea (Hymowitz and Boyd, 1977) and/or Africa (Madagascar or

Mauritius), and may have spread to Asia. It is now cultivated usually as a market garden crop in southern India, Myanmar, Malaysia, New Guinea, Indonesia, the Philippines, China and Thailand, and to a lesser extent in Africa ~ mainly in Ghana and Nigeria, and in the West Indies (Root Crops, 2023).

Winged bean is widely recognized by farmers and consumers in South Asia for its disease resistance *cum* variety of uses. Although mainly cultivated for root tubers, whole plant of winged bean is edible and has the potential to become a major multi-use food crop in the tropics of Asia, Africa and Latin America (Khan, 1982).

Botany

Winged bean is a herbaceous perennial vine twining to the left (Martin, 1978), growing to a height of 3–4m, generally taller and notably larger than the common bean (Etats-Unis, 2009), with moderately thick and generally green stem though sometimes boasting purple (en.wikipedia.org/wiki/Winged_bean), that is slightly ridged and grooved, and trifoliate leaves on long and stiff petioles. The leaflets are ovate, 7.5-15cm long with terminal one usually longer than the

Psophocarpus tetragonolobus

KENPEI. "*Psophocarpus tetragonolobus*, Osaka-fu Japan". Wikimedia Commons, 11 Aug. 2007, (CC BY-SA 2.5), commons. wikimedia.org/wiki/File:Psophocarpus_ tetragonolobus4.jpg. Accessed 19 May 2023.

Winged Bean Flower

Vinayaraj. "*Psophocarpus tetragonolobus*". Wikimedia Commons, 13 Oct. 2013, (CC BY-SA 4.0), commons.wikimedia. org/wiki/Category:Psophocarpus_ tetragonolobus#/media/File:Psophocarpus_ tetragonolobus_10.JPG. Accessed 19 May 2023.

laterals and attached to the petiole by a marked pulvinus (Root Crops, 2023). Shape of its leaves ranges from ovate to deltoid, ovate-lanceolate, lanceolate and long lanceolate (Khan, 1982). The green tone of the leaves also varies in various tones. Its chromosome number is 2n=2x=18.

The inflorescence is borne on an axillary raceme, up to 15cm in length, with 2-10 blue, white or lilac flowers. It is usually pollinated by insects, especially bees. The rectangular (sometimes appearing flat) bean pods, with characteristic serrated wings running down the four corners contain 5-20 seeds which can vary in colour from white, through varying shades of yellow and brown to black, and may also be mottled (Root Crops, 2023). Depending upon the genotype, the bean pod may be smooth or rough and cream, green, pink or purple in colour. It is typically 15–22 cm long having waxy skin and the young pods having partially translucent flesh. A fully ripe pod turns ash-brown and splits open to release the seeds (beans) that is often round though oval and rectangular also occur (Khan, 1982). The beans are similar to soybeans in both use and nutritional content (containing 29.8-39 per cent protein) (en.wikipedia.org/wiki/Winged_bean).

Winged Bean Pods

By STRONGlk7 - Own work, CC BY-SA 3.0,
https://commons.wikimedia.org/w/index.php?curid=21474925

Though the plant is perennial and produces new growth annually from shallow and persistent roots, it is usually grown as an annual for optimum results. Winged bean has numerous fibrous roots with the main lateral ones that run horizontally near the soil surface and usually become thickened and tuberous near the base of the plant after a few months; however, this does not occur in all the strains (Root Crops, 2023). Roughly cylindrical, starchy, fibrous and generally small tuberous roots are 8-12cm long, 2-4cm in diameter, and the weight of a single tuber is seldom more than 50g (Martin, 1978; Root Crops, 2023; Bassal *et al.*, 2020). Tubers bear brown and easy to peal skin, and white or cream-coloured flesh that is firm and moist, with a distinctive nutty, earthy flavour (Growables, 2023).

Varieties

Winged bean is a self-pollinating plant but mutations and occasional outcrossing may produce variations in the species (Khan, 1982). There are many varieties of winged bean varying in length, width and colour of pods. Various economic attributes of the crop were worked out to find out the relation of a particular variety in producing various economic traits for the greatest benefit to the farmers (Motis, 2019; Growables, 2023). The varieties evaluated were 'Bogor', 'Chimbu', 'Day Neutral', 'Flat', 'Ribbon', 'Siempre' and 'Square'.

Nutritional Value

All the parts of the plant including leaves, flowers, roots, young bean pods and seeds have high nutrient composition, especially proteins, vitamins and minerals; the starchy tubers have high percentage of proteins, *i.e.* up to 20 per cent (by weight) compared to other vegetables; the leaves and flowers containing 5-15 per cent proteins (by weight) and the seeds 32-37 per cent proteins which is similar to the amount of proteins found in soybeans though higher than that found in other beans (Sri Kantha and Erdman, 1984).

Tuberous Roots

The water content of the fresh tubers is 52-68 per cent. Dry matter composition is protein 13-20 per cent, fat 0.6-1.4 per cent, carbohydrate 63-77 per cent, fibre 1.5-21 per cent, ash 1.7-3.9 per cent, calcium 40mg/100g, iron 3mg/100g and phosphorus 64mg/100g. The wide range of values appears to be associated with variety and possibly with maturity of the root tubers. The carbohydrate is about 80 per cent starch and 20 per cent sugars. The non-protein nitrogen was found to be 7.4-15 per cent of the crude protein but in some other reports the values were found as high as 50 per cent. Sulphur-containing amino acids are low but the aspartic acid content is exceptionally high (Root Crops, 2023).

Seeds

Nutritional value per 100g of seeds of winged beans, raw and mature seeds is given below:

Energy	1,711 kJ (409 kcal)
Carbohydrates	41.7g
Dietary fiber	25.9g
Fat	16.3g
Saturated	2.3g
Monounsaturated	6g
Polyunsaturated	4.3g
Protein	29.65g
Vitamins	**Per cent DV† (Quantity)**
Vitamin A	0 IU
Thiamine (B1)	90 per cent (1.03mg)
Riboflavin (B2)	38 per cent (0.45mg)
Niacin (B3)	21 per cent (3.09mg)
Pantothenic acid (B5)	16 per cent (0.795mg)
Vitamin B6	13 per cent (0.175mg)
Folate (B9)	11 per cent (45μg)
Vitamin C	0 per cent (0mg)
Minerals	**per cent DV† (Quantity)**
Calcium	44 per cent (440mg)
Iron	103 per cent (13.44mg)
Magnesium	50 per cent (179mg)
Manganese	177 per cent (3.721mg)
Phosphorus	64 per cent (451mg)
Potassium	21 per cent (977mg)
Sodium	3 per cent (38mg)
Zinc	47 per cent (4.48mg)

Units: μg = micrograms; mg = milligrams; IU = International units†Percentages are roughly approximated using US recommendations for adults. Source: USDA FoodData Central

Source: en.wikipedia.org/wiki/Winged_bean

The seeds also contain 23-40 per cent carbohydrates (National Research Council, 1981; Misra *et al.*, 1987; Mohanty *et al.*, 2015), vitamins such as vitamins B1, B2, B3, B6, B9, C, A and E. The mature seeds contain 14-25 per cent fats by weight, of which 94 per cent remains in free form whereas the rest are complexed with carbohydrates and proteins (Mohanty *et al.*, 2015).

The seed of winged bean is rich in minerals including calcium, iron, phosphorous, potassium, sulphur, sodium, magnesium, zinc, manganese, boron, barium, copper, chromium and strontium (Misra *et al.*, 1987; King and

Puwastien, 1987; Jaffe and Korte, 1976). The seeds contain 54-75 per cent of unsaturated fatty acids, of which 38.6 per cent is mono-unsaturated and 36.9 per cent is polyunsaturated (Mohanty *et al.*, 2015; Varriano-Marston *et al.*, 1983), but no trans fatty acids were recorded. Winged bean oil contains 30 to 40 per cent saturated fatty acids which represent more than the content observed in soybean (Khor *et al.*, 1980). Oleic and linoleic acids represent nearly 50 per cent of unsaturated fatty acids. Unsaponifiable lipids are mainly represented by â-sitosterol (66.4 per cent) and stigmasterol (25.1 per cent) (Mohanty *et al.*, 2015). Moreover, winged bean oil is found qualitatively better than soybean oil because of its high oxidative and high thermal stabilities (Makeri *et al.*, 2016).

Phytochemicals

As per IMPPAT (2023), the phytochemicals present in different parts of the winged bean are given in the Table below. This work is licensed under the Creative Commons Attribution-NonCommercial 4.0 International License. To view a copy of this license, one will have to visit http://creativecommons.org/licenses/by-nc/4.0/ or send a letter to Creative Commons, PO Box 1866, Mountain View, CA 94042, USA.

Plant Part	*Phytochemical Name*	*Reference*
Fruit	Phaseollidin	ISBN:9788185042084
Fruit	Isomedicarpin	ISBN:9788185042084
Leaf	beta-Carotene	ISBN:9788172362461
Seed	Myristic acid	ISBN:9788172362461
Seed	Stearic acid	ISBN:9788172362461
Seed	Maritimetin	ISBN:9788185042114
Seed	N6-Methylagmatine	ISBN:9788172362461, ISBN:9788185042145
Seed	Docosanoic acid	ISBN:9788172362461
Seed	Palmitic acid	ISBN:9788172362461
Seed	Isovitexin	ISBN:9788185042114
Seed	Arachidic acid	ISBN:9788172362461
Seed	Oleic acid	ISBN:9788172362461
Seed	Vitexin	ISBN:9788185042114
Seed	Linolenic acid	ISBN:9788172362461
Seed	Linoleic acid	ISBN:9788172362461
Seedling	N6-Methylagmatine	ISBN:9788185042145

Plant Part	Phytochemical Name	Reference
	Ascorbic acid	DOI:10.15482/usda.adc/1239279
	4-Carvomenthenol	ISBN:9788185042114
	beta-Carotene	DOI:10.15482/usda.adc/1239279
	Betulinic acid	DOI:10.15482/usda.adc/1239279

Source: IMPPAT (2023).

Mantel Test Analysis of the mutual genotypic relationships among 24 genotypes of *P. tetragonolobus* gave a significant correlation (r = 0.839) between similarity matrices of the results obtained from the use of the RAPD and ISSR molecular markers. Quantification of total polyphenols, flavonoids and tannin revealed the highest percentage of occurrence of kaempferol (1.07- 790.5μg/g) and the lowest percentage of gallic acid (0.09-3.49μg/g) in the seeds (Rajan, 2018).

Evaluation of the physical properties, approximate composition of the seeds and the physico-chemical properties of fatty oil from immature, mature and fully mature seeds by GC-FID, GC/MS and 1 H-NMR revealed that the fatty oil from fully mature seeds contained high proportion of unsaturated fatty acids (75.5 per cent) compared to immature seeds which showed higher percentage of saturated fatty acid (61.3 per cent). Unsaponification matter (0.25 per cent) of fatty oil was identified as stigmasterol (66.4 per cent) and â-sitosterol (25.1 per cent). Total lipids of fully mature seeds were extracted and isolated as neutral, glyco- and phospholipids. Thus the fatty oil of fully mature seeds was enriched with mono-unsaturated fatty acids (38.6 per cent) and poly-unsaturated fatty acids (36.9 per cent) without trans-fatty acids thus meeting the standard of edible oil (Rajan, 2018). The beans are also rich in tocopherols and antioxidants that facilitate vitamin-A utilisation in the body (en.wikipedia.org/wiki/Winged_bean).

Anti-Nutritional Factors

Winged bean contains many anti-nutritional factors such as trypsin inhibitors, chymotrypsin inhibitors (WCI), hemagglutinins, amylase inhibitors, phytates, phytic acid, flatulence factors, hydrogen cyanide, saponins, tannins and other phenolic compounds (Kortt, 1984; Sri Kantha *et al.*, 1986). The seeds and root tubers have high levels of trypsin inhibitory activity (13,500 – 30,100 IU/ mg fresh weight) though it is inactivated by cooking (Root Crops, 2023).

The main phytochemicals of *Psophocarpus tetragonolubs* responsible of its anti-oxidant and anti-inflammatory activities are:

Metabolites	*Activity*	*Plant Part*	*Country*	*References*
Phytate	Anti-nutritional factors, affinity for specific blood cell antigens	Seeds	Malaysia	Sri Kantha *et al.*, 1986; Tan *et al.*, 1983
Tannin	Nonspecific enzyme inhibitors, hemagglutinin activity	Seeds	Malaysia	Tan *et al.*, 1983
Psophocarpin	Chemotrypsin inhibitory activity	Seeds, pods	India	Mit and Singh, 1988; Mit and Singh, 1986
Lectin	Hemagglutinin activity	Mainly in seeds, roots	India	Kortt, 1984; Manjunath and Madaiah, 1987; Kortt, 1985
Albumin 1 (WBA-1)	Kunitz-type trypsin inhibitors	Seeds	Malaysia	Kortt *et al.*, 1989
Phaseolin	α-amylase inhibitor	Seeds	Malaysia	Yao *et al.*, 2016

Source: www.mdpi.com/2223-7747/9/12/1730.

Uses

Culinary

All parts of the plant including leaves, flowers, roots, young unripe seed pods and seeds have high nutrient composition and each part except the ripe seeds can be eaten raw or cooked. Mature seeds are consumed only after cooking and mature seed pods become too fibrous to eat. Smoked pods, dried seeds, tubers (cooked and uncooked), and leaves have been sold in domestic markets in South East and South Asia (Khan, 1982).

The nutrient-rich tuberous roots have a distinctive nutty *cum* earthy flavour. Highland tribesmen in Papua New Guinea esteem it so highly that they hold winged bean sing-sings (feasts) at harvest time. Researchers in Thailand also have made tasty snacks from winged bean tubers sliced thin, fried and salted, or softened in sugar syrup (Growables, 2023).

Tender pods which are the most widely eaten part of the plant are best when eaten before they exceed 2.5cm in length. They are ready for harvest within three months of planting. The flowers are used to colour rice and pastry. The young leaves can be picked and prepared as a leaf vegetable similar to spinach (en.wikipedia.org/wiki/Winged_bean).

Boiled Winged Bean Roots Served as a Snack in Burma

By Wagaung, CC BY-SA 3.0, https://commons.wikimedia.org/w/index.php?curid=18506943

Its seeds are rich not only in protein but also in tocopherols ~ antioxidants facilitating vitamin-A utilisation in the body (en.wikipedia.org/wiki/Winged_bean). They can be made into milk when blended with water and an emulsifier. Winged bean milk is similar to soy milk but without the bean-rich flavour (Yang and Tan, 2011).

Other Uses

Similar to soyabean oil, its seeds can also be processed to produce oil for culinary purposes, illumination and for preparing soap (Growables, 2023). Expressed oil cakes are fed to livestock. Whole plant or a part is also fed to animals (Root Crops, 2023).

Growth

Winged bean can grow at least as fast as any other legume including soy. Plants start flowering 40-140 days after sowing, depending on the type and climatic conditions. Pods reach full-length about two weeks after pollination. Three weeks after pollination, the pod becomes fibrous and after six weeks mature seeds are ready for harvest (en.wikipedia.org/wiki/Winged_bean). Tuber development and flower production vary according to genotype and environmental factors and there are a few varieties not producing tuberous roots. This being a tropical plant flowers only at less than 12h day length though there are varieties with neutral day-length. Winged bean being a vining plant requires growing over a support for better performance and yield. Some examples of support systems include growing against exterior walls of houses, huts, buildings; supporting against larger perennial trees and shrubs; stakes placed in the ground vertically; and the structures made from posts and wires (Khan, 1982).

Cultivation

Propagation

Winged bean is usually propagated through seeds that remain viable for approximately one year. It can be transplanted or direct-seeded (Growables, 2023) since these do not require pre-treatment in tropical climate though scarification or soaking in water overnight before sowing enhances the germination rate. Seed soaking may be used in conjunction with scarification to increase the germination speed. Seedlings under natural field conditions germinate in a week.

In certain areas of Myanmar and the Philippines, it is grown as a perennial crop where tubers are left in the ground as such to produce fresh plants. A great deal of breeding work is currently underway to develop new cultivars to suit specific conditions and for tuber production (Root Crops, 2023).

Climate

Winged bean is a tropical crop therefore it is resistant to high temperature. It is grown between 20°N and 15°S latitude at elevations up to about 2,000m in the tropics (Root Crops, 2023). Though it can be grown at temperatures between 14 to 40°C, optimum being 18-30 °C (Growables, 2023) but the most ideal 25°C. Lower temperatures suppress germination and extremely high temperatures inhibit yield. Growing winged bean at lower temperatures can increase tuber production though leaf expansion rate remains higher in a warmer climate (Khan, 1982).

For highest yield the soil should remain moist throughout its life cycle. Winged bean production is optimal in humid climate though it is susceptible to moisture stress, *vis-a-vis* waterlogging. Yields remain higher if plants mature during the drier part of the growing season (Khan, 1982). A well-distributed rainfall in excess of 150cm per year is required and the plant thrives in areas with an annual rainfall of 250cm or more. It can be grown as a dry season crop provided there is adequate irrigation with proper drainage facilities. Despite its perennial nature and extensive root-system, it does not survive prolonged drought, however, drought-resistant cultivars are being developed (Root Crops, 2023).

Winged bean requires short days for normal flower induction as growing these under long photoperiod there is excessive vegetative growth at the expense of flowers. Recent experiments have shown that there is an interaction between day-length and day/night temperatures, as both the flowering and tuber growth are inhibited in 16h days but with 8h days tubers are formed regardless of temperatures though flowering is dependent upon the temperature regime used (Root Crops, 2023).

Soil

Winged bean is not very demanding in its soil requirements but there should be ample provision of drainage facilities. Waterlogging and salinity both are dangerous to this crop. Well-cultivated, rich and sandy loam soils are best for optimum yield of tubers as well as pods. On clay soils the tubers are usually small and lacking in flavour. It is frequently grown successfully in nitrogen-poor soils because of its exceptional ability of nodulation. However, recently in Nigeria, experimentally-grown winged beans have made comparatively slow growth accompanied by markedly chlorotic light-green foliage which suggests that the rate of nodulation is very dependent upon the availability of the most effective rhizobial strains, probably of the cowpea group. The manurial requirements of the winged bean have not been studied in detail though it responds favourably to nitrogen fertilization. When grown as a vegetable the routine application of standard NPK fertilizer at intervals of 14-21 days has been recommended (Root Crops, 2023).

Planting

Winged bean seeds are usually sown at the beginning of the rainy season. For pod and seed production, planting is usually done on the flat bed and the seeds are dibbled in holes about 2.5cm deep. The plant is generally provided with supports to twine over. When poles are used the plants may grow so tall that pod picking becomes difficult and the use of a trellis or wire fence 1-1.5m high is always better (Root Crops, 2023).

If the crop is being grown on a field scale for tuber production as is practice in Myanmar, the seeds are normally sown 5-7.5cm deep at 7.5-15cm along on about 20cm high ridges drawn 60cm apart and earthed up, stakes are frequently dispensed with, and the plants are left to ramble over the ground reaching a height of up to 30cm. Though use of stakes may slightly increase the yield of root tubers but staking cost may well offset any increase in yield (Root Crops, 2023). For covering one hectare of land some 500kg seeds are required which will give us some 1,50,000 plants.

Fertilizer

Well rotted compost some 10 quintals per hectare along with 80:80:80kg/ha of NPK fertilizers can be added before planting (Growables, 2023) when the soil is ready for planting. Winged bean can be grown without added fertilizer as the plant has a bacterium on the nodules of the roots that fixes nitrogen and allows the plant to absorb nitrogen. Factors that influence nitrogen fixation include *Rhizobium* strain, interactions between strain and host genotype and the available nutrients and soil pH (Khan, 1982).

Weeding

Seedlings make slow growth for the first 3-5 weeks, therefore 1-2 initial manual weedings are required. Efficient weed control is necessary until the plants are well established (Root Crops, 2023).

Interplanting and Deflowering

The winged bean is often interplanted with sweet potatoes, taro, bananas, sugarcane or any other vegetable crop and the removal of flowers increases root tuber production to a tune of four times (Root Crops, 2023).

Harvest and Postharvest

Tubers are harvested from 5 to 12 months after sowing depending upon variety used, prevailing environmental factors and cultivation practices (Growables, 2023). The root tubers are normally harvested when these attain 7.5-12cm length and 2.5-5cm diameter. Lifting is usually by fork so care should be taken to avoid any damage. The practice of growing plants on ridges facilitates lifting operation. Where grown on the flat beds, the ground is sometimes flooded to make digging easier and to reduce the possibility of injury. There is little information on storage of the root tubers as they are normally eaten immediately after harvest but it has been found that under normal tropical conditions deterioration is rapid (loss of moisture, loss of vitamin C and long cooking time) though at lower temperatures *cum* high humidity storage for a few weeks is possible provided that fungal growth is properly prevented (Root Crops, 2023).

Young leaves of large maturing plants can be used at any time but especially during the off-season when vegetative growth is abundant and the plants are not in bloom. The periodic harvest of up to 10 per cent of the foliage is not likely to reduce flowering, seed yield and fleshy roots. During pod production period, only light leaf harvesting is desired (Growables, 2023).

Pods are usually harvested 2-3 weeks after fruit set while they are still tender and not very fibrous though afterwards these become fibrous. They grow rapidly after the flowers are produced and attain proper maturing size by 3 weeks. As the pods begin to mature, these lose their shiny appearance and develop fibres. On Guam, the most popular local winged bean variety is cultivated seasonally or left in the field to grow year-round. Immediately after harvest, fresh immature winged bean pods should be cooled and stored at 10°C at 90 per cent relative humidity to slow down water loss and decay (Growables, 2023). The dried beans or seeds can be stored for long periods under cool dry conditions.

Yields

Root tuber yield is greatly influenced by growth conditions, altitude and variety used; and virtually all reported yields are from crops grown primarily for pod or seed. Malaysia has reported 2.5-6t/ha and the Philippines 2.3t/ha of tubers. In Indonesia, higher altitudes were stated to favour root tuber production as compared to pod production and in the cool highlands yields were recorded 6-16t/ha though in the lowland areas it was much less. Papua New Guinea also recorded higher yields from the highlands though in Malaysia 14t/ha from staked plants of selected strains in experimental plantings, however, in none of these cases flower removal was effected (Root Crops, 2023).

Pests and Diseases

Winged bean when grown in mixed market garden culture or shifting agriculture, it is generally free from serious pests and diseases though the crop is occasionally attacked by caterpillars, leaf miners, grasshoppers, spider mites and nematodes (Root Crops, 2023), so for these, appropriate treatment is required to check the menace. *Polyphagotarsonemus latus* and *Tetranychus virticae* mites are serious pests so these require to be checked through use of malathion, primiphos methyl or bifenthrin at 0.2 per cent sprayings 1-3 times fortnightly. Root knot nematodes have been reported to cause up to 70 per cent losses in tuberous roots. *Meloidogyne incognita* has been recorded as more aggressive than *M. javanica* (Root Crops, 2023) and these can be controlled by applying carbofuran granules in the soil and through crop rotation. Caterpillars of *Podalia* spp. have been reported from Brazil as a serious pest which can be controlled by spraying the crop with 0.2 per cent malathion. Other insects can also be controlled through 0.2 per cent malathion spraying.

Probably the most serious disease occurring in Papua New Guinea, Java, the Philippines and Malaysia is false rust caused by *Synchytrium psophocarpi* which may be controlled through use of copper fungicides. In plantations, *Cercospora* spp. causes leaf spots though not to the extent of producing defoliation therefore it would be better to leave it as such. Viral symptoms have been observed in Nigeria and the Ivory Coast so such plants should gently be removed along with the adhering soil and burnt or buried and their vectors should be controlled through insecticides. With increase in its plantation, the disease problem may also increase to alarming state so immediately when any such problem appears appropriate chemicals should be used.

References

1. Bassal, H., O. Merah, A.M. Ali, A. Hijazi and F. El Omar, 2020. *Psophocarpus tetragonolobus*: An underused species with multiple potential uses. *Plants*, **9**(12):1730. (https://doi.org/10.3390/plants9121730).
2. Echo community, 2023. Retrieved on 17 May, 2023. https://www.echocommunity.org/en/resources/3e995069-d3e7-4a2b-8188-a50f2d168193
3. en.wikipedia.org/wiki/Winged_bean
4. Etats-Unis, 2009. The Complete Guide to Edible Wild Plants, 57p.. United States Department of the Army. New York. Skyhorse Publishing, USA.
5. Growables: Grow Florida Edibles, 2023.. Retrieved on 19 May, 2023 from www.growables.org/informationVeg/BeanWinged.htm
6. Hymowitz, T. and J. Boyd, 1977. Ethnobotany and Agriculture Potential of the Winged Bean. *Econ. Bot.*, **31** (2): 180–88. (doi:10.1007/bf02866589. S2CID 5944034).
7. IMPPAT: Indian Medicinal Plants, 2023. Phytochemistry and Therapeutics. Retrieved on 18 May, 2023. cb.imsc.res.in/imppat/phytochemical/Psophocarpus per cent 20tetragonolobus
8. Jaffe, W.G. and R. Korte, 1976. Nutritional characteristics of the winged bean in rats. *Nutr. Rep. Int.*, **4**: 449–55. (www.mdpi.com/2223-7747/9/12/1730).
9. Khan, T. 1982. Winged Bean Production in the Tropics (1 page leaflet). Food and Agriculture Organization of the United Nations.
10. Khor, H.T., N.H. Tan and K.C. Wong, 1980. Winged bean seed: potential food source of the tropics. In: Proceedings of the 6th Malaysian Biochemical Society Conference, Kuala Lumpur, Malaya, held on 22–23 August 1980, pp. 157–162. (www.mdpi.com/2223-7747/9/12/1730).
11. King, R.D. and P. Puwastien, 1987. Effects of germination on the proximate composition and nutritional quality of winged bean (*Psophocarpus tetragonolobus*) seeds. *Food Sci.*, **5**: 106–8. (www.mdpi.com/2223-7747/9/12/1730).
12. Kortt, A.A. 1984. Purification and properties of the basic lectins from winged bean seed [*Psophocarpus tetragonolobus* (L.) DC]. *Eur. J. Biochem.*, **138**: 519–25.

13. Kortt, A. 1985. Characterization of the acidic lectins from winged bean seed {*Psophocarpus tetragonolobus* (L.) DC}. *Arch. Biochem. Biophys., 236*: 544–54.

14. Kortt, A.A., P.M. Strike and J.D. Jersey, 1989. Amino acid sequence of a crystalline seed albumin (winged bean albumin-1) from *Psophocarpus tetragonolobus* (L.) DC. Sequence similarity with Kunitz-type seed inhibitors and 7s storage globulins. *Eur. J. Biochem.*, **181**: 403–8.

15. Makeri, M.U., R. Karim, M.S. Adbulkarim, H.M. Ghazali, M.S. Miskandar and K. Muhammad 2016. Comparative analysis of the physiochemical, thermal and oxidative properties of winged bean and soybean oils. *Int. J. Food Prop.*, **19**: 2769–87. (www.mdpi.com/2223-7747/9/12/1730).

16. Manjunath, S. and S. Madaiah, 1987. Distribution of lectin activity at different stages in the tissues of winged bean [*Psophocarpus tetragonolobus* (L.) DC]. *Plant Sci.*, 53: 161–5.

17. Martin, Franklin W. 1978. Vegetables for the Hot, Humid Tropics Part 1. The Winged Bean, Psophocarpus tetragonolobus. Agricultural Research Service, U.S.D.A., Southern Region. (ia801207.us.archive.org/9/items/CAT79711173/CAT79711173.pdf.).

18. Misra, P.S., G. Misra, D. Prakash, R.D. Tripathi, A.R. Chaudhary and P.N. Mishra, 1987. Assay of some nutritional and anti-nutritional factors in different cultivars of winged bean *Psophocarpus tetragonolobus* (L.) DC seeds. *Plant Foods Hum. Nutr.*, **36**: 367–71. (www.mdpi.com/2223-7747/9/12/1730).

19. Mit, R. and M. Singh, 1986. Purification of a storage protein of *Psophocarpus tetragonolobus*. *Phytochem.*, **25**: 595–600.

20. Mit, R. and M. Singh, 1988. Psophocarpin B1, a storage protein of *Psophocarpus tetragonolobus*, has chymotrypsin inhibitory activity. *Phytochem.*, **27**: 31–4.

21. Mohanty, C.S., R.C. Pradhan, V. Singh, N. Singh, R. Pattanayak, O. Prakash, C.S. Chanotiya and P.K. Rout, 2015. Physicochemical analysis of *Psophocarpus tetragonolobus* (L.) DC seeds with fatty acids and total lipid compositions. *J. Food Sci. Technol.*, **52**: 3660–70. (www.mdpi.com/2223-7747/9/12/1730).

22. Motis, T. 2019. Winged Bean Revisited. Echo Research, Echo Community, edn.link/hwtqkj. Accessed 25 May 2019.

23. National Research Council. 1981. Winged Bean: A High-Protein Crop for the Tropics. The National Academies Press, Washington DC, USA. (www.mdpi.com/2223-7747/9/12/1730).

24. POWO, 2023. Plants of the World Online. Kew Royal Botanical gardens. (Retrieved 2023-05-15. https://powo.science.kew.org/taxon/urn:lsid:ipni.org:names:515970-1).

25. Rajan, D. 2018. A Review on Pharmacognostical, Phytochemical and Pharmacological aspects of *Psophocarpus tetragonalobus*. *Res. J. Pharmacognosy and Phytochem.*, **10**(4): 331-5. (doi: 10.5958/0975-4385.2018.00053.5).

26. Root Crops, 2023. Humanity Development Library 2.0. Retrieved on 16 May, 2023 *via* https://www.nzdl.org/cgi-bin/library.cgi.

27. Sri Kantha, S. and J.W. Erdman, 1984. The winged bean as an oil and protein source: A review. *J. Am. Oil Chem. Soc.*, **61**: 515–25. (www.mdpi.com/2223-7747/9/12/1730).

28. Sri Kantha, S, N.S. Hettiarachchy and J.J.W. Erdman, 1986. Nutrient, antinutrient contents and solubility profiles of nitrogen, phytic acid and selected minerals in winged bean flour. *Cereal Chem.*, **63**: 9–13.

29. Tan, N.H., Z.H.A. Rahim, H.T. Khor and K.C. Wong, 1983. Winged bean (*Psophocarpus tetragonolobus*) tannin level, phytate content and hemagglutinating activity. *J. Agric. Food Chem.*, **31**: 916–7.

30. Varriano-Marston, E., A. Beleia and C.C. Lai, 1983. Structural characteristics and fatty acid composition of *Psophocarpus tetragonolobus* seeds. *Ann. Bot.*, **51**: 631–40.

31. www.mdpi.com/2223-7747/9/12/1730

32. Yang, J. and H. Tan, 2011. Winged Bean Milk. International Conference on New Technology of Agricultural Engineering, Zibo, pp. 814–17. (doi:10.1109/ICAE.2011.5943916. ISBN 978-1-4244-9574-0. S2CID 22046459).

33. Yao, Y., Y. Hu, Y. Zhu, Y. Gao and G. Ren, 2016. Comparisons of phaseolin type and α-amylase inhibitor in common bean (*Phaseolus vulgaris* L.) in China. *Crop J.*, **4**: 68–72.

31

Woundwort (*Stachys* L.)

Family: Lamiaceae (mint family). **Subfamily**: Lamioideae.

Type species: *Stachys sylvatica* L.

Synonyms: *Aspasia* E.Mey. ex Pfeiff., *Betonica* L., *Bonamya* Neck., *Eriostemum* Steud, *Eriostomum* Hoffmanns. and Link, *Galeopsis* Hill 1756 (not L. 1753 nor Adans. 1763 nor Moench 1794), *Galeopsis* Moench 1794 (not L. 1753 nor Adans. 1763 nor Moench 1794 nor Hill 1756), *Lamiostachys* Krestovsk., *Menitskia* (Krestovsk.) Krestovsk., *Olisia* Spach, *Ortostachys* Fourr., *Stachyus* St.-Lag., *Tetrahitum* Hoffmanns. and Link, *Trixago* Haller, *Trixella* Fourr., and *Zietenia* Gled.

***Stachys sylvatica* (Type species)**
By AnRo0002 - Own work, CC0, https://commons.wikimedia.org/w/index.php?curid=27068898

Common names: Betany, Chinese artichoke (*Stachys affinis*), Florida betony (*Stachys floridana*), Hedgenettle, Heal-all, Self-Heal, Lamb's Ears, Marsh woundwort (*Stachys palustris*) and Woundwort.

Introduction and Origin

Scientific name *Stachys* is a Greek word meaning 'an ear of grain' (Quattrochi, 2000) which refers to the fact that the inflorescence is often a spike. *Stachys*, one of the largest temperate genus in the mint family comprises some 300-450 species distributed in Europe, Asia, Africa, Australasia and North America (en.wikipedia.org). Some species in the genus produce tuberous underground structures out of which certain are edible. The edible tuberous species of *Stachys* covered in this chapter include *S. affinis* (Chinese artichoke), *S. floridana* (Florida betony) and *S. palustris* (marsh woundwort). *Stachys affinis* (Chinese artichoke) is the most well known edible tuberous species not only in its genus but in whole of the mint family.

Stachys or other similar plants in other genus of the mint family such as *Lycopus* and *Plectranthus* (now known as *Coleus*) should be grown where they can establish as a perennial population because it is very hard to collect all of the tubers at harvest. Some of them can be pretty invasive if measures are not taken to grow them in confined areas. The large number of small tubers makes these plants nearly impossible to harvest completely and even small fragments of rhizome can easily produce new plants. Hard sided beds can be used for a perennial patch. Marsh woundwort is particularly good at escape but all these plants find the gaps between cement blocks or wood slats and try to make their escape into the larger garden. Small segments of rhizome can sprout into new plants and gets extended easily into the loose soils of cultivated ground. These plants can become difficult to control in almost any climate.

Botany

Stachys is a genus of shrubs and annual or perennial herbs that have the small, serrated leaves and angular stems in the mint family. The height of stems varies from 0.5 to 3m with simple, opposite and triangular leaves that are 1–14cm long with serrate margins and softly hairy in most species. The flowers are 1-2cm long and clustered in the axils of the leaves on the upper part of the stem. The corolla is 5-lobed with the top lobe forming a 'hood', varying from white to pink, purple, red or pale yellow (en.wikipedia.org). Chinese artichoke rarely flowers or sets seed (cultivariable.com). Flowers are primarily formed in warm *cum* dry weather (cultivariable.com). There have been some breeding programmes that managed to get seed from Chinese artichoke (cultivariable.com). Marsh woundwort and Florida betony set seeds easily although they mature over a long period of time and can be difficult to collect (cultivariable.com). It would be a good idea to tuck white plastic grocery bags around the base of the plant and collect the seeds that fall inside them. Hybrids of marsh woundwort exist in nature so crosses with other wild relatives may be a way to increase genetic diversity in this crop (Wilcock, 1974).

Some *Stachys* species in the genus produce edible tubers. Depending on the area that they grow in or if there has been little moisture the tubers may end up almost dehydrated which makes them inedible.

Nutritional Values

Although the edible tubers of most of the tuberous species in the mint family contain a smaller portion of starch, the primary carbohydrate in them is stachyose which is not digestible by humans so these tubers are relatively low in calories (cultivariable.com). As with most vegetables that contain indigestible sugars, excessive consumption may cause gastric trouble. This is usually not a problem with tuberous mints since it is generally consumed in smaller portions than other roots with indigestible sugars such as Jerusalem artichoke and yacon (cultivariable.com). The tubers of Chinese artichoke are considered to be high in iron though original source for nutritional values couldn't be recorded (cultivariable.com).

Uses

Several species in the genus have culinary and/or medicinal uses while some others are used as ornamental plants. *S. officinalis* (wood betony) was the most important medicinal herb to the Anglo-Saxons of early medieval England (en.wikipedia.org). If dried and powdered the roots of *S. affinis* are said to be anodyne and the aerial parts are made into a tea for colds and pneumonia. *S. palustris* (marsh woundwort), *S. arvensis* (field woundwort) and *S. sylvatica* (hedge woundwort) have wound healing capability (en.wikipedia.org) while *S. affinis* (Chinese artichoke), *S. floridana* (Florida betony), *S. palustris* and *S. hyssopifolia* Michx (hyssop-leaf hedge-nettle, rough hedge-nettle, hyssop-leaved hedgenettle and hyssop hedge-nettle) have edible tubers (cultivariable.com) and possibly aerial parts are also edible. Seeds of *S. scopulorum* are eaten by natives in the desert southwest of North America and *S. sylvatica* is consumed in Europe and in New York City.

Woolly betony (*S. byzantina*) is a popular decorative garden plant (en.wikipedia.org). The tuber as well as the aerial portion of several plants may have some medicinal and healing capabilities. If dried and powdered the roots of these plants are said to be anodyne and the aerial parts are made into a tea for colds and pneumonia.

The edible tubers of Chinese artichoke, Florida betony and marsh woundwort are traditionally pickled in Japan and that may be the only good way of preserving the tubers (cultivariable.com). They can potentially be dried and ground into flour (cultivariable.com). The tubers are quite delicious when sauteed with butter or oil, boiled, pickled, added to soup and stews or just eaten raw in salads.

Species

Among several tuberous species in the mint family the most well known is the *S. affinis* for culinary uses. In this section three edible tuberous species namely *S. affinis*, *S. floridana* and *S. palustris* have been discussed though several other species may also produce edible tubers. These temperate species almost have similar growth habits, cultivation details and culinary uses. The growth habit of these plants is through runner spreading indefinitely by rhizomes or stolons. The root pattern is rhizomatous with underground stems sending roots and shoots along their length.

Stachys affinis Bunge

Synonyms: *Stachys sieboldii* Miq. and *Stachys tuberifera* Naudin.

Common names: Artichoke betony, Chinese artichoke, Crosne, Japanese artichoke or Knotroot.

Introduction and Origin

The specific epithet *affinis* is Latin for 'neighbouring'. The common name Chinese or Japanese artichoke may be because it is believed to have been originated either in China or Japan. Its refined and slightly sweet flavour is reminiscent of artichoke. It is called 'knotroot' because its rhizomes resemble a succession of knots. As per the views of different authorities, the native range of *Stachys affinis* is Japan or China to North Myanmar (en.wikipedia.org; Anon., 1833; Gourmetpedia; Specialtyproduce), and is cultivated in several parts of Asia. It has been introduced to Europe in 1880s; in France, the crop was cultivated on a farm for the first time in Crosne village in 1882 (Bussard, 1909) where it became widely known at the beginning of 20th century though was abandoned from 1970s to 1990 due to problems with viruses and the plant's strong tendency to spread (Peron and Dubost, 1992). Now it is being cultivated for its edible rhizomes in many parts of Asia and Europe, though in small scale. Today it is common in France, Belgium and other European countries. This tuberous geophyte grows primarily in the temperate biome (Anon., 1833).

Botany and Genetics

Chinese artichoke is an erect, hairy and stoloniferous perennial herb (uses. plantnet) growing to a height of 30–120cm, often up to 50cm tall. The herb shares common features with other plants in the mint family, producing square stems

Stachys affinis
Source: CC BY-SA 3.0, https://commons.wikimedia.org/w/index.php?curid=8209

Close-up of flower of Knotroot
Source: **Anon. (1833)**

and softly-haired or pubescent foliage made up of small, rough, nettle-like, deep green and prominently veined leaves that are oppositely arranged on stem. The leaves can be ovate-cordate shaped with a width of 2.5–9.5cm or ovate-oblong with a width of 1.5–3.5cm. The leaves are separated into a leaf blade and a 1–2cm long petiole that becomes shorter towards the stem apex. The plant remains in leaf from May to November and in flower from July to August. Its chromosome number is 2n = 66-70 (uses.plantnet).

Its plant rarely flowers as possibly the flowering and stolonization being antagonistic processes. White, pink and red to purple flowers are borne on erect inflorescences above the foliage usually from spring to summer. This hermaphrodite species is entomophilous. It rarely produces seeds as is common for many plants in the mint family.

Stolons and tubers are formed 5-7 months after planting in the same way as in Irish potato (*Solanum tuberosum* L.) (uses.plantnet). Tubers are formed at the end of the stolons when temperature drops. Whitish tubers of the size of 5-8 × 1-2cm in great numbers just below the soil surface are produced by thickening of the top part of the stolons that have constrictions at the internodes into rows of rounded segments (uses.plantnet). In temperate regions, tubers remain dormant during winter and start sprouting in spring (uses.plantnet). The pale-beige to white-skinned, small, thin, elongated or slightly annulated tubers have white flesh and are shaped like strings of beads due to their successive swellings. Thick rhizomes have a curious segmented or jointed appearance somewhat similar to those of the puffy caterpillars, *i.e.* a succession of knots but there is also another variety that looks more like a shell.

Rhizomes are Puffy Caterpillar-like or a Succession of Knots
Source: CC BY-SA 3.0, "https://commons.wikimedia.org /w/index.php?curid=443119"

Shell-shaped Rhizome
Source: Veggiesinfo

Tuberous Rhizomes of *Stachys affinis*

Nutritive Value

Chinese artichoke tubers are rich in proteins, carbohydrates, iron and vitamins. The macronutrient profile of the tuber was found to be dominated by carbohydrates whereas potassium was the most abundant micro-nutrient. According to various sources over internet, per 100g of raw rhizomes of this plant contain the following nutrients:

Nutrients	*Nutrition Value (per cent DV of RDA)*
Calories	63.6 (3 per cent)
Carbohydrates	14.3g (5 per cent)
Fibre	10.3g (41 per cent)
Protein	3.5g (7 per cent)
Total Fat	0.4g (1 per cent)
Iron	0.7mg (4 per cent)

Nutrients	***Nutrition Value (per cent DV of RDA)***
Manganese	0.3mg (13 per cent)
Copper	0.2mg (8 per cent)
Calcium	25.2mg (3 per cent)
Magnesium	50.4mg (13 per cent)
Phosphorus	87.6mg (9 per cent)
Potassium	343mg (10 per cent)
Selenium	0.2mcg (N/A)
Zinc	0.5mg (3 per cent)
Folate	107mcg (27 per cent)
Vitamin B3	1.3mg (7 per cent)
Vitamin B5	0.3mg (3 per cent)
Vitamin B2	0.1mg (6 per cent)
Vitamin B1	0.1mg (4 per cent)
Vitamin A	15.6IU (N/A)
Vitamin B6	0.1mg (5 per cent)
Vitamin C	8.9mg (15 per cent)
Vitamin E	0.2mg (1 per cent)
Vitamin K	17.8mcg (22 per cent)
Choline	41.3mg(N/A)
Betaine	0.2mg (N/A)

Source: Veggiesinfo

Nutrients	***Value/100g***
Energy (kj/kcal)	215/50.8
Fat (g)	< 0.5
Saturated fat (g)	< 0.01
Carbohydrate (g)	9.1
Sugar (g)	0.2
Dietary fibre (g)	1.9
Protein (g)	2.1
Salt (g)	0.013
Minerals and vitamins	
Copper (mg)	0.21
Potassium (mg)	310
Folic acid (vitamin B9 in µg)	32.3

Source: Princedebretagne

Per 100g edible portion of Chinese artichoke rhizomes contain: water 78g, protein 4.3g, fat 0.2g, carbohydrates 14.6g and ash 1g. Energy value is 320kJ/100g. The main carbohydrate is the tetrasaccharide stachyose (gal-gal-sucrose) which is exclusively localized in the vacuoles. In dormant rhizomes 80 per cent of the dry weight is stachyose (18 per cent of the fresh weight). Other carbohydrates present are sucrose, raffinose and verbascose.

Source: uses.plantnet.

Phytochemicals

Although not much data are available on phytochemical profile of *Stachys affinis* rhizomes but internet highlights the wide functional food and nutraceutical potential much of which remains unexploited. Several biologically active compounds that may possess medicinal values have been reported. Vacuoles in the tubers of *S. affinis* are rich in stachyose (Greutert and Keller, 1993) which is a marker compound of the species and is up to 80-90 per cent in dry rhizomes (Yin *et al.*, 2006). Stachyose is a tetrasaccharide comprising galactose, glucose and fructose (Yin *et al.*, 2006). The newly found compounds ~ harpagide, 8-O-acetyl-harpagide, melittoside and 5-O-allosyloxy-aucubin are also reported to be present in several other species of *Stachys* genus.

Some compounds isolated from the tuber ethanolic extract and structurally elucidated by Nuclear Magnetic Resonance (NMR) spectroscopy and mass spectrometry (MS) are shown in the following table:

Compounds	*Pharmacological Characteristics*
Verbascoside	Anti-inflammatory, anti-hyperalgesic, antioxidant and anti-cancer
Leucosceptoside A	Estrogenic and antioxidant
Martynoside	Estrogenic and antioxidant
Harpagide	helps in osteoporosis treatment
8-O-acetyl-harpagide	Anti-bacterial
Melittoside	Anti-inflammatory and moderate antioxidant
5-O-allosyloxy-aucubin	Anti-inflammatory and moderate antioxidant
Succinic acid	Anticonvulsant, sedative and anxiolytic; helps in reducing glucose and cholesterol levels in blood; useful in treating diabetes mellitus and other metabolic disorders.
Stachyose	Capable of limiting the proliferation of some harmful bacteria; lowers plasma glucose levels; high potential as prebiotic ingredient in functional foods.

Source: nutraingredients-asia.com

Uses

Although all parts of the plant are edible but the rhizome is the preferred one. The leaves can be dried and made into a tea. Rhizome is used as a root vegetable having a crunchy texture similar to radish with a slightly sweet and nutty flavour similar to oyster plant or artichoke and can be eaten raw, pickled, dried or cooked. The rhizomes kept in a box at the store for more than a day or two may have no flavour at all. They can be prepared similarly to Jerusalem artichokes. The crunchy rhizomes are a good addition to salads. It should be cooked for short periods of time. Stir frying, steaming or brief boiling for 2-4 minutes gives good results.

In China, Japan and Korea the rhizome is often fermented to preserve it (cultivariable.com). This also increases the fraction of digestible sugars in the rhizome. Ingredients are sometimes added to colour the rhizomes. In Japan, perilla is added to colour them red. Its rhizome is a part of Osechi, cooked for celebrating Japanese New Year. In French cuisine, its cooked rhizome is often served alongside dishes named *japonaise* or Japanese-styled.

The dried and powered rhizomes of *Stachys affinis* are believed to be anodyne (pain killer) and the aerial parts are made into a tea for colds and pneumonia. Root extract has shown antimicrobial activity (Amahe, 2002). Inhibitory effects on acetylcholine esterase, monoamine oxidase and xanthine oxidase activities were observed in rat brains after 20 days of feeding with methanolic extracts of *S. affinis* (Anon., 2004). Ethanol extract from this plant also seems to have antitumour activity (Ryu *et al.*, 2002). In traditional Chinese medicine, whole plant is used as an agent to treat colds and pneumonia (Paton *et al.*, 1995). The phytochemical compounds present in the tubers have potential to be used in pharmacology as anti-bacterial, anti-inflammatory, anti-hyperalgesic, antioxidant, anti-cancer, estrogenic, anticonvulsant, sedative, anxiolytic and in diabetes treatment. It helps in reducing glucose.

Stachys floridana Shuttlew. ex Benth.

Common names: Florida betony, Florida hedgenettle, Rattlesnake weed and Wild artichoke.

Introduction and Origin

The specific epithet *floridana* means 'of Florida'. *Stachys floridana*, native to the United States (probably Florida) is commonly found throughout the Southeast as an introduced common weed (Czarnota and Murphy, 2012). It is found as far west as Texas and in California (en.wikipedia.org). The plant grows in disturbed habitat types such as roadsides, often on wet soils, and in turf and in beds of ornamental plants (en.wikipedia.org). It is a prolific producer of seeds but often undergoes vegetative reproduction *via* its rhizome and tubers (en.wikipedia.org). Small segments of rhizome can sprout into new plants and gets extended easily into the loose soils of cultivated ground. It is a common weed of residential and commercial land (en.wikipedia.org). Once established, Florida betony is difficult to eradicate by hand because the tubers and rhizomes break easily and small root fragments can regrow. Generally it has better yield as compared to its relative Chinese artichoke, and readily produces seed which means that it can be more easily improved through breeding (cultivariable.com).

Botany and Genetics

Florida betony is a perennial herb quite similar to Chinese artichoke but the rhizomes are usually a bit larger and grow 45-60cm tall (cultivariable.com) with square, hairy and erect stem. The oppositely arranged leaves have up to 5.5cm long blades borne on up to 3.5cm long petioles (en.wikipedia.org). Charlotte Lindqvist and Victor (2002) reported its chromosome number as 2n=34.

From the upper leaf axils, 3-6-clustered flowers appear. Calyx is tubular and hairy with pointed lobes. Corolla is 2-lipped, 1.3cm long and white to pink with purple spots or darker lines. A few centimetres long fruit is a schizocarp that splits in half. The seeds are about a millimeter long (en.wikipedia.org).

Florida betony produces pale-white segmented rhizomes that look similar to a rattlesnake's tail. Both tubers and rhizomes are free-branching and break easily. The pale-coloured tubers are usually up to 30cm long with 1-2cm in diameter (cultivariable.com) and are segmented in such a way that these resemble the rattle on the tail of a rattlesnake hence the common name 'rattlesnake weed'. The rhizome can reportedly grow up to 1m long in sandy soils (Czarnota and Murphy, 2012).

Uses

Its 'crispy and succulent' rhizomes are edible (Czarnota and Murphy, 2012) having crunchy texture and a bland, slightly sweet taste, though less nutty and more bland as compared to Chinese artichoke (cultivariable.com). Like other tuberous species of its genus it too is a source of a sugar called stachyose which is less sweet than sugar and is used as a bulk sweetener. It is also not completely digestible. Stachyose promotes friendly bacteria in the gut and can inhibit the growth of those bacteria that can cause pneumonia or vaginal infections. The tuber as well as the aerial portion of the plant may have some medicinal and healing capabilities as do many of other species of *Stachys*. The aboveground parts of young plants can be cooked like greens. They are, however, musty in flavour. Tea can be made from the dried leaves. The seeds are also edible.

Stachys palustris L.

Common names: Marsh woundwort, Clown's woundwort, Clown's heal-all, Marsh hedgenettle or Hedge-nettle (en.wikipedia.org).

Introduction and Origin

Marsh Woundwort

By Ivar Leidus - Own work CC BY-SA 3.0 and 4.0; "https://commons.wikimedia.org/w/index.php?curid=27165322 and https://commons.wikimedia.org/w/index.php?curid=115281952" respectively

The species epithet *palustris* is Latin for 'of the marsh' which indicates its common habitat (en.wikipedia.org). *Stachys palustris* is native to parts of Eurasia (en.wikipedia.org) and probably in North America (cultivariable.com; en.wikipedia.org) where it is believed by some authorities to be an introduced species. *S. palustris*, with an odourless rhizome is widespread in moist places in most of the northern hemisphere (merriam-webster.com). It can be found in the northeast quadrant of North America, basically the Mason Dixon Line north and as far west as Illinois and Manitoba. It is typically found growing near the shore of lakes, in marshes with alder trees, on the banks of ditches and streams, in damp meadows, in arable ground and in waste places (en.wikipedia.org). It is a difficult weed to get rid of because of its persistent tubers, especially in arable land (en.wikipedia.org).

Genetics

Stachys palustris is polyploid (8×) with a somatic number for British material of 2n=102 (Morton, 1973). Wilcock (1974) recorded 97–103, most frequent being

102. It is found in diploid, tetraploid and hexaploid forms. The diploids and tetraploids are mostly native to North America while the hexaploids to Europe (cultivariable.com; Kenneth Taylor and Phillip, 2011). *Stachys × ambigua*, the hybrid between *Stachys palustris* and *S. sylvatica* is most frequent in northern and western Britain (Kenneth Taylor and Phillip, 2011).

Botany

Similar to its relatives such as Chinese artichoke, Florida betony and even marsh woundwort, it is a perennial growing from a horizontal tuberous runner (en.wikipedia.org). It forms long-creeping rhizomes that produce small tubers at the apices in the autumn in the upper layer of the mineral soil horizon (Kenneth Taylor and Phillip, 2011). It is a monoecious and almost odourless pubescent herb (Kenneth Taylor and Phillip, 2011) growing to a height of 80-102cm (en. wikipedia.org; cultivariable.com) with square stems having opposite pairs of leaves that are almost stalkless, linearly lanceolate, slightly cordate at the base and toothed (en.wikipedia.org). The calyx has five sharply-pointed lobes (en. wikipedia.org). Purplish-red flowers are in terminal spikes with gaps in the lower part of the spike (en.wikipedia.org). In Britain and Ireland, even white-flowered plants are sometimes recorded. Flowers are arranged in whorls, each consisting of five fused petals, the corolla being two-lipped, the upper lip being gently hooded and the lower lip flat and three-lobed (en.wikipedia.org). There are four stamens, two long and two short and the fruit is a dry 4-chambered schizocarp (en.wikipedia.org). Marsh woundwort tubers may grow up to 46cm, although the most useful thicker part is usually only 10-15cm long while rest rather thin (cultivariable.com). It is still perfectly edible and may even be preferable for some dishes.

Uses

Rhizomes can be eaten raw, cooked or pickled. These can be cooked in a variety of ways, or dried and made into bread and the young shoots are cooked and eaten like asparagus. As the common name suggests, along with its close relatives ~ Field woundwort (*S. arvensis*) and Hedge woundwort (*S. sylvatica*), this plant is capable to promote the healing of wounds (en.wikipedia.org).

Propagation

Vegetative Propagation

The species discussed here can be propagated asexually with tubers, rhizomes or stem cuttings. The rhizomes are both effective as an overwintering organ and as an agent in biotype dispersal. Cuttings of the elongated shoots of *S. palustris* taken during the active growth period can be readily rooted.

Tubers

Whole tubers or cut pieces of the previous harvest are planted in fall or spring in mild climates. Tubers are cut much like to those of potatoes and each piece should have at least three eyes for best results. They are planted 5-8cm deep, preferring deeper planting for larger tubers. Some growers plant 2-3 tubers together in upright position, 2-15cm deep (uses.plantnet). Dormancy has been reported to vary considerably as marsh woundworts emerge earliest but in late winter; Chinese artichoke emerges in early spring; and Florida bettony doesn't wake up until mid-spring (cultivariable.com). The tubers begin to sprout at temperatures somewhere above 5°C. Plants take 5-7 months to develop their tubers.

Stem Cuttings

For stem cuttings, 10-15cm sections of stem can be planted by setting the bottom about 5cm deep in damp soil. After about 6-8 weeks when significant roots are formed the new plant can be hardened off and moved outside (cultivariable.com).

Sexual Propagation

Marsh woundwort and Florida betony set seeds easily but Chinese artichoke rarely. Hence the sexual propagation of Chinese artichoke is rare though there have been some breeding programmes that managed to get seeds from Chinese artichoke. The germination of Florida betony seeds and probably of Chinese artichoke seeds will take over several weeks at 21°C and should be started 4-6 weeks before the last frost. Marsh woundwort requires stratification at least of two months in moist-cold after which it will germinate over 2-3 months at 21°C. It would be better to begin stratification in November and then the seeds can be sown indoors in pots in February (cultivariable.com).

Seeds can be sown in spring in a cold frame and when they are large enough to handle the seedlings should be pricked out into individual pots to grow them on in the greenhouse for their first winter. They are then planted out into their permanent positions in late spring or early summer after the last expected frosts.

Growth and Development

The shoots usually begin to appear in mid- or late-March and the plant is fully in leaf at its peak of biomass by early June. Plant flowers from July to September, set seeds from July to October, and the shoots die down by the end of November. Seeds germinate near to the parent in the following spring (Kenneth Taylor and Phillip, 2011).

Climate

The tubers are generally hardy down to USDA zone 6. Chinese artichoke is typically hardy down to USDA zone 5 (zone 4 under mulch), Florida betony to USDA zone 6, and though marsh woundwort is unknown but is probably suitable for zone 5/6. The foliage will die by the time temperatures reach less than -4°C but the tubers are set fairly deep and survive shallow freezing pretty well. These plants are tolerant to wide temperature range but would be less hardy if grown in containers where they can't form tubers as deeply (cultivariable.com).

Marsh woundwort is the most tolerant of chilly conditions while Florida betony is the most heat tolerant though Chinese artichoke falls somewhere in between (cultivariable.com). Chinese artichoke thrives in cool subtropical and tropical high-elevation climates, generally areas with annual low temperatures of 8-15°C, annual high temperature of 20-27°C, annual rainfall of 800 to 2,300mm and a dry season of 5 months or less (iplantz.com).

These plants will happily grow in shade but they need full sun to get the energy required to produce good rhizome yields. None of these plants have any photoperiod dependencies, though it has been noted that Florida betony foliage grows fastest under warm temperatures (21°C) and long day conditions (McElroy *et al.*, 2004). Presumably, increased growth in long day conditions is simply due to the increased amount of sunlight. This information appears to be true for Chinese artichoke and marsh woundwort as well (cultivariable.com).

Soil

These tuberous mints, especially Chinese artichoke grow best in full to partial sun and have similar soil requirements to other plants in the mint family performing best on free-draining moisture-holding loam and sandy-loam soils that have been enriched with organic matter and adjusted to a slightly acid to a bit alkaline nature, generally with a pH of 6.6 to 7.5 (iplantz.com). Marsh woundwort prefers well-drained, wet and probably acidic soil. The weight of rhizomes produced in pots was found to be roughly 2-4 times greater in wet soil than dry (Wilcock, 1974). The fact that Florida betony is a noxious weed of lawns throughout the Southeast suggests that it probably tolerates a wide range of soil conditions (cultivariable.com).

Field Preparation and Planting

Land should be prepared for planting by ploughing, harrowing and adding organic manures to make it fertile. The crop responds well to fertilizer (uses. plantnet) so at the time of final bed preparation one-fourth of nitrogen and entire dose of phosphorus and half of potash should thoroughly be mixed in the soil.

Other three doses of nitrogen should be applied subsequently after every 30 days afterwards. Remaining half dose of potash should be applied after 2 months of planting. Nitrogen @ 120kg, phosphorus 30kg and potassium about 200kg are usually required per hectare of land. Plants are best grown on raised beds so that these do not spread out of control. Planting distances vary from 15 to 40cm in the row and from 30 to 80cm between the rows (uses.plantnet). A wider spacing gives more rhizomes per plant (uses.plantnet). Moreover, a wider row spacing will leave a comfortable walkway (cultivariable.com). Tighter spacing should only be used where water is abundant (cultivariable.com). Rhizomes can be planted directly in the field or first in pots and then transferred to the field in fall or spring. These plants can be grown in pots (one rhizome per pot) of three gallons or larger size (cultivariable.com) with a mixture of garden soil and good-quality compost. This is particularly recommended in areas with very heavy soil.

Although these plants are tolerant to wide temperature range and as these are not drought-tolerant so need consistent supply of **water** to harvest good yield. Plants seem to withstand even waterlogged conditions in the winter. The yield will suffer if the soil is allowed to dry out too frequently. Initially, the field should be kept weed-free until the canopy of the crop covers entire open area of plantation or just after planting whole planted area should be **mulched** preferably with organic materials (dried leaves, dry grasses, bagasse, rice bran, sawdust or compost) or black polyethylene sheets.

When the plants are 30-60cm tall they are **earthed up** around the base of the stem (uses.plantnet). Whenever, the rhizome is found exposed or when a rhizome has started peeping out the bunds are hilled up which will help developing of healthy rhizomes.

Thinning and Trimming

It is virtually impossible to collect all the rhizomes at harvesting time as some are still left behind to grow the following season. If these crops have been grown previously in the same field, they may require thinning to maintain the preferred spacing (cultivariable.com). Moreover, failure to thinning will result in overcrowding which will ultimately give poor yield of rhizomes, *vis-à-vis* these tubers will also be quite small. Thinning out can be done in the spring itself (cultivariable.com) as at this time the planted crop would also have sprouted. Further, the plants may require trimming to prevent flowering especially in case of Florida betony so that energy consumed in flowering may be diverted towards rhizome development (cultivariable.com).

In Malaysia, Chinese artichoke is planted in beds after cabbage or in rotation with other tuber crops such as potato and yam (uses.plantnet). These species can make good understorey plants for shrubs or trees. The only problem is that they will probably escape from control in such an environment and become very weedy (cultivariable.com).

Harvest

The tubers can be harvested and replanted at any time whilst they are dormant. They do start into growth fairly early in the year so it is better to have moved them by the end of March (cultivariable.com). After harvest, the rhizomes should be planted as soon as possible since it can be hard to store the rhizome in good condition. They can be planted out in March and harvested from October onwards. Although top growth is killed back by frost, the tubers are very hardy and can be left in the ground over winter to be harvested as required.

The rhizomes require 5-7 months to fully develop for harvest. The rhizomes can be rather deep. Florida betony rhizomes are formed typically 10-15cm deep but Chinese artichoke and marsh woundwort may form tubers anywhere from 18 to 46cm deep or sometimes even deeper where they run into obstructions like the sides of a raised bed (cultivariable.com). They are ready for harvest any time after the foliage dies back in the fall. In cool summer climates, rhizomes are generally harvested in fall and in frost-free climates they may be best harvested in spring (cultivariable.com). It is important that the soil is not frozen when harvesting. Harvesting is done manually and they (uses.plantnet) are usually harvested and replanted at the same time by placing one rhizome in the spot from which the plants have been harvested.

Any of the tubers missed being harvested will re-sprout the next growing season to form new clumps of plants (iplantz.com), and the plant may become a weed. In climates where the soil doesn't freeze to significant depth, these are harvested after maturity and then cleaned through the help of water (iplantz.com).

Yield

Under optimal conditions a Chinese artichoke plant may form 40-220 rhizomes depending on the planting material (uses.plantnet), growing condition and cultural practices. The yield is 8-20t/ha (uses.plantnet). Cultivation trials in Switzerland gave maximum yields of 43t/ha (uses.plantnet).

Storage

Rhizomes dehydrate and discolour rapidly in the open and should therefore be marketed soon after harvesting (uses.plantnet). These must be stored in dark under cool and high humid conditions (cultivariable.com). These can be stored in moist sand at 0-2°C or frozen at -18°C (uses.plantnet). For a short period they can be stored in a refrigerator by wrapping in paper (uses.plantnet). The rhizomes have thin skin and can be stored only for a few days, *i.e.* about a week in crisper drawer of a refrigerator without much decline in quality (cultivariable.com).

Rhizomes keep very poorly out of the ground and also tend to discolour. Because of this, they are best left in the ground and harvested as needed in climates where the ground doesn't freeze. In very cold climate or where animals may feed on the tubers, they can be kept lightly wetted and sealed in a plastic bag to be stored through the winter for replanting the following spring (cultivariable. com). Seeds of marsh woundwort and Florida betony have the storage life of at least 3 years at 10°C (cultivariable.com).

Insect-Pests and Diseases

Not much literature, especially under Asian condition is available on pests and diseases attacking this crop. Sometimes, spider mites and tarnished plant bug are reported to be serious on this crop (cultivariable.com). Certain caterpillars may be found eating away most of the foliage. Root aphids can also be a nuisance. Slugs and snails will attack both stems and tubers leaving irregular holes there. Damage by slugs and snails is most common in spring when plants begin to produce new seeds and shoot tips. Integrated pest management approaches should be adopted in keeping these pests at a minimum.

In Europe, Chinese artichoke can be infected by several viruses known from other crops (Chinese artichoke mosaic virus and potyvirus) spreading through aphids (uses.plantnet). The only remedy is to abandon the current plot, killing off the volunteers by covering the soil with plastic and starting again with new tubers in new ground (cultivariable.com). The crop is very sensitive to *Rhizoctonia solani* (uses.plantnet).

Reported fungal diseases are

☆ **Powdery mildew** (*Oidium* sp. and *Neoerysiphe galeopsidis*) usually infects shaded or semi-shaded plants. Their spores can spread with the winds. These thrive in warm (15-27°C), dry climates, though it does require fairly high relative humidity around the plant to spread. In cooler *cum* rainy areas it does not spread well and temperatures above 32°C is also not congenial for their spread.

  Symptoms: Infected plants look as if they have been dusted with flour, usually starting as circular powdery white spots which can appear on leaves and stems. Usually this covers the upper part of the leaves but may also grow on the undersides as well. Leaves turn yellow and dry out and may also twist, break or become disfigured (en.wikibooks.org1).

Management: Plants should be grown in open sunny locations, and thinning or plant shearing should be practiced to allow good air circulation. Effective organic fungicides such as sulphur, lime-sulphur, neem oil and potassium bicarbonate for treating powdery mildew should be applied (en.wikibooks.org1).

- ✰ **Downy mildew** [*Peronospora lamii* (Lib.) Corda] forms greyish-brown or greyish-lilac patches on the yellowed leaves in September and May. This can be controlled similar to the treatments given for powdery mildew.
- ✰ **Leaf spot** (*Phoma* sp.; *Ramularia stachydis* Sacc.; *Septoria stachydis* Fr.): Circular lesions found on the leaves of plants caused either by bacteria or fungi. Spores or bacteria overwinter on fallen leaves or buds. Growth is favoured by moist conditions on the leaf surface.

  Symptoms: Initial infections are in the form of small circular spots, usually with a differently coloured zones on the outer edge. Severe cases cause distorted leaves or blight (en.wikibooks.org1; en.wikibooks.org3).

  Management: Pruning or thinning to allow good air circulation and thorough cleaning in the autumn should be carried out to reduce the amount of spores. Neem oil and copper-based sprays are recommended (en.wikibooks.org1; en.wikibooks.org3).
- ✰ **Southern blights** (*Sclerotium rolfsii*):

  Symptoms: Affected plants show signs of leaf wilting, *vis-a-vis* soft *cum* rotten crowns, often followed by the death of the infected plant. When the fungus forms sclerotia as the asexual dormant body this becomes visible to the naked eye and looks like the white, yellow or brown mustard seeds (en.wikibooks.org1; en.wikibooks.org2).

  Management: Infected plants should be removed as soon as the infection is discovered; tools and shoes should be cleaned; the sclerotia only become active when on the surface of the soil, so a deep top-dressing with compost may help controlling the disease, and removal of the mulch layer in winter is also recommended so that sclerotia are exposed in the soil to winter dessication. Solarization can kill the fungus in the top layer of the soil in regions where solarization can bring about high enough temperatures (en.wikibooks.org1; en.wikibooks.org2). Methyl bromide and chloropicrin have been applied as soil fumigants, however, as both the chemicals are extremely dangerous so are banned in many regions. PCNB (Terraclor) at 225-450g a.i./300m^2 can be incorporated onto the soil surface to get rid of this fungus (ppdl.org). *Trichoderma harzianum*, *T. viride*, *Bacillus subtilis*, *Penicillium* spp. and *Gliocladium virens* are also effective (en.wikibooks.org1; en.wikibooks.org2).

References

1. Amahe, 2002. Antimicrobial activity of the hexane extract of *Stachys sieboldii* MIQ leaf. *J. Life Sci.*, **12**(6): 803–11. doi: 10.5352/jls.2002.12.6.803.
2. Anon. 2004. Antioxidant activities of *Stachys sieboldii* MIQ roots. *J. Life Sci.*, **14**(1): 1–7. doi: 10.5352/jls.2004.14.1.001.
3. Anon. 1833. Plants of the World Online (POWO). Royal Botanic Garden, Kew. First published in Enum. Pl. China Bor.: 51 (1833). https: //powo.science.kew.org/taxon/urn: lsid: ipni.org: names: 459268-1.
4. Bussard, L. 1909. Crosne du Japon. In: Culture Potagere et Culture Maraichere, pp. 183–4.
5. Charlotte Lindqvist and A. Victor, 2002. Origin of the Hawaiian endemic mints within North American *Stachys* (Lamiaceae). *Amer. J. Bot.*, **89**(10): 1709–24. https: //doi.org/10.3732/ajb.89.10.1709;.https: //bsapubs.onlinelibrary.wiley.com/doi/full/10.3732/ajb.89.10.1709
6. cultivariable.com: .www.cultivariable.com/instructions/root-crops/how-to-grow-chinese-artichoke-and-tuberous-mints/
7. Czarnota, M. and T. Murphy, 2012. Controlling Florida Betony in the Landscape. C 867-11. College of Agricultural and Environmental Sciences, University of Georgia, USA.
8. en.wikibooks.org1: https: //en.wikibooks.org/wiki/Horticulture/Stachys
9. en.wikibooks.org2: https: //en.wikibooks.org/wiki/Horticulture/Sclerotium_rolfsii
10. en.wikibooks.org3: https: //en.wikibooks.org/wiki/Horticulture/Leaf_Spots
11. en.wikipedia.org/wiki/Stachys_affinis
12. Gourmetpedia: http: //gourmetpedia.net/products/vegetables/knotroot/
13. Greutert, H. and F. Keller, 1993. Further Evidence for Stachyose and Sucrose/H+ Antiporters on the Tonoplast of Japanese Artichoke (*Stachys sieboldii*) Tubers. *Pl. Physiol.*, **101** (4): 1317–1322. doi: 10.1104/pp.101.4.1317.

14. iplantz.com: https: //www.iplantz.com/plant/1442/stachys-affinis/
15. Kenneth Taylor and Rowland Phillip, 2011. Biological Flora of the British Isles: *Stachys palustris* L. *J. Ecol.*, **99**: 1081–90. https: //doi.org/10.1111/j.1365-2745.2011.01849.x.
16. McElroy, S., Fred H. Yelverton, Joseph C. Neal and Thomas W. Rufty, Jr. 2004. Influence of photoperiod and temperature on vegetative growth and development of Florida betony (*Stachys floridana*). *Weed Sci.*, **52**(2): 267-70.
17. merriam-webster.com: https: //www.merriam-webster.com/dictionary/hedge per cent 20nettle
18. Morton, J.K. 1973. A cytological study of the British Labiatae (excluding *Mentha*). *Watsonia*, **9**: 239–46.
19. nutraingredients-asia.com: https: //www.nutraingredients-asia.com/Article/2017/02/26/Nutraceutical-benefits-of-Chinese-artichoke-underlined-for-people-with-diabetes-and-CVD#: ~: text=Chinese per cent 20artichoke per cent 20(S.,much per cent 20of per cent 20which per cent 20remains per cent 20untapped.
20. Paton, Alan, Wu Zheng-yi aned P.H. Raven, 1995. Flora of China Vol. 17: Verbenaceae through Solanaceae. *Kew Bul.*, **50** (4): 838. doi: 10.2307/4110257.
21. Peron, J.Y. and D. Dubost, 1992. Revalorization of lost vegetables: a contribution to preservation of genetic resources. In: *II intern. Symp. on Specialty and Exotic Veg. Crops, Acta Hort.* No. 318, 2, S. 685.
22. ppdl.org: http: //www.ppdl.org/dd/id/southern_blight-hosta.html - resistant species.
23. Princedebretagne: https: //www.princedebretagne.com/en/our-vegetables/chinese-artichokes-crosne.
24. Quattrocchi, U. 2000. CRC World Dictionary of Plant Names, vol. I, 91p. CRC Press, Boca Raton, New York, Washington, DC (USA) and London (UK).
25. Ryu, B.H., P. Bg and S.K. Song, 2002. Antitumor effects of the hexane extract of *Stachys sieboldii. Biotechnol. Bioeng.*, **17**(6): 520–24.
26. Specialtyproduce: https: //specialtyproduce.com/produce/Crosne_240.php.
27. uses.plantnet: .*Stachys.sieboldii.*(PROSEA).https: //uses.plantnet-project.org/en/Stachys_sieboldii_(PROSEA); https: //uses.plantnet-project.org/en/Stachys_sieboldii_(PROSEA).

28. Veggiesinfo: https: //veggiesinfo.com/chinese-artichoke-properties-and-its-uses/.

29. Wilcock, C.C. 1974. Population Variation, Differentiation and Reproductive Biology of *Stachys Palustris* L, *S sylvatica* L. and *S. x Ambigua* SM. *New Phytol.*, pp. 1233-41.

30. Yin, J., G. Yang, S. Wang and Y. Chen, 2006. Purification and determination of stachyose in Chinese artichoke (*Stachys sieboldii* Miq.) by high-performance liquid chromatography with evaporative light scattering detection. Talanta,**70** (1): 208–12. doi: 10.1016/j.talanta.2006.03.027.

32

Yacon [*Smallanthus sonchifolius* (Poepp.) H. Rob]

Family: Asteraceae.

Synonyms: *Helianthus esculentus* Warsz. ex Otto and Dietr., *Polymnia edulis* Wedd., *P. sonchifolia* Poepp.

Other common names: Bolivian sunroot, Peruvian ground apple, Poire de terre, Strawberry jicama, Yacon strawberry.

Introduction and Origin

The name yacon is derived from the words *yakku* (insipid) and *unu* (water), meaning tasteless water in Quechua language. Yacon is indigenous to the Andean highlands where it is traditionally grown in the cool conditions of the tropical and subtropical Andes of Peru, Bolivia, Colombia, Ecuador and Argentina at altitudes of 900–3,300m for its crisp, sweet-tasting and juicy tuberous roots. It has become popular in Japan and the Philippines, where the roots and leaves of yacon are widely used for its purported antihyperglycemic properties. The tuberous root is composed mostly of water and fructooligosaccharide. Unlike many other root vegetables domesticated by the indigenous peoples of the Andes (ulluco, oca and mashua), yacon is not photoperiod-sensitive and can produce a commercial yield in the subtropics as well as in mountainous regions.

Botany

Yacon, a close relative of the sunflower and Jerusalem artichoke, is a perennial herbaceous plant growing up to a height of 2.5m (Zardini, 1991) with two kinds of stems:

i) Thick, cylindrical or a little angular above-ground stems that are striated, densely pilous ~ especially on the upper part and with purple lines. These stems are succulent when young but tend to harden (lignify) and form a hole through their central axis as they get older and aquire a yellowish-orange colour when they become dry. They tend to radially incline from the centre to the outside at a distance equivalent to their height.

ii) Underground stems forming a system of short, thick and sympodial rhizomes which can be eaten when young but are generally used for vegetative propagation.

Its leaves are opposite, large, dark green, thin and variable-shaped with serrated margins (Humanity development library 2.0) where lower surface remains smooth though upper surface hairy (en.wikipedia.org). From the lower leaf axils additional stems arise and at the end of the growing season small, yellow or orange-yellow composite flowers are borne at the end of these stems. The bisexual flower head is a pseudanthium (composed of several florets), yellow or orange female ray florets are up to 1.2cm long while yellow-brown male disc florets are about 7mm long. Seeds are stored inside achenes measuring about 2.2-3.7mm and are dark brown in colour (en.wikipedia.org). Flowering time depends on the environmental conditions; under favourable conditions flowering begins 6–7 months after planting and peaks about two months later (en.wikipedia.org). Seed production is usually quite low and some ecotypes do not produce any seeds at all which is due to pollen sterility.

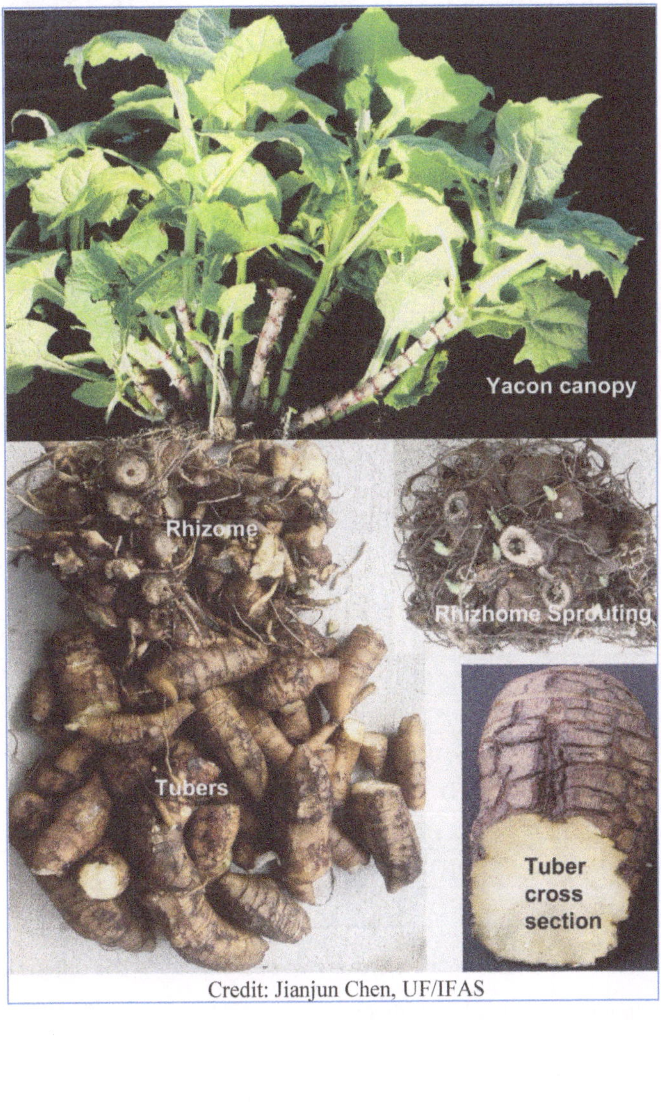

Credit: Jianjun Chen, UF/IFAS

Yacon has two types of underground tuberous growth ~ i) branching rhizomes directly at the base of the stem which can be eaten when young but are mainly used for propagation, and ii) up to 20 larger ellipsoid or cylindrical tuberous storage roots that are up to 25cm long and 10cm wide achieving a weight of 0.2–2kg with varying skin colours. Generally, in the Andes each tuberous root weighs 200-500g although these can reach up to 2kg. The rhizomes continually produce leafy shoots while storage roots are the principal economic product of the plant. The colour of peel and flesh ranges from white or cream to yellow, orange and red to pink to purple or brown, depending on the variety. Freezing temperatures cause the aboveground parts to die back but plants re-sprout from the rhizome under favourable temperature and moisture conditions (en. wikipedia.org). Optimal growing temperature range is 18-25°C (Dostert *et al.*, 2009).

The members within the species have different chromosome numbers, with individuals containing 2n=30, 32, 58, 87 and 116 chromosomes (Lorenzoni *et al.*, 2017).

Nutritional Values

Roots are very low in calories because most of the carbohydrate is in the form of indigestible fructooligosaccharide (FOS) which functions as a prebiotic. Its tubers contain high amounts of FOS and inulin (up to 60 per cent of DM) which are supposed to have positive effects on human health. The chemical composition of yacon varies depending on factors such as location, the variety, the growing season, harvest time and the post-harvest temperature (Valentova *et al.*, 2006).

Chemical Composition of Yacon Root

Components per cent	*Moscatto et al. (2006)*	*Lobo et al. (2007)*	*Itaya et al. (2002)*
Moisture	7.49±0.17	ND	8.09±1.74
Protein	6.48±0.15	2.64±0.07	4.50±1.26
Lipids	0.31±0.01	0.61±0.02	0.67±0.19
Ash	3.56±0.02	3.85±0.06	2.88±0.13
Insoluble fibre	ND	7.85±0.17	11.79±0.36
Carbohydrate	82.16	ND	ND
Calcium (mg/g)	ND	0.83±0.01	0.22±0.40
Magnesium (mg/g)	ND	0.62±0.09	0.40±0.00

Dry basis; ND=Not determined; Carbohydrate was estimated by the difference.

Source: https://en.wikipedia.org/wiki/Yacon

Carbohydrate Composition of Yacon Root

Carbohydrate per cent	*Moscatto et al. (2006)*	*Lobo et al. (2007)*	*Habib et al. (2011)*
Fructose	4.13	13.51	26.00
Glucose	1.96	8.97	10.01
Sucrose	3.25	13.42	10.00
FOS/fructans	ND	55.33	52.00
1-kestose (GF2)	8.19	ND	ND
Nystose (GF3)	5.36	ND	ND
Fructofuranosyl-nistose (GF4)	4.03	ND	ND

FOS: Fructooligosaccharides; ND: Not determined.

Source: https://en.wikipedia.org/wiki/Yacon

Yacon tubers consist mostly of water [about 70 per cent of fresh weight (Itaya *et al.*, 2002)] and carbohydrates. Therefore, the energy value is low. The dry matter is composed of 40-70 per cent of fructooligosaccharides. Inulin, a low-polymerization β(2-1)-oligosaccharide is the main fructooligosaccharide in yacon (Ojansivu *et al.*, 2011). These oligosaccharides are known to be nontoxic, non-digestible and sweet. There are also 15-40 per cent simple sugars as sucrose, fructose and glucose (Kapuler and Gurusiddiah, 1993). Most abundant minerals in yacon are calcium and potassium. In addition, yacon juice is rich in free essential amino acids (Kapuler and Gurusiddiah, 1993).

Chlorogenic acid and L-tryptophan are the major antioxidative compounds in yacon (Yan *et al.*, 1999). It also contains different phenolic compounds that enable epiphytic bacterial growth with specific metabolic properties inhibiting the attack of pathogens. Polyphenols found in yacon leaves and bark produce an acrid and astringent flavour as well as impart a typical odour. Polyphenols are also substrates for the enzymatic browning of damaged tissues in yacon root, giving it a greenish or black colour due to a condensation reaction of polyphenol compounds with amino acids and the enzymatic polymerization of polyphenols (Valentova, 2006).

Uses

Generally, yacon is cultivated in different countries for food and medicinal use. Mature storage roots are much sweeter as compared to young usable-sized roots. They can be eaten fresh, boiled, roasted or processed into beverages, jams, syrup, vinegar, flour, chips and juice (Paula *et al.*, 2015). These are also used as a source of natural sweeteners and syrups suitable for persons suffering from digestive problems. Yacon products have antidiabetic and hypoglycaemic effect. In the Andean folk medicine, it is used against liver and kidney diseases whereas in Bolivia it is used against diabetes and digestion problems (Paula *et al.*, 2015).

The rhizome is edible but it is not so sweet as to those of storage roots and is sometimes fibrous too. Moreover, its bitter skin is difficult to remove due to the convolutions. It can be used as animal feed.

Propagation

Yacon can be propagated by seed, rhizome, stem cutting and through tissue culture. Seed propagation is uncommon because not all varieties can produce abundant seeds. Moreover, plants produced from seeds take longer to mature than those grown from tubers or rhizomes (Dostert *et al.*, 2009). Stems can be easily rooted in a pot or nursery bed and then transplanted in the field. Tissue-cultured liners are available in Florida which can directly be planted in field. Rhizomes are the most commonly used planting material. Although the rhizomes store best in soil, they aren't frost resistant and it would be better to keep them safely indoors until planting time either in a cool and humid environment or in nursery beds/pots with growing media until ready to plant. Optimum temperature for sprouting is 20-25°C where up to 90 per cent sprouting is obtained.

Rhizomes can be propagated in three different ways:

i) Seedlings are divided from the mother plant and pre-cultivated in the greenhouse.
ii) Rhizome pieces are presprouted in the greenhouse in nursery bed or pots.
iii) Rhizome pieces can directly be planted in the main field which will decrease the propagation cost significantly. For better yield yacon is usually grown from chunks of the parent plant's subsurface crown of dormant rhizome. By planting the crown buds directly into the field, pre-cultivation in greenhouse would not be needed and mechanical planting by a potato planter could be possible. The rhizome pieces can be planted into the field earlier than the plantlets of the greenhouse.

Transplanting is done usually after the last frost. Sprouted rhizomes are transplanted in the field after the last frost, usually when the first pair of leaves is forming on the stem. Planting materials that have a few sets of leaves transplant better than dormant propagules in a cold and wet ground.

Growth and Development

The growth and development of yacon varies depending on its growing conditions and cultural practices. In the Andean region, yacon sprouts usually in 30-50 days after planting where for the 1st four months its growth is very slow, though stem diameter and leaf production reach a maximum in 170 days and then only root formation and coarsening accelerate (Seminario *et al.*, 2003). The reproductive period begins between 6 and 10 months after planting, when it blooms. The onset of flowering depends on the region where the plant is being grown, maturing earlier at lower altitudes (Santana and Cardoso, 2008). This stage is followed by a period in which the carbohydrate content in the rhizophores and the tuberous roots increase, leading to a senescence phase of the shoots and the numbness of the underground parts (Vilhena *et al.*, 2000). According to Oliveira and Nishimoto (2004), the tuberous roots and rhizophores should be harvested between the beginning of flowering and senescence. In the Andean regions yacon is harvested usually after blooming when the stems begin to wither away and bend. Depending on growing conditions, the growers must select a clone with shorter vegetative phase especially for Central Europe conditions.

Environmental Factors

Although the plant occurs in Pacific Coast regions up to 3,600m elevations but the best tuberous root production is obtained at altitudes ranging from 1,500 to 2,000m (Seminario *et al.*, 2003). It is day-length neutral and can be planted all year round in frost-free areas. It appears to be drought-tolerant and pest-free compared to other vegetable crops. For cold areas of Australia the rhizomes are started in styrofoam boxes in a greenhouse or on a warm verandah and planted out when frost is over. Yacon prefers the optimum temperatures of about 18-25°C but if sufficient water is available it can tolerate upto 40°C (Polreich, 2003). Yacon may grow up to twice of its normal height in areas with very long summer days and favourable weathers such as in parts of northern Canada and Alaska. Though it tolerates a little warmer condition as long as it does not dry out but copes rather poorly with wind so this is best planted in a protected spot, or grown in a block where the outside plants sacrificially protect the inside plants. Yacon is usually cultivated in its region of origin, *i.e.* Andes during rainy season though elsewhere it can also be grown anytime but with ample irrigation facilities *cum* cool weather. Planting it in the fall in a mountainous or lowland region favours yacon growth and development. Both fall and winter planting in the mountainous region allows root harvesting at the same time.

Cultivation

Yacon prefers well-drained, organic matter rich and friable soil with sufficient water for irrigation though its growth even in poor soils is also quite fast but on the cost of yield. Before planting, the field should be ploughed or dug thoroughly and mixed with sufficient compost, *i.e.* about 250q/ha. The rhizomes or presprouts are planted to about a 10cm depth (Tsukihashi *et al.*, 1996) with a row spacing of about 76cm and inter-row spacing of 1.2m. The rhizomes are placed in a hole or trench, cut side down, with buds at the top. They form roots at the base of each bud, not at the bottom of the cut, so the buds should not be allowed to protrude from the soil. If cuts are exposed, these may fail to root. In Korea, Ryu *et al.* (2001) studied certain spacings for planting but reported that among all the treatments, 80×60cm spacing which was widest in his studies was found better than others as this produced tallest plants with largest roots. Weeding is critical at early stage of growth as yacon establishes slow at this time though after start of the good growth it requires little care. Once when canopy is well developed there is no chance of developing of weed growth. As the storage roots grow, some of these may protrude from the soil when prior to planting the soil is hilled. It is best to cut free the lodged stems and hill up around the base of the plant to protect the rhizome. Many greens, radishes, turnips and peas make good early season companions for yacon.

Harvesting, Curing and Postharvest

Yacon is very flexible about harvest. It needs about five months in the ground to produce a reasonable yield but roots continue to grow until the plant enters senescence. If the soil starts to heave at the base of the plant a few early tubers can be harvested by carefully around the plant thus extending the harvest season. Tubers continue to sweeten as the plant dies back so the main harvest should only take place once all the top growth is dead. Harvesting should not be delayed too long, especially in areas that have mild winters, as the plant may start shooting when the weather warms up and the days get longer. The plant needs to be dug carefully to avoid damage and after separation from the central stem undamaged storage roots can be stored in a cool, dark and dry place with good air circulation for some months. The average sugar content of the tubers increases during storage because of starch conversion. These can also be exposed to sun for up to two weeks to accelerate sweetening process.

After digging, the rhizomes are kept out of the sun and covered with slightly damp sand, sawdust or cocopeat to stop them drying out and also to protect them from late frosts, and put aside for replanting in a dark and dry place. It is right time to plant when rhizomes begin to sprout in storage. Rhizomes are best stored intact with some small storage roots still attached. Stems can be trimmed leaving at least 5-7cm from the base and cleared off dirt by shaking or spraying water. Wet rhizome should be allowed to dry completely before putting in storage or dried in the sun and then stored otherwise mould may develop. If weather is humid during drying process, these can be kept under fans indoors for drying so that these may scape mould attack. Storage temperature should be close to 3°C with around 95 per cent humidity. If this condition for storage is not possible, the whole rhizome balls can be potted in the large containers. Broken roots as do not store well, these can be stored for longer by dipping the broken ends in beeswax or in some other food-safe wax. They have a respiratory rate similar to potatoes (Hermann *et al.,* 1997) and should last about as long in storage. These roots can be easily stored for 3-4 months at 10°C under humid conditions, and can last up to a year if stored at 2°C temperature.

Pests and Diseases

Insects such as **sunflower caterpillar** (*Chlosyne lacinia* Saundersii) feed on yacon leaves and may cause a significant damage. These insects can best be controlled by their natural enemies and trap plants (Silva *et al.,* 2015). Sunflowers can be planted as trap plants between yacon plants as insects are more enticed to these so insect population on yacon will be drastically reduced (en.wikipedia.org). **Root-knot nematodes** (*Meloidogyne incognita*) can lead to significant crop loss resulting in poor growth and roots with characteristic galls (Riedel, 1981; Mogor *et al.,* 2003). A safe procedure is to take shoot axillary buds as ideal explants and a disinfection treatment with a sodium hypochlorite aqueous solution (Mogor *et al.,* 2003).

Rhizoctonia sp. can lead to rot and discolouration in the root and crown of the yacon plant. Use of clean and healthy yacon offsets and rootstocks for propagation are important to reduce disease spreading (Fenille *et al.,* 2005). Many viruses also infect the yacon plant. Badnavirus (yacon necrotic mottle virus) causes necrosis, chlorosis, stunting and malformation of leaves. *Yucca bacilliform* virus leads to damage as chlorotic lesions on the leaves. The lesions disperse along the leaf veins and increase in intensity towards the tips whereby the lesions gradually turn necrotic (Bhat *et al.,* 2016). To get virus-free yacon, there should be regular spraying of insecticides such as pyrethrum, derris and insecticidal soaps to control virus-vectors. Virus infected plants should be uprooted and burnt to check further spread.

References

1. Bhat, A., Thomas Hohn and Ramasamy Selvarajan, 2016. Badnaviruses: the current global scenario. *Viruses,* **8** (6): 177.
2. Dostert, N., J. Roque, A. Cano, M. La Torre and M. Weigend, 2009. Factsheet – Botanical Data: Yacon – *Smallanthus sonchifolius.* https: //repositorio.

promperu.gob.pe/bitstream/handle/123456789/1343/Factsheet_botanical_data_yacon_2009_keyword_principal.pdf?sequence=1 and isAllowed=y

3. Fenille, R. C., M.B. Ciampi, N.L. Souza, A.K. Nakatani and E.E. Kuramae, 2005. Binucleate *Rhizoctonia* sp. AG G causing root rot in yacon (*Smallanthus sonchifolius*) in Brazil. *Pl. Path.*, **54**(3): 325–30.

4. Habib, Natalia C., Stella Maris Honore, Susana B. Genta and Sara S. Sanchez, 2011. Hypolipidemic effect of *Smallanthus sonchifolius* (yacon) roots on diabetic rats: biochemical approach. *Chemico-biolog. Inter.*, **194** (1): 31–9.

5. Hermann, M., I. Freire and C. Pazos, 1997. Compositional diversity of the yacon storage root. Impact on a changing world. *Program report*, **98**, 425-32.

6. https: //en.wikipedia.org/wiki/Yacon

7. Humanity development library 2.0. Root crops: Yacon (*Polymnia sonchifolia*). nzdl.org.

8. Itaya, Nair Massumi, Machado de Carvalho, Maria Angela and Rita De Cassia Leone Figueiredo-Ribeiro, 2002. Fructosyl transferase and hydrolase activities in rhizophores and tuberous roots upon growth of *Polymnia sonchifolia* (Asteraceae). *Physiologia Plantarum*, **116**(4): 451–9.

9. Kapuler, A.M and S. Gurusiddiah, 1993. The twenty protein amino acids free in the juices of our common vegetables and herbs. *J. Home Cons. Hort.* **1**(1): 3–18.

10. Lobo, Alexandre R., Celia Colli, Eliana P. Alvares and Tullia M.C.C. Filisetti, 2007. Effects of fructans-containing yacon (*Smallanthus sonchifolius* Poepp and Endl.) flour on caecum mucosal morphometry, calcium and magnesium balance, and bone calcium retention in growing rats. *British J. Nutrit.*, **97**(4): 776–85.

11. Lorenzoni, R.M., F. Menine, E. Marques Jr., F.L. Oliveira and T.C.B. Soares, 2017. Genetic diversity of yacon accessions using ISSR markers. http: //www.funpecrp.com.br/gmr/year2017/vol16-2/pdf/gmr-16-02-gmr.16029576.pdf.

12. Mogor, G., A.F. Mogor and G.P.P. Lima, 2003. Bud source, asepsis and benzylaminopurine (BAP) effect on yacon (*Polymnia sonchifolia*) micropropagation. *Acta Hort.*, No. 597, pp. 311–4.

13. Moscatto, Janaina Andrea, Dionisio Borsato, Evandro Bona, Antonio Sergio de Oliveira, and Maria Celia de Oliveira Hauly, 2006. The optimization of the formulation for a chocolate cake containing inulin and yacon meal. *Intern. J. Food Sci. Techn.*, **41**(2): 181–8.

14. Ojansivu, Ilkka, Celia Lucia Ferreira and Seppo Salminen, 2011. Yacon, a new source of prebiotic oligosaccharides with a history of safe use. *Trends in Food Sci. Tech.*, **22**(1): 40–46.

15. Oliveira, M. A. and E.K. Nishimoto, 2004. Avaliação do desenvolvimento de plantas de yacon (*Polymnia sonchifolia*) e caracterização dos carboidratos de reservas em HPLC. *Brazilian J. Food Tech.*, **7**(2): 215-20.

16. Paula, Hudsara Aparecida de Almeida, Monise Viana Abranches and Ferreira, Celia Lucia de Luces Fortes Ferreira, 2015. Yacon (*Smallanthus sonchifolius*): A food with multiple functions. *Crit. Rev. Food Sci. Nutrit.*, **55**(1): 32–40.

17. Polreich, S. 2003. Establishment of a Classification Scheme to Structure the Post-harvest Diversity of Yacon Storage Roots [*Smallanthus sonchifolius* (Poepp. and Endl.) H. Robinson]. Ph.D. Thesis. Brazil.

18. Riedel, Richard M. 1981. Introduction to plant nematology. *BioSci.*, **31** (11): 853.

19. Ryu, J.H., H.S. Doo, K.S. Lee, and S.Y. Park, 2001. Effect of plant density on growth responses and yield in yacon. *Korean J. Crop Sci.*, **46**(5): 407-10.

20. Santana, I. and M.H. Cardoso, 2008. Raiz tuberosa de yacon (*Smallanthus sonchifolius*): potencialidade de cultivo, aspectos tecnológicos e nutricionais. *Ciencia Rural*, **38**(3): 898-905.

21. Seminario, J., M. Valderrama and I. Manrique, 2003. El yacon: fundamentos para el aprovechamiento de un recurso promisorio. Centro Internacional de la Papa (CIP), Universidad Nacional de Cajamarca, Agencia Suiza para el Desarrollo y la Cooperacion (COSUDE), Lima, Peru.

22. Silva, Diego M.N., Fabio L. Oliveira, Leandro P. Dalvi, Dirceu Pratissoli, Wellington A. Erlacher and Mateus A.L. Quaresma, 2015. Occurrence of insects causing injuries to the yacon crop. *Horticultura Brasileira,* **33**(3): 394–7.

23. Tsukihashi, T., Hara H. Kawamura and Y. Taguchi, 1996. Influence of planting depth of the seed tubers on the growth and yield in the cultivation of yacon (*Polymnia sonchifolia*). *Jap. J. Farm Work Res.*, **31**(4): 257-64.

24. Valentova, Katerina, Ales Lebeda, Ivana Dolezalova, David Jirovsky, Breda Simonovska, Irena Vovk, Pavel Kosina, Nikol Gasmanova, Marta Dziechciarkova and Jitka Ulrichova, 2006. The biological and chemical variability of yacon. *J. agr. Food Chem.*, **54** (4): 1347–52.

25. Vilhena, S.M.C., F.L.A. Camara and S.T. Kakihara, 2000. O cultivo de yacon no Brasil. *Horticultura Brasileira*, **18**(1): 5-8.

26. Yan, Xiaojun, Masahiro Suzuki, Mayumi Ohnishi-Kameyama, Yasutoshi Sada, Tatao Nakanishi, Tadahiro Nagata, 1999. Extraction and identification of antioxidants in the roots of yacon (*Smallanthus sonchifolius*). *J. agr. Food Chem.*, **47**(11): 4711–3.

27. Zardini, Elsa, 1991. Ethnobotanical notes on yacon, *Polymnia sonchifolia* (Asteraceae). *Econ. Bot.*, **45**(1): 72–85.

33

Yams (*Dioscorea* L.)

Family: Dioscoreaceae.

Synonyms: *Borderea* Miegev.; *Botryosicyos* Hochst.; *Elephantodon* Salisb.; *Epipetrum* Phil.; *Hamatris* Salisb.; *Helmia* Kunth; *Higinbothamia* Uline; *Hyperocarpa* (Uline) G.M.Barroso, E.F.Guim. and Sucre; *Merione* Salisb.; *Nanarepenta* Matuda; *Oncorhiza* Pers.; *Oncus* Lour.; *Polynome* Salisb.; *Rhizemys* Raf.; *Ricophora* Mill.; *Sismondaea* Delponte; *Strophis* Salisb.; *Tamnus* Mill.; *Tamus* L.; *Testudinaria* Salisb. ex Burch; and *Ubium* J.F.Gmel.

Introduction and Origin

Dioscorea, native throughout the tropical and warm temperate regions of the world, is a genus of over 600 species of flowering plants, of which majority of the species are tropical, and only a few species occur in the temperate climates. They occur in diverse environments from forests to grasslands. Wild species grow best near canopy gaps in woodlands and rainforests. Several species that are commonly known as 'yams' have large starchy edible tubers, providing food and medicines to some 300 million people in low-income food-deficient tropical and subtropical countries. These are considered as under-utilized crops. These are particularly important in parts of Africa, Asia and Oceania. The most important yam cultivation zone is West Africa that produces about 93 per cent of the world's edible yams. The crop is also important in parts of eastern Africa, the Pacific area (including Japan), the Caribbean and tropical America. They are very important in all parts of Melanesia, Papua New Guinea, the Solomon Islands, New Hebrides and New Caledonia. Yams are also important in the high islands of Micronesia, in Ponape, Kusaie, Truk and Yap. In India, it is grown in almost all the states but major yam producing states are Kerala, West Bengal, Bihar, Odisha, Tamil Nadu, Asom, Rajasthan, Gujarat, Maharashtra and NE States. The major yams being grown in India are **greater yam** (*Dioscorea alata*), **white yam** (*D. rotundata*), **lesser yam** (*D. esculenta*) and **potato yam** (*D. bulbifera*). Apart from these, many wild yams such as *D. belophylla, D. glabra, D. hispida, D. intermedia, D. oppositifolia, D. pentaphylla, D. pubera, D. spicata, D. tomentosa, D. vexans, D. wallichii* and *D. wightii* are also used as vegetable by tribal population in India. The yam tubers come in a variety of forms owing to numerous cultivars and related species. Yam is recognized as the fourth most important tuber crop after potato, cassava and sweet potato, contributing to about 10 per cent of the total root and tuber production around the world. Since yam tubers can be stored for very long periods, they are also considered as famine food, especially during the food scarcity periods. Although yams are unique for their food, medicinal and economic values, their wider utilization is limited due to the presence of anti-nutritional compositions.

The genus *Dioscorea*, one of the earliest angiosperms, may have originated from SE Asia and Indo-Malayan region, and majority of the species are grown in SE Asia, Tropical America and West Africa (Kumar *et al.*, 2017). Origin of yams is uncertain and genetic information suggests that there may be more than one point of origin. For centuries, wild yams have been used medicinally to prevent menstrual cramps, stomach cramps and general pain. During 1950s, a plant-based estrogen called diosgenin was found in the roots of wild yams which was believed to aid in chemical defense against herbivores. During 1960s, diosgenin was used to create the first birth control pills. Moreover, some *Dioscorea* species that are rich in nutrients and antioxidants have been used by native individuals in their diets.

The genus *Dioscorea* has been divided into several sections. Most of the commercially cultivated *Dioscorea* species such as *D. alata*, *D. rotundata*, *D. opposita* L., *D. cayennensis* and *D. japonica* Thunb. are placed under Enantiophyllum section though other cultivated species such as *D. esculenta* is placed under Combilium section, *D. trifida* L.f. in Macrogynodium, and *D. dumetorum* (Kunth) Pax to Lasiophyton (Peter, 2007).

Etymology

The generic name given by monk Charles Plumier is attributed to the ancient Greek physician and botanist Dioscorides. The moist-fleshed varieties of sweet potato are often called yams in the United States, but the true yams are botanically distinct from sweet potato.

Botany

Morphological diversity of the above-ground parts of *Dioscorea* species. (a) *D. tokoro*, (b) *D. quinqueloba*, (c) *D. rotundata*, (d) a stem of *D. mangenotiana* with thorns, (e) flowers of *D. japonica*, (f) a bulbil of *D. bulbifera*

***Source*: "Sugihara *et al.* (2021). Population Genomics of Yams: Evolution and Domestication of Dioscorea Species"**

Yams are deciduous herbaceous perennial vines that provide edible tubers, though some of these monocots are invasive plants and may be considered a noxious weed outside the cultivated areas. The vines of most of the yams die back at the end of growing season and regrow on the return of favourable conditions, while tubers increase in size each year. The slender and annual vines are smooth or prickly twiners growing 15m or more in length by climbing over nearby plants, trees, rocks and other structures for physical support. Their touch receptors allow them to sense objects by sending out shoots to assess the suitable support. Depending on the species, the twining or coiling habit of the vines could be a clockwise (left to right) or counterclockwise (right to left) direction. Leaf growth generally gets slowed until the relative stem finds a support, thus allowing shoots to grow rapidly. Vines of some species produce calcium containing adhesive to better latch onto structures. The stem wings which help in twining of the vines are present in some cultivated species, *e.g. D. alata.*

The simple or compound and alternate or opposite leaves are either lobed or entire. The leaves of some *Dioscorea* species are cordate, simple or acuminate with long petiole, and some species may have leaves that are palmate or lobed having pointed tips. Some simple leaved right twining wild species are *D. oppositifolia* L., *D. hamiltonii* Hook.f., *D. pubera*, *D. wallichii* Hook.f., and *D. glabra* Roxb., and left twiner species with compound leaves are *D. hispida* Dennst. and *D. pentaphylla* (Behera *et al.*, 2009). Edible *Dioscorea* species have usually opposite leaves though poisonous ones have alternate leaves.

Dioscorea is mostly dioecious vine usually with small flowers and starch-rich tubers as the underground organ which in India is being grown since ancient times but is considered as under-utilized crop. Being a dioecious plant, yam blooms irregularly. The male and female flowers usually open at different times making the process of pollination very difficult. The small and unisexual flowers are borne in long clusters on the axillary spikes of the leaf axils, each having a greenish bell-shaped or flat perianth of six pieces, enclosing six or fewer stamens in the male flowers and surmounting a 3-celled 3-winged ovary in female flowers. The glabrous, sessile and spherical male flowers are borne axially or terminally. The capsule shape varies in different species (Behera *et al.*, 2009) and has numerous light and flattish or globose seeds. Most species produce winged seeds and 6-seeded capsular fruits, while some species have wingless seeds, samaroid or berry fruits (Caddick *et al.*, 2002; Noda *et al.*, 2020). Seeds are dormant for months after plant senescence. These can be germinated to produce small seedlings but with no economic significance except for breeding purposes.

The formation of new tuber and withering of the old one occurs simultaneously when the re-growth is initiated. The thick, large and fibrous tubers are generally a development of the base of the stem and may grow up to 1.5m deep into the soil, though most are placed in the top layer of the soil. Depending on the species, the plant can produce one singular tuber or several tubers and can be cylindrical, curved or lobed. They often have rough, thick and almost bark-like skin with varying colours ranging from dark brown to light pink. The colour of soft flesh of majority of edible yams ranges from white to yellow or orange, purple or pink and varies in taste from sweet to bitter to tasteless. Though the rough skin is difficult to peel, it easily gets softened after boiling or roasting. In some species yam tubers can reach up to 1.5m long. The tubers serve as underground starch storage organs and represent morphologically diverse structures. Several species have evolved various defence traits to protect themselves from burrowing/digging animals (e.g. *D. praehensilis* has crown roots with spines to protect tubers), fire (some species have thick corky barks covering the pachycaul) and herbivores. Most of the species produce secondary metabolites that serve a variety of functions including defense against herbivores. Some yam varieties have needle-shaped crystals of calcium oxalate raphides just under the skin of the tubers so they should be peeled thoroughly, but the needle-shaped crystals cab cause hand irritation. Similarly leaves should be double boiled before being consumed. Some varieties contain dioscorine, a toxic alkaloid that must be removed by slicing and leaching in water. Many *Dioscorea* species contain the steroid saponin disogenin.

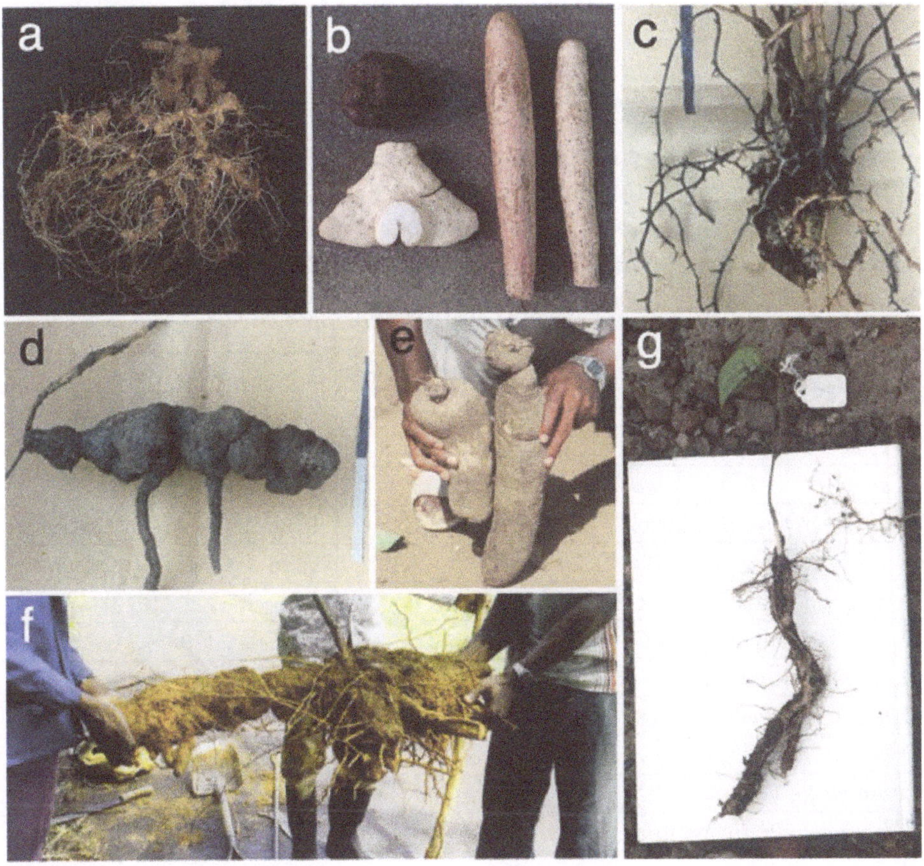

Rhizomes and tubers of *Dioscorea* species. (a) rhizomes of *D. tokoro*, (b) Tukuneimo group (left top), Ichoimo group (left bottom), Nagaimo group (right) in *D. polystachya*, (c) *D. praehensilis*, (d) *D. minutiflora*, (e) *D. rotundata* (left), *D. cayenensis* (right), (f) *D. mangenotiana*, (g) *D. abyssinica*

***Source*: "Sugihara *et al.* (2021). Population Genomics of Yams: Evolution and Domestication of Dioscorea Species"**

Many yam species, such as *D. bulbifera* (aerial yam), *D. alata*, *D. pentaphylla*, *D. pubera* and *D. tokoro* have aerial tubers or bulbils formed in the axils of leaves or bracts, which may become similar but smaller to underground tubers and can be eaten. In a few species, *e.g. Dioscorea bulbifera* (Opsophyton), the bulbils are the main storage organs. Bulbils are usually formed when the plants have been growing for a long time and are near the end of their growth for that year. Bulbils can sometimes make shoots and roots if a stem cutting is taken with the bulbil.

The size and shape of bulbils vary in different species, though their colour is generally grey to dark brown and have small tubercles over their surface (Murty and Purnima, 1983). Bulbils are mainly consumed as food, but also used as folk medicine in many cultures (Ikiriza *et al.*, 2019) and as planting materials.

Genetics

Several hybridization events have led to the changes in the ploidy levels of *Dioscorea* species. Basic chromosome number in the section Stenophora is x ¼ 10 but in the section Enantiophyllum it is x ¼20 (Scarcelli *et al.*, 2005; Arnau *et al.*, 2009). At least 33 per cent of the species in the genus *Dioscorea* have variable ploidy levels within the same species (Sugihara *et. al.*, 2021).

The Various Ploidy Levels of *Dioscorea* Species

Species	*Chromosome Number*	*References*
D. abyssinica Hochst.	2n ¼ 40	Miege (1954)
D. aculeata Webstar	2n ¼ 40	Martin and Ortiz (1963)
D. alata L.	2n ¼ 40, 60, 80	Arnau *et al.* (2009)
D. alata var. *purpurea* Roxb.	2n ¼ 80	Ramachandran (1968)
D. althaeoides Knuth	2n ¼ 20	Anon. (1976)
D. asclepiadea Prain and Burkill	2n ¼ 40	Takeuchi *et al.* (1970)
D. aspersa Prain and Burkill	2n ¼ 40	Chin *et al.* (1985)
D. balcanica Kosanin	2n ¼ 20	Miege (1954)
D. banzhuana Pei and Ting	2n ¼ 20	Anon. (1976)
D. batatas Decne.	2n ¼ 140, 144	Nakajima (1933), Smith (1937)
D. belophylla Voigt	2n ¼ 50, 80, 100	Raghavan (1958, 1959)
D. benthamii Prain and Burkill	2n ¼ c. 100	Chin *et al.* (1985)
D. bernoulliana Prain and Burkill	2n ¼ 36	Martin and Ortiz (1966)
D. biformifolia Pei and Ting	2n ¼ 20	Anon. (1976)
D. bulbifera L.	2n ¼ 36, 40, 54, 60, 80, 98–100	Smith (1937), Miege (1954), Raghavan (1958)
D. bulbifera var. *bulbifera* L.	2n ¼ 40	Paul and Debnath (2019)
D. caucasica Lipsky	2n ¼ 20	Meurman (1925)
D. cayenensis Lam.	2n ¼ 36, 54, 80, 140	Smith (1937), Miege (1954), Martin and Ortiz (1963)
D. ceratandra Uline	2n ¼ 36	Martin and Ortiz (1963)
D. chingii Prain and Burkill	2n ¼ 20	Anon. (1976)
D. cirrhosa Lour.	2n ¼ 40	Chin *et al.* (1985)

Species	*Chromosome Number*	*References*
D. cirrhosa var. *cylindrical* Ting and Chang	2n ¼ 40	Chin *et al.* (1985)
D. collettii Hook.f.	2n ¼ 20	Anon. (1976)
D. collettii var. *hypoglauca* (Palibin) Pei and Ting	2n ¼ 40	Anon. (1976)
D. composita Hemsl.	2n ¼ 36, 54	Martin and Ortiz (1963, 1966)
D. convolvulacea Schlect. and Cham.	2n ¼ 36	Martin and Ortiz (1963)
D. deltoidea Wall.	2n ¼ 20, 40	Raghavan (1958), Mehra and Sachdeva (1976)
D. discolour	2n ¼ 40	Smith (1937)
D. dumetorum Pax	2n ¼ 36, 40, 45, 54	Miege (1954), Baquar (1980)
D. escuintlensis Matuda	2n ¼ 36	Martin and Ortiz (1966)
D. esculenta Burkill	2n ¼ 40, 60, 80, 90, 100	Miege (1954), Cox *et al.* (1958), Raghavan (1958), Chin *et al.* (1985)
D. esculenta (Cherukizhengu)	2n ¼ 100	Ramachandran (1968)
D. esculenta (Nanakizhengu)	2n ¼ 90	Ramachandran (1968)
D. exalata Ting and Chang	2n ¼ 60, 80	Chin *et al.* (1985)
D. fargesii Franch.	2n ¼ 64	Smith (1937)
D. floribunda Mart. and Gal.	2n ¼ 36, 54, 72, 144	Martin and Ortiz (1963, 1966)
D. fordii Prain and Burkill	2n ¼ 40	Chin *et al.* (1985)
D. friedrichsthalii R.Knuth	2n ¼ 36	Martin and Ortiz (1963)
D. futschuensis Uline	2n ¼ 40	Anon. (1976)
D. glabra Roxb.	2n ¼ 40	Mehra and Sachdeva (1976)
D. gracillima Miq.	2n ¼ 20, 40	Nakajima (1933), Takeuchi *et al.* (1970)
D. hamiltonii	2n ¼ 40	Paul and Debnath (2019)
D. hastata J. Miege	2n > 120	Miege (1954)
D. hemsleyi Prain and Burkill	2n ¼ 60	Pei *et al.* (1979)
D. henryi (Prain and Burkill) Ting	2n ¼ 40	Chin *et al.* (1985)
D. hirtiflora Benth.	2n ¼ 40	Miege (1954)
D. hispida Dennst.	2n ¼ 40	Raghavan (1958)
D. hondurensis R. Knuth	2n ¼ 36	Martin and Ortiz (1966)
D. izuensis Akahori	2n ¼ 20	Takeuchi *et al.* (1970)
D. japonica Thunb.	2n ¼ 40, 80, 100	Nakajima (1933), Araki *et al.* (1983), Chin *et al.* (1985)

Species	*Chromosome Number*	*References*
D. kamoonensis Kunth	2n ¼ 40, 60	Chin *et al.* (1985)
D. macroura Harms.	2n ¼ 40	Smith (1937)
D. mangenotiana J. Miege	2n ¼ 72, 80	Miege (1954)
D. melanophyma Prain and Burkill	2n ¼ 40	Chin *et al.* (1985)
D. mexicana Guillemin	2n ¼ 36	Martin and Ortiz (1966)
D. minutiflora Engl.	2n > 120	Miege (1954)
D. nentaphylla L.	2n ¼ 120	Chin *et al.* (1985)
D. nipponica Makino	2n ¼ 20, 40	Takeuchi *et al.* (1970), Kiangsu Institute of Botany (1976)
D. nipponica var. *rosthornii* Prain and Burkill	2n ¼ 40	Anon. (1976)
D. nitens Prain and Burkill	2n ¼ 60	Chin *et al.* (1985)
D. nummularia Lam.	2n ¼ 60, 100, 120	Lebot *et al.* (2017)
D. opposita Thunb.	2n ¼ 140	Araki *et al.* (1983)
D. oppositifolia L.	2n ¼ 40, 140, 138–142	Smith (1937), Raghavan (1958), Chin *et al.* (1985)
D. oppositifolia var *linnaei* Prain and Burkill	2n ¼ 40	Ramachandran (1968)
D. paniculata Michx	2n ¼ 36	Martin and Ortiz (1966)
D. panthaica Prain and Burkill	2n ¼ 40	Anon. (1976)
D. parviflora C. T. Ting	2n ¼ 20	Pei *et al.* (1979)
D. pentaphylla L.	2n ¼ 40, 50, 60, 80, 120, 144	Smith (1937), Raghavan (1958, 1959), *Chin et al.* (1985), Paul and Debnath (2019)
D. pentaphylla var. *jacquemontii* Prain and Burkill	2n ¼ 40, 80	Raghavan (1958)
D. pentaphylla var. *linnaei* Prain and Burkill	2n ¼ 40, 80	Raghavan (1958)
D. pentaphylla var. *rheedei* Prain and Burkill	2n ¼ 40	Ramachandran (1968)
D. persimilis Prain and Burkill	2n ¼ 40	Chin *et al.* (1985)
D. poilanei Prain and Burkill	2n ¼ 20	Anon. (1976)
D. polygonoides Humb. and Bonpl.	2n ¼ 36, 54	Martin and Ortiz (1963, 1966)
D. polystachya Turcz. (Ichoimo group)	2n ¼ 100	Babil *et al.* (2013)
D. polystachya Turcz. (Nagaimo group)	2n ¼ 140	Babil *et al.* (2013)

Species	*Chromosome Number*	*References*
D. polystachya Turcz. (Tsukuneimo group)	2n ¼ 100	Babil *et al.* (2013)
D. praehensilis Benth.	2n ¼ 40	Miege (1954)
D. preussii Pax	2n ¼ 40	Miege (1954)
D. pubera Blume	2n ¼ 40	Raghavan (1958)
D. pyrenaica Bord.	2n ¼ 24	Miege (1954)
D. quaternata Gmel.	2n ¼ 36, 54	Jensen (1937), Martin and Ortiz (1966)
D. quinqueloba Thunb.	2n ¼ 20	Smith (1937)
D. reticulata C. Gay	2n ¼ 61	Smith (1937)
D. rotundata Poir.	2n ¼ 40, 60	Martin and Ortiz (1963), Baquar (1980)
D. sansibarensis Pax.	2n ¼ 40	Miege (1954)
D. sativa L.	2n ¼ 40	Sharma and De (1956)
D. sativa Thunb.	2n ¼ 80	Ramachandran (1962)
D. septemloba Thunb.	2n ¼ 20, 40	Takeuchi *et al.* (1970), Anon. (1976)
D. septemloba var. *sititoana*	2n ¼ 40	Takeuchi *et al.* (1970)
D. shimperana Hochst ex Kunth	2n ¼ 80	Baquar (1980)
D. simulans Prain and Burkill	2n ¼ 20	Anon. (1976)
D. sinuata Vell.	2n ¼ 24, 34 (male), 36 (female)	Meurman (1925), Suessenguth (1921)
D. smilacifolia de Wild.	2n > 120	Miege (1954)
D. spiculiflora	2n ¼ 36	Martin and Ortiz (1963)
D. spinosa Roxb.	2n ¼ 90	Ramachandran (1962)
D. subcalva Prain and Burkill	2n ¼ 60	Pei *et al.* (1979)
D. subcalva var. *submollis* (Knuth) Ling and Ting	2n ¼ 60	Chin *et al.* (1985)
D. tentaculigera Prain and Burkill	2n ¼ 40	Pei *et al.* (1979)
D. tenuipes Franch. and Savat.	2n ¼ 20, 30, 40	Takeuchi *et al.* (1970)
D. tokoro Makino	2n ¼ 20	Nakajima (1933)
D. tomentosa Koenig	2n ¼ 40, 60	Raghavan (1959)
D. trifida	2n ¼ 40, 60[b], 80	Bousalem *et al.* (2006, 2010)
D. villosa L.	2n ¼ 60	Smith (1937)
D. *wallichii* Hook.f.	2n ¼ 40	Raghavan (1959)
D. yunnanensis Prain and Burkill	2n ¼ 40	Chin *et al.* (1985)

Species	*Chromosome Number*	*References*
D. zingiberensis Wright	2n ¼ 20, 30, 40	Anon. (1976), Xiaoqin *et al.* (2003)
Tamus communis L.[c]	2n ¼ 48	Meurman (1925)
Tamus edulis Lowe[c]	2n ¼ 96	Borgen (1974)

Source: Sugihara *et al.* (2021). Population Genomics of Yams: Evolution and Domestication of Dioscorea Species.

a: Species name used in the references

b: 2C DNA content suggests the existence of the triploid *D. trifida* (Bousalem *et al.*, 2010).

c: Synonym of *D. communis* (L.) Caddick and Wilkin.

Chemical Composition

Diosgenin, a steroid sapogenin (an aglycone), is a chemical substance found in the tubers of certain wild yams, including a variant of 'Kokoro' yam (variety of *D. rotundata*)) and other *Dioscorea* species, such as *D. nipponica* (*Maiden, 1889*). Apart from diosgenin, other alkaloids, such as dioscorin and dioscin are also found. Yam contains phytosterols, alkaloids and tannin and is a rich source of starch. Other substances found are ash, protein, beta-carotene, ascorbic acid, niacin, riboflavin, thiamine, Al, Ca, Ch, Co, Fe, Mg, Mn, K, P, Se, Si, Na, Sn (tin) and Zn.

Species	*Phytochemicals*	*Reference*
D. alata	Phenolics, phenol, flavonoid, flavonol, phytates/phytic acid, saponin, oxalates, alkaloid, tannins, allantoin, dioscin, diosgenin, dioscorin, hydrogen cyanide	Wu *et al.*, 2016; Polycarp *et al.*, 2012; Estiasih *et al.*,2013; Shajeela *et al.*, 2011; Chaudhary *et al.*, 2018
D. bulbifera	Carotenoid, phenolics, phenol, polyphenol, flavonoid, terpenoid, saponin, steroid, alkaloid, tannins, phytates/phytic acid, oxalates, hydrogen cyanide	Bhandari and Kawabata 2004; Polycarp *et al.*, 2012; Shajeela *et al.*, 2011; Chaudhary *et al.*, 2018
D. belophylla	Saponins, alkaloids, flavonoids, tannins and phenols	Poornima and Ravishankar, 2009
D. cayenensis	Phenolics, phenol, saponin, alkaloid, tannins, phytates/phytic acid, oxalates, dioscin	Bhandari and Kawabata 2004; Polycarp *et al.*, 2012; Muzac-Tucker *et al.*, 1993
D. deltoid	Polyphenol	Bhandari *et al.*, 2004

Species	*Phytochemicals*	*Reference*
D. dumetorum	Phenols, flavonoid, alkaloid, tannins, phytates/phytic acid, oxalates, dioscorine	Polycarp *et al.*, 2012
D. esculenta	Phenolics, tannins, phytates/phytic acid, oxalates, hydrogen cyanide	Polycarp *et al.*, 2012; Muzac-Tucker *et al.*, 1993
D. fordii	Allantoin, dioscin	Wu *et al.*, 2016
D. glabra	Phenol, flavonoid	Padhan *et al.*, 2020
D. hamiltonii	Phenol, flavonoid	Padhan *et al.*, 2020
D. hirtiflora	Phenol, flavonoid	Sonibare and Abegunde, 2012a; Price *et al.*, 2016
D. hirsute	Dioscorine	Price *et al.*, 2016
D. hispida	Dioscorine, phenol, flavonoid	Padhan *et al.*, 2020
D. japonica	Phenols, flavonoilds, glycans	Wadkar *et al.*, 2014; Chiu *et al.*, 2013
D . mangenotiana	Dioscin	Kwon *et al.*, 2015
D. oppositifolia	Phenolics, phenol, flavonoid, tannins, oxalates, hydrogen cyanide	Mohan and Kalidas, 2010; Shajeela *et al.*, 2011; Padhan *et al.*, 2020
D. panthaica	Saponins	Dong *et al.*, 2001
D. persimilis	Allantoin, dioscin	Wu *et al.*, 2016
D. pentaphylla	Phenolics, tannins, oxalates, hydrogen cyanide, phenol, flavonoid	Mohan and Kalidas, 2010; Shajeela *et al.*, 2011; Padhan *et al.*, 2020
D. polystachya	Flavones, polyphenols, allantoin, dioscin	Wu *et al.*, 2016; Chen *et al.*, 2017
D. praehensalis	Tannins, phytates/phytic acid, oxalates,	Polycarp *et al.*, 2012;
D. preussii	Saponins	Tabopda *et al.*, 2014
D. pubera	Phenol, flavonoid	Padhan *et al.*, 2020
D. rotundata	Phenolics, phenol, tannins, phytates/phytic acid, oxalates, saponin, alkaloid, hydrocyanatem dioscin	Polycarp *et al.*, 2012; Muzac-Tucker *et al.*, 1993; Kwon *et al.*, 2015
D. sansibarensis	Dioscorine	Price *et al.*, 2016
D. spicata	Phenolics, tannins, oxalates, hydrogen cyanide	Mohan and Kalidas, 2010; Shajeela *et al.*, 2011;
D. tomentosa	Phenolics, tannins, oxalates, hydrogen cyanide	Mohan and Kalidas, 2010; Shajeela *et al.*, 2011;
D. trifida	Phenolics	Muzac-Tucker *et al.*, 1993
D. triphylla	Polyphenol	Bhandari and Kawabata, 2004

Species	Phytochemicals	Reference
D. versicolour	Polyphenol	Bhandari and Kawabata, 2004
D. villosa	Flavonoid, phenol, saponin, alkaloid, tannins, Phytates/Phytic acid, oxalates	Zhu, 2015
D. wallichi	Phenolics, tannins, oxalates, hydrogen cyanide, phenol, flavonoid	Shajeela *et al.*, 2011; Padhan *et al.*, 2020

D. polystachya: Chinese yam formerly known as *D. opposita* and *D. batatas*. NR = not reported.

Nutritional Values

Tubers of several species of *Dioscorea* are good source of food due to their high starch content and calorie value. The most widely cultivated edible species of yams are white yam (*Dioscorea rotundata* Poir.), greater yam (*Dioscorea alata* L.) and lesser yam (*Dioscorea esculenta* (Lour.) Burk.). Raw yam has 10 per cent or more of the Daily Value (DV), limited to potassium, vitamin B6, manganese, thiamine, dietary fibre and vitamin C (Uwaegbute and Obatolu, 1998). Although it is rich in starch and has high calorie value, yam is a relatively low-protein food and is not a good source of essential amino acids. Hence, it is advised to supplement a yam dominated diet with more protein-rich foods. The quantity and quality of protein in roots and tubers is lower than other food-staples. However, the potassium level in raw yam is highest as compared to other 10 major staple foods of the world. Yam supplies 118 calories per 100g and generally has a lower glycemic index, and about 54 per cent of glucose per 150g serving, compared to potato products (Anon., 2008). The relative nutritional and antinutritional contents in cooked yam are remarkably different from that of raw form of these staples.

Uses

Yams are used differently in different parts of the world. Yams are consumed in a variety of preparations such as flour or whole vegetable pieces across their range of distribution in Asia, Africa, North America, Central America, the Caribbean, South America, and Oceania (Anon. 2016). The tubers of most of the yams are eaten baked, boiled, or fried much like potatoes, and certain others as sweet potatoes. They are consumed after cooking, *i.e.* by frying, boiling or roasting. The green parts of some plants can be cooked and consumed as vegetable. Yams may also be used to produce flour or starch. Most of the raw yam tubers are not edible and cannot be digested. These must be sprouted or prepared and cooked for human consumption. Various natural substances that may be present in yams, especially in African yams, can cause illness if consumed raw, hence they must be taken only after cooking as then these become safe (Nweke *et al.*, 2013). An

Approximate Compositions of some Species of *Dioscorea*

Species	Proximate Composition (Percentage)						Reference
	Moisture	Crude Protein	Crude Fat	Crude Fibre	Ash	Starch	
D. alata	64.9–87.8	0.6–18.7	0.23–5.28	0.75–11.0	0.69–8.81	15.6–84.3	Wu *et al.*, 2016; Bekele and Bekele, 2018; Chu and Figero, 1991; Shanthakumari *et al.*, 2008; Shajeela, 2011; Chaudhury *et al.*, 2018
D. abyssinica	Not reported	3.13–5.37	0.31–1.22	1.94–4.91	2.31–3.58	NR	Bekele and Bekele, 2018
D. bulbifera	61.6–92.5	0.89–15.8	0.30–8.13	0.61–18.2	0.05–8.15	12.5–62.7	Bekele and Bekele 2018; Chu and Figueiredo-Ribeiro, 1991; Shanthakumari *et al.*, 2008; Shajeela, 2011
D. cayenensis	62.2–89.4	2.62–6.63	0.27–7.86	0.17–3.26	0.63–5.48	80.75	Bekele and Bekele, 2018; Chu and Figueiredo-Ribeiro, 1991
D. delicata	92.7	0.41	NR	4.87	NR	0.54	Chu and Figueiredo-Ribeiro, 1991
D. deltoidea	80.2	1.6	0.2	1.5	0.6	NR	Bhandari *et al.*, 2003
D. dodecaneura	68.4	1.50	NR	NR	NR	18.46	Chu and Figueiredo-Ribeiro, 1991
D. dumetorum	64.3–90.2	0.19–10.3	0.37–3.65	0.82–5.65	2.17–7.79	17.0–63.34	Chu and Figueiredo-Ribeiro, 1991; Polycarp *et al.*, 2012; Ogbo and Agu, 2014; Ojo *et al.*, 2017
D. esculenta	50.65–86.67	5.60 –10.50	0.08–2.58	1.23–7.82	0.25–8.50	17.25	Chu and Figueiredo-Ribeiro, 1991; Shanthakumari *et al.*, 2008; Polycarp *et al.*, 2012
D. fordii	NR	9.8–10.2	NR	0.92–1.14	NR	75.7–77.1	Wu *et al.*, 2016
D. hamiltonii	78.73	4.37	10.2	4.15	8.70	NR	Mohan and Kalidas, 2010
D. hispida	15.8–37.8	1.13–6.20	1.99–9.36	NR	0.29–1.24	11.5	Saleha *et al.*, 2018

Species	Proximate Composition (Percentage)						Reference
	Moisture	Crude Protein	Crude Fat	Crude Fibre	Ash	Starch	
D. laxiflora	82.0	0.26	NR	2.34	NR	8.92	Chu and Figueiredo-Ribeiro, 1991
D. nipponica	NR	NR	NR	NR	NR	35.4	Jiang *et al.*, 2012
D. olfersiana	84.6	0.42	NR	9.53	NR	0.54	Chu and Figueiredo-Ribeiro, 1991
D. oppositifolia	78.5–92.1	7.00–13.54	4.40–7.42	4.92–8.47	2.60–6.38	NR	Mohan and Kalidas, 2010; Shanthakumari *et al.*, 2008; Shajeela *et al.*, 2011
D. pentaphylla	90.1–93.1	6.48–9.18	4.01–6.24	5.14–7.24	3.36–4.64	NR	Mohan and Kalidas, 2010; Shanthakumari *et al.*, 2008; Shajeela *et al.*, 2011
D. persimilis	NR	7.70–8.20	NR	0.88–0.92	NR	68.2–72.2	Wu *et al.*, 2016
D. piperifolia	55.4–74.8	2.27–4.38	NR	NR	NR	18.2–26.1	Chu and Figueiredo-Ribeiro, 1991
D. polystachya	NR	6.30–12.2	NR	0.99–1.50	NR	60.7–72.5	Wu *et al.*, 2016; Jiang *et al.*, 2012
D. praehensilis	64.1	3.64–5.38	0.26–7.83	1.41–3.21	2.13–4.90	NR	Bekele and Bekele, 2018; Polycarp *et al.*, 2012
D. pyrifolia	NR	1.34	NR	NR	0.88	NR	Sharlina *et al.*, 2017
D. remotiflora	78.18	1.91	0.47	1.22	0.85	NR	Soto *et al.*, 2014
D. rotundata	54.5–75.2	0.09–8.28	0.09–3.39	0.41–4.33	1.03–4.92	22.0–80.8	Bekele and Bekele, 2018; Chu and Figueiredo-Ribeiro, 1991; Polycarp *et al.*, 2012; Ezeabara and Anona, 2018; Adeyeye *et al.*, 2014; Ahmadu *et al.*, 2018; Alinnor and Akalezi., 2010
D. sanpaulesis	69.2	0.77	NR	10.3	NR	2.62	Chu and Figueiredo-Ribeiro, 1991
D. sinuate	75.6	2.32	NR	NR	NR	8.00	Chu and Figueiredo-Ribeiro, 1991
D. spicata	81.5–89.3	6.38–8.20	3.26–4.78	4.67–6.31	5.18–5.20	NR	Mohan and Kalidas, 2010; Shajeela *et al.*, 2011

Species	Proximate Composition (Percentage)						Reference
	Moisture	Crude Protein	Crude Fat	Crude Fibre	Ash	Starch	
D. steriscus	72.5	0.83	NR	16.8	2.06	9.02	Wu *et al.*, 2016
D. subhastata	89.0	0.59	NR	0.95	NR	3.69	Chu and Figueiredo-Ribeiro, 1991
D. tomentosa	84.5–93.7	5.25–9.54	2.86–6.84	3.21–4.38	2.48–6.53	NR	Mohan and Kalidas, 2010; Shanthakumari *et al.*, 2008; Shajeela *et al.*, 2011
D. trifida	69.4–81.3	0.38–6.79	0.03–0.30	NR	0.2–3.37	7.94–64.0	Chu and Figueiredo-Ribeiro, 1991; Perez *et al.*, 2011
D. triphylla	76.9	2.3	0.2	0.6	0.6	NR	Bhandari *et al.*, 2003
D. versicolor	80.1	1.7	0.2	1.1	0.5	NR	Bhandari *et al.*, 2003
D. villosa	76.4	2.21	6.01	3.50	3.13	NR	Afiukwa *et al.*, 2013
D. wallichi	71.1–76.4	10.5–10.8	1.18–3.34	7.48–9.23	6.36–8.42	NR	Shanthakumari *et al.*, 2008; Shajeela *et al.*, 2011

Source: https://www.ncbi.nlm.nih.gov/pmc/articles/PMC7555206/

Approximate Composition of Minerals in *Dioscorea* Species

Species	No. of Varieties	Minerals (mg/100g)									Reference
		K	*Na*	*P*	*Ca*	*Mg*	*Cu*	*Fe*	*Mn*	*Zn*	
D. abyssinica [1]	13	NR	NR	5.1–56.5	31.02–118.8	NR	NR	20.3–69.7	NR	0.48–0.77	Bekele *et al.*, 2018
D. alata [1]	7	1157–2016	52–82.7	117–194	62.5–78.0	64.0–74.6	6.4–6.9	9.9–10.9	3.1–4.3	3.4–4.3	Wanasundera and Ravindran, 1994
D. alata [1]	7	240–400	190–380	100–340	20.04–80.2	24.31–97.2	NR	NR	NR	NR	Udensi *et al.*, 2008
D. alata [1]	9	NR	NR	NR	31.64–45.3	32.68–47.8	0.42–0.48	0.83–2.2	NR	0.82–2.6	Wu *et al.*, 2016
D. alata [1]	20	1055-2010	8.30–13.1	NR	26.0–53.5	39.0–59.5	NR	NR	NR	1.01–1.8	Dufie *et al.*, 2013
D. alata [3]	1	476.8 ± 0.1	68.9 ± 0.02	163.7 ± 0.10	285.8 ± 0.02	116.3 ± 0.69	NR	2.48 ± 0.02	NR	2.12 ± 0.00	Ezeabara and Anona, 2018
D. alata [3]	2	622.5–742.5	62,5–95.0	219.0–239	6.50–16.50	40.0–41.5	0.10–0.15	1.50–2.00	2.15–2.20	6.65–6.80	Polycarp *et al.*, 2012
D. alata [1]	16	1055–2010	8.4–13.1	87.8–190.0	26.0–41.0	39.0–58.0	1.23–1.57	NR	0.48–2.21	1.01–1.41	Baah *et al.*, 2009
D. alata [1]	4	NR	NR	26.59–49.12	11.24–120.0	NR	NR	17.75–51.1	NR	0.38–1.18	Bekele and Bekele, 2018
D. alata [1]	1	3932.9 ± 0.16	75.4 ± 0.02	NR	3032.1 ± 0.25	120.7 ± 0.005	1.216 ± 0.001	124.3 ± 0.004	1.33 ± 0.001	5.7 ± 0.001	Chaudhary *et al.*, 2018
D. alata [1]	1	5.25 ± 2.12	0.35 ± 0.0	NR	0.22 ± 0.99	0.65 ± 0.71	NR	0.75 ± 0.73	NR	NR	Ojo *et al.*, 2017

Species	No. of Varieties	Minerals (mg/100g)									Reference
		K	Na	P	Ca	Mg	Cu	Fe	Mn	Zn	
D. alata [1]	1	786.3 ± 0.14	44.56 ± 0.3	140.14 ± 0.14	448.36 ± 0.11	656.31 ± 0.07	11.20 ± 0.14	24.30 ± 0.19	6.36 ± 0.21	2.26 ± 0.01	Shajeela *et al.*, 2011
D. buibifera [1]	12	NR	NR	8.72–55.26	15.74–121.3	NR	NR	20.26–90.9	NR	0.4–8.33	Bekele and Bekele, 2018
D. buibifera [1]	1	NR	NR	0.521	1410.0	250	NR	NR	NR	NR	Ezeocha *et al.*, 2014
D. buibifera [1]	1	1554.4 ± 0.36	78.24 ± 0.07	154.42 ± 0.53	338.15 ± 0.09	396.20 ± 1.07	2.14 ± 0.04	19.20 ± 0.20	9.40 ± 0.14	1.48 ± 0.03	Shajeela *et al.*, 2011
D. buibifera [3]	1	525.8 ± 1.41	87.8 ± 0.10	159.5 ± 0.04	378.5 ± 0.10	128.7 ± 0.04	NR	3.14 ± 0.02	NR	2.79 ± 0.01	Ezeabara and Anona, 2018
D. buibifera [2]	1	560 ± 49	17.8 ± 9.8	61.61 ± 0.8	29.3 ± 4.8	25.9 ± 9.2	0.21 ± 0.03	2.92 ± 0.3	0.35 ± 0.03	0.53 ± 0.06	Bhandari *et al.*, 2003
D. buibifera [3]	2	1250–1475	92.5–102.5	223.5–224.5	103–116.5	76.5–83.5	0.20	6.00–6.50	1.30–1.35	6.10–6.35	Polycarp *et al.*, 2012
D. cayenensis [1]	2	NR	NR	19.15–26.12	6.3–27.6	NR	NR	17.2–27.95	NR	0.74–0.75	Bekele and Bekele, 2018
D. cayenensis [2]	1	262.3 ± 0.25	8.53 ± 0.05	19.5 ± 0.10	22.53 ± 0.13	61.53 ± 0.25	NR	0.79 ± 0.02	NR	0.39 ± 0.01	Adepoju *et al.*, 2018
D. cayenensis [3]	1	523.8 ± 0.04	76.8 ± 0.03	167.8 ± 0.02	345.8 ± 0.01	120.2 ± 0.55	NR	2.50 ± 0.08	NR	2.18 ± 0.02	Ezeabara and Anona, 2018
D. cayenensis [3]	2	700–825	62.5–70.0	164.5–190.5	74.5–80.0	57.5–38.0	0.20	5.0–5.5	1.2–1.25	5.45–5.85	Polycarp *et al.*, 2012
D. deltoidea [2]	1	340 ± 51	9.12 ± 1.6	33.1 ± 0.6	46.9 ± 6.2	22.8 ± 7.1	0.10 ± 0.0	1.85 ± 1.0	0.31 ± 0.02	0.22 ± 0.04	Bhandari *et al.*, 2003

Species	No. of Varieties	Minerals (mg/100g)									Reference
		K	Na	P	Ca	Mg	Cu	Fe	Mn	Zn	
D. dumetorum [1]	1	7.03 ± 0.78	0.41 ± 0.14	NR	0.81 ± 0.21	0.95 ± 0.71	NR	0.07 ± 0.14	NR	NR	Ojo *et al.*, 2017
D. dumetorum [3]	2	670–772	72.5–77.5	269–286	27.5–29.5	61.5	0.10	2.0–2.50	2.50–2.65	5.80	Polycarp *et al.*, 2012
D. dumetorum [3]	2	0.03	0.02	NR	0.19–0.21	0.65–0.72	1.36–1.48	0.13–0.16	0.34–0.38	0.03–0.18	Oyeyinka *et al.*, 2018
D. dumetorum [3]		NR	NR	151	57.8	NR	NR	8.89	NR	NR	Afoakwa *et al.*, 2001
D. esculenta [3]	2	765–795	87.5–92.5	273.5–294.5	20.5–27.0	67.5–73.0	0.10	2.0	2.70–2.95	6.20–7.80	Polycarp *et al.*, 2012
D. esculenta [1]	1	1594.3 ± 1.34	86.40 ± 0.14	138.10 ± 0.14	314.01 ± 0.33	436.06 ± 0.54	3.40 ± 0.01	11.48 ± 0.11	5.46 ± 0.11	1.76 ± 0.04	Shajeela *et al.*, 2011
D. fordii [1]	3	NR	NR	NR	28.56–30.05	34.58–35.63	0.45–0.51	1.82–2.02	NR	1.79–1.85	Wu *et al.*, 2016
D. oppositifolia [1]	1	1431 ± 1.56	102.2 ± 0.54	78.2 ± 0.08	680.6 ± 0.82	432.5 ± 1.11	2.74 ± 0.03	22.0 ± 0.08	6.34 ± 0.01	3.24 ± 0.08	Mohan and Kalidas, 2010
D. oppositifolia [1]	2	1534–1624	124–168.2	114.1–124.1	294.2–646.2	540.1–634.1	7.62–14.56	32.16–40.76	7.42–9.04	1.56–6.26	Shajeela *et al.*, 2011
D. pentaphylla [1]	1	1322 ± 2.40	95.2 ± 0.12	96.1 ± 0.06	632.1 ± 0.22	380.0 ± 0.74	12.60 ± 0.14	103.48 ± 0.94	1.32 ± 0.01	3.10 ± 0.01	Mohan and Kalidas, 2010
D. pentaphylla [1]	1	1441.0 ± 0.98	96.20 ± 0.63	158.18 ± 0.21	444.24 ± 0.09	532.12 ± 0.56	13.26 ± 0.05	66.32 ± 0.14	3.46 ± 0.21	3.42 ± 0.01	Shajeela *et al.*, 2011
D. persimilis [1]	3	NR	NR	NR	46.55–47.64	46.70–47.42	0.382–0.423	1.73–1.93	NR	1.32–1.45	Wu *et al.*, 2016

Species	No. of Varieties	Minerals (mg/100g)									Reference
		K	Na	P	Ca	Mg	Cu	Fe	Mn	Zn	
D. polystachya [1]	10	NR	NR	NR	39.73–55.82	33.26–54.47	0.35–0.54	1.43–2.58	NR	0.99–2.27	Wu *et al.*, 2016
D. praehensilis [1]	5	NR	NR	20.9–39.0	13.1–118.2	NR	NR	18.36–76.4	NR	0.4–1.09	Bekele and Bekela, 2018
D. praehensilis [3]	1	1000 ± 21.2	80.0 ± 7.07	200.5 ± 0.71	79.5 ± 3.54	43.5 ± 0.71	0.40 ± 0.14	9.0 ± 0.0	0.95 ± 0.07	5.4 ± 0.57	Polycarp *et al.*, 2012
D. remotiflora [2]	1	4891 ± 25	79 ± 6	720 ± 20	242 ± 14	250 ± 10	3.3 ± 0.2	12.4 ± 0.5	4.1 ± 0.2	7.1 ± 0.3	Soto *et al.*, 2014
D. rotundata [3]	2	475–900	70.0–87.5	158–211.5	91.5–103.3	35.5–53.0	0.20–0.25	5.0–6.75	1.15–1.80	6.30–6.80	Polycarp *et al.*, 2012
D. rotundata [1]	6	NR	NR	26.96–40.21	22.77–114.4	NR	NR	17.75–78.3	NR	0.35–1.02	Bekele and Bekele, 2018
D. rotundata [1]	1	1591	10.4	NR	31.0	51.0	NR	NR	NR	1.23	Dufie *et al.*, 2013
D. rotundata [3]	3	9.00–71.00	NR	22.00–35.00	2.00–4.00	11.00	NR	1.00	NR	1.00	Ahmadu *et al.*, 2018
D. rotundata [3]	1	530.7 ± 0.10	80.75 ± 0.14	168.7 ± 0.01	278.8 ± 0.15	125.7 ± 0.08	NR	2.88 ± 0.02	NR	2.34 ± 0.00	Ezeabara and Anona, 2018
D. rotundata [3]	1	209.13 ± 0.03	185.2 ± 0.05	54.00 ± 0.04	132.02 ± 0.04	45.90 ± 0.02	10.06 ± 0.05	81.85 ± 0.01	NR	5.46 ± 0.02	Alinnor *et al.*, 2010
D. spicata [1]	1	1255 ± 0.48	52.2 ± 0.11	86.1 ± 0.11	172.0 ± 0.21	112.4 ± 0.32	0.78 ± 0.21	22.36 ± 0.38	0.98 ± 0.14	4.18 ± 0.13	Mohan and Kalidas, 2010
D. spicata [1]	1	1136 ± 0.74	66.34 ± 0.54	166.30 ± 0.27	234.10 ± 0.58	324.16 ± 0.24	7.41 ± 0.11	24.10 ± 0.26	6.70 ± 0.14	2.56 ± 0.04	Shajeela *et al.*, 2011
D. tomentosa [1]	1	1354 ± 1.34	32.2 ± 0.18	96.1 ± 0.04	272.1 ± 1.01	120.4 ± 0.08	1.34 ± 0.01	24.56 ± 0.04	1.32 ± 0.04	5.20 ± 0.03	Mohan and Kalidas, 2010

Species	No. of Varieties	Minerals (mg/100g)									Reference
		K	Na	P	Ca	Mg	Cu	Fe	Mn	Zn	
D. tomentosa [1]	1	1245.6 ± 1.14	46.14 ± 0.30	104.06 ± 0.09	266.36 ± 0.16	321.04 ± 0.14	2.46 ± 0.14	28.50 ± 0.07	2.10 ± 0.11	5.40 ± 0.02	Shajeela *et al.*, 2011
D. trifida [1]	3	830–1350	NR	50.0–120.0	40.0	40.0–50.0	0.67–1.19	NR	NR	0.62–1.79	Perez *et al.*, 2011
D. triphylla [2]	1	317 ± 32	4.15 ± 0.7	56.6 ± 0.1	39.7 ± 8.1	27.3 ± 5.6	0.18 ± 0.05	1.00 ± 0.05	0.25 ± 0.07	0.39 ± 0.1	Bhandari *et al.*, 2003
D. versicolor [2]	1	250 ± 4	4.91 ± 2.5	40.8 ± 0.2	14.3 ± 1.8	18.3 ± 3.8	0.18 ± 0.02	0.39 ± 0.1	0.14 ± 0.0	0.3 ± 0.06	Bhandari *et al.*, 2003
D. villosa [3]	1	145.33 ± 1.15	5.40 ± 0.10	43.82 ± 0.49	28.06 ± 4.01	9.47 ± 0.23	NR	NR	0.032 ± 0.0	0.26 ± 0.0	Afiukwa *et al.*, 2013
D. wallichi [1]	1	1361.7 ± 1.01	63.01 ± 0.27	106.40 ± 0.11	748.31 ± 0.32	578.06 ± 0.19	2.46 ± 0.08	20.14 ± 0.04	3.31 ± 0.05	6.66 ± 0.01	Shajeela *et al.*, 2011

NR = not reported, [1]dry weight, [2]fresh weight, [3]Sample type: not specified, *D. polystachya*: Chinese yam formerly known as *D. opposita* and *D. batatas*.

Source: https://www.ncbi.nlm.nih.gov/pmc/articles/PMC7555206/

exception to the cooking rule is the 'mountain or Chinese yam' (*D. polystachya*) which in Japan is known as *nagaimo*. It can be further classified into *ichoimo* or *yamatoimo*, depending on the root shape. Mountain yam is usually eaten raw (sliced or grated) after only a relatively minimal preparation that involves soaking the whole tubers in a vinegar-water solution to neutralize irritant oxalate crystals found in their skin. The raw vegetable is starchy and flavourless, mucilaginous when grated, and may be eaten plain as a side dish, or added to noodles.

Boiled tubers can be mashed with or without oil, seasoned with spices and then can be used either alone or with other vegetables in various preparations such as *pakoras, bharta, chaat*, cutlets, *etc.* It can be finely sliced and deep-fried before or after seasoning with spices. It can be sliced into finger-like pieces, seasoned with spices and deep fried to prepare finger chips similar to French fries. Sliced tubers can be used to prepare vegetable dishes. Yam tubers can be steamed and then cooked with spices. In Southern India, the vegetable is a popular accompaniment to rice dishes and curry.

Medicinal Uses

Tubers of different *Dioscorea* species are used as a cure for different diseases and ailments including cough, cold, stomach-ache, leprosy, burns, fungal infections, dysentery, skin diseases, rheumatism, arthritis, cancer, *etc.* in several formulations, and even for birth control (Kumar *et al.*, 2017). Diosgenin extracted from the tubers of several species of *Dioscorea* are used commercially in pharmaceutical industry for commercial synthesis of cortisone, pregnenolone, progesterone and other steroidal products (Marker and Krueger, 1940). Such preparations were used in early combined oral contraceptive pills (Djerassi, 1992). Diosgenin is also used for the development of cortisone treatments of arthritis. It has been reported that diosgenin has effect on diabetic neuropathy (Kang *et al.*, 2011). The unmodified steroid has estrogenic activity (Liu *et al.*, 2005).

Barbasco (*Dioscorea composita* L.f.), native to Mexico, is one of the several species used as a source of diosgenin. Russell Earl Marker, an American chemist who invented the octane rating system, developed the extraction and manufacture mechanism of hormones from *D. mexicana* but later discovered that *D. composita* has much higher content of diosgenin than *D. mexicana*. Medicinal yam or mule's-hoof (*D. floribunda* M.Martens and Galeotti, syn. *D. barclayi* R.Knuth), found from central Mexico to northern Central America, is another species grown commercially for its diosgenin content.

D. zingiberensis C.H. Wright (syn. *Dioscorea henryi* Uline ex Diels), endemic to China, is used extensively in traditional Chinese medicines. The tubers of *D.*

zingiberensis are used for isolation of diosgenin. Steroidal saponins are believed to be the active principles. More than 70 compounds from tubers of *Dioscorea* have been identified. A wide spectrum of biological effects including cardiovascular, anti-thrombosis, hyperlipidemia, neuroprotection, anti-inflammatory and anthelmintic effect have been verified (Zhang *et al.*, 2018). The **wild yam** (*D. dregeana* (Kunth) T.Durand and Schinz, 1894), native to eastern South Africa and Eswatini, has slightly thorny stems, and is commonly used and traded as a traditional medicine or *muti*.

Ethnobotanical Values of Common *Dioscorea* Species of SBR, India

Botanical Name	*Ethno-botanical/ Pharmacological Values*	*Supporting Literature*
D. alata L.	Tuber powder is used to cure piles.	Jadhav *et al.*, 2011
	Tubers are eaten raw twice a day until weakness is reduced.	Kamble *et al.*, 2010
	Juice of tuber is used to kill stomach worm.	Samanta and Biswas, 2009
D. belophylla (Prain) Voigt ex Haines	Tuber juice with hot water is given to treat fever, malaria, headache, and dysentery.	Srivastava and Nyishi, 2010
D. bulbifera L.	Raw tuber is eaten to enhance appetite.	Mishra *et al.*, 2008
	Tuber is good for intestinal colic, relieving dysmenorrhoea, reducing acidity, against rheumatoid arthritis, to relieve intense inflammation in the acute phase, in spasmodic asthma, for menopausal problems, for labor pain and the prevention of early miscarriage, for hernia, relieving the pain of childbirth.	Nayak *et al.*, 2004; Patil and Patil, 2005
	Leaf paste is used against skin infections.	Girach *et al.*, 1999
	Bulbils are used to reduce throat pain.	Mbiantcha *et al.*, 2011
	Boiled tubers are taken orally to reduce body heat.	Singh *et al.*, 2009
	Used against boils and dysentery.	Nag, 1999
	Tuber powder mix with butter is given to check diarrhoea.	Jadhav *et al.*, 2011
	Tuber is used as refrigerant to reduce body heat during summer.	Dutta, 2015
	Tuber is used to treat skin infection.	Tiwari and Pande, 2006

Botanical Name	***Ethno-botanical/ Pharmacological Values***	***Supporting Literature***
	Tuber is used to treat bronchial cough as well as antiseptic.	Bhatt and Negi, 2006
	Tuber is useful for acidity and ulcers.	Dutta, 2015
	Crushed tender shoots/twigs are rubbed on wet hair to remove dandruff.	Dutta, 2015
	Tuber paste mixed with cow milk is taken orally for treating cough and asthma.	Teron, 2011
	Tuber is used to treat typhoid with *Curcuma aromatica*	Jain *et al.*, 2008
	Tubers are used in ulcer, piles, syphilis and dysentery, and the powder against hair lice.	Abhyankar and Upadhyay, 2011
	About 10g of tuber powder is given once a day for 5–6 days after menses as contraceptive.	Kamble *et al.*, 2010
	Tubers are boiled after processing and given for abdominal pains. The tubers are dried and pea-sized pieces are cut and given in early morning with water for 3 days to cure piles.	Abhyankar and Upadhyay, 2011
	The roasted and mashed tubers are eaten with salt to cure cough.	Singh *et al.*, 2009
D. dumetorum (Kunth) Pax	Tuber juice is used to make arrow poison.	Edison *et al.*, 2006
	Tuber is used against jaundice.	Edison *et al.*, 2006
D. esculenta (Lour.) Burkill	Tubers are used for treatment of chest pain, nervous disorders, and swellings.	Edison *et al.*, 2006
	Tuber paste is used to relieve pain and to treat boils, dysentery and swellings.	Dutta, 2015
D. hamiltonii Hook.f.	Tuber is used for treatment of stomach-ache.	Edison *et al.*, 2006
	Tuber is eaten for poor appetite.	Edison *et al.*, 2006
	Crushed tubers are given as body refrigerant during summer seasons, and is also good for treating diarrhoea.	Sharma and Bastakoti, 2009
D. hirtiflora Benth.	Tuber is used to treat gonorrhoea.	Sonibare and Abegunde, 2012b
D. hispida Dennst.	Water of soaked tuber is used as medicine for eyes.	Meena and Yadav, 2011
	Tuber is used as fish poison.	Nashriyah *et al.*, 2011
	Sap of tuber is pasted around the affected parts and covered with cloths for about one night to treat peeling of skin of feet.	Sharma and Bastakoti, 2009

Botanical Name	*Ethno-botanical/ Pharmacological Values*	*Supporting Literature*
	Tuber is used to treat vomiting, indigestion, possesses narcotic properties and fresh tuber taken as purgative.	Dutta, 2015
	Tubers are roasted and pounded and its paste is applied on wounds and injuries.	Kamble *et al.*, 2010
D. kamoonensis Kunth	Tubers are used in the treatment of arthritis and rheumatism.	Edison *et al.*, 2006
D. oppositifolia L.	Tuber is boiled with *D. uniflorus* and given to women once a day for nearly a month after delivery to revive their strength.	Mishra *et al.*, 2008
	Oral administration of tuber powder mixed with honey is used for increasing sperm count.	Sharma and Bastakoti, 2009
	Leaf paste is used as antiseptic for ulcers.	Sheikh *et al.*, 2013
	Powered tuber mixed with cow urine is applied on scorpion bite.	Nashriyah *et al.*, 2011
	Leaves are mixed with leaves of clematis and 2–3 drops of juice put in the nose of affected person to get relief after sneezing in fits and epilepsy.	Kamble *et al.*, 2010
D. pentaphylla L.	Tubers are applied on swelling of joints and as a tonic to improve body immunity.	Edison *et al.*, 2006
	Tubers are used for stomach pain.	Choudhary *et al.*, 2008
	Crushed mass of tuber is given to cattle when they become sick by eating green leaves of maize.	Sharma and Bastakoti, 2009
	Tuber is used as tonic, *vis-a-vis* to cure stomach troubles and rheumatic swellings.	Dutta, 2015
	Inflorescence is used as vegetables for body weakness.	Kamble *et al.*, 2010
	Tubers are useful to allay pain and swelling.	
D. pubera Blume	Bulbils are cooked and taken to cure colic pain.	Sheikh *et al.*, 2013
D. spinosa Burm.	Tubers are edible in the district Mayurbhanj of Odisha, India.	Behera, 2006
D. wallichii Hook.f.	Tubers are roasted and eaten for flatulence.	Dutta, 2015
	Tuber is used in stomach pain.	Rout and Panda, 2010

Source: https://www.ncbi.nlm.nih.gov/pmc/articles/PMC5306286/

Toxicity

Usually the edible *Dioscorea* species have opposite leaves though poisonous ones alternate. Most edible and mature varieties of cultivated yam do not contain toxic compounds though exceptions are there. Polyphenols or tannin-like bitter compounds tend to accumulate in immature tuber tissues of white and yellow yams. Wild forms of bitter yams (*Dioscorea dumetorum*) contain little or high levels of toxins such as dihydrodioscorine which tastes bitter. They are not normally eaten except at times of local food scarcity. The bitter compounds in these yams are water-soluble alkaloids and can be detoxified by soaking in a vessel of salt water, in cold or hot fresh water or in a stream. Ingestion of these bitter compounds may produce severe and distressing symptoms and in severe case can even prove lethal.

D. alata supplements may be allergic to some people and may have adverse effects in people taking estrogens, anticoagulant drugs or during pregnancy and breast-feeding (drugs.com, 2020). *D. alata* has relatively high levels of oxalates (486–781mg/100g dry matter) (Wanasundera and Ravindran, 1994), which are associated with antinutritional effects and kidney stone formation (Otegbayo *et al.*, 2017).

Aerial or potato yams (*D. bulbifera*) have antinutritional factors. Uncultivated forms such as those found growing wild in Florida can be poisonous. In Asia, detoxification methods used for bitter cultivars of this yam include water extraction, fermentation and roasting of the grated tuber. The bitter compounds in these yams include diosbulbin and possibly saponins such as diosgenin (Jesus *et al.*, 2016). In Indonesia, an extract of air potato is used in the preparation of arrow poison (Oke, 1990).

Species

Out of around 600 species, approximately 50 species (including *D. hirtiflora* subsp. *pedicellata* and *D. japonica*) are consumed as wild-harvested staples or famine food, 7-10 including *D. rotundata*, *D. cayennensis*, *D. alata*, *D. polystachya*, *D. bulbifera*, *D. esculenta*, *D. dumetorum* and *D. trifida* are cultivated species so all these along with other 3 species [*D. alata* L., *D. cayennensis* Lam. and *D. rotundata*) constitute a major staple food in most parts of Africa and Austronesian cultures. *D. pubera* Blume, native to the Indo-China region, is another wild species distributed throughout the temperate and tropical Americas, China, wet regions of Himalayas, Central Nepal, Western Malaysia and Bhutan (Kumar *et al.*, 2017). *D. villosa* L. (wild yam), the most well-known species, is native to North America (Avula *et al.*, 2014). It is most diversified and extensively distributed species throughout tropical Asia and the Pacific.

From India, more than 50 different *Dioscorea* species have been reported in the states of Asom, Tamil Nadu, Kerala, Bihar, Odisha, West Bengal, Sikkim, Rajasthan, Gujarat and Maharashtra out of which maximum of 19 species are from Asom, and *D. belophylla* (Prain) Voigt ex Haines, *D. wightii* Hook.f. and *D. spicata* Roth are endemic to Western Ghats (Kumar *et al.*, 2017). The wild species *D. prazeri* Prain and Burkill and *D. deltoidea* Wall. ex Griseb. are found at high altitudes (Saikia *et al.*, 2010).

There are about 13 *Dioscorea* species found in Similipal Biosphere Reserve (SBR); the most common ones being *D. bulbifera* L. (Pita aalu), *D. pentaphylla* L. (Panja Sanga), *D. hispida* Dennst. (Banya aalu), *D. alata* L. (Khamba aalu), *D. oppositifolia* L. (Paani aalu), *D. pubera* Blume (Kukai Sanga), *etc.* About 10 species are known to be bitter in taste or unpalatable when taken raw. The local people mostly soak the tubers overnight in water or leave them overnight in stream before boiling to remove the bitterness (Bhandari and Kawabata, 2005; Kumar *et al.*, 2012). Central Tuber Crops Research Institute, Sreekaryam, Thiruvananthapuram (Kerala) has developed several improved varieties of major cultivated yam species.

Major Cultivated *Dioscorea* Species

Dioscorea alata L. (Greater yam)

Common names: Greater yam, Purple yam, Water yam, Winged yam and Yam. **Other names**: Guyana arrowroot, Ten-months yam, Ube, Violet yam, White yam or Yam. In Okinawa (Japan): *Beniimo*; and in **India**: Ratalu, Khambha Aalu, *etc.*

Synonyms: *Dioscorea atropurpurea* Roxb., *D. colocasiifolia* Pax, *D. eburina* Lour., *D. eburnea* Lour., *D. globosa* Roxb., *D. javanica* Queva, *D. purpurea* Roxb., *D. rubella* Roxb., *D. sapinii* De Wild., *D. sativa* Munro, *D. vulgaris* Miq., *Elephantodon eburnea* (Lour.) Salisb. and *Polynome alata* (L.) Salisb.

D. alata foliage, flowers, whole tubers and sliced tuber

Greater yam originating in SE Asia (in tropical Myanmar and Thailand) is the most economically important species (Tamiru, 2006). Being dioecious, its male and female cultivars are available. It has become naturalized throughout tropical South America, Africa, Australia and the SE U.S. It has the largest distribution worldwide of any cultivated yam, though not grown in the same quantities as the African yams. Known to humans since ancient times, this yam is now being cultivated in Asia, the Pacific islands, Africa and the West Indies (Anon., 2016). It is a highly invasive plant outside the cultivated areas. In India, it is distributed throughout the country, though rich in Odisha, Kerala and Andhra Pradesh where it is grown commercially. It is also grown in M.P., U.P., West Bengal, T.N. and NEH region as a subsidiary food crop. The right twining stems of greater yam have 4-5 wings that are sometimes reduced to ribs. Bulbils are globose, ovoid or obpyriform and occasionally with rootlets. The colour of tubers is usually a vivid violet-purple to bright lavender, but some varieties may be cream to plain white. Leaves are 7-18×4-10cm size, opposite or rarely subopposite, broadly ovate or deltoid-ovate, the base is cordate or subsagittate, apex acuminate, primary veins are 7, and the petioles are up to 8cm long and winged. These usually produce one to several cylindrical tubers which are polymorphic and variously branched or lobed. It may be sometimes confused with taro and the sweet potato (*Ipomoea batatas*) due to their colour, taste and culinary use similarities. However, purple yam tends to have a moister texture than sweet potatoes and also has higher anthocyanin content as compared to sweet potatoes. They can otherwise be used interchangeably in most recipes. It also produces bulbils in sufficient numbers though not so prolifically as to *D. bulbifera*. *D. alata* is easy to grow and harvest as it requires minimal processing. Each plant may produce 1-3 tubers.

Varieties

Some varieties developed by CTCRI and other Indian centres are:

Greater yam (common Hindi name *Ratalu*) is commercially grown in Andhra Pradesh, Odisha and Kerala. Its varieties are:

i) **'Sree Karthika'** for Kerala is a high yielding variety having long *cum* oval tubers with excellent cooking quality. Maturity period is 9 months and average yield is 30t/ha.

ii) **'Sree Keerthi'** for Kerala is a high yielding variety with large sized tubers, good cooking quality and excellent taste. It is a clonal selection from germplasm. These climb to a height of 4-5m. Its maturity duration is 9-10 months with starch content of 20-22 per cent on fresh weight basis. It matures in 270-300 days with average tuber yield of 25-30t/ha. It is suitable for intercropping with banana and can be grown in mature coconut gardens.

iii) **'Sree Neelima'** for Kerala is a high yielder (35t/ha) with good culinary and nutritive quality and light purple flesh colour. It matures in 270-300 days.

Sree Keerthi

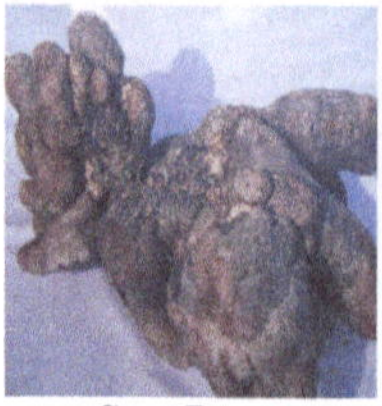
Sree Roopa

Sree Karthika

Sree Shilpa

Sree Swathy

iv) **'Sree Roopa'** is a high yielding selection for Kerala, having medium-sized tubers with excellent cooking quality. This climbs to a height of 4-5m. Its starch content is 16-18 per cent and protein 1-2 per cent on fresh weight basis. It matures in 270-300 days with average tuber yield of 25-30t/ha.

v) **'Sree Shilpa'** for Kerala is the first intervarietal hybrid between 'D-140' and 'Sree Keerthi' of *Dioscorea alata* with medium-sized oval tubers which are of good cooking quality as the flesh is also white. Its leaves are cordate with overlapping lobes, slightly cupping lamina and with purple petiole at both the ends. It matures in 9-10 months with average yield of 28t/ha though it has potential of giving up to 40t/ha tuber yield. It is easily harvestable.

vi) **'Sree Swathy'** is a variety for Kerala released for its high yield (30t/ha), good culinary and nutritive qualities and moderate tolerance to anthracnose disease. It matures in 270-300 days.

vii) **'Orissa Elite'**is a clonal selection for Odisha bearing long cylindrical tubers. It has excellent cooking quality. It is field-tolerant to leaf spot, scales and mealy bugs. This gives an average tuber yield of 25t/ha. It matures in 270-300 days.

viii) **'Indu'** is a high yielding (39.39t/ha) variety developed by K.A.U. for Kerala. Its tubers are digitate and have brownish-black skin and white to pale flesh. It matures in 270-300 days.

ix) **'Co 1'** matures in 270-300 days with average tuber yield of 40t/ha. It is developed for Tamil Nadu.

x) **'Konkan Ghorkhand'** recommended for Maharashtra, maturing in 270-300 days with average tuber yield of 20t/ha.

Phytochemical Composition

Along with four known compounds (*RRR*-α-tocopherol, coenzyme Q_9, cycloartane, and 1-feruloylglycerol), two new compounds (hydro-Q_9 chromene, and γ-tocopherol-9) were isolated (Cheng *et al.*, 2007). These compounds had phenolic hydroxyl group in common with the known phytoestrogens and have also antioxidant activity.

Nutritional Values and Uses

Greater yam tubers, depending on varieties or location, have approximately 70 per cent moisture, 28 per cent starch, 0.5 per cent sugars, 0.1 per cent fat, 1-3 per cent crude protein, 0.5-1.5 per cent crude fibre and 6-12mg/100g vitamin C. The mildly sweet, earthy and nutty taste of purple yam (*D. alata*) is somewhat similar to sweet potato or taro. The violet cultivars with high amount of anthocyanins can be used as a food colourant. These yams can be processed to extract starch. It can be used to prepare many desserts, soups, vegetables, and as an ingredient to colour or flavour pastries and cakes, icecreams, milk, donuts, tarts, jam, *etc.* It is often eaten boiled, baked, fried or as stir-fried chips. The purple yam is also eaten in India, where it is also called the violet yam. Species may be called by the regional name 'taradi', which can refer to *D. belophylla*, *D. deltoidea* (Rana *et al.*, 2015) and *D. bulbifera*. Digging and selling 'taradi' is a major source of income in the region of Palampur (H.P., India). The species were used in the promotion of the health of postmenopausal women.

Dioscorea bulbifera L. (Potato yam/Air potato)

Common names: Aerial potato/yam, Air potato/yam, Bitter yam, Cheeky yam, Parsnip yam, Potato yam and Up yam; in **Hindi**: Garaadoo/Garadu; in **Sanskrit**: varahi.

Synonyms: *Dioscorea anthropophagorum* A.Chev., *D. crispata* Roxb., *D. heterophylla* Roxb., *D. hoffa* Cordem., *D. hofika* Jum. and H.Perrier, *D. korrorensis* R.Knuth, *D. latifolia* Benth., *D. longipetiolata* Baudon, *D. perrieri* R.Knuth, *D. pulchella* Roxb., *D. rogersii* Prain and Burkill, *D. tamifolia* Salisb., *D. tenuiflora* Schltdl., *Helmia bulbifera* (L.) Kunth and *Polynome bulbifera* (L.) Salisb.

Potato yam (*D. bulbifera*), native to Asia, tropical Africa and Northern Australia, is the most popular wild species (Kumar *et al.*, 2017) and is mostly seen in moist deciduous forests and also in plains. It is also widely cultivated and has become naturalized in many regions including Latin America, the West Indies, the SE

Aerial potato

United States and various oceanic islands (Anon., 2017) with slight differences between those found everywhere. In India, it is distributed mainly in the forest patches of lower Eastern Ghats. This quick-growing left-twining vine bearing broad alternate leaves can reach over 18m long. It has two types of storage organs, both capable of producing new plants: i) bulbils in the leaf axils that are able to sprout at a very early stage and are the primary means of spread and reproduction, and ii) small and oblong potato-like tubers beneath the ground that are also capable of re-sprouting. The plant produces small white flowers though these can not be seen at all the places where it grows. The flowering and fruiting occurs mainly during September to October. It produces a crop after only four months of growth and continues producing so for whole life of the vine or as long as two years. Moreover, the bulbils are easy to harvest and cook.

Phytochemical Composition

The mannose-binding lectin from bulbils of *D. bulbifera* was studied for its clinical potential in HIV and cancer research (Sharma *et al.*, 2017). The main bioactive compounds are reported to be flavonoids, clerodane diterpenoids, steroidal saponins and phenolic compounds. Liu *et al.* (2011) isolated two new compounds (bibenzyl type-2,5,22 ,52 -tetrahydroxy-32 -methoxybibenzyl and diarylheptanone containing a dihydrophenanthrene moiety named as diobulbinone A), along with 16 known compounds. Nine norclerodane diterpenoids were isolated from the tubers including two new compounds (diosbulbin N and diosbulbin P), and a naturally occurring compound (diosbulbin O) along with six known diosbulbins A-D, F and G (Tang *et al.*, 2014). Another study also showed the molecular changes of liver dysfunction and revealed the overall metabolic and physiological mechanisms of the subchronic toxic effect of *Dioscorea bulbifera* tuber (Zhao *et al.*, 2018).

Nutrition and Uses

Air potato tubers contain 14.74 per cent moisture, 2.56 per cent ash, 0.35 per cent fibre, 73.62 per cent carbohydrate and 7.47 per cent protein. It also shows the presence of some minerals such as Ca, Mg, K, P and Na with some phytochemicals like saponin, tannin, flavonoids and alkaloids. The tubers or bulbils of edible varieties often have a bitter taste which can be removed by boiling and can then be prepared in the same way as other yams, potatoes and sweet potatoes. Aerial tubers are boiled, baked or fried for consumption. The aerial tubers of this species contain toxic substances including the alkaloid dioscorine which can be neutralized by thorough cooking.

Dioscorea cayennensis Lam. and *D.c.* subsp. *rotunda* (Poir.) J. Miege

Common names of *D. cayennensis*: Guinea yam, yellow Guinea yam and yellow yam.

Common names of *D.c.* subsp. *rotunda*: Guinea yam, West African yam and White yam.

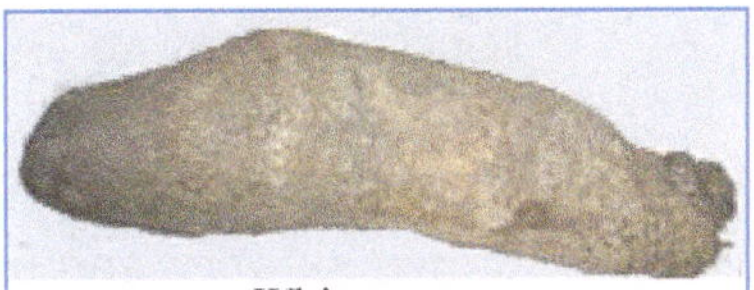

White yam

The **yellow yam** (*D. cayennensis*) and the **white yam** (*D.c.* subsp. *rotunda* or *D. rotundata*), native to Africa, are the most important cultivated yams. Yellow yam is a widely consumed West African domesticated crop. Its brown and smooth tuber is roughly cylindrical where flesh is usually white and firm. The yellow flesh in yellow yam is due to the presence of carotenoids. In outer appearance both seem to be similar though tuber skin of yellow yam is usually a bit firmer and less extensively grooved. The yellow yam has a longer period of vegetation and a shorter dormancy than white yam. The vines of these plants can be 10-12m long and the weight of their tubers is usually 2.5-5kg each, though can weigh as much as 25kg. Tubers are harvested in 7-12 months. 'Kokoro', a variety of white yam, can give good tuber yield even on degraded soil. This variety is important in making dried yam chips and is abundantly found in western Nigeria, Benin and Togo.

Varieties

Some white yam varieties developed by CTCRI are:

- ☆ **'Sree Dhanya'** is a white yam (*D. rotundata*) variety for Kerala which matures in 250-270 days with average tuber yield of 20t/ha. Its cylindrical tubers contain 23.3 per cent starch. It is the first dwarf bushy variety not requiring staking and can be accommodated in closer spacing.

Sree Subhra

Sree Priya

Sree Dhanya

- ☆ **'Sree Priya'** is a drought-tolerant white yam variety for Kerala which matures in 270-300 days with average tuber yield of 35-40t/ha producing

2-3 tubers per plant. It is a seedling selection from the African variety 'Umidika', bearing cylindrical tubers with smooth surface and excellent cooking quality having excellent flavour. It is quite amenable to intercropping in mature coconut gardens as well as with banana. On fresh weight basis, its starch content is 20-21 per cent and protein 2-3 per cent.

- ☆ **'Sree Subhra'** released for cultivation for Kerala is a drought tolerant selection from an African variety. It matures in 270-300 days with average tuber yield of 35-40t/ha, having excellent cooking quality with starch content of 21-23 per cent on fresh weight basis. Its tubers are brown-skinned and partially hairy with cylindrical shape.

Phytochemical Composition

Along with the known spirostanol saponins described as methyl protodioscin, asperoside and prosapogenin A of dioscin, a new furostanol glycoside isolated from the methanolic extract of the tuber of *Dioscorea cayenensis* growing in Cameroon was 26-*O*-β -D-glucopyranosyl-3β,26-dihydroxy-20,22- seco-25(*R*)-furost-5-en-20,22-dione-3-*O*-α-L-rhamnopyranosyl-(1→4)-α-L-rhamnopyranosyl-(1→4)-[α-L-rhamnopyranosyl-(1→2)]-β-D-glucopyranoside (Sautour *et al.*, 2004). Prosapogenin A of dioscin having a spirostan skeleton showed antifungal activity against *Candida* species with MIC values between 6.25 and 25mg/ml (Sautour *et al.*, 2004). Two new fatty acid-spirostan steroid glycoside esters, progenin III palmitate and progenin III linoleate were isolated from the methanol extract of the tubers of *Dioscorea cayenensis* (Ali *et al.*, 2013).

Nutrition and Uses

Yellow yam tubers are used as food and as a remedy for treating burns and fevers. Raw *Dioscorea cayenensis* tubers are reported to contain 66.79g moisture, 2.62g crude protein, 0.27g lipid, 0.17g fibre, 0.63g ash, 29.69g carbohydrates, 262.3mg potassium, 61.53mg magnesium, 0.79mg iron, 0.39mg zinc and 108.26kcal energy with insignificant vitamin content per 100g of edible portion (Adepoju *et al.*, 2018).

Dioscorea dumetorum (Kunth) Pax

Common names: Bitter yam, Cluster yam or Three-leaved yam.

Wild bitter yam vines (*D. dumetorum*)

The **bitter yam** (*D. dumetorum*) is cultivated mainly in the Bight of Biafra region from SE Nigeria to Gabon as well in the Ubangi-Shari region of inland Central Africa, having distinctive trifoliate leaves similar to the Indian 3-leaved yam, *D. hispida* (Blench, 2006). Its left-twining stems grow up to 10m long patterned with sparsely velvety or rather densely with 1-3mm long prickles. The trifoliate alternate leaves are up to 25cm long, usually with a few scattered prickles.

It's very toxic so must be leached in water for days and thoroughly cooked before eating (Blench, 2006). The toxicity of wild form is due to the high content of water soluble dihydrodioscorine.

Dioscorea esculenta (Lour.) Burkill

Common name: Lesser yam: In Hindi Suthani.

The **lesser yam** (*D. esculenta*), a native to SE Asia, probably China, is one of the first cultivated and 3rd most cultivated yam species in SE Asia, though very little in other parts of the world. Most of its varieties have small tubers, with vines that usually have 2-4cm long black compound spines. These spiny vines seldom reach more than 3m in length. The soft and heart-shaped leaves are 5–8cm long and 6–8cm wide. In India, the lesser yam is cultivated in Kerala, Goa and Konkan parts of Maharashtra as well as in U.P. to be used during fasting. It is known as 'nana kizhangu', 'nheruvalli kizhangu' or 'cheru kizhangu' in Malayalam. In Goa it is called 'Kaate Kanaga'. Like *D. alata*, it too requires minimal processing, unlike the other bitterer yam species.

Some **varieties** of lesser yam/Chinese yam; common Hindi name Suthani or Kangar, developed by CTCRI are:

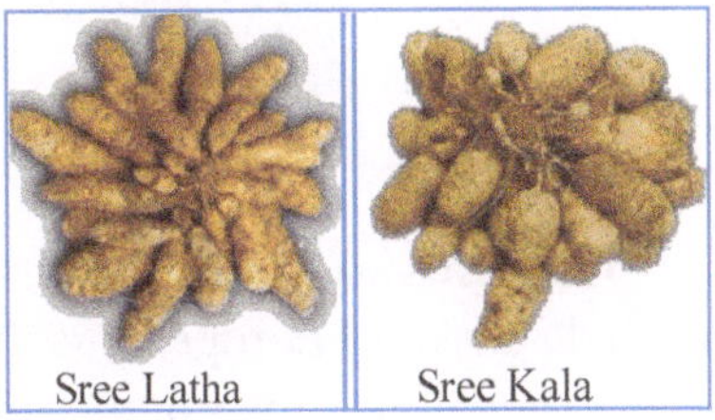

- ☆ **'Sree Latha'** is released for cultivation in Andhra Pradesh, Asom, Kerala, Maharashtra, West Bengal and Bihar and is widely adaptable variety of lesser yam (*D. esculenta*) which can climb up to the height of 3m. This matures in 225-240 days with an average tuber yield of 20-25t/ha with starch content of 24.4 per cent. Its tubers have shelf-life of 75-90 days. It is a selection from an indigenous germplasm. It is a good-flavoured non-bitter variety with fair cooking quality and of acceptable taste.
- ☆ **'Sree Kala'**recommended for cultivation in Kerala is an early maturing (210-240 days) variety of lesser yam with an average tuber yield of 20t/

ha. It has oval tuber with excellent culinary quality. It is a selection from exotic germplasm named 'Kombi'.

- ☆ **'Konkan Kanchan'** released for Maharashtra region matures in 190-210 days with an average tuber yield of 18-20t/ha.

Phytochemical Composition

Fiteen steroidal saponins have been isolated from the powdered tuber and leaf of lesser yam where 4 of them belonged to furostanol-type saponin and 11 as spirostane-type saponins (Lee *et al.,* 2017).

Nutrition and Uses

Lesser yam tuber has almost similar percentage of compositions as to that of greater yam but is also reported containing vitamins A, B_1 and B_2. The tubers are eaten baked, boiled or fried much like potatoes, and are traditionally used locally in the treatment of various diseases.

Dioscorea polystachya Turcz.

Common names: Chinese potato, Chinese yam, Cinnamon-vine/Mountain yam; In Manipuri: Ha.

Synonyms: *Dioscorea batatas* Decne., *D. decaisneana* Carriere, *D. doryphora* Hance, *D. potaninii* Prain and Burkill, *D. pseudobatatas* (Hauman) Herter, *D. rosthornii* Diels, and *D. swinhoei* Rolfe.

Leaf, large tuber and aerial tubers of Chinese yam

Chinese yam (*D. polystachya*), native to East Asia, was introduced to Japan, United States and Europe. This species is more tolerant to frost and cooler climates than other yams. The tubers can be eaten raw or cooked. It is cultivated largely in Asia and sometimes used in alternative medicine. Three groups of this species in cultivation are recognized in Japan: the nagaimo (common long, cylindrical type tubers), the ichoimo (ginkgo-leaf yam with flat, palmate shape tubers), and the tsukune imo (meatball yam with round or globular tubers) (Thompson, 2014). The clockwise twinning vines of Chinese yam generally grow 3–5m long but can be longer. Up to 11cm long and wide leaves are lobed at the base and larger ones may have lobed edges. It may be alternately or oppositely arranged or borne in whorls. About 2cm long, warty and rounded bulbils appear in the leaf axils which are sometimes informally referred to as 'yam berries'. New plants sprout from the bulbils which are also edible but not considered as palatable. The flowers are cinnamon-scented. The plant produces one or more spindle-shaped or cylindrical up to 1m long tubers.

The Chinese yam's growing cycle is approximately one year. It should be planted between winter and spring using smaller tubers, top cut of bigger tubers or through cuttings of branches as planting materials. The first two methods can produce 20cm or even longer tubers. The latter produces smaller tubers of around 10cm size that are usually replanted for the next year. After about 7 to 9 months of replanting of tubers when the aboveground portion starts drying the tubers are ready for harvest.

Dioscorea rotundata **(White yam)** recommended for cultivation in Kerala has three important varieties, *viz.* 'Sri Dhanya' which matures in 250-270 days with average tuber yield of 25t/ha; as well as 'Sri Priya' and 'Sree Subhra' which mature in 10 months with average tuber yield of 35t/ha each.

Dioscorea trifida L.f.

Common names: Cush-cush yam, Indian yam and Yampee.

Cush-cush yam (*Dioscorea trifida*), native to the Caribbean and C and S America, is an important, good flavoured and easy to cultivate New World yam, mainly used for foods in native places. Usually the plant produces up to 12 vines growing upto 3m or more in length. The shape and size of tubers may vary though they are generally up to 20cm long by 8cm wide, spherical, cylindrical or club-shaped. The starchy tuber has a thin and smooth skin marked with some cracks having various colours including white, purple and black. Leaves being 23-30cm long have blades divided into pointed lobes and are borne on long and winged petioles. Green flowers that are borne in the axils have 6 tiny tepals. Fruit is a winged, lightly hairy capsule up to about 2.7cm long. Growth cycle of cush-cush yam is not very much affected by the seasonal changes as compared to other yams. This may be due to its origin in tropical rainforest conditions. The crop is cultivated like the potato but must be given a strong trellis for support to twining vines. It is propagated by planting small tubers or cut tuber pieces (cut setts) and can be harvested in 10 to 11 months.

D. trifida

The tuber can be consumed like potatoes. It can be baked, boiled, fried or prepared as vegetables. It can be mashed or used in soups. The tuber contains about 38 per cent waxy starch that lacks amylose and can be used as a binder and thickener in food processing.

Wild Edible Species

Dioscorea hirtiflora Benth. subsp. *pedicellata* Milne-Redh

Common names: Lusala, Busala or Lwidi.

Lusala, an important wild edible tuber native to Tropical Africa, is widely harvested seasonally (March–September) for consumption and/or trade from natural forests in Southern Zambia. Research on propagation of this subspecies to alleviate the threat from wild harvest or deforestation has been successful (Zulu *et al.*, 2020), though domestication is likely to be hampered by the long period required for vines to emerge and establish as the tubers remain dormant for 6-7 months. More sustainable foraging including re-planting in natural forests is recommended to balance consumption of *lusala* in the region and to promote its long-term conservation (Zulu *et al.*, 2020). It is found at altitudes of 200–1,700m with 3 to 8m long vines which produce 1–6 cylindrical tubers of up to 5cm in diameter (Wilkin, 2001). At the onset of the rainy season, new vines often emerge from the dormant tubers.

Lusala can be propagated from tubers, mini-tubers, minisetts and nodal vine cuttings for domestication, for conservation of genetic resources and for multiplication.

Dioscorea japonica Thunb. 1784

Common names: East Asian mountain yam and Japanese mountain yam. In Japanese language: *Yamaimo* (Mountain yam); in Chinese: *Ye shanyao* (Wild Chinese/ Wild yam) or *Riben shuyu* (Japanese yam); and In Korean: *Cham ma* and *Dang ma*.

Japanese yam

Japanese mountain yam is an edible wild yam native to Japan, Korea, China, Taiwan and Asom (India). It has cold-hardy trait which differentiates it from Chinese yam or cinnamon vine. Its 30-90cm long tan to white skinned tubers are usually buried 30-90cm underground. A single perennial climbing vine grows from each wild yam that may grow 3m long. Vine's elongated leaves look like other heart-shaped sweet potato leaves. Out of several proposed botanical varieties, the four accepted varieties are: i) *D.j.* var. *japonica* ~ native to Japan (Ryukyu, Bonin), Korea, Taiwan, China (Anhui, Fujian, Guangdong, Guangxi, Guizhou, Hubei, Hunan, Jiangsu, Jiangxi, Sichuan, Zhejiang; ii) *D.j.* var. *nagarum* Prain and Burkill native to Asom (India); iii) *D.j.* var. *oldhamii* R.Knuth native to China (Guangdong, Guangxi) and Taiwan; and iv) *D.j.* var. *pilifera* C.T.Ting and M.C.Chang a native to China (Anhui, Fujian, Guangxi, Guizhou, Hubei, Hunan, Jiangsu, Jiangxi, Zhejiang). Yamaimo's edible parts include the tuber and air potatoes but tubers require some processing before consumption.

Phytochemical Composition

D. japonica contains the antimutagenic compounds eudesmol and paeonol. Phytochemical studies revealed presence of active hypoglycemic compounds (dioscorans A–F), sesquiterpene and acetophenone along with the steroidal constituents including spirostane, furostane and cholestane types (Kim *et al.*, 2011).

Nutrition and Uses

The tubers of *D. japonica*, known as 'San Yak' in Korea are used as food similar to potatoes and as medicine. These contain about 1.9 per cent protein, 20 per cent carbohydrate, 0.1 per cent fat and 1 per cent ash. Air tubers can be roasted and snacked on like edamame. Air as well as underground tubers have a neutral flavour that often act as a binder for various foods such as soba noodles. In Japanese cuisine, both the Japanese yam and the introduced Chinese yam are used interchangeably in dishes and recipes (Itoh, 2015). It is one of the few yams that can be eaten raw similar to the Chinese yam (*D. polystachya*). *Jinenjo* (wild yam) native to fields and mountains in Japan is another kind of *D. japonica* that is mainly used as an ingredient in soba noodles.

Tubers are prescribed in the treatment of diarrhoea, enteritis, enuresis and spermatorrhoea. These are also dried and cut into shavings to be used as a tonic. The plant was used in traditional Chinese medicine to control/eliminate diarrhoea, diluting sputum, improving anorexia and moisturizing skin (Kim *et al.*, 2011).

Edible Species of Minor Importance

Dioscorea abyssinica Hochst. ex Kunth

D. abyssinica, with native range from W. Tropical Africa to Eritrea, has non-spiny and right-twinning stems growing up to 4m long from a tuberous rootstock. Its starchy edible tubers are either harvested from the wild or cultivated; however, they are difficult to obtain due to the depth at which they grow in the soil. It is usually propagated by seed. This species is believed to be one of the parents of *Dioscorea cayenensis*.

Dioscorea alatipes Burkhill and H.Perrier

D. alatipes, indigenous to the Morondava Prefecture in Madagascar, is a wild species that can be found growing in forested areas on sandy soils or on limestone substrate, and is managed for human consumption as part of the SuLaMa (Sustainable Landmanagement in south-western Madagascar) Project. It is listed as vulnerable on the IUCN Red List in 2017 (Kennerley and Winkin, 2017) due to overharvesting of its tubers, *vis-à-vis* by extension of farmland areas in the area. This species is usually confused with *D. bako* which is a food source of the indigenous people of the region (Wilkin *et al.*, 2008).

Dioscorea altissima Lam.

Common Name: Dunguey.

Synonyms: *Dioscorea balsapuertensis* R.Knuth, *D. calcarea* R.Knuth, *D. chondrocarpa* Griseb., *D. hoehneana* R.Knuth, *D. maranonensis* R.Knuth, *D. nitida* R.Knuth, *D. poeppigii* Kunth, *D. rajanioides* Uline ex R.Knuth, *D. revillae* Ayala, *D. riparia* Kunth and R.H.Schomb. and *D. samydea* Griseb.

Dunguey (*Dioscorea altissima*) is indigenous to forested areas of S. America, *viz.* Brazil, Bolivia and Peru, north through Central America to Panama and the Caribbean. It has 10m or even more longer vines produced from a tuberous rootstock. The plant is harvested mainly from the wild for its edible tubers that are consumed only after cooking, but is also cultivated on a small scale locally in Brazil.

Dioscorea deltoidea Wall. ex Griseb.

Common name: Nepal yam.

Synonym: *Tamus nepalensis* Jacquem. ex Prain and Burkill.

Nepal yam, native to the Himalayas, is found growing in forests and humus-rich soils. Its tubers contain diosgenin and are harvested by local people as a treatment for a variety of conditions, including gastrointestinal disorders and intestinal worms. Its boiled and baked tubers are eaten after thorough washing.

D. deltoidea foliage and Inflorescence

Dioscorea elephantipes (L'Her.) Engl.

Common names: Elephant's foot or Hottentot bread.

Elephant's foot, native to the dry interior of South Africa, grows naturally in the arid inland regions of the Cape, stretching from the centre of the Northern Cape, south to the Clanwilliam and Cederberg area, and eastwards as far as Grahamstown. The species is mainly a winter grower having slender, leafy and climbing shoots. It loses its leaves in the summer when it goes through a dry dormancy period. The dark-spotted greenish-yellow flowers can be seen in winters, *i.e.* May or June in South Africa. It has large, partially buried and very slow growing starchy tuberous stem that often reaches a considerable size of more than 3m in circumference with a height of nearly 1m aboveground. The tuber is covered on the outside with thick, hard and corky plates.

When this plant grows naturally in thick undergrowths on rocky slopes it requires extremely well-drained soil with a large mineral component. In cultivation in temperate areas, it can tolerate temperatures upto -4°C. In its natural environment, the caudex is usually in shade beneath thicket vegetation and only the leafy tendrils reach up to the partial or full sunlight. Therefore, the caudex is sensitive to prolonged exposure to heat and full sunlight. Green vine tendrils however, thrive if they are able to reach partial or full sunlight. It can receive regular watering when a new growth appears from the caudex until the growth withers and dies back (summer dormancy) whence the watering should become quite rare until the next new growth appears.

Before being eaten, the tubers of this plant require thorough processing to remove its toxic compounds so that these may be consumed similar to potatoes or other yams.

Dioscorea fordii Prain and Burkill

Dioscorea fordii is indigenous to the Fujian, Guangdong, Guangxi, Hunan and Zhejiang provinces of China where it is typically found in forest areas or on disturbed ground besides roads or rivers (*Dioscorea fordii* in 'Flora of China'). It has oppositely arranged leaves and the large starchy tubers growing vertically into the soil. It is cultivated in China for food as well as to be used in traditional Chinese medicine. *D. fordii* contains higher starch though lower protein contents as compared to the well known Chinese yam (*D. polystachya*). On dry weight basis, the tubers contain approximately 76.5 per cent starch, 9.9 per cent protein and 1.03 per cent fibre (Wu *et al.*, 2016).

Dioscorea hamiltonii Hook.f.

Local name: Kaluvenni.

D. hamiltonii

Dioscorea hamiltonii, distributed in Indo-Myanmar region, is mostly seen in semi-evergreen and moist deciduous forests. The plant has angled and glabrous stem with ovate or lanceolate leaves. Flowering and fruiting occur during October-December. Tubers having medicinal properties are cooked for consumption. It has been used for the treatment of stomach-ache and for poor appetite. The crushed tubers are given as body refrigerant during summer and also for treating diarrhoea.

Dioscorea hastifolia Nees

Other names: Adjigo yam, Warram or Warrine.

D. hastifolia

The species name *hastifolia* literally translates to 'spear leaf' due to the spear-shaped foliage of the plant. *Dioscorea hastifolia*, native to SW Australia bears up to 3m long stem with spearheaded leaves, yellow flowers and green triangular fruits. *D. hastifolia* is one of the hardiest of the yams, tolerating drought and sandy soil but for best results it should be grown in loamy soil with proper mulching and support for vines. The flowers appear during May-July producing triangular fruits that mature in October. The plant is cultivated by Australian Aboriginals for its long, thin and white tubers that are used as a source of carbohydrates. The plant is harvested in late winter, just after the yellow flowers turn to fruits. The tuber sections can be used to grow more plants. With careful planning, up to three crops of this species can be taken in one year.

Its tubers are largely consumed by the local aborigines for food (Maiden 1889). The tubers can be eaten raw, roasted with butter and/or a pinch of salt or as an ingredient in a variety of savoury and sweet dishes like stews, pies, curries, dumplings, cakes, *etc.*

Dioscorea hispida Dennst

Common names: Asiatic bitter yam, Indian three-leaved yam; locally as Banya aalu or Kottunoora.

Dioscorea hispida, native to SE Asia, is climbing herb with a fibrous root system and trifoliate leaves but no bulbils. It is quite common in Odisha,

D. hispida

Jharkhand, Andhra Pradesh and Telangana and is distributed throughout the mountainous areas of the country. Fresh tubers are toxic due to presence of high content of dioscorine alkaloid and calcium oxalate raphides; hence careful processing is required to make it edible. The toxin in the tubers is often extracted to be used as bait for animals, to eliminate fish from shrimp pond cultures or for other devious purposes. It's time consuming detoxification is done by finely slicing and washing the fresh or boiled tubers several times in salt water or in running water for a few days until the toxins have leached out. Then it is dehydrated and cooked.

Dioscorea intermedia Thw.

D. intermedia

Local name: Shoddikalasu.

Dioscorea intermedia is distributed in the evergreen forest from Sri Lanka to Southern India in south Western Ghats. The plant has elliptic or oblanceolate leaf with cylindrically shaped tubers. Flowering and fruiting occur from October to January. The tuber and bulbils are edible though bulbils are generally not collected for food. It is used as a remedy for stomach-ache and consumed for poor appetite.

Dioscorea oppositifolia L. syn. *Dioscorea opposita*

Common names: Air potato; Local name: Kavalakalasu.

Dioscorea oppositifolia is native to the subcontinent of India but has been consistently misapplied to Chinese yam. The plant is distributed in Indo-Malaysian region and China. Alternate or opposite leaves of this deciduous perennial creeping *cum* climbing vine are long-pointed with a heart-shaped base. The margins, leaf-stalk and stems are purplish to red in colour. This species flowers and fruits from August to October. Small bell-shaped yellowish-white flowers arise from the axils of the leaves. The male flowers are in bundles and spikes or panicles at the end of the branches. Flowers have a spicy fragrance similar to cinnamon. Bulbils (small aerial tubers) are produced in the axils of the leaves.

Though tubers and bulbils are edible but bulbils are usually not collected for food. The tuber is also used as herbal tonic. It stimulates the stomach and spleen and has an effect on the lungs and kidneys. The tubers are medicinally used for chronic diarrhea, asthma, dry coughs, frequent or uncontrollable urination and diabetes.

Dioscorea ovinala Baker

Dioscorea ovinala is native mostly to the dry deciduous forests of West Madagascar. It is important to the Malagasy people as food, medicine and as a source of starch (Kennerley and Wilkin, 2017).

Dioscorea pentaphylla L.

Common names: Five-leaf yam; also called as Noora kizhangu or Panja Sanga by local people in India.

Synonyms: *Botryosicyos pentaphyllus* (L.) Hochst.; *Dioscorea changjiangensis* F.W.Xing and Z.X.Li; *D. codonopsidifolia* Kamik.; *D. digitata* Mill.; *D. globifera* R.Knuth; *D. jacquemontii* Hook.f.; *D. kleiniana* Kunth; *D. spinosa* Burm.; *D. triphylla* L.; *Hamatris triphylla* (L.) Salisb.; *Ubium quadrifarium* J.F.Gmel.; and *Ubium scandens* J.St.-Hil.

D. pentaphyla

Five leaf yam is native to Tropical Asia and Eastern Polynesia, and is distributed in SE Asia, Tropical Asia and North America (Kumar *et al.*, 2017). In India, it is distributed in Deccan, Western Himalayas, Rajasthan, Nasik (Maharashtra) and Odisha states of the country. The populations of this species are rich in Eastern and Western Ghats. It is a prickly and counterclockwise twining vine growing to a length of upto 10m. The alternately arranged leaves are compound, divided into 3 to 5 leaflets each up to 10cm long. It produces about 1cm long horseshoe-shaped bulbils. New plants can sprout from the bulbils as well as from tubers. Flowers are borne in spikes. Flowering and fruiting have been observed from August to November. The tubers with hairy, black skin and white flesh are occasionally used as food after proper cooking. Tubers may weigh some 1.3kg and may be buried over a metre underground. Male inflorescences and young leaves can also be consumed after cooking. The extract of the plant is traditionally used for boils and swellings.

Dioscorea praehensilis Benth.

Dioscorea praehensilis, native to rainforests and seasonal tropical forests of Africa, is the wild progenitor of the West African domesticated crops *Dioscorea rotundata* and *Dioscorea cayennensis* (Blench, 2006). It is a long stemmed vine that uses trees or other means of support to climb up to the canopy in search of direct sunlight. Its edible tubers reach their maximum starch reserves during the dry season. Like several other species, it too renews its stems every year at the start of the rainy season.

Dioscorea pubera Blume

Local name: Arikizhngu, Kukai Sanga.

D. pubera

Dioscorea pubera, native to the Indo-China region, is another wild species distributed throughout the temperate and tropical Americas, China, wet regions of Himalayas, Central Nepal, Western Malaysia and Bhutan (Kumar *et al.*, 2017). These right-twinning vines are mostly seen in semi-evergreen forests. In India, it is distributed mainly in areas of the Himalayas, Sikkim, Asom, Northern Bengal, Western Ghats, Wayanad and Udupi in Kerala, Jharkhand, Bihar, Upper Gangetic Plain, and Andhra Pradesh. The leaves are very rarely alternate, broadly elliptic, abruptly acuminate, with glabrous and coriaceous margins. Closely packed flowers, *vis-a-vis* fruiting occur during February-March. The raw tubers of this species are bitter in taste *cum* unpalatable. The rural and local people who use them as food supplements make them edible with different traditional practices. The nutritive and antioxidant content not only enrich the diet of the local and rural people, it also has ethnomedicinal importance. The tubers are used for birth control and for the treatment of skin infections. Tubers and vegetative parts of this species are used in multiple traditional medicinal formulations.

Dioscorea spicata Roth

Tubers of *D. spicata*

Local name: Kavalakizhangu.

Dioscorea spicata, native to the Indian subcontinent, is a wild species mostly seen in Shola and semi-evergreen forests of South India, Sri Lanka, and SE Bangladesh. Opposite or subopposite (rarely alternate) leaves are elliptically lanceolate, up to 10cm long and 5cm wide. Small cylindrical tubers are edible. Flowering and fruiting season is during September to December. The tubers are used as a food source. It is placed in the section Enantiophyllum.

Dioscorea tomentosa J. Koenig ex Spreng.

D. tomentosa

Local name: Noolamkizhangu.

Dioscorea tomentosa, native to India and Sri Lanka, is distributed in semi-evergreen, moist and dry deciduous forests. This left-twinner has tomentose stems and trifoliate tomentose leaves having sub-equal broadly ovate leaflets which are oblique at base. The flowering and fruiting occur mainly during July to December.

The tubers are thoroughly washed and steam-cooked, boiled or roasted before consumption as food. The tubers are believed to clean the digestive system and promote the hair growth. It is also used in the synthesis of many steroidal drugs as corticosteroids, sex hormones and oral contraceptives. It has also been used in rheumatism.

Dioscorea transversa R.Br.

Synonyms: *Dioscorea punctata* R.Br.

Common names: Long yam and Pencil yam.

Pencil yam

Dioscorea transversa is native to eastern and northern Australia. The plant has shiny heart-shaped leaves with 5-7 prominent veins. Round, green or pink seed pods become brown papery when dry. One form of the species that does not have bulbils occurs in eastern rainforest and wet sclerophyll forest, while the other form that occurs in open northern forests has small bulbils, along with the large inground tubers. The long slender tubers are a staple food of Australian aborigines and are eaten after cooking, usually in ground ovens (Govaerts *et al.*, 2007).

Dioscorea wallichii Hook.f.

Dioscorea wallichii is native to Bangladesh, China, India, Malaysia, Myanmar and Thailand. It has single, about 1m long *cum* 3-6cm wide cylindrical edible tuber that is white when young, becoming yellow and stringy once aged. The tubers are eaten after boiling and washing.

Propagation

Although most of the yam species can be grown by sowing their seeds though it is of no use to the farmers but for a breeder. The plants grown from seeds need not be like the parent plants, the yield is usually less, the size and quality of tubers is not up to the mark, and moreover it may contain poisonous dioscorine. Farmers propagate yams clonally by small tubers, cuttings of large tubers, aerial tubers (bulbils) or sometimes using stem cuttings. The most accepted propagation method is through small seed tubers and tuber cuttings, though small seed tubers are ideal for planting as they have no exposed parenchyma for rapid infection by fungi.

Usually the tubers from previous harvest are selected and stored in a well-ventilated and dry place until planting, and then if required they are cut into 100-150g pieces in case of lesser yam and 259-300g pieces in case of greater yam and planted directly in the prepared field when the cut surfaces are dry, *i.e.* usually after 3 days. If these cuttings are to be first raised in beds or nursery before planting in the main field, they should be planted well before so that the plantlets get developed before planting time. Most yams have a definite period of dormancy, but this may be broken by the use of ethylene chlorohydrin (C_2H_5ClO).

Seed Yam Production Methods

There are various methods of seed yam production.

Traditional Method

Farmers usually set aside a portion of their harvest for planting the next year. It is approximately 30 per cent of the total yield. The yam crop stays in the field for 8 to 10 months and when dormancy is broken, *i.e.* after another 2 to 3 months after harvest, the tubers are ready to grow. Though traditional methods of production provide the assurance of obtaining seed tubers by farmers, it is generally characterized by low multiplication ratios (Alvarez and Hahn, 1984) with uneven and prolonged sprouting periods when planted (Otoo *et al.*, 1987).

Double Harvesting Method

The plants of early maturing species or variety are harvested twice. First harvest is done for food usually between 4 and 5 months after shoot emergence when the shoot of the mother plant is still green. The tubers are harvested from the base of the vine without disturbing the above ground portion. This is done by opening the mounds from one side and re-moulding them after detaching the tubers. The production of new tubers get initiated from the base of the vine, which often bulks into an amorphous shape with protuberances or digitations that can be detached and planted separately at harvest. At the end of the season (about 1–3 months after the first harvest) when the plants senesce, the second harvest is done for regenerated tubers to be used solely as seed (Onwueme, 1973). It is almost certain that regenerated tubers are slightly or not at all infested by nematodes (such as *Scutellonema bradys* and *Pratylenchus coffeae*) and are the better planting material than normal setts from single harvests (Degras, 1993), but the method is labour intensive and only suited to early maturing varieties of yams.

Cut Setts

The large tubers are cut into tops, middles and bottoms, and size of these cut setts may vary depending on the species, for example in *D. rotundata* or in other yams with large tubers each piece is usually 200 to 500g. The larger the sett, the earlier and greater is the rate of sprouting. These setts are planted after

2-3 days when cut surfaces are dry. It is advisable to treat the cut surfaces with wood ash and/or with recommended chemicals before planting. Only the head portion would have a sprout and it establishes better and faster. Since middle and bottom portions do not have pre-formed sprouts, shoots from these setts take longer to emerge after planting as compared to the setts of top portion. Depending on the variety and agro-ecology, tubers from the cut setts are harvested after 7 to 12 months. Setts that sprout earlier have a longer growth period because the crop is harvested when the dry season sets in and plants senesce. In this method, there is greater chance of fungi infection and nematode infestation. Some of the modifications or variations of cut setts method of seed production are ~ cut setts intercropping, minisett and microsett methods.

- ☆ **Cut setts intercropping**: Seed yams are produced by cutting small tuber pieces (80–100g) from post-dormant 'mother' tubers weighing 500–1000g and planted directly in the field (Okonmah, 1980). These setts are usually planted along with other crops such as maize, cassava, cocoyam and vegetables. It would be better to dust the cut surfaces of tubers with wood ash and/or treat the setts with recommended chemicals before planting (Aighewi, 1998) to protect them from fungal rots after planting. The inter-plant distance and depth of planting depend on the density and type of companion crops and the moisture content of the soil. The seed tubers produced by this system are usually planted whole and have better establishment than when cut setts are used as seed.
- ☆ **Minisett and microsett methods**: These techniques are beneficial to dedicated seed yam producers. To maximize the setts from a single tuber and the need for a reasonable proportion of seed yams in the yield, the size of cut setts is further reduced to around 25g in minisett method (Kalu *et al.*, 1989) and to about 10g or less in microsett method though some people believe that larger than 25g minisett sizes would be better (George, 1990). However, Okoli and Akoroda (1995) stated that there is wastage of edible tubers being used for planting as 5–30g setts could produce the recommended seed yam sizes of 150–300g. Even 30g setts of 5mm thickness which are normally discarded as yam peels produce mean tuber weights above 200g (Aighewi, 1998). Using these methods, greater number of setts can be produced from the post-dormant 'mother' tubers. The major aim of the techniques is to increase the quantity and quality of seed tubers available to farmers. These developments are based on the principle that any section of the tuber is capable of developing buds and sprouting provided it has a portion of the periderm (Onwueme, 1973). The freshly cut minisetts

can be treated using wood ash or fungicide *cum* insecticide, and then spread out in light shade to dry for 1–2 days (Anon., 1986). While the use of wood ash as a substitute for fungicide and insecticide has been recommended, its effectiveness has been variable (Otoo, 1992). According to Alvarez and Hahn (1984), microsett propagation with the aid of phytohormones increased the multiplication ratio of the traditional system from 1:4 to 1:90.

- During growing season the treated minisetts can be directly planted in the field on ridges at a depth of 9–12cm with a plant spacing of 25–30cm × 100cm (4 stands per m^2). Sprouting and tuber yield of direct planted setts are strongly affected by variety (Ayankanmi *et al.*, 2005; Aighewi, 1998). The condition of the soil at the time of planting also affects the survival of propagules. It is usually recommended that they should be first pre-sprouted in a nursery before transplanting because pre-sprouting in a pest- and disease-free medium can help maximum survival of minisetts though it is more labour intensive (Okoli and Akoroda, 1995).
- The minisetts and microsett techniques can be ranked best with regard to the number and cost of mother tubers needed for seed yam production, and the rate of multiplication (Anon., 1985). Though enormous amounts of yam could be saved using these methods, preparation of setts could be labour intensive. Moreover, farmers may not be able to conveniently handle the small setts, and not all species or varieties may be suitable for use with this technique. Also, inadequate seed treatment (fungicide/insecticide application) before and after planting (mulching/ irrigation) may reduce the rate of tuber yield.

Small Whole Tubers

Whole seed tubers sorted out from a ware crop are ideal for planting as they have no exposed parenchyma for rapid infection by fungi. These small tubers are planted at the beginning of the dry season. In addition to whole seed tubers produced by double harvesting, some varieties produce 1–3 small tubers together with a large ware tuber that is used for food. The small tubers are sorted and reserved for planting. Sometimes the tubers have exceptionally reduced size due to infection of certain viruses so these should be used with caution.

Vine Cuttings

This method of propagation improves the economic value of the crop by allowing the entire tuber to be saved for food. Rooted vine cuttings of 20cm length with 1 to 3 nodes are taken from healthy plants for tuber production. It will be better if the cuttings are taken from plants aged between 30 and 60 days after shoot emergence, and before the start of tuber development (Okonmah, 1980; Kikuno *et al.*, 2007). These cuttings are capable of producing mini-tubers of 50-600g after about 8 months of planting, giving a 1:30 propagation ratio. Cuttings are rooted in a high humidity chamber or carbonized rice husk before transplanting into soil or planting directly in top soil in a garden. The setts produced using this method would be almost free of nematodes and other soil-borne pathogens if a sterilized medium or pest-free soil has been used for propagation. Large quantities of small size minitubers that are appropriate for international exchange of germplasm can be produced using vine cuttings, as well as normal seed size tubers for production of ware yam.

Tissue Culture

In this technique, disease-free plantlets are grown rapidly in test tubes on a nutrient medium. Selected clones are cleaned from virus diseases through meristem culture combined with thermotherapy. Each plantlet is further cut into several sections. Each new cutting is then planted in a new test tube. This can be repeated until the desired number of plantlets is obtained. Plantlets are then removed from the tubes and grown under sterile conditions in artificial media containing growth substances such as sugars, vitamins and plant growth regulators, and multiplied to reach target numbers. Thereafter, they are acclimatized and grown in pots with sterile soil or other media where they complete their entire growth cycle to give whole seed tubers of high quality after 5–6 months. Tissue culture plays an important role in the production and international exchange of disease-free planting materials of yams in the form of plantlets and potentially as microtubers. The aseptic steps taken in tissue culture produce almost disease-free planting materials, though there is also a risk of re-infection once the seeds are transferred from the aseptic conditions of laboratories to fields where there is high disease pressure. Some other drawbacks of this technique include ~ high costs, the need for skilled personnel and specialized equipment, and more crop cycles are required to produce large quantities of suitable seed size tubers for ware yam production.

Growth and Development

The crop yield depends on how and where the setts are planted, such as quality of setts, sizes of mounds, interplant spacing, provision of stakes for the resultant plants, yam species, and tuber sizes desired at harvest. Budding occurs in 5 to 10 weeks, depending on the species. Most edible yams normally reach maturity 8-11 months after planting though some are first harvested after 5 months of planting. The growth period ends at the end of the rainy season. The yam growth can be divided into 3 phases which develop over 12 months and repeat indefinitely during the life of this perennial plant. First is the vegetative growth phase that begins 5-10 weeks after planting. In this phase the outbreak and development of stems, leaves and roots take place. During the first two months, stems have few or no leaves. The soil should be rich in potassium for the formation of tubers. Several species such as *D. alata* and *D. rotundata* need stakes or support for vines to climb. Second phase is the reproductive growth in which tubers and bulbils thicken whereas the growth of stems and leaves remain stable. The plant blooms and then its tubers ripen. They are harvested in the next phase of dormancy or resting phase. The third and the final phase is resting or tuber maturity phase, preceding the flowering of the plant, which coincides with the dry season. The tubers get matured and the aerial parts get dried or decayed. This final stage ends with the sprouting of new tubers and beginning of the growing season.

The areas where yams are to be grown should be carefully selected so that the produce remains disease-free. Although environmental factors such as temperature, moisture, light and soil type influence the growth and quality of yams, other factors such as species and varieties, seed setts, planter operation, planting *cum* harvesting time, application of nutrition, irrigation, timely pest management, *etc.* are other important factors for maximizing the healthy tuber yield. Most profitable yield can be attained only when all factors are at optimum levels and at proper time. White, yellow, and water yams typically produce a single large tuber per year, generally weighing 5 to 10kg. In several yam growing countries, the agronomical advancement has significantly increased the tuber yield.

Environmental Factors

Most of the species of *Dioscorea* have a wide adaptability to diverse agro-climatic conditions. The growth and dormant phases of yam crops correspond respectively to the wet season and the dry season. Yams grow well if there is a dry season lasting 2-5 months followed by a wet season. Most yams require a humid tropical environment with an annual rainfall between 1,200mm and 1,500mm, distributed uniformly throughout the growing season though it can tolerate 500-700mm of annual rainfall but water stress causes poor yield. Yams are mostly planted at the end of the dry season and before the tubers begin to make long sprouts. Where the rains last 8-10 months, usually the planting takes place just before or at the beginning of the rains. Where the rains last less than 8 months, early planting up to 3 months before the rains can give increase in yield. Yams prefer full sun or partial shade and are mainly grown in tropical and subtropical climates. Day length greater than 12 hours favours vine growth though short-day conditions the tuber development. Depending on the species, the average temperatures for yams are 18-34°C, best being 25-31°C (Sheela, 2019), however, at mean temperatures of less than 20°C the growth is adversely affected and when temperature is around 15-17°C the dormancy is prolonged. The optimum temperature for growth of most yams is between 25°C and 30°C and do not grow well at temperatures below 22°C. Yams are not at all tolerant to frost though non-tropical species *D. opposita* and *D. japonica* can tolerate freezing temperatures upto -4°C. Day length greater than 12h at initial stages while short day conditions during tuber initiation favour tuber production satisfactorily.

Soil Type

Yams in general prefer moist but well-drained light friable sandy or gravelly soils rich in organic matter with a pH of 5.5 to 6.5 in full sun or partial shade. Too clayey soils are not suitable for yam cultivation. Too compact soils lead to tuber malformation. Very wet soils should be avoided as this promotes tuber rot. However, white yam can come up well in heavy soils, can tolerate high clay content though not the water-logged conditions. *D. alata* has good adaptation to low soil fertility but little tolerant to soil acidity.

Land Preparation and Planting Method

Land should be prepared for planting by three deep ploughings and with addition of well-rotten FYM @ 150-200 quintals per hectare. All the foreign materials should be taken out and then the land should be levelled, followed by lay out. Seed tubers should be planted in trenches to a depth of around 15cm with interplant spacing of about 30cm and 1.5m between rows. Most farmers plant the tubers in the top of the mound at a depth of 5 to 10cm and cover it with soil. The soil is often mounded around plants or ridged to aid drainage and to facilitate tuber development. It is common practice to stake plants with a 2–4m support to allow them to climb and ensure that all parts of the plant receive adequate sunlight.

Planting is done usually manually preferably before the rainy season between March-April and may continue up to June as these are relatively tolerant to dry conditions. Harvesting is usually done between November and January. Though seed tubers (bulbils, small tubers and cut setts) or nursery-generated plantlets can be directly planted in a well prepared field, they should be disinfected using proper treatments for large productions. Planting on the flat beds in areas such as river flood plains (where the soil is deep and soft), in trenches or holes just below the soil surface or on mounds (along the sides of ridges or raised beds) is the most widespread method. Setts are oftenly planted deep to avoid drying out of the young shoots. Most common spacing is 1.2×1.2m, however wider spacing in usual cropping lowers the yield. The recommended spacings for yams are 90×90cm (12,345 plants/ha) in case of *D. alata* and *D. rotundata*; and 75×75cm (18,000 plants/ha) in case of *D. esculenta* (Sheela, 2019). In case of lesser yam some 1,800-2,700kg and in case of greater yam and white yam 3,000-3,700kg of seed materials are required for planting one hectare of land. When planting inside coconut plantation, leaving a radius of 2m from the base of palms, at 60cm spacing the yams are planted where its requirement is about 9,000 plants. When planted among Nendran and Robusta banana plantations, in the inter-spaces 3 rows of yams can be planted which will require 8,000 plants in one hectare of land. Immediately after planting, the yams are irrigated. During 14th and 20th week of planting there should be good moisture in the field though waterlogging is at all not desired.

Manure, Fertilizers and Irrigation

It is important to enrich the soil in organic matter to make it fertile. The thickening of the tubers requires a generous amount of potassium. Organic manures such as well-rotten farmyard manure or compost @ of at least 10t/ha or green manure should be added at the time of field preparation. Though inorganic fertilizers may be used to get a greater yield, the amounts may vary according to the country, region, or even the soils in the same field. N:P:K chemical fertilizers @ 80:60:80kg/ha can be applied for optimum yield. Full dose of P and half doses of N and K should be applied within a week of sprouting and the remaining doses of N and K at one month of applying first dose. However, in fertile soils, N:P:K @ 60:40:60kg/ha for greater yam and 40:40:40kg/ha for lesser yam are quite promising. As these are relatively tolerant to drought so under such conditions larger tuber setts should be planted so that the plants during drought may survive by the moisture already present in the tubers. For uniform sprouting, after planting, the yams are immediately watered. During 14th and 20th weeks, the yam field should have sufficient moisture (Sheela, 2019). However, waterlogging is dangerous. Yams require 100cm of water distributed evenly throughout the growing season.

Interculture

Yam plants should be mulched after planting to prevent plants from drying out and to conserve moisture. Failure to mulch the plants will result in drastic decrease in yield. Home gardeners and small-scale growers can cover the growing area with mulches such as straw or plastic sheets.

When there is too much sun or the light is too strong, most farmers cover the mound with grass so that soil remains moistened and the plants are saved from

sun burn. This way the rain will not wash away the soil exposing the tubers. In savannah country where there is a long dry period, stakes or supports are not used. The aerial stems trail on the ground covering it completely thus preventing the weed growth and thus protecting the plants from drying out.

Earthing up of the soil also brings about weed control besides improving the physical condition of the soil. Regular weeding and earthing up should be performed to keep the soil loose and to destroy the weeds. Exposed tubers must be covered else they would become greenish and promote root aeration. For a good harvest, hoeing must be done 2-3 times during early stages of growth. The mounds must be remade at the same time. Later, the abundant vegetation of the yams prevents the growth of weeds.

Stakes, trellises, strings attached to horizontal ropes, corn stalk and nearby trees can be used as a **support** that is necessary for good plant growth and tuber development though some species such as *D. alata* and *D. esculenta* are adapted to trailing on the ground without support. Removing of some tubers that are growing too close together should be done to promote thickening of the tubers.

Crop Rotation

A good crop rotation maintains soil fertility, reduces soil erosion and depletion, and suppresses insect and disease outbreaks. Rotating with other crops on a 3 or 4 year cycle is the standard procedure to prevent outbreak of insect-pest attack and disease infections. It is best to plant yams at the beginning of the rotation as a first-year crop after clearing the land. If yams are grown after a long fallow, they find plenty of mineral salts in the soil so yielding good tubers. Yams can also be intercropped with other seasonal crops such as cereals and vegetables.

Common Pests and Diseases

Yam crops face pressure from a range of insect-pests, fungal and viral diseases as well as nematodes.

Mealybugs (*Rastrococcus* spp.) are flattened oval to round disc-like soft-bodied grayish-white insects up to 4mm long. They are covered in waxy substance. Its nymphs and adults both are found clustered on branches or leaves of several plants sucking sap of the plants. In their severe infestation the plants become weak and yellow. As these excrete honeydew so ants are also enticed, *vis-à-vis* fungi colonise such organs as is symptomized by occurrence of sooty mould there. These can be controlled by pruning out heavily infested branches, by their natural enemies or by using horticultural oils or soapy solutions. Use of chemical pesticides may also decrease populations of natural enemies leading to mealybug outbreaks.

White scale insect (*Aspidiella hartii*) infestation may cause poor foliage growth and tubers may show delay in sprouting or may fail to sprout. Small white scales on leaves and tubers are seen from field to storage. Severe infestation may lead to tuber shrivelling. Preventive measures include dipping of the planting material in methyl demeton 0.25 per cent and using scale free seed tuber for planting.

Nematodes

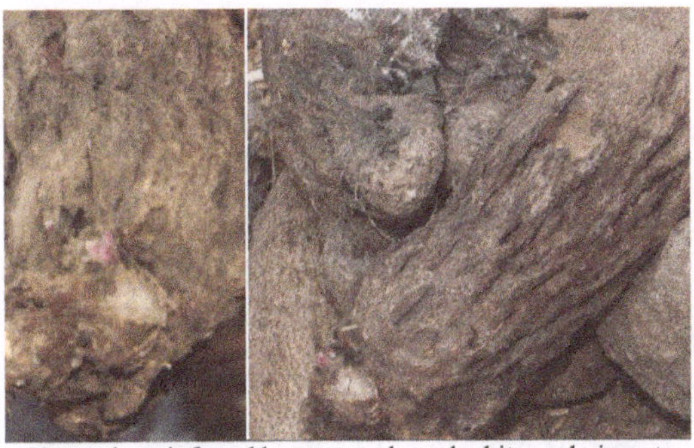

Yam tubers infested by nematode and white scale insect

Galled tubers with crazy roots

Yam tubers infested by white scale insects

Dry rot disease is caused by yam nematode (*Scutellonema bradys*). The infested tubers have no aboveground symptom but often show dry rot which initially appears as cream and light yellow lesions just below the outer skin but without any external symptom. It then spreads deeper (maximum up to 2cm). At later stage the rot becomes light and dark brown to black in colour and tubers may show external cracks. Entry of fungus through these wounds causes further decay of tubers in storage. The cultural practices to control nematodes include use of healthy tubers/setts for planting and smearing tubers with wood ash or cow dung. Crop rotation with non-host or antagonist crops like groundnut, sorghum, maize, chilli pepper, *etc.* can be performed.

Root-knot nematode [*Meloidogyne incognita* (Kofoid and White)] causes size-reduction of tubers so there is substantial loss in tuber yield. The symptoms include stunted growth, leaf yellowing and galling on tubers and feeder roots. It can be controlled by deep summer ploughing and solarization of soils, *vis-a-vis* by following crop rotation with non-host crops such as groundnut and maize.

Anthracnose symptoms on foliage

Anthracnose is a worldwide fungal disease caused by *Colletotrichum gleosporoides which* overwinters in plant debris. The symptoms include small dark brown spots or black lesions on leaves which may be surrounded by a chlorotic halo; leaf necrosis; dieback of stem; withered leaves and scorched appearance. It is usually managed by planting yam varieties resistant to anthracnose such as 'TDA 291' or 'TDA 297'. For its effective eradication, CTCRI recommended field sanitation; use of resistant varieties as developed by CTCRI; spraying with mancozeb 0.2 per cent and prophylatic spray of Bordeaux mixture.

Yam mosaic infected leaves

Yam mosaic disease is caused by yam mosaic potyvirus transmitted by aphids and tubers/setts. This disease is mostly associated with yam mild mosaic virus, yam badnaviruses and cucumber mosaic virus thus the management gets more complicated. The symptoms include yellow and green patterns (called mosaics) between the veins or a narrow green strip bordering the veins (called vein banding). In severe cases, the leaves become long, thin and strap-shaped (called shoe-string symptom) and whole plant becomes stunted. Plant may produce few small tubers with less starch content. Some plants may recover from the virus infection soon after first symptom but virus may survive in plant and reduce the vigour. It can be prevented using healthy disease-free tubers/setts for planting and by keeping the fields free from weeds. The crop debris should be destroyed.

Harvest

Usually, the yam crop yields 20–25t/ha of tubers depending on the species, area and time of planting, planting methods, soil fertility *cum* soil pH, irrigation, and cultural practices employed by the farmers. Depending on the location and species, yams are harvested between November to January or sometimes in February after 7-10 or at most after 12 months of planting. Several yams including greater yam and white yam become ready for harvest by 8-10 months of planting. Lesser yam takes 8-9 months for attaining maturity. The time of harvest is characterized by wilting of aerial parts when the stems become slightly turgid and the leaves turn brownish. The distal part of the tuber becomes brown. Farmers usually harvest the tubers manually using spades or other tools. The commercial producers harvest the crop both ways, *i.e.* mechanically as well as manually. Mechinical harvest is time and labour saving but results in more damage to crops whereas manual harvest is time consuming and labour intensive though little damage to tubers. Aerial tubers or bulbils are harvested by manual plucking from the vines. Harvesting may be quite difficult in mixed cropping and when stakings are used. When the tubers grow in areas where other tree roots are present, especially in forested areas, harvesting of tuber involves the additional step of freeing them from other roots but this often causes tuber damage.

Postharvest

After harvesting, yams should be sorted into boxes or baskets to cure them before storage. Curing can be performed by placing the tubers in single layers in a well-aerated, warm and dark place at least for two weeks. This process will allow bruising and wounds on the yams to heal, and sealing out rot bacteria.

Storage

Early varieties do not store well and should be consumed immediately after lifting. Late varieties such as *D. alata* may be stored for 5 or 6 months under a roof in dry and ventilated location with protection from direct sunlight, rats and other rodents. There are more chances of rot if the tubers are heaped up too much. Inspection and removal of rotting tuber and any sprouts that gets developed is necessary. Sprouting rapidly increases a tuber's respiration rates, and accelerates the rate at which its food value decreases. The ideal temperature range to store yam tubers is 14-16°C along with high technology-controlled humidity and climatic conditions. Most countries that grow yams as a staple food are too poor to afford high-technology storage systems. Storing yam at low temperature reduces the respiration rates, however, temperatures below 12°C may damage the tubers through chilling, causing a breakdown of internal tissues, increasing water loss and yam's susceptibility to various decay fungi. The chilling injury symptoms are not always noticeable when the tubers are still in cold storage. The chilling injury can be clearly seen when the tubers are restored to ambient temperatures.

The yam breeds adapted to arid climate are easier to store because they tend to stay in a dormant low-respiration stage much longer than yam breeds adapted to humid tropical lands where they do not need dormancy. The water yam, white yam or lesser yams have much longer dormancy periods than yellow yam and cush-cush yam.

References

1. Abhyankar R. K. and R. Upadhyay, 2011. Ethno-medicinal studies of tubers of Hoshangabad, M.P. *Pharmacol. Life Sci.*, **1**: 57–9.

2. Adepoju, O.T., O. Boyejo and P.O. Adeniji, 2018. Effects of processing methods on nutrient and antinutrient composition of yellow yam (*Dioscorea cayenensis*) products. *Food Chem.*, **238**: 160–65.
3. Adeyeye, S.A.O. and O.O. Jacob, 2014. Quality assessment and acceptability of pounded yam from different varieties of yam. *Nat. Sci.*,**12**: 115–9.
4. Afiukwa, C.A., O. Ogah, O.P.C. Ugwu, J.O. Oguguo, F.U. Ali and E.C. Ossai, 2013. Nutritional and antinutritional characterization of two wild yam species from Abakaliki, Southeast Nigeria. *Res. J. Pharm. Biol. Chem. Sci.*, **4**: 840–48.
5. Afoakwa, E.O. and S. Sefa-Dedeh, 2001. Chemical composition and quality changes occurring in *Dioscorea dumetorum* pax tubers after harvest. *Food Chem.*, **75**: 85–91.
6. Ahmadu, U., O. Agbomeji, M. Yahya and A. Odeku, 2018. Physicochemical and material properties of starches from three cultivars of *Dioscorea rotundata. Agric. Nat. Resour.*, **52**: 79–83.
7. Aighewi, B.A. 1998. Seed yam (*Dioscorea rotundata* Poir.) production and quality in selected yam zones of Nigeria. Ph.D. Thesis, 252p. University of Ibadan, Ibadan, Nigeria.
8. Ali, Z., T.J. Smillie and I.A. Khan, 2013. Two spirostan steroid glycoside fatty esters from *Dioscorea cayenensis. Nat. Prod. Commun.*, **8**: 323–6.
9. Alinnor, I.J. and C.O. Akalezi, 2010. Proximate and mineral compositions of *Dioscorea rotundata* (white yam) and *Colocasia esculenta* (white cocoyam). *Pak. J. Nutr.*, **9**: 998–1001.
10. Alvarez, M.N. and S.K. Hahn, 1984. Seed yam production. In: Tropical Root Crops: Production and Uses in Africa (eds Terry, E.R., E.V. Doku, O.B. Arene and N.M. Mahungu), pp. 129–132. *Proc. 2nd triennial symp. of ISTRC-AB, IDRC-221e*, Ottawa, Canada.
11. Anon., 1976. Studies on Chinese *Dioscorea* sect. *Stenophora* Pr et Burk and their chromosome numbers. *Acta Phytotax Sin.*, **14**(1): 65–72.
12. Anon. 1985. Root and tuber improvement program. International Institute of Tropical Agriculture (IITA). Research Highlights 1981–1984, Ibadan, Nigeria.
13. Anon. 1986. Annual report of the root and tuber improvement program. International Institute of Tropical Agriculture (IITA), Ibadan, Nigeria.
14. Anon. 2008. Glycemic Index and Glycemic Load for 100+ Foods. Harvard Health Publications. Harvard Medical School, Great Britain.

15. Anon. 2016. *Dioscorea alata* (white yam). Centre for Agriculture and Biosciences International (CABI). Retrieved 5 December 2017.

16. Araki, H., T. Harada and T. Yakuwa, 1983. Some characteristics of interspecific hybrids between *Dioscorea japonica* Thunb. and *Dioscorea opposita* Thunb. *J. Japan. Soc. Hort. Sci.*, **52**(2): 153–8.

17. Arnau, G., A. Nemorin, E. Maledon and K. Abraham, 2009. Revision of ploidy status of *Dioscorea alata* L. (Dioscoreaceae) by cytogenetic and microsatellite segregation analysis. *Theor. Appl. Genet.*, **118**(7): 1239–49.

18. Avula, B., Y.H. Wang, Z. Ali, T.J. Smillie and I.A. Khan, 2014. Chemical fingerprint analysis and quantitative determination of steroidal compounds from *Dioscorea villosa*, *Dioscorea* species, and dietary supplements using UHPLC-ELSD. *Biomed. Chromatogr.*, **28**: 281–94. doi: 10.1002/bmc.3019.

19. Ayankanmi, T., H. Shiwachi and R. Asiedu, 2005. Sprouting and yield of yam (*Dioscorea* spp.) minisetts in relation to sett size, soil moisture and agro-ecology. *Trop. Sci.*, **45**: 23–7.

20. Baah, F.D., B. Maziya-Dixon, R. Asiedu, L. Oduro and W.O. Ellis, 2009. Nutritional and biochemical composition of *D. alata* (*Dioscorea* spp.) tubers. *J. Food agric. Environ.*, **7**: 373–8.

21. Babil, P., S. Kondo, H. Iwata, S. Kushikawa and H. Shiwachi, 2013. Intra-specific ploidy variations in cultivated Chinese yam (*Dioscorea polystachya* Turcz.). *Trop. Agr. Develop.*, **57**(3): 101–7.

22. Baquar, S.R. 1980. Chromosome behaviour in Nigerian yams (Dioscorea). *Genetica,* **54**(1): 1–9.

23. Behera, K.K. 2006. Ethnomedicinal plants used by the tribals of Similipal Bioreserve Orissa, India: a pilot study. *Ethnobot. Leaflet,* No. 10, pp. 149–73.

24. Behera, K.K., S. Sahoo and A. Prusti, 2009. Relative agronomic performance of different *Dioscorea* species found in different parts of Orissa. *Nat. Sci.*, **7** (3): 23–35.

25. Bekele, A. and E. Bekele, 2018. Proximate and mineral composition variability in Ethiopian yam (*Dioscorea* spp.). *J. Food Nutr. Sci.*, **6**: 12–7.

26. Bhandari, M.R., T. Kasai and j. Kawabata, 2003. Nutritional evaluation of wild yam (*Dioscorea* spp.) tubers of Nepal. *Food Chem.*, **82**: 619–23.

27. Bhandari, M.R. and J. Kawabata, 2004. Organic acid, phenolic content and antioxidant activity of wild yam (*Dioscorea* spp.) tubers of Nepal. *Food Chem.*, **88**: 163–8.

28. Bhandari, M.R. and J. Kawabata, 2005. Bitterness and toxicity in wild yam (*Dioscorea* spp.) tubers of Nepal plant foods. *Hum. Nutr.*, 60: 129–35.
29. Bhatt, V.P. and G.C.S. Negi, 2006. Ethnomedicinal plant resources of Jaunsari tribe of Garhwal Himalaya, Uttranchal. *Indian J. Trad. Knowl.*, **5**: 331–5.
30. Blench, R. 2006. Archaeology, language and the African past. Altamira Press.
31. Borgen, L. 1974. Chromosome numbers of Macaronesian flowering plants. II. *Norwegian. J. Bot.* **21**: 195– 210.
32. Bousalem, M., G. Arnau, I. Hochu, R. Arnolin, V. Viader and S. Santoni, 2006. Microsatellite segregation analysis and cytogenetic evidence for tetrasomic inheritance in the American yam *Dioscorea trifida* and a new basic chromosome number in the Dioscoreae. *Theor. Appl. Genet.*, **113** (3): 439–51.
33. Bousalem, M., V. Viader, C. Mariac, R.M. Gomez, I. Hochu and S. Santoni, 2010. Evidence of diploidy in the wild Amerindian yam, a putative progenitor of the endangered species *Dioscorea trifida* (Dioscoreaceae). *Genome*, **53**(5): 371–83.
34. Caddick, L.R., P. Wilkin, P.J. Rudall, T.A.J. Hedderson and M.W. Chase, 2002. Yams reclassified: a recircumscription of Dioscoreaceae and Dioscoreales. *Taxon*, **51**(1): 103–14.
35. Chaudhury, S., C.H. Rahaman, H. Singh, K. Chaudhuri, B. Pillai and T. Seal, 2018. *Dioscorea alata*: a potent wild edible plant consumed by the Lodha Tribal community of West Bengal, India. *J. Pharmacogn. Phytochem.*, **7**: 654–63.
36. Chen, X., X. Li, X. Mao, H. Huang, T. Wang, Z. Qu, J. Miao and W. Gao, 2017. Effects of drying processes on starch-related physicochemical properties, bioactive components and antioxidant properties of yam flours. *Food Chem.*, **224**: 224–32.
37. Cheng, W.Y., Y.H. Kuo and C.J. Huang, 2007. Isolation and identification of novel estrogenic compounds in yam tuber (*Dioscorea alata* cv. Tainung No 2) *J. Agr. Food Chem.*,**55**: 7350–58.
38. Chin, H.C., M.C. Chang, P.P. Ling, C.T. Ting and F.P. Dou, 1985. A cytotaxonomic study on Chinese *Dioscorea* L. The chromosome numbers and their relation to the origin and evolution of the genus. *Acta Phytotax. Sinica*, **23**(1): 11–8.
39. Chiu, C.S., J.S. Deng, H.Y. Chang, Y.C. Chen, M.M. Lee, W.C. Hou, C.Y. Lee, S.S. Huang and G.J. Huang, 2013. Antioxidant and anti-inflammatory

properties of Taiwanese yam (*Dioscorea japonica* Thunb. var. *pseudojaponica* (Hayata) Yamam.) and its reference compounds. *Food Chem.*, **141**: 1087–96.

40. Choudhary, K., M. Singh and U. Pillai, 2008. Ethno-botanical survey of Rajasthan-An update. *Am. Eurasian J. Bot.* **1:** 38–45.

41. Chu, E.P. and R.C.L. Figueiredo-Ribeiro, 1991. Native and exotic species of *Dioscorea* used as food in Brazil. *Econ. Bot.*, **45**: 467–79.

42. Cox, D.K., A.H. Corzo, E. Matuda and J.G.G. Durán, 1958. Estudio de las *Dioscoreas mexicanas*. I. *Dioscorea spiculiflora* Hemsl. *Bol. Soc. Bot. Mex.*, **22**: 12–27.

43. Degras, L. 1993. The Yam: A Tropical Crop (ed. Réné, C.). Macmillan Press Ltd, London, 409p.

44. *Dioscorea fordii* in Flora of China @ efloras.org". www.efloras.org.

45. Djerassi, C. 1992. Steroid research at Syntex: the pill and cortisone. *Steroids*, **57**(12): 631–41. doi: 10.1016/0039-128X(92)90016-3. PMID 1481227. S2CID 5933910.

46. Dong, M., X.Z. Feng, L.J. Wu, B.X. Wang and T. Ikejima, 2001. Two new steroidal saponins from the rhizomes of *Dioscorea panthaica* and their cytotoxic activity. *Planta Med.*, **67**: 853–7.

47. *Drugs.com. 24 July 2020.* Retrieved 15 March 2021 (Wild yam).

48. Dufie, W.M.F., I. Oduro, W.O. Ellis, R. Asiedu and B. Maziya-Dixon, 2013. Potential health benefits of water yam (*Dioscorea alata*). *Food Funct.*, **4**: 1496–1501.

49. Duke, J.A., L. Judith and DuCellier, 1993. Handbook of Alternative Cash Crops. CRC Press, USA.

50. Dutta, B. 2015. Food and medicinal values of certain species of *Dioscorea* with special reference to Asom. *J. Pharmacog. Phytochem.*, **3:** 15–8.

51. Edison, S., M. Unnikrishnan, B. Vimala, S.V. Pillai, M.N. Sheela and M.T. Sreekumari, 2006. Biodiversity of tropical tuber crops in India. National Biodiversity Authority, Chennai (India).

52. Estiasih, T., M.W. Apriliyanti, A. Afriliana and J. Kusnadi, 2013. Physicochemical and bioactive characteristics of purple and yellow water yam (*Dioscorea alata*) tubers. *Int. J. PharmTech Res.*, **5**: 1691–1701.

53. Ezeabara, C.A. and R.O. Anona, 2018. Comparative analyses of phytochemical and nutritional compositions of four species of *Dioscorea*. *Acta Sci. Nutr. Health*, **2**: 90–94.

54. Ezeocha, V.C., J.S. Nwogha, A.N. Ohuoba and L.I. Chukwu, 2014. Evaluation of poultry manure application rates on the nutrient composition of *Dioscorea bulbifera* (Aerial yam) Niger. *Food J. Off.*, **32**: 92–6.

55. George, J. 1990. Effect of minisett sizes and nursery media on the sprouting of yams. *J. Root Crops*, **16**(2): 71–5.

56. Girach, R.D., A.A. Shaik, S.S. Singh and M. Ahmad, 1999. The medicinal flora of Similiphar forests, Orissa state, India. *J. Ethnopharmacol.*, **65**: 165–72.

57. Govaerts, R., P. Wilkin and R.M.K. Saunders, 2007. World Checklist of Dioscoreales. Yams and their allies: pp. 1-65. The Board of Trustees of the Royal Botanic Gardens, Kew, England.

58. Ikiriza, H., P.E. Ogwang, E.L. Peter, O. Hedmon, C.U. Tolo and M. Abubaker, 2019. *Dioscorea bulbifera*, a highly threatened African medicinal plant, a review. *Cogent Biol.*, **5**(1): 16-31.

59. Itoh, M. 2015. Yamaimo: Japan's slimy mountain yam. *The Japan Times* (15/07/2015). Retrieved 07-06-2019.

60. Jadhav, V.D., S.D. Mahadkar and S.R. Valvi, 2011. Documentation and ethnobotanical survey of wild edible plants from Kolhapur District. *Rec. Res. Sci. Technol.*, **3:** 58–63.

61. Jain, A., S.S. Katewa, P. Galav and A. Nag, 2008. Some therapeutic uses of biodiversity among the tribal of Rajasthan. *Indian J. Tradit. Know.*, **7:** 256–62.

62. Jensen, H.W. 1937. Meiosis in several species of dioecious Monocotyledoneae I. The possibility of sex- chromosomes. *Cytologia*, **1**: 96–103.

63. Jesus, M., A.P. Martins, E. Gallardo and S. Silvestre, 2016. Diosgenin: Recent highlights on pharmacology and analytical methodology. *J Analyt. Meth. Chem.*, **16**: 1–16.

64. Jiang, Q., W. Gao, X. Li, Y. Xia, H. Wang, S. Wu, L. Huang, C. Liu and P. Xiao, 2012. Characterization of starches isolated from five different *Dioscorea* L. species. *Food Hydrocoll.*, **29**: 35–41.

65. Kalu, B.A., J.C. Norman, V.R. Pal and D.K. Adedzwa, 1989. Seed yam multiplication by the mini-sett technique in three yam species in a tropical guinea savanna location. *Exper. Agric.*, **25:** 181–8.

66. Kamble, S.Y., S.R. Patil, P.S. Sawant, S. Sawant, S.G. Pawar and E.A. Singh, 2010. Studies on plants used in traditional medicines by Bhilla tribe of Maharashtra. *Indian J. Tradit. Know.*, **9**: 591–8.

67. Kang, T.H., E. Moon, B.N. Hong, S.Z. Choi, M. Son, J.H. Park and S.Y. Kim, 2011. *Biol Pharm Bull.*, **34**(9): 1493-8.

68. Kennerley, J.A. and P. Wilkin, 2017. *Dioscorea alatipes*. IUCN Red List of Threatened Species.

69. Kennerley, J.A., P. Wilkin and S. Contu, 2017. *Dioscorea ovinala*. IUCN Red List of Threatened Species.

70. Kikuno, H., R. Matsumoto, H. Shiwachi, H. Youohara and R. Asiedu, 2007. Comparative effects of explants sources and age of plant on rooting, shooting and tuber formation of vine cutting of yams. *Jap. J. Trop. Agric.*, **51**(2): 71–2.

71. Kim, N., S.H. Kim, Y.J. Kim, J.K. Kim, M.K. Nam, H. Rhim, S.K. Yoon, S.Z. Choi, M. Son and S.Y. Kim, 2011. Neurotrophic activity of DA-9801, a mixture extract of *Dioscorea japonica* Thunb. and *Dioscorea nipponica* Makino, *in-vitro*. *J. Ethnopharmacol.*, **137**: 312–9.

72. Kumar, S., G. Das, H.S. Shin and J.K. Patra, 2017. *Dioscorea* spp (a wild edible tuber): a study on its ethnopharmacological potential and traditional use by the local people of Similipal Biosphere Reserve, India. *Front. Pharmacol.*, **8**: 52.

73. Kumar, S., P.K. Jena and P.K. Tripathy, 2012. Study of wild edible plants among tribal groups of Similipal Biosphere Reserve forest, Odisha, India; with special reference to *Dioscorea* species. *Int. J. Biol. Technol.*, **3:** 11–9.

74. Kumar, S., G. Das, H.S. Shin and J.K. Patra, 2017. *Dioscorea* spp. (A Wild Edible Tuber): A Study on its Ethnopharmacological Potential and Traditional Use by the Local People of Similipal Biosphere Reserve, India. *Front. Pharmacol.*, **8**: 52. doi: 10.3389/fphar.2017.00052.

75. Kwon, Y.K., E.Y. Jie, A. Sartie, D.J. Kim, J.R. Liu, B.W. Min and S.W. Kim, 2015. Rapid metabolic discrimination and prediction of dioscin content from African yam tubers using fourier transform-infrared spectroscopy combined with multivariate analysis. *Food Chem.*, **166**: 389–96.

76. Lebot, V., R. Malapa and K. Abraham, 2017. The Pacific yam (*Dioscorea nummularia* Lam.), an under- exploited tuber crop from Melanesia. *Genet. Resour. Crop Evol.*, **64**(1): 217–35.

77. Lee, H.J., B. Watanabe, M. Nakayasu, M. Onjo, Y. Sugimoto and M. Mizutani, 2017. Novel steroidal saponins from *Dioscorea esculenta* (Togedokoro). *ýBiosci. Biotechnol. Biochem.*, **81**: 2253–60.

78. Liu, M.J., Z. Wang, Y. Ju, R.N. Wong and Q.Y. Wu, 2005. Diosgenin induces cell cycle arrest and apoptosis in human leukemia K562 cells with the disruption of Ca2+ homeostasis. *Cancer Chemoth. Pharmac.*, **55**(1): 79–90.

79. Liu, H., K.W. Tsim, G.X. Chou, J.M. Wang, L.L. Ji and Z.T. Wang, 2011. Phenolic compounds from the rhizomes of *Dioscorea bulbifera*. *Chem. Biodivers.*, **8**: 2110–16.
80. Maiden, J.H. 1889. The Useful Native Plants of Australia including Tasmania. Turner and Henderson Publishers. Sydney (Australia).
81. Marker, R.E. and J. Krueger, 1940. Sterols. CXII. Sapogenins. XLI. The Preparation of Trillin and its Conversion to Progesterone. *J. Amer. Chem. Soc.*, **62**(12): 3349–50.
82. Martin, F.W. and S. Ortiz, 1963. Chromosome numbers and behavior in some species of *Dioscorea*. *Cytologia*, **28**(1): 96–101.
83. Martin, F.W. and S. Ortiz, 1966. New chromosome numbers in some *Dioscorea* species. *Cytologia*, **31**(1): 105–7.
84. Mbiantcha, M., A. Kamanyi, R.B. Teponno, A.L. Tapondjou, P. Watcho and T.B. Nguelefack, 2011. Analgesic and anti-inflammatory properties of extracts from the bulbils of *Dioscorea bulbifera* L. var *sativa* (Dioscoreaceae) in mice and rats. *Evid-Based Compl. Alt.*, **9**: 129-35.
85. Meena, K.L. and B.L. Yadav, 2011. Some ethnomedicinal plants used by the Garasia tribe of district Sirohi, Rajasthan. *Indian J. Trad. Know.*, **10**, 354–7.
86. Mehra, P.N. and S.K. Sachdeva, 1976. Cytological observations on some W. Himalayan Monocots IV. Several families. *Cytologia*, **41**(1): 31–53.
87. Meurman, O. 1925. The chromosomal behavior of some dioecious plants and their relatives with special reference to the sex chromosomes. *Soc. Scient. Fenn. Comm. Biol. II.*, **3**: 1–105.
88. Miege, J. 1954. Nombres chromosomiques et repartition geographique de quelques plantes tropicales et equatoriales. *Rev. Cytol. Biol. Vég. Bot.*, **15**(4): 312–48.
89. Mishra, R.K., V.P. Upadhyay and R.C. Mohanty, 2008. Vegetation ecology of the Similipal Biosphere Reserve, Orissa, India. *Appl. Ecol. Environ. Res.*, **6:** 89–99.
90. Mishra, S., S. Swain, S.S. Chaudhary and T. Ray, 2008. Wild edible tubers (*Dioscorea* spp.) and their contribution to the food security of tribes of Jaypore tract, Orissa, India. *Plant Genet. Resour.*, **156:** 63–7.
91. Mohan, V.R. and C. Kalidass, 2010. Nutritional and antinutritional evaluation of some unconventional wild edible plants. *Trop. Subtrop. Agroecosystems.*, **12**: 495–506.

92. Murty, Y.S. and Purnima, 1983. Morphology, anatomy and development of bulbil in some dioscoreas. *Proc. Indian Acad. Sci.* (*Plant Sci.*), **92**(6): 443–9.

93. Muzac-Tucker, I., H.N. Asemota and M.H. Ahmad, 1993. Biochemical composition and storage of Jamaican yams (*Dioscorea* sp.). *J. Sci. Food Agric.*, **62**: 219–24.

94. Nag, A. 1999. A Study of the Contribution of Some Wild Food Plants to the Diet of Tribals of South East Rajasthan. Ph.D. Thesis, Mohanlal Sukhadia University, Udaipur.

95. Nakajima, G. 1933. Chromosome numbers in some angiosperms. *Jap. J. Genet.*, **9**(1): 1–5.

96. Nashriyah, M., M.Y.N. Athiqah, H.S. Amin, N. Norhayati, A.W.M. Azhar and M. Khairil, 2011. Ethnobotany and distribution of wild edible tubers in Pulau Redang and nearby islands of Tereengganu, Malaysia. *Int. J. Biol. Vert. Agric. Food Eng.*, **5**: 110–13.

97. Nayak, S., S.K. Behera and M.K. Misra, 2004. Ethno-medico botanical survey of Kalahandi District of Odisha. *Indian J. Trad. Know.*, **3**: 72–9.

98. Noda, H., J. Yamashita, S. Fuse, R. Pooma, M. Poopath and H. Tobe, 2020. A large-scale phylogenetic analysis of *Dioscorea* (Dioscoreaceae), with reference to character evolution and subgeneric recognition. *Acta Phytotax. Geobot.*, **71**(2): 103–28.

99. Nweke, Felix; Aidoo, Robert and Okoye, Benjamin, 2013. Yam Consumption Patterns in West Africa. Bill and Melinda Gates Foundation. Retrieved 5 December 2017.

100. Ogbo, A.F. and K. Agu, 2014. Proximate composition, physiological changes during storage, and shelf life of some Nigerian varieties of yams (*Dioscorea* species). *J. Sci. Res. Rep.*, **3**: 553–62.

101. Ojo, A., J.O. Odedeji, S.A. Oladapo and O.O. Olosunde, 2017. Comparative studies on the chemical compositions of cereal (wheat and corn) and tubers (water yam and bitter yam) flours. *Int. J. Innov. Sci. Eng. Technol.*, **4**: 369–85.

102. Oke, O.L. 1990. Roots, tubers, plantains and bananas in human nutrition (eds Redhead, J. and, M.A. Hussain). FAO. Rome, Italy.

103. Okoli, O.O. and M.O. Akoroda, 1995. Providing seed tubers for the production of food yams. *Africa J. Root and Tuber Crops*, **1**(1): 1–6.

104. Okonmah, L.U. 1980. Rapid multiplication of yams (Manual Sieries No. 5). International Institute of Tropical Agriculture. Ibadan, Nigeria.

105. Onwueme, I.C. 1973. The Tropical Tuber Crop: Yams, Cassava, Sweet potato, Cocoyam. English Language Book Society. Wiley and Sons, Chichester, England.

106. Otegbayo, B.O., D.J. Oguniyan, B.A. Olunlade, O.O. Oroniran and O.E. Atobatele, 2017. Characterizing genotypic variation in biochemical composition, anti-nutritional and mineral bioavailability of some Nigerian yam (*Dioscorea* spp.) land races. *J. Food Sci. Techn.*, **55** (1): 205–216.

107. Otoo, J. A. 1992. Substitutes for chemicals, sawdust and plastic mulch in improved seed yam production. Promotion of root crop-based industries: An incentive for research and development (pp. 281–284). *Proc. 4th Triennial Symp. of ISTRC-AB* (eds Akoroda, M.O. and O. B. Arene), Kinhasa.

108. Otoo, J.A., D.S.O. Osiru, S.Y. Ng and S.K. Hahn, 1987. Improved technology for seed yam production. International Institute of Tropical Agriculture, Ibadan, Nigeria, 56p.

109. Padhan, B., M. Biswas and D. Panda, 2020. Nutritional, anti-nutritional and physico-functional properties of wild edible yam (*Dioscorea* spp.) tubers from Koraput, India. *Food Biosci.*, **34**: 15-27.

110. Patil, M.V. and D.A. Patil, 2005. Ethnomedicinal practices of Nasik district, Maharashtra. *Indian J. Tradit. Know.*, **4:** 287–90.

111. Paul, C. and B. Debnath, 2019. A report on new chromosome number of three *Dioscorea* species. *Plant Sci. Today*, **6**(2): 147–50.

112. Pei, C., C.T. Ting, H.C. Chin, P. Su, S.Y. Tang and H.C. Chang, 1979. A preliminary systematic study of *Dioscorea* L. sect. *Stenophora* Uline. *Acta Phytotax. Sin.*, **17**(3): 61–72.

113. Perez, E., O. Gibert, A. Rolland-Sabaté, Y. Jiménez, T. Sánchez, A. Giraldo, B. Pontoire, S. Guilois, M.C. Lahon and M. Reynes, 2011. Physicochemical, functional, and macromolecular properties of waxy yam starches discovered from 'mapuey' (*Dioscorea trifida*) genotypes in the Venezuelan Amazon. *J. agric. Food Chem.*, **59**: 263–73.

114. Peter, K.V. 2007. Tuber Crops. National Book Trust, New Delhi, 124p.

115. Polycarp, D., E.O. Afoakwa, A.S. Budu and E. Otoo, 2012. Characterization of chemical composition and anti-nutrition factors in seven species within the Ghanaian yam (*Dioscorea*) germplasm. *Int. Food Res. J.*, **19**: 985–92.

116. Poornima, G.N. and R.V. Ravishankar, 2009. Evaluation of phytonutrients and vitamin contents in a wild yam, *Dioscorea belophylla* (Prain) Haines. *Afr. J. Biotechnol.*, **8**: 971–3.

117. Price, E.J., P. Wilkin, V. Sarasan and P.D. Fraser, 2016. Metabolite profiling of *Dioscorea* (yam) species reveals underutilised biodiversity and renewable sources for high-value compounds. *Sci. Rep.*, **6**: 29-36.

118. Raghavan, R.S. 1958. A chromosome survey of Indian dioscoreas. *Proc. Indian Acad. Sci.* B.**48**(1): 59- 63.

119. Raghavan, R.S. 1959. A note on some south Indian species of the genus *Dioscorea*. *Curr Sci.*, **28** (8): 337–8.

120. Ramachandran, K. 1962. Studies on the cytology and sex determination of the Dioscoreaceae. *J. Indian bot. Soc.*, **41**: 93–8.

121. Ramachandran, K. 1968. Cytological studies in Dioscoreaceae. *Cytologia*, **33**(3–4): 401–10.

122. Rana, M., A. Kabra, R. Kabra, M. Rana and V. Dhatwalia, 2015. Plant Species used by locals as Ethano- Medicine in Gohar Tehsil, Distt. Mandi Region of North Western Himalaya. *Pharma Tutor.*, **3**(4): 47-52.

123. Rout, S.D. and S.K. Panda, 2010. Ethnomedicinal plant resources of Mayurbhanj district, Orissa. *Indian J. Tradit. Know.*, **9:** 68–72.

124. Saikia, B., H. Tag and A.K. Das, 2010. *Dioscorea brandisi* Prain et Burkill (Dioscoreaceae) – A new record for India. *Pleione.*, **4**(1): 160–61.

125. Saleha, C., N. Saidi, S. Murniana, S. Rasnovi and T.M. Iqbalsyah, 2018. Nutritional composition of *Dioscorea hispida* from different locations around Leuser ecosystem area. *J. Nat.*, **18**: 1–6.

126. Samanta, A.K. and K.K. Biswas, 2009. Climbing plants with special reference to their medicinal importance from Midnapore Town and its adjoining areas. *J. Econom. Taxonom. Bot.*, **33:** 180–88.

127. Sautour, M., A.C. Mitaine-Offer, T. Miyamoto, A. Dongmo and M.A. Lacaille-Dubois, 2004. Antifungal steroid saponins from *Dioscorea cayenensis*. *Planta Med.*, **70**: 90–92.

128. Scarcelli, N., O. Daïnou, C. Agbangla, S. Tostain and J.L. Pham, 2005. Segregation patterns of isozyme loci and microsatellite markers show the diploidy of African yam *Dioscorea rotundata* (2n=40). *Theor. Appl. Genet.*, **111**(2): 226–32.

129. Shajeela, P.S., V.R. Mohan, L.L. Jesudas and P.T. Soris, 2011. Nutritional and antinutritional evaluation of wild yam (*Dioscorea* spp.). *Trop. Subtrop. Agroecosystems,* **14**: 723–30.

130. Shanthakumari, S., V.R. Mohan and J. de Britto, 2008. Nutritional evaluation and elimination of toxic principles in wild yam (*Dioscorea* spp.). *Trop. Subtrop. Agroecosystems,* **8**: 319–25.

131. Sharlina, M.S.E., W.A. Yaacob, A. Mat, S. Fazry, S. Joe, S. Abdullah, A. Noordin and M. Kumaran, 2017. Physicochemical properties of starch from *Dioscorea pyrifolia* tubers. *Food Chem.*, **220**: 225–32.

132. Sharma, A.K. and D.N. De, 1956. Polyploidy in *Dioscorea. Genetica*, **28**(1): 112–20.

133. Sharma, L.N. and R. Bastakoti, 2009. Ethnobotany of *Dioscorea* L. with emphasis on food value in Chepang communities in Dhading District, central Nepal. *Botanica Orientalis. J. Plant Sci.*, **6:** 12–7.

134. Sharma, M., V. Hotpet, B. Sindhura, A, Kamalanathan, B.M. Swamy and S.R. Inamdar, 2017. Purification, characterization and biological significance of mannose binding lectin from *Dioscorea bulbifera* bulbils. *Int. J. Biol. Macromol.*, **102**: 1146–55.

135. Sheela, M.N. 2019. Yams. In: Handbook of Horticulture, vol.1, pp. 680-82. ICAR, New Delhi.

136. Sheikh, N., Y. Kumar, A.K. Misra and L. Pfoze, 2013. Phytochemical screening to validate the ethnobotanical importance of root tubers of *Dioscorea* species of Meghalaya, North East India. *J. Med. Plant Stud.*, **1:** 62–9.

137. Singh, N., Y.P.S. Pangtey, S. Khatoon and A.K.S. Rawat, 2009. Some ethnobotanical plants of Ranikhet region, Uttaranchal. *J. Econ. Taxon. Bot.*, **33:** 198–204.

138. Smith, B.W. 1937. Notes on the cytology and distribution of the Dioscoreaceae. *Bull. Torrey Bot. Club*, **64**(4): 189–97.

139. Sonibare, M.A. and R.B. Abegunde, 2012a. *In vitro* antimicrobial and antioxidant analysis of *Dioscorea dumetorum* (Kunth) Pax and *Dioscorea hirtiflora* (Linn.) and their bioactive metabolites from Nigeria. *J. Appl. Biosci.*, **51**: 3583–90.

140. Sonibare, M.A. and R.B. Abegunde, 2012b. Ethnobotanical study of medicinal plants used by the Laniba village people in South Western Nigeria. *Afr. J. Pharm. Pharmacol.*, **6:** 1726–32.

141. Soto, L.J.M., J.V. González, A.B. Nicanor, L.G. Cruz and J.Y. Fernández, 2014. Chemical characterisation and nutritional evaluation of mountain yam (*Dioscorea remotiflora* Kunth) tubers. *Adv. Biores.*, **5**: 153–60.

142. Srivastava, R.C. and C. Nyishi, 2010. Traditional knowledge of Nyishi (Daffla) tribe of Arunachal Pradesh. *Indian J. Tradit. Know.*, **9:** 26–37.

143. Suessenguth, 1921. Bemerkungen zur meiotischen und somatischen Kernteilung bei einigen Monokotylen. *Flora Allg Bot Ztg.*, **114**(3–4): 313–28.

144. Sugihara, Y., A. Kudoh, M. Tamiru, H. Takagi, S. Natsume, M. Shimizu, A. Abe, R. Asiedu, A. Asfaw, P. Adebola and R. Terauchi, 2021. Population Genomics of Yams: Evolution and Domestication of *Dioscorea* Species. *Jap. J. Trop. Agric.*, **65**(2): 58-62.

145. Tabopda, T.K., A.C. Mitaine-Offer, C. Tanaka, T. Miyamoto, J.F. Mirjolet, O. Duchamp, B.T. Ngadjui and M.A. Lacaille-Dubois, 2014. Steroidal saponins from *Dioscorea preussii. Fitoterapia*, **97**: 198–203.

146. Takeuchi, Y., T. Iwao and A. Akahori, 1970. Chromosome numbers of some Japanese *Dioscorea* species. *Acta Phytotax Geobot.*, **24**(4–6): 168–73.

147. Tamiru, M. 2006. Assessing Diversity in Yams (*Dioscorea* spp.) from Ethiopia Based on Morphology, AFLP Markers and Tuber Quality, and Farmers' Management of Landraces. Ph.D. Thesis, Georg August University, Goettingen.

148. Tang, Y., Y.B. Xue, L. Zhou, J.W. Zhang, G.M. Yao, Z.W. Luo, G. Du and Y.H. Zhang, 2014. New norclerodane diterpenoids from the tubers of *Dioscorea bulbifera. Chem. Pharm. Bull.*, **62**: 719–24.

149. Teron, R. 2011. Studies on Ethnobotany of Karbi-Anglong District, Asom: Trans-cultural Dynamism in Traditional Knowledge. Ph.D. Thesis, Guwahati University, Asom.

150. Thompson, A.K. 2014. Chinese yam, Fruits and Vegetables: Harvesting, Handling and Storage. John Wiley and Sons, 1429p.

151. Tiwari, L. and P.C. Pande, 2006. Indigenous veterinary practices of Darma Valley of Pithoragrah district, Uttranchal. *Indian J. Tradit. Know.*, **5:** 201–6.

152. Udensi, E.A., H.O. Oselebe and O.O. Iweala, 2008. Investigation of chemical composition and functional properties of water yam (*Dioscorea alata*): Effect of varietal differences. *Pakistan J. Nutr.*, **7**: 342–4.

153. Uwaegbute, Osho and Obatolu, 1998. Postharvest technology and commodity marketing: *Proceedings of a Postharvest Conference.* International Institute of Tropical Agriculture, 172p.

154. Wadkar, K.A., C.S. Magdum, S.S. Patil and N.S. Naikwade, 2014. Anti-diabetic potential and Indian medicinal plants. *J. Herb. Med. Toxicol.*, **2:** 45–50.

155. Wanasundera, J.P.D. and G. Ravindran, 1994. Nutritional assessment of yam (*Dioscorea alata*) tubers. *Plant Foods Hum. Nutr.*, **46**(1): 33–9.

156. Wilkin, P. 2001. Dioscoreaceae of South-central Africa. *Kew Bull.*, **56**: 361–404.

157. Wilkin, P., Mamy Tiana Rajaonah, Vololoniaina Harimanga Jeannoda, Annette Hladik, Victor Louis Jeannoda and Claude Marcel Hladik, 2008. An endangered new species of edible yam (*Dioscorea*, Dioscoreaceae) from Western Madagascar and its conservation. *Kew Bull.*, **63** (1): 113–20.

158. Wu, Z.G., W. Jiang, M. Nitin, X.Q. Bao, S.L. Chen and Z.M. Tao, 2016. Characterizing diversity based on nutritional and bioactive compositions of yam germplasm (*Dioscorea* spp.) commonly cultivated in China. *J. Food Drug Anal.*, **24**: 367–75.

159. Xiaoqin, Z, L. Guolu anf L.A. Xiaolin, 2003. A comparison among natural variations of *Dioscorea zingiberensis. J. Trop. Subtrop. Bot.*, **11**(3): 267–70.

160. Zhang, X., M. Jin, N. Tadesse, J. Dang, T. Zhou, H. Zhang, S. Wang, Z. Guo and Y. Ito, 2018. *Dioscorea zingiberensis* C. H. Wright: An overview on its traditional use, phytochemistry, pharmacology, clinical applications, quality control, and toxicity. *J. Ethnopharmac.*, **220**: 283–93.

161. Zhao, D.S., Z.T. Wu, Z.Q. Li, L.L. Wang, L.L. Jiang, W. Shi, P. Li and H.J. Li, 2018. Liver-specific metabolomics characterizes the hepatotoxicity of *Dioscorea bulbifera* rhizome in rats by integration of GC-MS and 1H-NMR. *J. Ethnopharmacol.*, **226**: 111–9.

162. Zhu, F. 2015. Isolation, composition, structure, properties, modifications, and uses of yam starch. *Compr. Rev. Food Sci. Food Saf.*, **14**: 357–86.

163. Zulu, D., R. Ellis and A. Culham, 2020. Propagation of lusala (*Dioscorea hirtiflora*), a wild yam, for *in situ* and *ex situ* conservation and potential domestication. *Exper. Agr.* ISSN 0014-4797 https: //doi.org/10.1017/ S0014479720000083.

34

Yellow Nutsedge (*Cyperus esculentus* L.)

Family: Cyperaceae (Sedge family).

Common names: Atadwe, Chufa, Chufa flatsedge, Earth/Ground almond, Earth nut, Rush nut, Tiger nut and Yellow nutsedge or Zulu nut. In **Hindi**: Chichoda or Kaseru; in **Bengali**: Motha.

Synonyms: *Chlorocyperus aureus* (K.Richt.) Palla ex Kneuck., *C. phymatodes* (Muhl.) Palla, *Cyperus aureus* Ten, *C. aureus* (L.) Nyman, *C. bahiensis* Steud., *C. buchananii* Boeckeler, *C. callistus* Ridl., *C. chrysostachys* Boeckeler, *C. cubensis* Steud., *C. damiettensis* A.Dietr., *C.esculentus* var. *angustispicatus* Britton, *C.e.* f. *angustispicatus* (Britton) Fernald, *C.e.* subsp. *aureus* K.Richt., *C.e.* var. *cyclolepis* Boeckeler ex Kük., *C.e.* f. *evolutus* C.B.Clarke, *C.e.* var. *heermannii* (Buckley) Britton, *C.e.* var. *helodes* (Schrad. ex Nees) C.B.Clarke, *C.e.* var. *leptostachyus* Boeckeler, *C.e.* var. *lutescens* (Torr. and Hook.) Kük. ex Osten, *C.e.* var. *lutescens* (Torr. and Hook.) Kük., *C.e.* var. *macrostachyus* Boeckeler, *C.e.* f. *macrostachyus* (Boeckeler) Fernald, *C.e.* var. *phymatodes* (Muhl.) Kük., *C.e.* f. *princeps* C.B.Clarke, *C.e.* var. *sativus* Boeckeler, *C.e.* var. *sprucei* C.B.Clarke, *C. fresenii* Steud., *C. fulvescens* Liebm., *C. gracilescens* Schult., *C. gracilis* Link, *C. heermannii* Buckley, *C. helodes* Schrad. ex Nees, *C. hydra* Kunth, *C. lutescens* Torr. and Hook., *C. melanorhizus* Delile, *C. nervosus* Bertol., *C. officinalis* T.Nees, *C. pallidus* Savi, *C. phymatodes* Muhl., *C.p.* var. *heermannii* (Buckley) S.Watson, *C. repens* Elliott, *C. ruficomus* Buckley, *C. scirpoides* R.Br., *C. sieberianus* Link, *C. tenoreanus* Schult., *C. tenorei* C.Presl, *C. tenorianus* Roem. and Schult., *C. tuberosus* Pursh,

C. variabilis Salzm. ex Steud., *Pterocyperus esculentus* (L.) Opiz, and *Pycreus esculentus* (L.) Hayek.

Introduction and Origin

The name *Cyperus* derives from the Greek 'kyperos'. The word *esculentus* comes from Latin words *esca* (food) and *ulentus* (full of, abounding in). **Chufa**, probably originated in the Mediterranean area and western Asia, from where it has spread to many other parts of the world where it is often found in the wet soils such as paddies, and peanut farms as well as well-irrigated lawns and golf courses during warm weather. It can grow in a very wide range of climatic conditions and occurs in the tropics, subtropics and warm temperate regions. The plant got its common name 'earth almonds' and 'tiger nuts' due to its hard shell and the stripes on tubers. Chufa tubers are eaten as vegetables, made into sweets or used to produce 'horchata' of the Valencia region. Due to its clonal nature, *C. esculentus* can take advantage of soil disturbances caused by anthropogenic or natural forces (Renne Ian and Tracy Benjamin, 2006; Oldfield Callie and Evans Jonathan, 2016).

Although cultivated for its edible tubers in several countries, it is an invasive species outside its native range and is stated to be the 15th worst weed in the world. *Cyperus esculentus* and *C. rotundus* are considered the most serious weeds of Africa (Little, 1965). These two species, grouped as a single entity, were placed on a list of the 10 worst weeds of field crops in the United States by the editors of Crops and Soils Magazine (Lewis and Worsiiam, 1970).

Botany

There are reports that chromosome numbers in *Cyperus esculentus* are n=54 or 108 (Hicks, 1929; Federov, 1969; Rath and Patnaik, 1978; Roalson, 2008) and 2n=96 and 108. Grass-like **Chufa** is an erect annual or perennial sedge growing to a height of usually 30-90cm with slender, long, narrow (3–10mm wide) and dark-green leaves arranged in three rows around the solitary triangular stem growing from a tuber. The plant develops a series of shoots and stem tubers connected by brown wiry rhizomes which are strengthened by lignification of the inner cortex. Basal tubers grow from rhizome tips producing shoot growth and new plants. The tough and fibrous foliage is often mistaken for a grass. The roots form an extensive and complex system of fine and fibrous roots and small, scaly, hard and spherical rhizomes.

Spikelets are linear to narrowly elliptic with pointed tips having 8 to 35 florets, colour varying from straw- to gold-brown. Spikelets are 5mm-3cm long bearing a cluster of flat *cum* oval seeds surrounded by four hanging leaf-like bracts positioned at 90° from each other. They can produce up to 2420 seeds per plant. *C. esculentus* is wind-pollinated and requires cross pollination as it is self–incompatible (en.wikipedia.org).

During a single growing season one plant can produce hundreds to thousands of small rhizomes that are 0.3-2.5cm (generally 1-2cm) in diameter, and are borne at intervals along the rhizomes. Its colour varies from yellow to brown and black. Rhizomes have very thin skins and the flesh is slightly yellowish-white in the newly-formed ones but darken with increasing maturity and the flavour is sweet and nutty.

Nutritional Value

Rhizomes are rich in both starch and oil though variable in composition. Dry matter of fresh thizomes is 70-90 per cent. Average analytical figures of the dry matter are residual moisture 9.3 per cent, protein 8.6 per cent, fat 21.8 per cent, carbohydrate 48 per cent, ash 1.7 per cent, Mg 0.1 mg/100g, P 211.5mg/100g and K 0.5mg/100g. Out of a total of 48 per cent carbohydrate, 33.4 per cent is starch and 14.6 per cent total sugars. Quinones occur and have been used as an aid to the classification of the genus *Cyperus.*

Uses

Dried tiger nut has a smooth, tender, sweet and nutty taste compared to the more bitter-tasting tuber of the related *Cyperus rotundus* (purple nutsedge). Dried tiger nuts have a hard texture and soaking is required to render them more easily edible and to ensure acceptable sensory quality. The tubers can be eaten raw, baked as a vegetable, roasted like groundnuts or grated and used to make icecream, *vis-a-vis* dried and ground into flour, sherbets or a milky beverage which in Spain and Latin American countries is known as 'horchata' (en.wikipedia.org).

Flour

Flour of roasted tiger nut can be added to biscuits and other bakery products as well as in making oil, soap and starch extracts. It is also used for production of nougat, jam, beer and as a flavouring agent in icecream. The flour contains approximately 3.4 per cent protein, 27 per cent fat, 38 per cent starch and 2.5 per cent ash and can be mixed with wheat flour in baking.

Milk

Tiger nut 'milk' has been tried as an alternative source of milk in fermented products such as yogurt production. To a certain extent it is a useful replacement of dairy milk in the diet of people intolerant to lactose (Sanchez-Zapata *et al.*, 2012).

Horchata Beverage

In Spain 'horchata' continues a popular beverage in spite of severe competition from carbonated drinks. As per the *Consejo Regulador de Chufa de Valencia* 2022 (Regulating Council for Valencia's Tiger Nuts), the nutritional composition per100ml of the Spanish beverage 'horchata de chufas' is energy content around 66kcal, proteins around 0.5g, carbohydrates over 10g with starch at least 1.9g and fats at least 2g.

Oil

Rhizomes contain 20-36 per cent of non-drying pleasantly flavoured yellow oil similar to olive or sweet almond oil. It is used in Spain and Italy for culinary purposes and for manufacture of soap. The rhizome oil contains 17-18 per cent saturated (approximately 12 per cent palmitic acid and 5 per cent stearic acid) and 82-83 per cent unsaturated (about 75 per cent oleic acid and 6 per cent linoleic acid) fatty acids (Elias and Dykeman, 2009; en.wikipedia.org) though this varies to some extent with growing conditions. Tiger nut oil is considered to be high quality oil which can be used naturally with salads or for deep frying. Chufa oil is also resistant to oxidative changes.

Starch

Rhizomes are potentially a rich source of starch which may be extracted after the oil has been removed from the tubers. Its starch is a white flavourless product and when heated in water forms a transparent gelatinous paste. The starch has approximately 9 per cent moisture, 0.3 per cent nitrogenous material *cum* traces of fat, 89.8 per cent starch, 0.3 per cent cellulose and 0.5 per cent ash.

Subsidiary Uses

Chufas can be used as animal feed and in parts of the southern USA these are grown as pig feed. They are sometimes used in certain types of confectionery often as a substitute for almonds. The ground ones are sometimes used as a substitute or adulterant of coffee and cocoa.

Alcohol

The tubers can be used for production of alcohol through fermentation. In Sicily, a cultivar with a very high sucrose content is grown and used commercially for this purpose.

Propagation

The plant is reproduced by seeds, creeping rhizomes and tubers. In cultivation, usually tubers are used for propagation. They are soaked in water for 24-36 hours and then planted by hand or through drill. Sprouting can be stimulated by treatment with ethylene at 3-10ppm in air or by ethephon (ethrel) at 10-100ppm in water.

Growth and Development

Many rhizomes initiate when tubers sprout and this ends in a basal tuber near the soil surface. These basal tubers initiate the stems and leaves aboveground and fibrous roots underground. The roots form an extensive and complex system of fine, fibrous roots and scaly rhizomes with small and hard spherical tubers with basal bulbs attached. Tubers develop about 6–8 weeks after seedling emergence, growing quickly during July and August and maturing in 90–110 days. All parts die with cool temperatures except the tubers which resprout the following spring when soil temperatures remain above 6°C (Stoller, 1981) and these tubers can resprout up to several years later (Stoller and Wax, 1973).

Although **environmental factors** such as temperature, moisture level, light and soil type influence the growth and quality, other factors such as ground preparation, harrowing and ploughing, varieties, quality of planting material, planting method and date, planter operation, irrigation, nutrition, pest management, harvest date, *etc.* are equally responsible and can be controlled by the grower. The most profitable yields can be attained only when all the factors are at optimum levels. Chufas normally take 3-4 months to reach maturity.

Although the plant can be grown in relatively cool climates, optimum yields are obtained with moderately high temperatures throughout the growing season and a moderate well-distributed rainfall. Low temperature, shade and light intensity can inhibit flowering. Tuber initiation is inhibited by high levels of nitrogen, long photoperiods and high levels of gibberellic acid. Flower initiation occurs under photoperiods of 12-14h/day. Although chufas are fairly drought resistant, commercial crops receive copious and repeated irrigations during dry summer season as in Valencia (Spain) in order to ensure maximum yield.

Chufas prefer light sandy loams with pH range of 5.0-7.5 (Anon., 2003) but can be grown on any well-drained soil. Tubers can develop in soil depths around 30cm, but most occur in the top or upper part. They tolerate many adverse soil conditions including periods of drought and flooding and survive soil temperatures as low as –6°C. Alluvial sands containing relatively high quantities of Mn, S, Ca, Mg and B are particularly suitable. The densest populations of *C. esculentus* are often found in low-lying wetlands (Ross Merrill and Lembi Carole, 2008).

There is little information on the precise fertiliser requirements of chufas but nitrogen is often limiting under natural conditions. The application of a 6:6:8 complete (NPK) fertilizer at the rate of 1,000 kg/ha has been recommended. In many areas growers give a heavy dressing of FYM (if available) and wood ash before or immediately after planting. In Florida, however, the crop has been found to respond very erratically to applications of fertilizers.

Planting

Planting is usually done between April and May followed by irrigation every week until they are harvested in November and December. Field spacing for chufas vary greatly depending on the soil conditions, local cultural methods and the purpose for which the crop is being grown. Chufa is generally planted on previously tilled flat soils with ridges facilitating irrigation. Seeds or tubers are planted manually on these ridges which are around 60-90cm apart. Distances between seeds may vary from 15 to 30cm with some 2.5-8cm seeding depth. A typical seeding rate for chufa is about 120kg of tubers/ha (Pascual-Seva *et al.*, 2012, 2013).

Harvesting and Handling

Harvesting is usually done in November or December when plants begin to die back. The entire plant that is pulled out of the ground is laid on the soil and allowed to dry for 1-3 days before the tubers are separated for storage in thin layers in sheds. Tubers for human consumption are washed in running water and then dried either artificially or in the sun which can take upto three months and then they are graded and stored.

The drying prevents rot or other bacterial infections and ensures a longer shelf life. The temperatures and humidity levels have to be monitored very carefully during this period. The tiger nuts have to be turned every day to ensure uniform drying.

Tuber Yield and Storage

The average tuber yield can be anywhere between 10 and 19t/ha. During storage, the starch content of the tubers decreases and the reducing sugar content (invert sugar) increases (Coskuner *et al.,* 2002). Before consumption, the tubers are rehydrated by soaking overnight without losing the crispy texture.

Pests and Diseases

In Florida the negro bug (*Thyreocoris pulicaria*) has been reported to damage the crop where larvae develop inside the tubers. Proper crop rotation is the best means of control. Chufas are seldom affected seriously by diseases.

Other Edible Species

Various other species of *Cyperus* are also eaten though to a smaller extent. Some *Cyperus* species are used in folk medicines. Roots of Near East species were a component of *kyphi* ~ a medical incense of Ancient Egypt. Tubers of *C. rotundus* (purple nut-sedge) are used in *kampo.*

Cyperus articulatus L. 1753

Common names: Priprioca, Jointed flatsedge, Guinea rush or Adrue (Wren, 1923). **Chromosome no.** n=56, 57.

Synonyms: *Chlorocyperus articulatus* (L.) Rikli, *C. cordobensis* Palla, *Cyperus articulatus* Benth., *C.a. var. articulatus, C.a. var. conglomeratus* Britton, *C.a. var. erythrostachys* Graebn., *C.a. var. fistulosus* Kuk., *C.a. f. longispiculosus* Kuk., *C.a. var. multiflorus* Kuk., *C.a. var. nodosus* (Humb. and Bonpl. ex Willd.) Kuk., *C. autumnalis* Pursh, *C. borbonicus* Steud., *C. cordobensis* (Palla) Hicken, *C. corymbosus var. pangorei* Rottb., *C.c. var. subnodosus* (Nees and Meyen) Kuk. ex Osten, *C.c. var. subnodosus* (Nees and Meyen) Kuk., *C. fistulosus* Ehrenb. ex Boeckeler, *C. gymnos* Schult., *C. interceptus* Steud., *C. niloticus* Forssk., *C. nodosus* Humb. and Bonpl. ex Willd., *C.n. var. aphyllus* Boeckeler, *C.n. var. subnodosus* (Nees and Meyen) Boeckeler, *C. pertenuis* Roxb., *C. subarticulatus* Nees and Meyen, *C. subnodosus* Nees and Meyen, and *Papyrus pangorei* Nees.

Priprioca being widespread across tropical and subtropical regions in Africa, southern Asia, northern Australia, the SE United States, the West Indies and Latin America grows in water or near rivers, streams, lakes and swamps with a hyperhydrate (emergent aquatic) or possibly tenagophyte (submerged juvenile and terrestrial adult) growth pattern (Cook Christoper, 1996; Uphof, 1968). While it is closely related to highly invasive sedges such as purple nut sedge (*C. rotundus*), priprioca is less prolific and competitive than its relative.

Uses

It is one of the traditional spices of the Amazon region and its reddish essential oil is used commercially both by the cosmetic industry and as flavouring for food (Azambuja, 2017). Interest is increasing in the larger, fast-growing species as crops for paper and bio-fuel production. Related to other nut sedges (such as tigernut), priprioca roots release a light, woody and spicy fragrance with floral notes. It is also used medicinally in local tradition. The essential oil has antifungal properties and its main components are mustakone, alpha-pinene, beta-pinene, caryophyllene oxide, trans-pinocarveol, myrtenal/myrtenol, ledol, cyperotundone and alpha-cyperone, though the oil is very complex and no single element plays a dominant role over the others. According to local Amazonian medical tradition, priprioca may be harmful to pregnant women as it may cause adverse effects including possibly the miscarriage (Azambuja, 2017).

Cyperus bulbosus Vahl

Common names: Australian bush onion, Nalgoo or Wild onion though at all it is not related to the onion or other members of Alliaceae.

Australian bush onion is found across Africa, the Middle East, Indian subcontinent, SE Asia and Australia. Although considered an agricultural weed in several areas, it is a component of Australian bushfood.

Uses

The small tubers formed on short stolons are a traditional bush tucker food for Australian aboriginal peoples. They are dug up on creek banks when their aerial parts have dried out. They are eaten raw or cooked in the hot earth by the fire and are said to have a pleasant nutty taste. The husk can also be eaten. It propagates *via* rhizomes ending at bulbs.

Cyperus rotundus L.
(Chromosome numbers 2n=16, 26, 96, 108 and 152)

Common names: Coco-grass, Java grass, Nut grass, Purple nutsedge, Red nutsedge and Khmer *kravanh chruk*.

The word *rotundus* is from Latin, meaning 'round'. The names 'nut grass' and 'nut sedge', shared with the related species *Cyperus esculentus* are derived from its tubers which to some extent resemble nuts though botanically they have nothing to do with nuts.

Purple nutsedge is native to Africa, southern and central Europe (north to France and Austria), and southern Asia. It prefers dry conditions but will tolerate moist soils and often grows in wastelands and in crop fields (Martin Robert and Pol Chanthy, 2009). Young plants initially form white and fleshy rhizomes up to 2.5cm in dimension, in chains. Some rhizomes grow upward in the soil and then form a bulb-like structure from which new shoots and roots grow, and from the new roots new rhizomes grow. Other rhizomes grow horizontally or downward and form dark reddish-brown tubers or chains of tubers.

Cyperus rotundus is one of the most invasive known weeds having spread out to a worldwide distribution in tropical and temperate regions.

Chemical Constituents

Several chemical substances have been identified in *C. rotundus* such as cadalene, cyprotene, flavonoids, sesquiterpenes, terpenoids, mustakone, isocyperol, acyperone, rotundene, valecine (Imam *et al.*, 2004), kaempferol, luteolin, quercetin, patchoulenone, isopatchoulenone, sugeonyl acetate, cellulose triacetate and sugebiol (Ying and Bing, 2016). A sesquiterpene, rotundone called so because it was originally extracted from the tuber of this plant, is responsible for the spicy aroma of black pepper and the peppery taste of certain Australian Shiraz wines (Siebert *et al.*, 2008).

Extract from leaves and tubers of *C. rotundus* increases the adventitious rooting of different species. These extracts contain a large amount of auxins and phenolic compounds that promote the rooting of cuttings and seedlings (Rezende *et al.*, 2013; Yamashita *et al.*, 2017).

Uses

Purple nutsedge is a staple carbohydrate in tropical regions for modern hunter-gatherers and is a famine food in some agrarian cultures in famine-stricken areas, especially in Rajasthan, despite bitter taste of its tubers (Bhandari, 1974). The tubers of *C. rotundus* are also edible, but for this purpose it is not cultivated like *C. esculentus*. The tubers are collected from the wild in times of need.

In **Ayurveda**, the plant is used for medicinal purposes in treating fevers, digestive system disorders, dysmenorrhea and other maladies. Modern alternative medicine recommends using the plant to treat nausea, fever and inflammation; for pain reduction; for muscle relaxation and for many other disorders (Imam *et al.*, 2004).

References

1. Anon. 2003. Weeds in the West project: Status of introduced plants in Southern Arizona Parks, Factsheets for *Cyperus esculentus* L. Tucson, Arizona, USA.
2. Azambuja, W. 2017. Priprioca Essential Oil - Óleos Essenciais O Guia do Brasil. *Óleos Essenciais Brazil.* Archived from the original on 13 September 2017.
3. Bhandari, M,M. 1974. Famine Foods in the Rajasthan Desert. *Econ. Bot.*, **28** (1): 78.
4. Consejo Regulador de Chufa de Valencia. Chufadevalencia.org. 2002-12-31.
5. Cook Christopher, D. K. 1996. Cyperaceae. Aquatic and Wetland Plants of India: a reference book and identification manual for the vascular plants found in permanent or seasonal fresh water in the subcontinent of India south of the Himalayas. Oxford University Press, New York, 105p.
6. Coþkuner, Y., Ercan Recai, Karababa Erþan and Nazlýcan Ahmet Nedim, 2002. Physical and chemical properties of chufa (*Cyperus esculentus* L) tubers grown in the Çukurova region of Turkey. *J. Sci. Food and Agric.*, **82** (6): 625–31.
7. Elias Thomas, S. and A. Dykeman Peter, 2009. Edible Wild Plants: A North American Field Guide to Over 200 Natural Foods, 130p. Sterling Press, New York, USA.
8. Federov, A. 1969. Chromosome Numbers in Flowering Plants. Academy of Sciences of the USSR. Leningrad, Russia.

9. Hicks, G.C. 1929. Cytological studies in *Cyperus*, *Eleocharis*, *Dulichium* and *Eriophorum*. *Bot. Gaz.*, **88**: 132–49.
10. https: //en.wikipedia.org/wiki/*Sagittaria*
11. https: //uses.plantnet-project.org/e/index.php?title=Cyperus_(PROSEA) and oldid=333106.
12. Imam, H., Sofi, G., Seikh, A. and Lone, A. 2004. The incredible benefits of Nagarmotha (*Cyperus rotundus*). *Intern. J. Nutr., Pharmac., Neurolog. Dis.*, **4** (1): 23–27.
13. Lewis, W. M. and A.D. Worsiiam, 1970. Nutsedge. *Crops and Soils Mag.* (M: tr4t—L6).
14. Little, E.C.S. 1965. Iriloes, fiom, a fiaibon offiber III. PANS, Sect. C. *Wepdi Confim*, **l1**: l1-13.
15. Martin Robert and Pol Chanthy, 2009, Weeds of Upland Cambodia. Archived on 2014-02-25 at the Wayback Machine, ACIAR Monagraph 141, Canberra, Australia.
16. Oldfield Callie, A. and Evans Jonathan, P. 2016. Twelve years of repeated wild hog activity promotes population maintenance of an invasive clonal plant in a coastal dune ecosystem. *Ecol. and Evol.* **6**(8): 2569–78.
17. Pascual-Seva, N., A. San Bautista, S. López Galarza, J.V. Maroto and B. Pascual, 2012. Yield and Irrigation Water Use Efficiency for Ridge-and Bed-cultivated Chufa (*Cyperus esculentus* L. var. *Sativus Boeck*). *Acta Hort.*, No. 936, pp.125-32.
18. Pascual-Seva, N., A. San Bautista, S. López Galarza, J.V. Maroto and B. Pascual, 2013. Furrow-irrigated *chufa* crops in Valencia (Spain). *Spanish J. agric. Res.*, **11**(1): 258-67. http: //revistas.inia.es/index.php/sjar/article/view/3385/1803.
19. Rath, S.P. and S.N. Patnaik, 1978. Cytologic studies in Cyperaceae with special reference to its taxonomy II. *Cytologia*, **43**: 643–53.
20. Renne, Ian J. and F. Tracy Benjamin, 2006. Disturbance persistence in managed grasslands: shifts in aboveground community structure and the weed seed bank. *Pl. Ecol.*, **190** (1): 71–80.
21. Rezende, F.P.F., K.C. Zuffellato-Ribas and H.S. Koehler, 2013. Aplicaçao de extratos de folhas e tuberbulos de *Cyperus rotundus* L. E de auxinas sintéticas na estaquia caulinar de *Duranta repens* L" [Application of extracts of leaves and tubers of *Cyperus rotundus* L. and of synthetic auxins in the cutting stem of *Duranta repens* L]. *Revista Brasileira de Plantas Medicinais* (in Portuguese), **15** (4 suppl 1): 639.

22. Roalson, E.H. 2008. A synopsis of chromosome number variation in the Cyperaceae. *Bot. Rev.*, **74**: 209–393.

23. Ross Merrill, A. and A. Lembi Carole, 2008. *Applied Weed Science* (3rd ed.), 322p. Prentice Hall, England,

24. Sanchez-Zapata, E., J. Fernandez-Lopez and J. Angel Perez-Alvarez, 2012. Tiger Nut (*Cyperus esculentus*) Commercialization: Health Aspects, Composition, Properties, and Food Applications. *Comprehensive Reviews in Food Science and Food Safety*, **11**(4): 366–77.

25. Siebert, T.E., C.M. Wood, G. Elsey and P. P. Alan, 2008. Determination of rotundone, the pepper aroma impact compound, in grapes and wine. *J. agric. Food Chem.*, **56** (10): 3745–8.

26. Stoller, E.W. 1981. Yellow Nut Sedge: A Menace in the Corn Belt (Leaflet No. 1642). US Department of Agriculture, Agricultural Research Service, USA.

27. Stoller, E. W. and L.M. Wax, 1973. Yellow nutsedge shoot emergence and tuber longevity. *Weed Sci.*, **21** (1): 76–81.

28. Uphof, Johannes Cornelis Theodorus, 1968. Dictionary of Economic Plants (second ed. Rev. by J. Cramer), 169p. New York, USA.

29. wcsp.science.kew.org.

30. Wren, R.C. 1923. Potter's Cylopedia of Botanical Drugs and Preparations, 4p. Potter and Clark, London.

31. Yamashita, O.M., G.W.P. Azevedo, W.M. Peres, G.Q. David, M.A.C. de Carvalho and P.S. Koga, 2017. Seedling production of fruit and ornamental species to the use of weed hormone (*Cyperus rotundus*). *Nucleus*, **14**: 279–88.

32. Ying, J. and X. Bing, 2016. Chemical constituents of *Cyperus rotundus* L. and their inhibitory effects on uterine fibroids. *African Health Sci.*, **16** (4): 1000–1006.

35

Zombi pea (*Vigna vexillata* (L.) A.Rich)

Family: Fabaceae. **Subfamily**: Faboideae.

Other names: Wild cowpea and Narrow-leaved wild sweetpea (Eng.) (pza.sanbi.org).

Synonyms: *Dolichos cylindricus* Desv.; *D. vexillatus* (L.) Kunth; *Phaseolus capensis* sensu Thunb.; *P. glycinaeformus* Weinm.; *P. humifusus* Savi; *P. pulniensis* Wight; *P. quadriflorus* A. Rich.; *P. sepiarius* Dalzell; *P. vexillatus* L.; *Plectrotropis angustifolia* Schumach and Thonn.; *P. hirsuta* Schum. and Thonn.; *Strophostyles capensis* E. Mey.; *Vigna capensis* (Thunb.) Burtt Davy; *V. carinalis* Benth.; *V. crinita* A. Rich.; *V. davyi* Bolus; *V. dinteri* Harms; *V. dolichoneura* Harms; *V. golungensis* Baker; *V. hirta* Hook.; *V. lobatifolia* Baker; *V. phaseoloides* Baker; *V. scabra* Sond.; *V. senegalensis* A. Chev.; *V. thonningii* Hook. f; *V. tuberosa* A. Richard (Quattrocchi, 2016)

Introduction and Origin

The genus *Vigna* is named in honour of Italian botanist Domenico Vigna (*d.* 1647), who was a professor of botany and Director of the Botanical Garden of Pisa. The specific epithet *vexillata* means 'having a standard petal', referring to the widely spreading standard petal (pza.sanbi.org). *Vigna vexillata*, a highly heterozygous legume (Dachapak *et al.*, 2017) is an underutilized crop that can be found in diverse areas of Africa, America, Australia and Asia (Tomooka *et al.*, 2011). It is cultivated for edible storage roots and seeds. Earlier observations have shown that some varieties of zombi pea adapt well to environmental stresses including infertile soil (Karuniawan *et al.*, 2006), alkaline soil (Lawn and Watkinson, 2002), drought (Roecklein and Leung, 1987) and waterlogging conditions (MIller and Williams, 1981). Zombi pea is naturally found in grasslands with dispersed trees, thickets, bushlands, in deserted cultivated land and even forests at altitudes ranging from 0–2500m and prefers southern slopes (pza.sanbi.org). Plants can be found in the deep shade of trees in the wild (tropical.theferns.info).

Flower of *Vigna vexillata*

By Jayesh Patil from Mumbai, India - Zombi Pea, CC BY 2.0, https://commons.wikimedia.org/w/index.php?curid=32507018

Vigna vexillata probably originated in the Old World tropics with centres of diversity in SE Africa and SE Asia (tropical.theferns.info). It is a pan-tropical herbaceous legume distributed widely throughout tropical and sub-tropical regions of Africa, Asia, Australia and America in its wild form (Duke, 1991). The species often develops storage roots that are edible and consumed by people in Africa, India, Australia and SE Asia (Lawn and Cottrell, 1988; Sasikumar and Sardana, 1988; Karuniawan *et al.*, 2006). In Namibia, it is an important food crop where it is commonly harvested from the wild for local use (tropical.theferns.info). Its use has spread beyond its native range and the plant is now sometimes cultivated for its edible tubers. The cultivated form of *V. vexillata* is found in a limited areas of Africa and Asia (Bali, Indonesia) where both seeds and tuberous roots are consumed (Amkul *et al.*, 2019).

Botany

Zombi pea with chromosome number 2n=2x=22, is a strong perennial twiner with fusiform tuberous roots producing stems usually covered with brownish silky hairs, or trichomes. Its oval-shaped and tip-pointed dark green leaflets have appressed trichomes on both surfaces. Leaflets appear in threes with terminal leaflet being 7.5–15cm long. Sweetly scented pink or purplish to yellow flowers are 2.5cm long borne out on 7.5–30cm long 2-4-flowered peduncles with the keel prolonged into an uncurved beak (Quattrocchi, 2016). Linear, recurved and silky pod is 7.5–9cm long (Palaniswami and Peter, 2008). The plant produces a number of tubers on thread-like roots, lying quite deep in the soil (tropical.theferns.info).

The seeds of *V. vexillata* have a significant defense mechanism and resistance to pest attacks due to the presence of a dipeptide para-aminophenylalaline which can fight off predators such as parasitic beetles (pza.sanbi.org).

V. vexillata var. *macrosperma*

Source: cropped from Wettberg et al. (2010)

Wild and Cultivated Varieties

Though wild distribution of zombi pea includes all the five continents (Lawn and James, 1991; Pienaar and Kok, 1991; Sasikumar and Sardana, 1988) but cultivated varieties have been reported only in Sudan and Indonesia (Karuniawan *et al.*, 2006). These two cultivated forms are not cross compatible suggesting that the wild form has substantial genetic variation within it and is likely a species complex. Sudanese cultivated form which is moderately tolerant to drought, salt and certain biotic stresses is generally cultivated for seeds. However, in Bali, where varieties are more susceptible to many kinds of stresses, it is generally cultivated for its tubers (Wettberg *et al.*, 2020).

Infraspecific Varieties

- ☆ *Vigna vexillata* var. *angustifolia* (Schumach. and Thonn.) Baker: native to Tropical and Subtropical Old World; grows primarily in the seasonally dry tropical biome.
- ☆ *Vigna vexillata* var. *davyi* (Bolus) B.J.Pienaar: native to S. Africa; grows primarily in the subtropical biome.
- ☆ *Vigna vexillata* var. *macrosperma* Marechal, Mascherpa and Stainier: cultivated variety native to Indo-China; grows primarily in the wet tropical biome.
- ☆ *Vigna vexillata* var. *ovata* (E.Mey.) B.J.Pienaar: native to S. Africa; grows primarily in the subtropical biome.
- ☆ *Vigna vexillata* var. *sepiaria* (Dalzell) Babu and S.K.Sharma: native to India; grows primarily in the seasonally dry tropical biome.

- ☆ *Vigna vexillata* var. *tsusimensis* Matsum.: native to S. Korea, Japan (Tsushima); grows primarily in the temperate biome.
- ☆ *Vigna vexillata* var. *vexillata*: native to Tropics and Subtropics; grows primarily in the seasonally dry tropical biome.
- ☆ *Vigna vexillata* var. *wightii* (Benth. ex Bedd.) Babu and S.K.Sharma: native to India; grows primarily in the seasonally dry tropical biome.
- ☆ *Vigna vexillata* var. *youngiana* F.M.Bailey: native to Australia; grows primarily in the seasonally dry tropical biome.

Source: (POWO, 2023)

Nutritional Value

The tubers of *V. vexillata* as well as seeds contain 15 per cent protein content which is much greater than the contents of potato, yam and cassava (Chandel *et al.*, 1972; tropical.theferns.info). Across all the regions where it is consumed, it often serves an important food security role, particularly in times of famine.

Uses

The tubers have a soft, easy to peel skin and creamy, tasty flesh which can be eaten raw or cooked. These can be boiled or roasted like sweet potato or cassava. Young leaves can be cooked and eaten as a leafy vegetable. Leaves and young shoots are cut into small pieces and cooked with salt and chilli, and garnished with mustard seeds, curry leaves and onions in oil. Young pods are cooked and eaten as a vegetable. Seeds are cooked and consumed as a pulse in India. It is also grown as a green manure and ground cover crop especially in poor soils (tropical.theferns.info). Its seeds contain L-Dopa, a medicine for parkinsons disease (pharmanest.net; Gangarao Battu *et al.*, 2011).

Cultivation

In India, zombi pea flourishes at an altitude of 1200–1500m in the foothills of the Himalayas and in the hills of eastern and north-eastern India. In Australia, it grows in the north where long dry season is common and also with a monsoon rain of 1250–1500mm (eol.org; PROSEA, 2023). The plant can survive and perform well in drought-prone areas (pza.sanbi.org).

Optimal mean annual temperature for zombi pea is 20–28°C though is tolerant to 15-32°C temperatures. It prefers a mean annual rainfall of 1300–2100mm but also survives in areas with lower or higher rainfall (650-4200mm). Zombi pea prefers full sun exposure with well-drained light or medium soils.

It can also survive in water-logged, poor soils or in areas prone to flooding but requires more time to grow in too much water. Though it prefers slightly acidic soil, *i.e.* 5.5-6.5 pH range but tolerates soil pH level even up to 8 (tropical.theferns.info; pza.sanbi.org).

The plant can be grown in the garden beds as well as in spacious containers filled with a commercial potting soil mixture. The plant is propagated from seeds that take about 10 days to germinate. Though not necessary but it is advisable to scarify the seeds before for better results. The process of scarification is used to soften the seed coat with a tool such as a scalpel. Other forms of scarification include freeze-thawing, heat and mechanical *e.g.* sandpaper rubbing and acid scarification (pza.sanbi.org).

A small amount of water should be sprinkled to moisten the soil before planting in the garden bed. Seeds are sown 1cm deep with the seed eyes facing downwards. After sowing, soil and water should be sprinkled to cover the seeds. The stems of mature plants tend to twine and find support on surrounding vegetation. The 1-1.5m long sticks can be used as support on which the crop can twine. Irrigation twice per week is sufficient (pza.sanbi.org).

Flowering occurs 3-4 months after planting, and the seedpods ripen about 1 month later and at this time the tubers are also ready for harvesting. Bhattacharyya (1984) reported *Vigna vexillata* yielding 7t/ha of root tubers.

References

1. Amkul, K., L. Wang, P. Somta, S. Wang and X. Cheng, 2019. Construction of a high density linkage map and genome dissection of bruchid resistance in zombi pea [*Vigna vexillata* (L.) A. Rich]. *Sci Rep.*, **9**: 117-19. (https://doi.org/10.1038/s41598-019-48239-5).
2. Bhattacharyya, P.K. 1984. Grow *Vigna vexillata* for protein-rich tuber-cum-pulse crop in North-eastern hill region. *Seeds Farms*, **10**: 33–6.
3. Chandel, K.P.S., R.K. Arora and B.S. Joshi, 1972. *Vigna capensis* Walp. (*V. vexillata*) an edible root legume. *Curr. Sci.*, **41**: 537.
4. Dachapak, S., P. Somta, S. Poonchaivilaisak, T. Yimram and P. Srinivas, 2017. Genetic diversity and structure of the zombi pea (*Vigna vexillata* (L.) A. Rich) gene pool based on SSR marker analysis. *Genetica*, **145**: 189–200.
5. Duke, J.A. 1981. *Vigna unguiculata* (L.) Walp. spp. *unguiculata*. In: Legumes of World Economic Importance (ed. Okeson, O. N.), pp. 303–5, Plenum Press, England.
6. eol.org/pages/655241/articles

7. Gangarao, Battu, Ch.K.V.L.S.N. Anjana Male, T. Haripriya, Venna Naga Malleswari and S.K. Reeshma, 2011. A Phytopharmacological Review on Vigna species. *Pharmanest - An International J. Adv. on Pharmac. Sci.*, **2** (1) (www.pharmanest.net. ISSN 0976 - 3090 (Print) 2231 - 0541 (Online). https://pharmanest.net/journal_pharmanest/uploads/1/414_pdf.pdf).

8. Karuniawan, A., A. Iswandi, P.R. Kale, J. Heinzemann and W.J. Gruneberg, 2006. *Vigna vexillata* (L.) A. Rich. cultivated as a root crop in Bali and Timor. *Genet. Resour. Crop Evol.*, **53**: 213–7.

9. Lawn, R.J. and A. Cottrell, 1988. Wild mungbean and its relatives in Australia. *Biologist*, **35**: 267–73.

10. Lawn, R.J. and A.T. James, 1991. Inheritance of selected traits in accessions of *Vigna vexillata* (L.) A. Rich of Australian and African origin. *Aust. J. Bot.*, **39**: 415-29.

11. Lawn, R.J. and A.R. Watkinson, 2002. Habitats, morphological diversity and distribution of the genus *Vigna savi* in Australia. *Aust. J. agric. Res.*, **53**: 1305–16.

12. Miller, I.L. and W.T. Williams, 1981. Tolerance of some tropical legumes to six months of stimulated waterlogging. *Trop. Glassland*, **15**: 39–41.

13. Palaniswami, M.S. and K.V. Peter, 2008. Tuber and Root Crops, 67p. Horticulture Science Series, Vol. 9. New India Publishing, New India.

14. Pienaar, B.J. and P.D.F. Kok, 1991. The *Vigna vexillata* complex (Fabaceae) in southern Africa. *S. Afr. Tydskr. Plantk.*, **57**(5):236-45.

15. POWO, 2023. Plants of the World Online – Kew Royal Botanical Garden. Retrieved 12 May, 2013 via https://powo.science.kew.org/taxon/urn:lsid:ipni.org:names:525561-1.

16. PROSEA: Plant Resources of South-East Asia, Vol. 11. Auxiliary Plants. Yayasan Obor, Indonesia, 261p. Retrieved on 11 May, 2023.

17. pza.sanbi.org/vigna-vexillata-var-vexillata

18. Quattrocchi, U. 2016. CRC World Dictionary of Medicinal and Poisonous Plants: Common Names, Scientific Names, Eponyms, Synonyms, and Etymology (reprint ed.), 3909p. CRC Press, USA.

19. Roecklein, J.C. and P.S. Leung, 1987. A Profile of Economic Plants. Transaction Publishers: Piscataway/New Brunswick, NJ, USA.

20. Sasikumar, B. and S. Sardana, 1988. *Vigna vexillata* (Fabaceae), a pulse cum tuber crop of northeastern hill region of India. *Econ. Bot.*, **42**(2): 292.

21. Tomooka, N., A. Kaga, T. Isemura and D. Vaughan, 2011. *Vigna*. In: Wild Crop Relatives: Genomic and Breeding Resources: Legume Crops and Forages (ed. Kole, C.). Springer, Berlin/Heidelberg, Germany, pp. 291–311.
22. tropical.theferns.info/viewtropical.php?id=Vigna+vexillata
23. Wettberg, Eric B. von, Ken Naito, Andi Kur and Ndiko Ludidi, 2020. The unusual biogeography of zombi pea, *Vigna vexillata*. *Legume Perspectives*: The journal of the International Legume Society, 19: 8-9. (www.researchgate.net/profile/Taofeek-Adegboyega/publication/348430504_Winged_bean_Psophocarpus_tetragonolobus/links/5ffea468a6fdccdcb84d8bf1/Winged-bean-Psophocarpus-tetragonolobus.pdf).

Index – Common and Regional Names

Lotus, 214, 277-287, 414

Index – Botanical Names

D

E

F

G

R

S

Index – Family and Subfamily

Printed at Repro India Ltd.